Springer Complexity

Springer Complexity is a publication program, cutting across all traditional disciplines of sciences as well as engineering, economics, medicine, psychology and computer sciences, which is aimed at researchers, students and practitioners working in the field of complex systems. Complex Systems are systems that comprise many interacting parts with the ability to generate a new quality of macroscopic collective behavior through self-organization, e.g., the spontaneous formation of temporal, spatial or functional structures. This recognition, that the collective behavior of the whole system cannot be simply inferred from the understanding of the behavior of the individual components, has led to various new concepts and sophisticated tools of complexity. The main concepts and tools – with sometimes overlapping contents and methodologies – are the theories of self-organization, complex systems, synergetics, dynamical systems, turbulence, catastrophes, instabilities, nonlinearity, stochastic processes, chaos, neural networks, cellular automata, adaptive systems, and genetic algorithms.

The topics treated within Springer Complexity are as diverse as lasers or fluids in physics, machine cutting phenomena of workpieces or electric circuits with feedback in engineering, growth of crystals or pattern formation in chemistry, morphogenesis in biology, brain function in neurology, behavior of stock exchange rates in economics, or the formation of public opinion in sociology. All these seemingly quite different kinds of structure formation have a number of important features and underlying structures in common. These deep structural similarities can be exploited to transfer analytical methods and understanding from one field to another. The Springer Complexity program therefore seeks to foster cross-fertilization between the disciplines and a dialogue between theoreticians and experimentalists for a deeper understanding of the general structure and behavior of complex systems.

The program consists of individual books, books series such as "Springer Series in Synergetics", "Institute of Nonlinear Science", "Physics of Neural Networks", and "Understanding Complex Systems", as well as various journals.

Springer Series in Synergetics

SSSyn – An Interdisciplinary Series on Complex Systems

The success of the Springer Series in Synergetics has been made possible by the contributions of outstanding authors who presented their quite often pioneering results to the science community well beyond the borders of a special discipline. Indeed, interdisciplinarity is one of the main features of this series. But interdisciplinarity is not enough: The main goal is the search for common features of self-organizing systems in a great variety of seemingly quite different systems, or, still more precisely speaking, the search for general principles underlying the spontaneous formation of spatial, temporal or functional structures. The topics treated may be as diverse as lasers and fluids in physics, pattern formation in chemistry, morphogenesis in biology, brain functions in neurology or self-organization in a city. As is witnessed by several volumes, great attention is being paid to the pivotal interplay between deterministic and stochastic processes, as well as to the dialogue between theoreticians and experimentalists. All this has contributed to a remarkable cross-fertilization between disciplines and to a deeper understanding of complex systems. The timeliness and potential of such an approach are also mirrored – among other indicators – by numerous interdisciplinary workshops and conferences all over the world.

C. W. Gardiner P. Zoller

Quantum Noise

A Handbook of Markovian and Non-Markovian Quantum Stochastic Methods with Applications to Quantum Optics

Third Edition
With 59 Figures

Professor Crispin W. Gardiner
School of Chemical and Physical Sciences
Victoria University
Wellington, New Zealand

Professor Peter Zoller
Institute for Theoretical Physics
University of Innsbruck
Technikerstrasse 25
6020 Innsbruck, Austria

ISSN 0172-7389

ISBN 3-540-22301-0 Third Edition Springer Berlin Heidelberg New York

ISBN 3-540-66571-4 Second Edition Springer Berlin Heidelberg New York

Library of Congress Control Number: 2004108253

Springer is a part of Springer Science+Business Media
springeronline.com

Printed in Germany

Typesetting: by the author
Cover design: *design & production*, Heidelberg
Printed on acid-free paper 55/3141/mh 5 4 3 2 1

In Memory of

Nellie Muriel Olive Gardiner

1908–1989

Preface to the Third Edition

The methods in this book were developed for use in quantum optics, but they do have a much wider relevance. The basic concepts, related to the manipulation of quantum states in the presence of noise, are fundamental to any study of a wide range of physical systems, and the methods can often be appropiately adapted.

A particular example is the field of Bose-Einstein condensation, which has a long history as a part of condensed matter physics. With the dramatic experimental creation of alkali atom condensates in 1995, a whole body of theoretical work on them has arisen, and the field is now in the process of diversifying into the physics of cold atoms, including degenerate cold Fermi gases. This is now leading to a more unified view of the overall field of quantum statistics, of which quantum optics and condensed matter physics form significant subfields.

Another field which is more directly connected to quantum optics is that of quantum information theory and quantum computing, for which there are several different proposed experimental implementations, and a very extensive literature on its purely theoretical aspects. Connected to this is the concept of quantum state engineering: The engineering of interesting and useful quantum states. Currently, the frontier is moving towards building larger composite systems of a few atoms and photons, while still maintaining complete quantum control of the individual particles

These developments have been very extensive, and their full treatment would not be possible without expanding the book unreasonably. Therefore, for this third edition, we have limited changes to the provision of a supplementary chapter with a guide to developments in the field since completion of the second edition in 1999, with particular attention to the above fields and, of course, we have also corrected various misprints and minor errors. Proofreading and checking a book such as this is always a major task, and we would like to thank Ashton Bradley and Piyush Jain for their assistance in this process. The work on the supplementary chapter was partially funded by the Marsden Fund of the Royal Society of New Zealand under contract PVT202.

Wellington and Innsbruck
May 2004

Crispin Gardiner
Peter Zoller

Preface to the Second Edition

Since the first edition of this book was completed some eight years ago, the two authors of this new edition have developed a very fruitful collaboration, which has led us naturally to collaborate on the revision of *Quantum Noise*. The emphasis on quantum optical methods remains, but the significant advances in the theoretical methods of quantum optics in the intervening period have made it necessary to add a considerable amount of new material, which has resulted in the addition of two new chapters.

The stochastic Schrödinger equation is the subject of Chap.11. The wave of work in the early 1990s on this topic arose out of the need to compute the quite complex problems involved in laser cooling of atoms. The development of these methods, coupled with the growing availability of low cost fast workstations brought about a changed point of view in theoretical methods. One can nowadays seriously contemplate investigating systems with a very large number of degrees of freedom, without making the very drastic approximations which were the hallmark of the early days of quantum optics.

The concept of cascaded quantum systems—the subject of a new Chap.12—became essential with the growing capacity to create and use non-classical states of light. The subject matter of this chapter covers some applications of these methods, including one application to the transmission of quantum information.

The remainder of the book is not greatly changed. The main addition is new material on the applicability of positive-P function methods. The positive P-function has proved very useful in quantum optics, even though there were examples known in which it could be demonstrated that it gave incorrect results. The first edition of this book touched on the problem, but did not offer a resolution. We are now able to present clear guidelines for it use, together with a definite explanation of the causes of difficulties.

The process of revision and adding more material was helped considerably by many colleagues. In particular we want to thank James Anglin, Rob Ballagh, Thomas Busch, Ignacio Cirac, Simon Gardiner, Dieter Jaksch and Klaus Gheri for advice and proofreading, and as well all those other colleagues who have worked with us on this subject over the past few years; in particular Hans Briegel, Peter Drummond, Ralph Dum, Klaus Ellinger, Steven van Enk, Andrea Eschmann, Alexei Gilchrist, Murray Holland, Stefan Marksteiner, Monika Marte, Peter Marte, Bill Munro, Scott Parkins, Thomas Pellizzari, and Richard Taïeb.

Both of us have had a long relationship with Dan Walls, whose untimely death came as we were finishing this book. Dan's influence on quantum optics over the last thirty years is unparalleled, and this book reflects that influence. He was instrumental in introducing both of us to this field, and for this and for the privilege of having known and worked with him we are deeply grateful.

Wellington and Innsbruck
May 1999

Crispin Gardiner
Peter Zoller

Preface to the First Edition

The term "Quantum Noise" covers a number of separate concepts, and has a variety of manifestations in different realms of physics. The principal experimental areas now covered by the term arise in quantum optics and in the study of Josephson junctions. Although there has been some interaction between these two fields, it is essentially true to say that the two fields have developed independently, and have their own body of techniques for theoretical investigation. This book is largely about the quantum optical field. By this I mean that the techniques presented here are based on the methods used in quantum optics—I certainly do not mean that these methods are applicable only to quantum optics.

My aim in this book is to give a systematic and consistent exposition of those quantum stochastic methods which have developed over the last thirty years in quantum optics. In my previous book, "A Handbook of Stochastic Methods", I concluded with a chapter on quantum mechanical Markov processes. This was only a brief outline of those methods which harmonized most naturally with the classical theory of Markov processes. Since that time the realization has grown that the study of *squeezing* and other more exotic properties of light fields required a much more careful description of quantum noise, and in particular the relation between inputs and outputs to quantum optical systems. After Matthew Collett and myself worked out such a formalism for describing the production of squeezed light beams, we realized that the same basic approach had within it the foundation for a complete description of quantum noise. This book is the result of this realization.

This book is not merely a collection of methods drawn from the literature—there is a significant amount of new work, and the interconnections between the various descriptions are new. The structure of the book is as illustrated in the following diagram.

The introduction and Chap.2 on quantum statistics set the background for the real development of the theory, which takes place mainly in Chap.3. This chapter develops the quantum Langevin equation of a system in interaction with a heat bath of harmonic oscillators in a form which is rather general, and which can be applied. The *adjoint equation*, which is analogous to a Schrödinger picture version of the quantum Langevin equation is also introduced, and it is shown how all the standard quantum optical techniques can be derived as limits of the adjoint equation. The great advantage of the adjoint equation comes from the remarkable fact that the quantum noise which arises can be exactly represented by a c-number stochastic process. To emphasize this point, a final section in Chap.3 shows how to apply it to the problem of macroscopic quantum coherence in a low temperature two level system, which can perhaps be realized with a SQUID—that is, it is an application of a quantum optical method to a problem in superconductivity. This section, written in collaboration

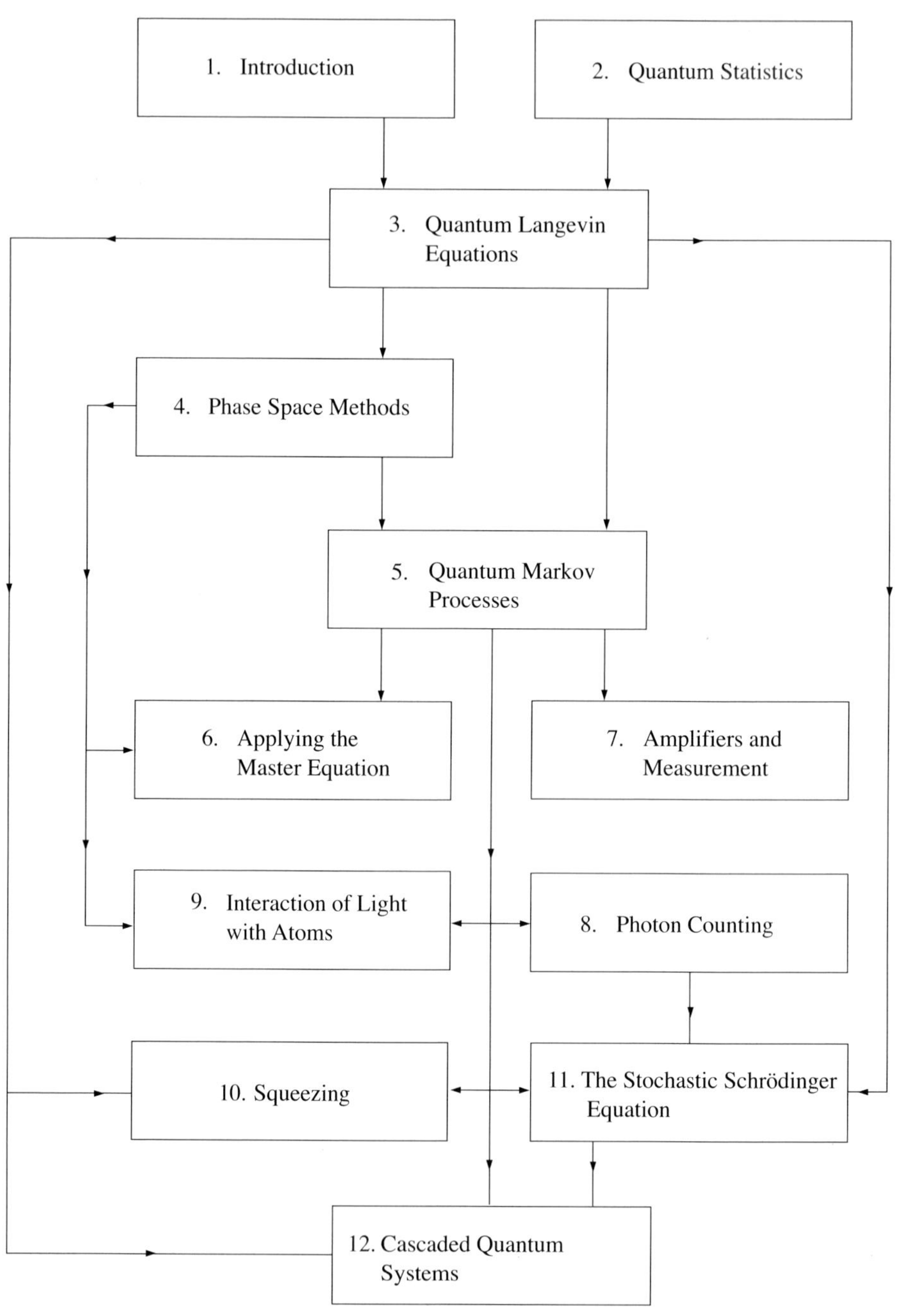
1. Introduction
2. Quantum Statistics
3. Quantum Langevin Equations
4. Phase Space Methods
5. Quantum Markov Processes
6. Applying the Master Equation
7. Amplifiers and Measurement
9. Interaction of Light with Atoms
8. Photon Counting
10. Squeezing
11. The Stochastic Schrödinger Equation
12. Cascaded Quantum Systems

with Scott Parkins, shows how quantum noise can be numerically simulated to give slightly more general results than are achievable from the path integral methods usually employed.

The fourth chapter is essentially a compendium of phase space methods, by which c-number representations of quantum processes can be developed. It contains a systematic explanation of coherent states, and the three principal phase space methods, the P-, Q- and Wigner representations.

Chapters 5, 6, and 7 form a group. Chapter 5 contains the theory of the Markovian limit of systems driven by quantum noise, and also shows that this limit occurs in a wider range of systems than those described in Chap.3. It contains, for the first time in any book, a dual description similar to that which occurs in classical stochastic processes. Thus we have the master equation, which corresponds classically to the Fokker-Planck equation, and *quantum* stochastic differential equations, which are in exact correspondence to their classical counterparts. Basically, these are all results of a quantum white noise limit, though in the case of linear systems, non-white noise can be handled. Thus Chap.5 contains all that is necessary for a theory of quantum Markov processes.

It is perhaps appropriate to add a word of caution here. The white noise limit is very simple and elegant and has led to some very rigorous mathematical theory. There have been claims that quantum mechanics as usually presented (i.e., using only unitary time evolution) is less general than the theory of quantum Markov processes, and even that because of this conventional quantum mechanics is incomplete. However there is no physical evidence that nature makes use of this more general theory—rather, it appears that the only way in which quantum Markov processes turn up in practice is as a *limiting* description of the motion of a subsystem of a larger system, from which all knowledge of a large number of degrees of freedom is eliminated. This is certainly the point of view I take throughout the book, though I acknowledge that in the field of quantum measurement theory there is a widespread opinion that the theory is not complete. To emphasize—*everything in this book can be understood as a reformulation or a limit of conventional quantum mechanics*.

Chapter 6 is very largely concerned with joining together the results of the previous two chapters, and in developing the ideas of generalized P-representations, introduced by Peter Drummond and myself ten years ago. These methods have gradually become standard techniques, even though there are still some fundamental existence questions to be answered.

On the other hand Chap.7 depends largely on quantum stochastic differential equations, and presents a way of formulating amplifier theory with inputs and outputs, which is the logical continuation of the 1982 formulation of Carl Caves. I feel that this point of view can be developed much more, and could lead to a "quantum engineering" formulation of amplifiers and related quantum systems. The section on measurement is included here because it is another application of the quantum stochastic differential equation approach, and because, in the model described, it is shown that an understanding of the von Neumann collapse of the wavefunction can be understood as the result of quantum noise and amplifier gain.

My chapter on photon counting is an almost entirely new formulation, in which the techniques of quantum Markov processes are utilized to develop a fully spatially dependent description of the photodetection process. Of course no new physical results arise from this, but I believe the derivation of the conditions under which the quantum Mandel formula is valid, and the relation to the theory of continuous measurements is very illuminating. The latter part of the chapter puts together a way of looking at photodetectors from the point of view of inputs and outputs—that is, we look at input photons and output electrons. Almost incidentally, a Fermionic form of quantum white noise is introduced to describe electron fields.

Chapter 9 is a rather brief summary of the interaction of light and the two level atom, including gas laser theory and optical bistability. The presentation is again unconventional, since it is an application of the quantum Markov process description developed in the earlier parts of the book.

The final chapter is a brief summary of some aspects of squeezed light. This field is still developing, and I did not feel a more intensive discussion would be appropriate.

In order to understand this book, it is necessary to have some knowledge of field quantization, and a thorough knowledge of non-relativistic quantum mechanics. I have chosen not to put in a description of classical stochastic processes, since this is well covered in my previous book, “A Handbook of Stochastic Methods”, to specific sections of which I shall frequently refer by using the abbreviation S.M. In that sense, this book does demand a lot of preparation, but I think that is unavoidable. To understand the full range of physical noise phenomena requires a thorough understanding of both the classical and the quantum fields, and I hope this book and my previous one will provide that.

I have not designed this book primarily as a textbook for a course, though I have included some exercises, which are not necessarily always easy. However, I have given a short course of eight two-hour lectures at the University of Linz, in which I covered the material in Chaps. 1,3 (two lectures), Chap.4, Chap.5 and the first three sections of Chap.9. A very suitable course of about 24 one hour lectures could be made from selections from “Handbook of Stochastic Methods”, say Chaps. 1–7, and Chap.9, and from “Quantum Noise”, Chaps. 3–7, and Chap.9. Other material would be optional. For a balanced view, the student should also attend a course on quantum optics with a more applied point of view, including information on experiments.

Acknowledgments: The material in this book, as well as the realization that such a book was possible, arose because of the work done by Matthew Collett in his M.Sc. thesis in 1983, which resulted in the first description of the input-output formalism, and the first correct description of travelling wave squeezed light. This was a period of remarkable productivity, and I wish to acknowledge here the central rôle of Matthew's thinking in the form that this book now takes.

The parts of the book dealing with simulations of the adjoint equation were largely carried out by Scott Parkins, who has also assisted me immensely by reading and checking the proofs. Andrew Smith developed the parts of Chap.9 involving unconventional phase space methods, and Moira Steyn-Ross carried out some of the work on quantum Brownian motion in Chap.3.

My most heartfelt thanks go to Heidi Eschmann, who gladly took up the challenge of producing a manuscript in and, as the reader can see, has managed some typesetting of very great complexity with great success.

The book was written over a period of five years, during which time I have travelled extensively, and benefited greatly from the points of view of many colleagues, including Nico van Kampen, Carl Caves, Jeff Kimble, Howard Carmichael, Gerard Milburn, Peter Drummond, Urbaan Titulaer, Peter Zoller, Marc Levenson, Bob Shelby, Dick Slusher, Bernard Yurke, Fritz Haake, Robert Graham, and my quantum optics colleagues here in New Zealand, Margaret Reid, Matthew Collett, and Dan Walls, who first introduced me to the field.

I also wish to thank my colleagues here at the University of Waikato, in particular Bruce Liley for his constant support of the project, and Alastair Steyn-Ross and Lawrence D'Oliveiro for their assistance in understanding the mysteries of the Macintosh.

I would also like to thank Hermann Haken, for including this book in the Springer Series in Synergetics, and Helmut Lotsch for his constant forbearance of the inevitable delays of such a book.

Finally, let me express my thanks to Helen May and our daughter Nell, who have lived with this project and supported it for the last five years.

Hamilton, New Zealand *Crispin Gardiner*
May 1991

Contents

1. A Historical Introduction

Quantum mechanics has had a statistical aspect since the formulation of the probability interpretation by *Born* [1.1] in 1926, and in spite of repeated attempts to re-interpret this intrinsic statistical aspect as the consequence of the observer's incomplete access to information which is in principle accessible (as is the case classically), Born's probability interpretation remains with us to this day. There is as yet no experiment which has shown any disagreement whatsoever with the basic principles of quantum mechanics, even though some of its predictions are quite counter intuitive.

In simple experiments, such as scattering experiments, the quantum mechanics of few body problems—essentially Schrödinger's equation—is sufficient to give a complete description. But even in the archetypal quantum mechanical concept of atomic energy levels and spectral lines the simple few body view is no longer adequate. Spectral lines are not sharp. The origin of the spectral line width is the coupling of the atom to the electromagnetic field, which has infinitely many degrees of freedom, and it is only because of this infinity that the irreversible phenomenon of atomic decay takes place.

Thus the description of atomic decay embodies the two fundamental building blocks of the theory of quantum noise—the intrinsic statistical aspect of quantum mechanics, and the statistical aspect which arises from our inability to specify each of all of the infinite number of degrees of freedom of the electromagnetic field. *Weisskopf* and *Wigner* [1.2] gave the first description of atomic decay and the consequent existence of a spectral line width as long ago as 1930, and their paper can be regarded as the beginning of the theory of quantum noise, though the concept of noise *per se* did not really enter their formulation. The purely quantum mechanical aspect of noise arises from Heisenberg's uncertainty principle, which we shall now investigate in some detail.

1.1 Heisenberg's Uncertainty Principle

Heisenberg's uncertainty principle states that it is impossible to measure simultaneously two canonically conjugate variables such as position x and momentum p, with arbitrary precision. Explicitly,

$$\Delta x \Delta p \geq \hbar/2. \tag{1.1.1}$$

From Heisenberg's principle alone, it is clear that we simply cannot measure all of the variables of a system precisely. Repeated measurements on the same system will yield values of x and p which fluctuate about certain mean values, with uncertainties Δx and Δp. It is instructive to derive (1.1.1), for this yields a connection with classical noise theory. Let us consider two variables x and y, and define

$$\begin{aligned}\delta x &= x - \langle x \rangle \\ \delta y &= y - \langle y \rangle\end{aligned} \tag{1.1.2}$$

where x and y are quantum mechanical operators, and $\langle \quad \rangle$ is the average over quantum wavefunctions as well as any other statistical elements. Then we know that for any operator A, with Hermitian conjugate $A^\dagger$

$$\langle AA^\dagger \rangle \geq 0. \tag{1.1.3}$$

Let us put $A = \delta x + \lambda e^{i\theta} \delta y$ and substitute in (1.1.3). We get

$$\langle \delta x^2 \rangle + \lambda(\cos\theta \langle [\delta x, \delta y]_+ \rangle - i \sin\theta \langle [\delta x, \delta y] \rangle) + \lambda^2 \langle \delta y^2 \rangle \geq 0. \tag{1.1.4}$$

The criterion for (1.1.4), regarded as a quadratic in λ, to be non-negative is

$$\left(\cos\theta \langle [\delta x, \delta y]_+ \rangle - i \sin\theta \langle [\delta x, \delta y] \rangle\right)^2 \leq 4 \langle \delta x^2 \rangle \langle \delta y^2 \rangle. \tag{1.1.5}$$

(Note that the mean of a commutator is imaginary, so that $i\langle [\delta x, \delta y] \rangle$ is real). This must be true for all θ, so maximizing the left hand side yields

$$\langle \delta x^2 \rangle \langle \delta y^2 \rangle \geq \tfrac{1}{4} |\langle [\delta x, \delta y] \rangle|^2 + \tfrac{1}{4} \langle [\delta x, \delta y]_+ \rangle^2. \tag{1.1.6}$$

We can now see a classical term and a quantum term on the right hand side. The second term involves the covariance

$$\sigma_{xy} = \tfrac{1}{2} \langle \delta x \delta y + \delta y \delta x \rangle. \tag{1.1.7}$$

In a classical theory the products would commute, and we would recover the usual definition of a covariance. In the classical case we would then have

$$\langle \delta x^2 \rangle \langle \delta y^2 \rangle \geq \sigma_{xy}^2 \tag{1.1.8}$$

and this expresses the fact that if x and y are correlated, then the product of the variances must satisfy (1.1.8) as illustrated in Fig.1.1. This second term, being one which arises classically, has an origin in the coupling of the simple system to an external reservoir, as mentioned in the introduction to this chapter. Quantum effects can still be manifested in this term through the quantum mechanical nature of the reservoir, so although it has a classical analogue, it cannot be called a purely classical term. The first term is purely quantum, since it vanishes if the operators commute. We can see that even if x and y are uncorrelated, there is a minimum product of their variances. In the case of position x and momentum p we know

$$[x, p] = i\hbar \tag{1.1.9}$$

so that the inequality (1.1.6) becomes in this case

$$\Delta x \Delta p \geq \sqrt{\tfrac{1}{4}\hbar^2 + \sigma_{xp}^2}\,. \tag{1.1.10}$$

If the correlation σ_{xp} vanishes then we recover the Heisenberg principle, as stated in (1.1.1).

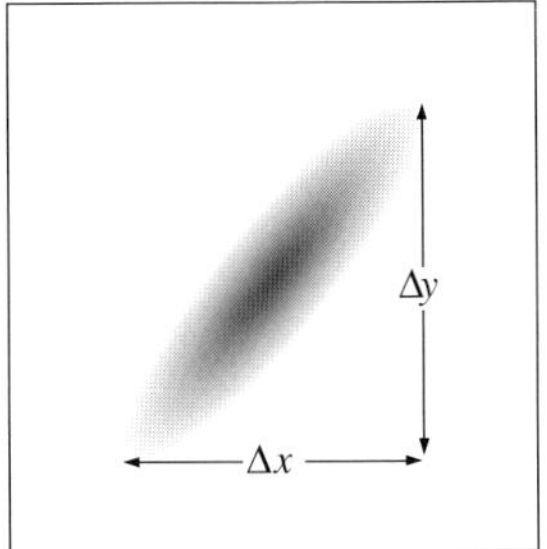

Fig. 1.1 Diagram illustrating how the existence of a covariance requires that the product of variances not be zero.

1.1.1 The Equation of Motion and Repeated Measurements

We can now ask how Heisenberg's principle will manifest itself as quantum noise. Like all noise, quantum noise is a time dependent phenomenon. Without noise, repeated measurements of a physical quantity would be expected to yield a smoothly varying function of time. To analyse such a time dependent situation one must have some equations of motion, so we consider a specific example; repeated measurements at time intervals t of the position of a free particle, in the absence of any thermal noise. The Heisenberg equations of motion are of the same form as the classical Hamilton equations :

$$\left.\begin{aligned} m\dot{x}(t) &= p(t) \\ \dot{p}(t) &= 0 \end{aligned}\right\} \tag{1.1.11}$$

so that

$$x(t+\tau) = p(t)\frac{\tau}{m} + x(t) \tag{1.1.12}$$

from which one can deduce

$$\langle \delta x(t+\tau)^2 \rangle = \langle \delta x(t)^2 \rangle + \left(\frac{\tau}{m}\right)^2 \langle \delta p(t)^2 \rangle + \left(\frac{\tau}{m}\right) \langle [\delta x, \delta p]_+ \rangle . \tag{1.1.13}$$

We note some interesting consequences of (1.1.13).

i) A precise measurement of $x(t)$ results, from Heisenberg's principle, in

$$\langle \delta p(t)^2 \rangle \to \infty \tag{1.1.14}$$

and hence also

$$\langle \delta x(t+\tau)^2 \rangle \to \infty . \tag{1.1.15}$$

The measurement acts back on the system in such a way as to make any further precise measurement of x impossible. The infinite uncertainty in $p(t)$ means that the velocity of the particle is infinitely arbitrary, and hence so is the future position.

ii) The best we could do would perhaps be to measure with uncertainties in $x(t)$ and $p(t)$ which were comparable with each other; for example Heisenberg's principle permits

$$\langle \delta x(t)^2 \rangle = \hbar\tau/(2m), \qquad \langle \delta p(t)^2 \rangle = m\hbar/(2\tau) \tag{1.1.16}$$

which yields

$$\langle \delta x(t+\tau)^2 \rangle = \hbar\tau/m. \tag{1.1.17}$$

Hence the measurement at time $t+\tau$ has twice the variance of that at time t.

iii) It should be borne in mind that to measure with the uncertainties given by (1.1.16) requires that the initial state be a pure quantum-mechanical state. How to carry out any of these measurements in practice is non-trivial, though it is theoretically possible.

iv) Not all quantum measurements have this property. For example, (1.1.11) clearly shows that

$$p(t+\tau) = p(t) \tag{1.1.18}$$

and hence repeated measurements of the momentum do not introduce noise. This is the basis of back action evading or quantum non-demolition measurements [1.3].

1.2 The Spectrum of Quantum Noise

Nyquist's theorem [1.4] as experimentally verified by *Johnson* [1.5] established that a resistor R develops a noise voltage $E(t)$ across its ends, whose value can be written classically as

$$\langle E(t+\tau)E(t) \rangle = \int_{-\infty}^{\infty} \mathrm{e}^{\mathrm{i}\omega\tau} S(\omega)\, d\omega \tag{1.2.1}$$

where

$$S(\omega) = RkT/\pi. \tag{1.2.2}$$

(k = Boltzmann's constant, T = absolute temperature.)

The origin of these fluctuations is well understood: they are simply the fluctuations inherent in maintaining a Boltzmann distribution of the appropriate canonical variables in the electric circuit. A stochastic analysis is given in S.M. 5.3.6(d).[1] Even in 1928 Nyquist considered what should happen at high frequencies, such that $\hbar\omega \gg kT$, since it was known that in the case of the black body radiation, the effect of quantization was to yield the Planck spectrum, which would be equivalent in this case to

$$S(\omega) = \frac{R\hbar\omega}{\pi[\exp(\hbar\omega/kT) - 1]}. \tag{1.2.3}$$

[1] The abbreviation S.M. is used for the book, "A Handbook of Stochastic Methods" by C.W. Gardiner, as noted in the preface.

This function is flat for $\hbar\omega \ll kT$, but for $\hbar\omega > kT$, it rapidly approaches zero. If (1.2.3) is used instead of (1.2.2), the correlation function (1.2.1) becomes a well-behaved approximation to a delta function. Thus Nyquist concluded that the physically unrealistic flat noise spectrum would be eliminated by the incorporation of quantum mechanics.

A more serious quantum treatment did not come until 1951, when *Callen* and *Welton* [1.6] gave a form of reasoning which suggested that the true correlation function should have a quantum mechanical spectral density given by

$$S(\omega) = \frac{R\hbar\omega}{2\pi}\coth\left(\frac{\hbar\omega}{2kT}\right) \tag{1.2.4}$$

$$= \frac{R}{\pi}\left[\tfrac{1}{2}\hbar\omega + \frac{\hbar\omega}{\exp(\hbar\omega/kT) - 1}\right]. \tag{1.2.5}$$

This result shows that Nyquist was correct in surmising that the thermal spectrum would be cut off when $\hbar\omega > kT$. However the total spectrum is not cut off—rather, it rises linearly with increasing ω because of the first term in (1.2.5), which arises from the zero point fluctuations in the harmonic oscillators which model the microscopic structure of the resistive element. It might be thought that effects arising from zero point fluctuations would be unobservable, since it is well attested by experiment that black body radiation is observed to have a Planck spectrum, corresponding only to the second term in (1.2.5)—we cannot detect the zero point spectrum when we measure the spectrum by means involving the absorption of photons from a radiation field.

However, serious doubts can be raised about the status of this quantum mechanical version of $S(\omega)$. Callen and Welton's derivation is not flawless, and furthermore they computed only the mean square voltage $\langle E(t)^2\rangle$ in the case that one has a frequency dependent resistor $R(\omega)$, so that

$$\langle E(t)^2\rangle = \frac{1}{2\pi}\int_{-\infty}^{\infty} d\omega\, \hbar\omega R(\omega)\coth\left(\frac{\hbar\omega}{2kT}\right) \tag{1.2.6}$$

and if $R(\omega)$ is sufficiently rapidly decreasing, this integral converges. But, most importantly, the meaning of the product specified classically in (1.2.1) for the quantum mechanical situation, where $E(t)$ and $E(t+\tau)$ are operators which cannot be expected to commute, has not been specified. *Ford et al.* [1.7] carried out a more well defined procedure, and brought up this matter of the choice of quantum mechanical product. They suggested the use of the *normal product*, (see Sect.4.3.1) which has the effect of omitting the linearly rising term in (1.2.5), and therefore seems more physical.

However this can only be a guess; a correct answer to the problem can only come by a detailed modelling of what is actually measured. We consider therefore two pieces of experimental evidence.

a) Measurement of the Spectrum of Black Body Radiation by Absorption: As just mentioned, in this we get a Planck spectrum. Furthermore, we will show later in

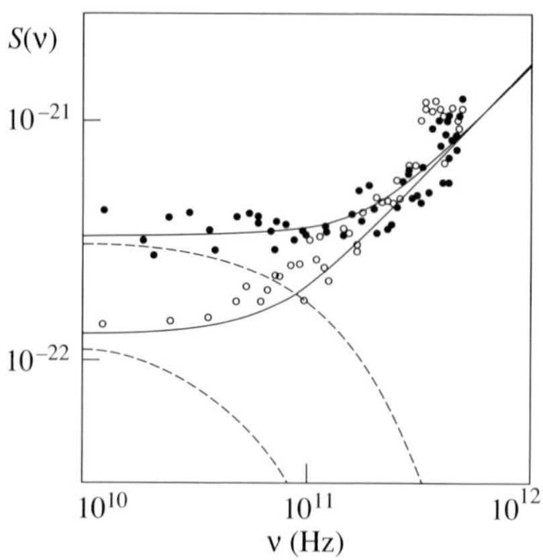

Fig. 1.2 Measured spectral density of noise current in the experiment of *Koch et al.* at 4.2 K (solid circles) and 1.6 K (hollow circles). The solid lines are the prediction of (1.2.5), while the dashed lines correspond to the Planck spectrum (1.2.3)

Chap.8 that in absorption measurements, it is the normal product that is measured, and hence the Planck spectrum is expected.

b) Heterodyne Measurement of the Spectrum of Quantum Noise: *Koch et al.* [1.8] chose to carry out measurements without a complete theory. They assumed that an electric circuit behaves classically, except that the classical flat spectrum is replaced by the expression (1.2.4), and they rather convincingly showed that this expression is correct—that the linearly rising term, which Clarke has termed "f-noise", does exist and is measurable, as is shown in Fig.1.2. The observed spectral density therefore does actually possess both terms. The first term, proportional to $\hbar\omega$, is temperature independent and is therefore purely quantum mechanical—it is truly "quantum noise".

However, the second "thermal" term is also influenced by quantum mechanics, since it does contain Planck's constant. There is no escape from the details of quantum mechanics. Pragmatically, the results of this experiment show that there is considerable validity in a treatment based on a classical noise theory with a modified "quantum" spectrum. For example, according to this kind of treatment, one would treat an LRC circuit by the equations

$$\frac{dQ}{dt} = I \tag{1.2.7}$$

$$L\frac{dI}{dt} = -\frac{Q}{C} - IR + E(t) \tag{1.2.8}$$

in which

$$\langle E(t)E(t')\rangle = \frac{R}{2\pi}\int_{-\infty}^{\infty} \mathrm{e}^{\mathrm{i}\omega(t-t')}\hbar\omega \coth\left(\frac{\hbar\omega}{2kT}\right) d\omega \tag{1.2.9}$$

and where Q, the charge on the capacitor C, I, the current through the circuit, and $E(t)$ are viewed as classical non-operator quantities. This kind of theory is widely used in the treatment of macroscopic superconductivity problems [1.9].

In the limit of high temperature it reduces to the classical white noise theories, as long as one is not interested in very high frequency phenomena. This is an example of a damped harmonic oscillator, and we will show in Sect.3.4.3 that a true quantum theory of the harmonic oscillator can be cast into this form exactly. But the introduction of any amount of anharmonicity destroys this rosy picture and serious discrepancies arise. The major difference that arises is one noted by *Dirac* [1.10]. The motion of a harmonic oscillator has only one frequency—no harmonics occur. Quantum mechanically, the harmonic oscillator has evenly spaced energy levels such that in a transition from one to the next, radiation of a corresponding frequency is emitted. There is therefore no a priori contradiction between the two viewpoints. However in the case of an anharmonic oscillator there is classically a fundamental frequency and an infinite number of harmonics. These do *not* match up with the frequencies of the possible transitions between the energy levels of the quantum-mechanical version—there is no way in which in the small noise limit one will obtain agreement between these two pictures.

1.3 Emission and Absorption of Light

Having introduced the subject of energy levels in the last section, it now seems appropriate to ask what is the relationship between these energy levels and quantum noise. Historically, energy levels occur in their most easily observable form in atoms and molecules, which interact with the radiation field, experiencing transitions from one energy level to the other as they absorb and emit light. This picture has the essentials given in the beginning of this chapter, of a small system with few degrees of freedom (an atom) interacting with a system having very many degrees of freedom (the radiation field).

In 1917 *Einstein* [1.11] introduced a stochastic method of treating emission and absorption of *thermal* light, with the principal result that he was able to show that the Planck distribution formula was necessarily required if molecules which emitted and absorbed the light had a Maxwellian velocity distribution. Einstein's equation remains with us today, though in a more general form.

We suppose that the molecules have (internal) energy levels $\epsilon_1, \epsilon_2, \epsilon_3, \cdots$. Statistical mechanics says that in equilibrium, the relative frequency of the state n is given by

$$w_n = p_n \mathrm{e}^{-\epsilon_n/kT} \tag{1.3.1}$$

where k is Boltzmann's constant, T is the temperature, and p_n represents the "weight" of the state, i.e., the number of different states with the same energy ϵ_n.

Suppose that n and m are two states whose energies satisfy

$$\epsilon_m > \epsilon_n . \tag{1.3.2}$$

The transition $m \rightarrow n$ is possible provided energy is emitted into the radiation field, and the transition $n \rightarrow m$ is possible if energy is absorbed from the radiation field. Einstein made two hypotheses :

a) Emitted Radiation: On the basis that it is known that oscillators radiate when excited, whether or not a radiation field is present, one assumes that the probability per unit time of making the transition $m \to n$ can be written

$$dW = A_m^n \, dt \tag{1.3.3}$$

where A_m^n are certain coefficients. This is also in exact correspondence with the law of radioactive decay.

b) Effect of Incident Radiation: Suppose a molecule is in the presence of a radiation field. Then the energy of the molecule can either increase or decrease, depending on the relative phase of the oscillator and the radiation field. Correspondingly, two hypotheses are introduced: The probability per unit time for the transition $n \to m$ (absorption) is

$$dW = B_n^m \, \bar{\rho} \, dt \tag{1.3.4}$$

and for the transition $m \to n$ (emission)

$$dW = B_m^n \bar{\rho} \, dt \tag{1.3.5}$$

where $\bar{\rho}$ is the energy density (per unit volume per unit frequency) of the radiation field, and B_m^n, B_n^m are certain constants, which we do not specify for the moment.

One can now derive the Planck radiation formula. In equilibrium, the rate of transitions $n \to m$ must balance that of the transitions $m \to n$. Using the expressions (1.3.1) and (1.3.3–5) one sees that

$$p_n e^{-\epsilon_n/kT} B_n^m \bar{\rho} = p_m e^{-\epsilon_m/kT} \left(B_m^n \bar{\rho} + A_m^n \right) . \tag{1.3.6}$$

If we assume that $\bar{\rho}$ can become arbitrarily large by increasing the temperature, then we may derive from (1.3.6) (by noting that the exponentials then approach unity)

$$p_n B_n^m = p_m B_m^n . \tag{1.3.7}$$

Using (1.3.7) we can solve (1.3.6) for $\bar{\rho}$, to find

$$\bar{\rho} = \frac{A_m^n / B_m^n}{\exp\left[(\epsilon_m - \epsilon_n)/kT \right] - 1} \tag{1.3.8}$$

which is almost Planck's radiation law. To complete the result, Einstein used Wien's displacement law, in the form that says that $\bar{\rho}$ can be written in the form

$$\bar{\rho} = \nu^3 f(\nu/T), \tag{1.3.9}$$

from which it is clear that

$$A_m^n / B_m^n = \alpha \nu^3 \tag{1.3.10}$$

and

$$\epsilon_m - \epsilon_n = h\nu . \tag{1.3.11}$$

The constants α and h are not determined by this reasoning—their determination requires both experimental information and a more precise theory of radiation. Even less so are the coefficients A_m^n and B_m^n determined. However identifying h with Planck's constant we find that

$$\bar{\rho}(\nu) = \alpha\nu^3/(\exp(h\nu/kT) - 1) \tag{1.3.12}$$

which is of course Planck's radiation law.

c) Commentary: Einstein went on further to show that the conservation of momentum requires that the molecules have the Maxwellian distribution of velocities, if it is assumed that *each photon is emitted in a particular direction*, rather than as a spherical wave, as might have been expected classically. From a modern quantum-mechanical point of view this is not acceptable. The emission of photons can be computed either as emission of plane waves (corresponding to quanta with a definite direction) or as emission of spherical waves. Since each is a superposition of the others, there is no difference. However the concept of superposition of probability amplitudes, which is essential for this equivalence, plays no part in Einstein's formulation; only probabilities enter into the argument, and the specific quantum-mechanical connection only arises from the assumption that energy levels *exist* and that the transitions between energy levels occur when electromagnetic energy is absorbed or emitted. This means that Einstein's derivation is consistent with a purely probabilistic quantum theory, in which, however, photons are to be regarded as *particles* travelling in a definite direction. The question of wave-particle duality, which is such a central part of quantum mechanics, is nowhere to be seen.

This does not invalidate Einstein's arguments, but the question of why photons *must* be treated as particles travelling in a definite direction is not answered. We will see in Sect.3.6.3 that purely quantum-mechanical arguments give the same results, though not with the simplicity of Einstein's methods.

There is a long tradition of this kind of "probabilistic" quantum mechanics. After the development of modern quantum theory in the 1920s, the use of time-dependent perturbation theory as developed by *Dirac* [1.10] and transformed by *Fermi* into his "golden rule number two" [1.12] was used as the basis for multitudes of calculations by probabilistic methods. The basic technique was to use quantum mechanics to compute probabilities for transitions (such as absorption and emission of photons) and then to use these probabilities in classical stochastic equations for the probability of occupation of quantum mechanical energy levels. Using such methods, it is not at all difficult to compute the coefficients A_m^n and B_m^n in Einstein's equations, and indeed to derive the whole equation, subject to one major assumption. This assumption was first made explicit by *Pauli* [1.13], and it was called the "repeated random phase assumption". It is assumed that the phase relations between wavefunctions are always (repeatedly) randomized, so that all one has to deal with are the probabilities, which are given by the squares of the wavefunctions. How this is achieved is not explained, but the success of many calculations based on it shows that it often has considerable validity.

The arrival of the laser in the 1960s changed this picture. The repeated random phase assumption does not hold for the highly coherent fields that can be produced in a laser, and the whole field of quantum noise, emission and absorption theory, and optical coherence was transformed under its influence. A highly effective theory of "quantum Markov processes" was developed for application in the field of quantum optics, and has generated a renaissance in the theory of stochastic processes. The development of this point of view will occupy much of this book.

1.4 Consistency Requirements for Quantum Noise Theory

In setting up theories of noise and damping one does not have complete freedom—to be acceptable, theories must be consistent with the laws of physics, and in this field there are two principal consistency requirements which have to be dealt with. Firstly, there is the requirement that the well verified results of statistical mechanics be reproduced, and secondly, the requirement that canonical commutation relations be preserved by any equation of motion.

1.4.1 Consistency with Statistical Mechanics

Methods based on generalizations of the "quasiclassical" approach of Sect. 1.2 and of the stochastic approach of Sect. 1.3 must be compatible with each other. In particular, the requirement of consistency with statistical mechanics, as enunciated by Mark Kac must be shown to be satisfied. Kac observed that in the classical theory of noise, we can show that the motion of a particle in a potential governed by the Langevin equation

$$m\ddot{x} = -V'(x) - \gamma\dot{x} + \sqrt{2\gamma kT}\;\xi(t) \tag{1.4.1}$$

in which

$$\langle \xi(t)\xi(t')\rangle = \delta(t-t') \tag{1.4.2}$$

yields the stationary distribution function

$$p_s(x,\dot{x}) = \mathcal{N}\exp\left(-\frac{m\dot{x}^2}{2kT} - \frac{V(x)}{kT}\right) \tag{1.4.3}$$

which is the familiar Boltzmann distribution of equilibrium statistical mechanics.

Any satisfactory theory of quantum noise must give the corresponding quantum result, though some care is needed in formulating what the corresponding quantum result should be: In the words of *Kac* and *Benguria* [1.14],

"The position of a quantum mechanical particle subject to an external potential $V(x)$ and in equilibrium with a heat bath of absolute temperature T should be distributed according to the probability density

$$\frac{\sum_{n=1}^{\infty}\exp\left(-E_n/kT\right)|\psi_n(x)|^2}{\sum_{n=1}^{\infty}\exp\left(-E_n/kT\right)} \tag{1.4.4}$$

where the E_n and ψ_n are respectively the eigenvalues and the normalized eigenfunctions of the Schrödinger equation with potential $V(x)$. One might thus suspect that as $t \to \infty$ the probability density of $x(t)$ should approach the canonical density (1.4.4). But this cannot be so for $\gamma \neq 0$, because for nonvanishing friction one would expect shifting and broadening of the spectral lines. What one should therefore expect is that only in the additional limit $\gamma \to 0$ the limiting distribution of $x(t)$ as $t \to \infty$ should be (1.4.4)."

In a limited form, Benguria and Kac were able to prove the above result by using a model of quantum noise developed previously by *Ford et al.* [1.7] which we shall describe in Chap.3. There we will see that the canonical distribution can be derived directly from a quantum Langevin equation of the kind (1.4.1), by means of a route which yields a generalization of the kind of method used by Einstein—but by a route quite different from that used by Kac and Benguria.

The essence of this model is the rather simple result that, appropriately generalized, (1.4.1) represents a satisfactory quantum theory of Brownian motion of a particle in a potential, provided that:

i) The quantities x, $\dot{x}$, $\ddot{x}$, are regarded as Heisenberg operators, obeying the appropriate canonical commutation relations.

ii) The noise quantity $\xi(t)$ is regarded as an operator, with the commutator

$$[\xi(t), \xi(t')] = 2\mathrm{i}\hbar\gamma \frac{d}{dt}\delta(t - t') \tag{1.4.5}$$

and the mean anticommutator

$$\langle\, [\xi(t), \xi(t')]_+ \rangle = \frac{\gamma\hbar}{\pi} \int_{-\infty}^{\infty} d\omega\, \omega \coth\left(\hbar\omega/2kT\right) \mathrm{e}^{\mathrm{i}\omega(t-t')} . \tag{1.4.6}$$

This latter quantity corresponds exactly to the spectrum of quantum noise (1.2.4). We neglect the technical problems associated with the rising spectrum, which we will go into in Sects.3.4.2, 3.4.3

The limit $h \to 0$ clearly gives a smooth transition to the classical Langevin equation, in the sense that in this limit all commutators vanish and the equation of motion becomes an equation for c-numbers driven by a noise term. Because of the limit,

$$\lim_{\hbar\to 0} \hbar\omega \coth\left(\hbar\omega/2kT\right) = 2kT \tag{1.4.7}$$

this noise term possesses a flat spectrum.

1.4.2 Consistency with Quantum Mechanics

If (1.4.1), regarded as quantum equations with the noise $\xi(t)$ satisfying the conditions (1.4.5, 6), are solved, we will obtain solutions for $x(t)$ and $p(t) \equiv m\dot{x}(t)$ which must obey the canonical commutation relations

$$[x(t), p(t)] = \mathrm{i}\hbar \tag{1.4.8}$$

for all times. The method of derivation of the equations used by Ford *et al.* [1.7] makes it clear that this will be the case. At this stage, however, let us show how a simplified model can illustrate how the existence of noise can ensure that the commutation relations are preserved.

If we consider a harmonic oscillator, of natural frequency ω, then the Heisenberg equations for the *destruction* and *creation* operators

$$a(t) = \frac{\omega x(t) + \mathrm{i}p(t)}{\sqrt{2\hbar\omega}}, \qquad a^\dagger(t) = \frac{\omega x(t) - \mathrm{i}p(t)}{\sqrt{2\hbar\omega}} \tag{1.4.9}$$

are, in the absence of noise and damping,

$$\dot{a}(t) = -\mathrm{i}\omega a(t), \qquad \dot{a}^\dagger(t) = \mathrm{i}\omega a^\dagger(t). \tag{1.4.10}$$

A simplified way to introduce damping and noise is to straightforwardly add appropriate terms to the equation, thus;

$$\dot{a}(t) = -\mathrm{i}\omega a(t) - \gamma a(t) + \sqrt{2\gamma}\, \varGamma(t). \tag{1.4.11}$$

Correspondingly, we also assume some simple delta-correlated commutation relations and correlation function of the noise term $\varGamma(t)$, namely

$$\begin{aligned} [\varGamma(t), \varGamma^\dagger(t')] &= D\delta(t-t') \\ \langle \varGamma^\dagger(t)\varGamma(t')\rangle &= \bar{N}\delta(t-t'). \end{aligned} \tag{1.4.12}$$

Here the delta functions indicate that some kind of quantum generalization of the classical delta correlated white noise process is being contemplated, and D and $\bar{N}$ are numbers which remain to be determined. Solving (1.4.11), we get

$$a(t) = \mathrm{e}^{-(\mathrm{i}\omega+\gamma)t}a(0) + \sqrt{2\gamma}\int_0^t dt'\, \mathrm{e}^{-(\mathrm{i}\omega+\gamma)(t-t')}\varGamma(t') \tag{1.4.13}$$

and if we assume that

$$[a(0), \varGamma^\dagger(t)] = [a^\dagger(0), \varGamma(t)] = 0 \tag{1.4.14}$$

and of course

$$[a(0), a^\dagger(0)] = 1, \tag{1.4.15}$$

then the commutator can be evaluated as

$$[a(t), a^\dagger(t)] = \mathrm{e}^{-2\gamma t} + D\left(1 - \mathrm{e}^{-2\gamma t}\right) \tag{1.4.16}$$

so the commutator will be preserved if

$$D = 1. \tag{1.4.17}$$

The mean energy of the oscillator is similarly found to be

$$\hbar\omega\langle a^\dagger(t)a(t)\rangle = \hbar\omega\mathrm{e}^{-2\gamma t} + \hbar\omega\bar{N}\left(1 - \mathrm{e}^{-2\gamma t}\right) \tag{1.4.18}$$

so that the quantity $\bar{N}$ represents the eventual mean occupation number of the oscillator, and for consistency with the Planck distribution this must require

$$\bar{N} = 1/[\exp(\hbar\omega/kT) - 1]. \tag{1.4.19}$$

It can be seen then that, subject to a set of rather sweeping assumptions (1.4.12, 14), a quantum mechanically consistent picture can be built up. Although this is only intended as an illustration and cannot pretend to give a fully realistic picture of the physics, we will show in Sect.3.4.5 that it can be derived as a high frequency limit of a more physical model, namely, the quantum version of (1.4.1), and indeed from the quantum electrodynamics of an atom interacting with the radiation field. In such a model $\Gamma(t)$ is essentially the operator of the electric field evaluated at the position of the atom. It is therefore quite an appropriate model for quantum optics, for which it was in fact first developed.

However the philosophy originally used was much more like that presented in this section—to assume a damping-noise equation like (1.4.11) and to deduce noise and damping parameters from the preservation of commutation relations and the requirement that the mean energy of the oscillator be that given by statistical mechanics. At these high frequencies this technique works well because there are considerable simplifications available in high frequency systems, as will be shown in Chap.3 and Chap.4. What it shows is what all physicists know—physics has sufficient internal consistency to enable one quite often to guess the right answer to a problem in favourable circumstances. Such guesses are often illuminating, but there is more to physics than guessing.

1.5 Quantum Stochastic Processes and the Master Equation

The approach to quantum noise based on quantum Langevin equations has its roots in the Heisenberg equations of motion. It is well known that in all quantum mechanical situations the Heisenberg equations of motion are not easy to deal with, mainly because of their non-linear operator nature. The Schrödinger equation is always the preferred alternative for practical calculations. The major advantage of the Heisenberg equations arises from their closer resemblance to the corresponding classical equations, which can be of advantage in trying to understand particular problems and in general theoretical formulations. This situation is amply illustrated by the material in this chapter—for example the quantum version of the Langevin equation (1.4.1) is at least easily described. In contrast, it does not seem at all obvious how to generalize the Schrödinger equation to take account of noise.

The appropriate approach was developed in the late 1960s by quantum optical workers, in particular by *Haken, Lax* and *Gordon et al.* [1.15], and this amounts to a generalization of the approach of Einstein discussed in Sect. 1.3. From the rate laws (1.3.3–5), it is clear that there is a probabilistic law governing the probability function $P_n(t)$ which gives the probability that the atom is in the state n; this will be

$$\dot{P}_n(t) = \sum_{m>n}(A^n_m + B^n_m\bar{\rho})P_m(t) + \sum_{m<n} B^n_m\bar{\rho}P_m(t) - \sum_{m<n}(A^m_n + B^m_n\bar{\rho})P_n(t) - \sum_{m>n} B^m_n\bar{\rho}P_n(t) \tag{1.5.1}$$

which is in the form of the classical master equation, S.M. 3.5.1, in which the positive terms represent gain of probability from transitions into the state n, and the negative terms represent loss of probability by transitions from the state n. (Here, as in Sect. 1.3, the symbol $\bar{\rho}$ is used for the energy density of the electromagnetic field to avoid confusion with the use of the symbol ρ for the density operator in the remainder of this book.)

This kind of equation is called the *Pauli master equation*, after *Pauli* who first introduced it [1.13]. However it has the serious defect of dealing only with probabilities, whereas a proper quantum mechanical theory should deal with the *density operator* for the system as described in Sect.2.1. The density operator was introduced by *von Neumann* [1.16] in order to treat situations where there is a statistical element to the physics other than that which arises directly from quantum mechanics, and is thus appropriate in the kind of situation relevant to emission and absorption of light, where the light field is in a thermal state. We will describe how this is to be done in a simplified derivation of the generalized master equation for the interaction of light with an atom.

1.5.1 The Two Level Atom in a Thermal Radiation Field

The physics of the problem of a thermal radiation field interaction with an atom can be viewed as that of an atom interacting with a *random electric field*—but this field must be *operator valued*. Let us consider the simplest possible atom—an atom with only two energy levels. This can be described simply in terms of matrix wavefunctions, labelled by a for the lower level, and b for the upper level. The energy eigenstates are

$$u(a) = \begin{pmatrix} 0 \\ 1 \end{pmatrix} \quad , \quad u(b) = \begin{pmatrix} 1 \\ 0 \end{pmatrix} \tag{1.5.2}$$

and the Hamiltonian is (in the Schrödinger picture)

$$H_{\text{Atom}} = \tfrac{1}{2}\hbar\Omega \begin{pmatrix} 1 & 0 \\ 0 & -1 \end{pmatrix} \equiv \tfrac{1}{2}\hbar\Omega\sigma_z. \tag{1.5.3}$$

We have set the zero of energy to be midway between the upper and lower levels, so that Ω is the frequency of the radiation that the atom will emit or absorb. The description of the interaction with a radiation field can be most simply dealt with by introducing the interaction Hamiltonian

$$H_{\text{Int}} = g(\sigma^+ + \sigma^-)E. \tag{1.5.4}$$

Here E is an operator representing the electric field at the position of the atom, and

$$\sigma^+ = \begin{pmatrix} 0 & 1 \\ 0 & 0 \end{pmatrix}, \qquad \sigma^- = \begin{pmatrix} 0 & 0 \\ 1 & 0 \end{pmatrix} \tag{1.5.5}$$

are operators which allow this field to induce transitions between levels a and b. Finally there must be a Hamiltonian for the radiation field, H_{Rad}, which we shall not specify in detail. The total Hamiltonian is

$$H = H_{\text{Atom}} + H_{\text{Int}} + H_{\text{Rad}}. \tag{1.5.6}$$

The interaction is assumed to be weak, so that to a first approximation, the atom and field do not interact. The appropriate method of solution to von Neumann's equation is to go to the *interaction picture* whose density operator is in terms of the Schrödinger picture density operator for the atom-field system

$$\rho_{\text{Int}}(t) = \exp\left(-\frac{H_{\text{Atom}} + H_{\text{Rad}}}{\mathrm{i}\hbar}t\right)\rho_S(t)\exp\left(\frac{H_{\text{Atom}} + H_{\text{Rad}}}{\mathrm{i}\hbar}t\right). \tag{1.5.7}$$

From von Neumann's equation for the Schrödinger picture density operator

$$\mathrm{i}\hbar\dot{\rho}_S(t) = [H, \rho_S(t)] \tag{1.5.8}$$

follows the interaction picture equation of motion

$$\mathrm{i}\hbar\dot{\rho}_{\text{Int}}(t) = [H_{\text{Int}}(t), \rho_{\text{Int}}(t)] \tag{1.5.9}$$

where

$$H_{\text{Int}}(t) = \exp\left(-\frac{H_{\text{Atom}} + H_{\text{Rad}}}{\mathrm{i}\hbar}t\right)H_{\text{Int}}\exp\left(\frac{H_{\text{Atom}} + H_{\text{Rad}}}{\mathrm{i}\hbar}t\right). \tag{1.5.10}$$

Since H_{Atom} and H_{Rad} represent different degrees of freedom, they commute, and we can then write

$$H_{\text{Int}}(t) = g\left(\sigma^+\mathrm{e}^{\mathrm{i}\Omega t} + \sigma^-\mathrm{e}^{-\mathrm{i}\Omega t}\right)E(t), \tag{1.5.11}$$

where $E(t)$ is defined by

$$E(t) = \exp\left(-\frac{H_{\text{Rad}}}{\mathrm{i}\hbar}t\right)E\exp\left(\frac{H_{\text{Rad}}}{\mathrm{i}\hbar}t\right). \tag{1.5.12}$$

At this stage we can make a "randomness" ansatz for $E(t)$, the electric field operator in the interaction picture. This operator will contain a wide range of frequencies, but the only ones which will give any significant effect in the equation of motion (1.5.9) will be those which almost match the frequencies $\pm\Omega$, so that there will be a part of $H_{\text{Int}}(t)$ which is sufficiently slowly varying for a significant change to $\rho_{\text{Int}}(t)$ to be built up. The terms corresponding to frequencies significantly different from $\pm\Omega$ will oscillate so rapidly that their net effect will be almost zero. We can even evaluate the correlations of $E(t)$, $E(t')$, under this approximation. Let us take the correlations to be given by the same formula as for $\xi(t)$ in (1.4.5, 6) but omit the factor γ, which refers specifically to the interaction. Thus we assume that the electric field operator behaves like our "quantum noise". Then we can write the correlation function in terms of the sum of the commutator and anticommutator terms given by (1.4.5, 6) as

$$\langle E(t)E(t')\rangle = \frac{\hbar}{\pi}\int_{-\infty}^{\infty} d\omega \left(-\omega + \omega \coth\left(\frac{\hbar\omega}{2kT}\right)\right) e^{i\omega(t-t')} \tag{1.5.13}$$

Since only the components near $\pm\Omega$ interact significantly, we may approximate the spectral factors depending on ω by their values at $\pm\Omega$, so

$$e^{i\Omega(t-t')}\langle E(t)E(t')\rangle \sim \frac{\hbar}{\pi}\left(\Omega + \Omega \coth\left(\frac{\hbar\Omega}{2kT}\right)\right)\int_{-\infty}^{\infty} d\omega\, e^{i\omega(t-t')}$$

$$= 4\hbar\Omega\left(\bar{N}(\Omega) + 1\right)\delta(t-t') \tag{1.5.14}$$

and similarly

$$e^{-i\Omega(t-t')}\langle E(t)E(t')\rangle \sim 4\hbar\Omega\bar{N}(\Omega)\delta(t-t') \tag{1.5.15}$$

where $\bar{N}(\Omega)$ is the Planck function (1.4.19). These correlation functions are of exactly the same form as used in the arguments for quantum mechanical consistency of Sect. 1.4.2 so we may make the replacements

$$E(t)e^{i\Omega t} \to 2\sqrt{\hbar\Omega}\,\Gamma(t), \qquad E(t)e^{-i\Omega t} \to 2\sqrt{\hbar\Omega}\,\Gamma^\dagger(t), \tag{1.5.16}$$

where $\Gamma(t)$, $\Gamma^\dagger(t)$ are the quantities of Sect. 1.4.2 with the properties (1.4.12).

The equation of motion for the interaction picture density operator thus becomes

$$i\hbar\dot{\rho}_{\mathrm{Int}} = 2g\sqrt{\hbar\Omega}\left[\sigma^+\Gamma(t) + \sigma^-\Gamma^\dagger(t), \rho_{\mathrm{Int}}\right]. \tag{1.5.17}$$

This is now a kind of quantum white noise equation, but it contains more information than we actually want, since ρ_{Int} is the density operator for the radiation field as well as the atom: we want to obtain an equation for a *reduced* density operator, which refers only to atomic variables—this would be defined by tracing out over the radiation variables i.e.,

$$\tilde{\rho}_{\mathrm{Int}} \equiv \mathrm{Tr}_{\mathrm{Rad}}\left\{\rho_{\mathrm{Int}}\right\}. \tag{1.5.18}$$

To find an equation for $\tilde{\rho}_{\mathrm{Int}}$ we need to trace out (1.5.13) over the radiation field, and will hence need to evaluate terms like

$$\mathrm{Tr}_{\mathrm{Rad}}\left\{\Gamma(t)\rho_{\mathrm{Int}}\right\}. \tag{1.5.19}$$

In the classical theory of white noise differential equations, the very irregularity of the white noise functions (which arises directly from the idea that the noise at one instant of time is absolutely independent of the noise at any other instant, no matter how short the time interval between the two instants) leads to considerable technical difficulties if we assume the equations are truly *differential* equations. The reason is quite straightforward—what is usually assumed to be white noise is the derivative of a non-differentiable, but continuous, random function; the Wiener process. The way out of this logical anomaly in the classical case is to integrate the differential equation and turn it into an integral equation. To do the same thing here we define the integrals of the quantum white noise quantities by

$$B(t,t_0) = \int_{t_0}^{t} \Gamma(t)\,dt, \qquad B^\dagger(t,t_0) = \int_{t_0}^{t} \Gamma^\dagger(t)\,dt. \tag{1.5.20}$$

The differential equation (1.5.17) is then written as the integral equation

$$\rho_{\text{Int}}(t) - \rho_{\text{Int}}(t_0) = -2ig\sqrt{\frac{\Omega}{\hbar}} \int_{t_0}^{t} \left[\sigma^+ dB(t) + \sigma^- dB^\dagger(t), \rho_{\text{Int}}(t)\right]. \tag{1.5.21}$$

The definition of the integral is taken in exactly the same form as that of the classical Stratonovich integral, S.M. 4.3.6, which leads to an approximate form for the situation where

$$t \to t + \delta t, \qquad t_0 \to t - \delta t \tag{1.5.22}$$

and

$$\delta B(t) \equiv B(t+\delta t, t-\delta t), \qquad \delta B^\dagger(t) \equiv B^\dagger(t+\delta t, t-\delta t) \tag{1.5.23}$$

in the form

$$\begin{aligned}\rho_{\text{Int}}(t+\delta t) &- \rho_{\text{Int}}(t-\delta t)\\ &= -ig\sqrt{\frac{\Omega}{\hbar}}\,[\sigma^+\delta B(t) + \sigma^-\delta B^\dagger(t), \rho_{\text{Int}}(t+\delta t) + \rho_{\text{Int}}(t-\delta t)].\end{aligned} \tag{1.5.24}$$

We now want to take the trace over the radiation field. Some terms will obviously vanish, for example $\text{Tr}_{\text{Rad}}\,\{\delta B(t)\rho_{\text{Int}}(t-\delta t)\}$, since from the equation of motion in the form (1.5.24) it is clear that $\rho_{\text{Int}}(t-\delta t)$ must be independent of $\delta B(t)$. The same is not the case with terms involving $\rho(t+\delta t)$, so we can substitute for these using (1.5.24) itself, which yields

$$\begin{aligned}\rho_{\text{Int}}(t+\delta t) - \rho_{\text{Int}}(t-\delta t) &= -2ig\sqrt{\frac{\Omega}{\hbar}}\,\left[\sigma^+\delta B(t) + \sigma^-\delta B^\dagger(t), \rho_{\text{Int}}(t-\delta t)\right]\\ &- \frac{g^2\Omega}{\hbar}\Big[\sigma^+\delta B(t) + \sigma^-\delta B^\dagger(t), \left[\sigma^+\delta B(t) + \sigma^-\delta B^\dagger(t),\right.\\ &\qquad\qquad \left.\rho_{\text{Int}}(t+\delta t) + \rho_{\text{Int}}(t-\delta t)\right]\Big].\end{aligned} \tag{1.5.25}$$

Taking the trace over the radiation field, all terms on the right hand side of the first line vanish, because of the independence of the noise terms of the density operator at the time $t - \delta t$. In the next line we still have the same problem, but we can substitute again, and without much difficulty show that the resulting higher order terms vanish sufficiently rapidly as $\delta t \to 0$ to be neglected. Thus we are left to evaluate terms like

$$\mathrm{Tr}_{\mathrm{Rad}}\left\{\delta B(t)\rho_{\mathrm{Int}}(t-\delta t)\delta B^{\dagger}(t)\right\} \tag{1.5.26}$$

Since the interaction is assumed weak, we approximate

$$\rho_{\mathrm{Int}}(t-\delta t) \approx \tilde{\rho}_{\mathrm{Int}}(t-\delta t)\otimes\rho_{\mathrm{Rad}}(t-\delta t) \approx \tilde{\rho}_{\mathrm{Int}}(t)\otimes\rho_{\mathrm{Rad}}(t). \tag{1.5.27}$$

Thus we can evaluate as follows;

$$\mathrm{Tr}_{\mathrm{Rad}}\left\{\delta B(t)\rho_{\mathrm{Int}}(t)\delta B^{\dagger}(t)\right\}$$

$$= \tilde{\rho}_{\mathrm{Int}}(t)\int_{t-\delta t}^{t+\delta t} dt' \int_{t-\delta t}^{t+\delta t} dt''\,\mathrm{Tr}_{\mathrm{Rad}}\left\{\Gamma(t')\rho_{\mathrm{Rad}}(t)\Gamma^{\dagger}(t'')\right\} \tag{1.5.28}$$

$$= \tilde{\rho}_{\mathrm{Int}}(t)\int_{t-\delta t}^{t+\delta t} dt' \int_{t-\delta t}^{t+\delta t} dt''\,\langle\Gamma^{\dagger}(t'')\Gamma(t')\rangle \tag{1.5.29}$$

$$= 2\tilde{\rho}_{\mathrm{Int}}(t)\bar{N}(\Omega)\delta t. \tag{1.5.30}$$

Putting all these together in (1.5.25), and taking the limit $\delta t \to 0$, we finally obtain the *master equation* for the reduced density matrix in the form

$$\begin{aligned}\frac{d\tilde{\rho}_{\mathrm{Int}}(t)}{dt} &= \frac{2g^2\Omega}{\hbar}\left(\bar{N}(\Omega)+1\right)\left\{2\sigma^-\tilde{\rho}_{\mathrm{Int}}\sigma^+ - \tilde{\rho}_{\mathrm{Int}}\sigma^+\sigma^- - \sigma^+\sigma^-\tilde{\rho}_{\mathrm{Int}}\right\}\\ &+\frac{2g^2\Omega}{\hbar}\bar{N}(\Omega)\left\{2\sigma^+\tilde{\rho}_{\mathrm{Int}}\sigma^- - \tilde{\rho}_{\mathrm{Int}}\sigma^-\sigma^+ - \sigma^-\sigma^+\tilde{\rho}_{\mathrm{Int}}\right\}.\end{aligned} \tag{1.5.31}$$

1.5.2 Relationship to the Pauli Master Equation

The reduced density matrix has the property that its diagonal elements are the probabilities of occupying the relevant states a or b. It is not difficult, by using the definitions (1.5.5), to show that we have the two equations for the diagonal matrix elements

$$P_a(t) \equiv \langle a|\tilde{\rho}_{\mathrm{Int}}(t)|a\rangle\,, \qquad P_b(t) \equiv \langle b|\tilde{\rho}_{\mathrm{Int}}(t)|b\rangle\,, \tag{1.5.32}$$

$$\frac{dP_a}{dt} = \frac{4g^2\Omega}{\hbar}\left\{-\bar{N}(\Omega)P_a + \left(\bar{N}(\Omega)+1\right)P_b\right\} \tag{1.5.33}$$

$$\frac{dP_b}{dt} = \frac{4g^2\Omega}{\hbar}\left\{\bar{N}(\Omega)P_a - \left(\bar{N}(\Omega)+1\right)P_b\right\} \tag{1.5.34}$$

and these equations are exactly of the form of the Pauli master equation (1.5.1). There are a number of points to note.

a) Stationary Solutions: It is straightforward that these are

$$P_a^{\mathrm{s}} = \frac{\bar{N}(\Omega)+1}{2\bar{N}(\Omega)+1} = \frac{1}{1+\exp(-\hbar\Omega/kT)} \tag{1.5.35}$$

$$P_b^{\mathrm{s}} = \frac{\bar{N}(\Omega)}{2\bar{N}(\Omega)+1} = \frac{\exp(-\hbar\Omega/kT)}{1+\exp(-\hbar\Omega/kT)}. \tag{1.5.36}$$

These are of course just the probabilities that Einstein used to derive the theory presented in Sect. 1.3

b) Stimulated and Spontaneous Processes: On comparing the Pauli master equation with (1.5.33,34), we can see that $\bar{\rho}$ and $\bar{N}(\Omega)$ must be proportional. Put more carefully, we can see that there are terms proportional to $\bar{N}(\Omega)$, identified as *stimulated emission and absorption terms* and there are also the *spontaneous emission terms*, which are present even when $\bar{N}(\Omega)$ is zero. These terms arise out of the non-commutativity of the noise terms, $\Gamma(t)$, $\Gamma^\dagger(t)$, which ultimately arises because it is assumed that $E(t)$ is an operator quantity whose commutation relations are given by a form like (1.4.5). Thus these terms, added almost arbitrarily by Einstein, are a natural consequence of the quantum nature of the effective noise. Indeed, one can assume that the radiation field is a classical random quantity with the proper spectrum of quantum noise (1.2.4) and carry out a similar analysis. The result is a similar equation to (1.5.33), but instead of $\bar{N}(\Omega)$ and $\bar{N}(\Omega)+1$, one obtains in both cases $\bar{N}(\Omega)+\frac{1}{2}$, which is wrong physics. Einstein's result amounts to assuming classical noise, but with a Planck spectrum, and then arbitrarily adding the spontaneous terms.

c) Off-Diagonal Elements: The density operator $\bar{\rho}$ has two off-diagonal elements, which are complex conjugates of each other. Following the same methods as before, this gives the equation

$$\frac{d}{dt}\langle a|\rho_{\mathrm{Int}}(t)|b\rangle = -\frac{2g^2\Omega}{\hbar}(2\bar{N}+1)\langle a|\bar{\rho}_{\mathrm{Int}}(t)|b\rangle \tag{1.5.37}$$

so that in the stationary state, this off-diagonal element vanishes. However this does not mean that such off-diagonal elements have no physical relevance. They represent the quantum-mechanical coherence, and what (1.5.37) implies is that, under the influence of random noise, this dies off with time.

d) Application of a Classical Light Field: Suppose, however, we introduce a slightly more intricate situation—an atom in a thermal light field being illuminated by a laser, modelled by a classical field, which we introduce into the Hamiltonian (1.5.4) by

$$E \to E + \mathcal{E}\mathrm{e}^{\mathrm{i}\Omega t} + \mathcal{E}^*\mathrm{e}^{-\mathrm{i}\Omega t} \tag{1.5.38}$$

so that E is the *operator* representing the quantized radiation field, while the remaining term is a classical c-number. (This procedure turns out to be in fact a perfectly correct quantum mechanical way of representing an ideal laser field (see Chap.4) and even in the absence of explicit justification, it is intuitively obvious that if the light field is strong enough, it will behave classically.)

The net effect of making this change is to give a master equation in the form

$$\begin{aligned}\frac{d\tilde{\rho}_{\text{Int}}(t)}{dt} &= \frac{2g^2\Omega}{\hbar}\left(\bar{N}(\Omega)+1\right)\{2\sigma^-\tilde{\rho}_{\text{Int}}\sigma^+ - \tilde{\rho}_{\text{Int}}\sigma^+\sigma^- - \sigma^+\sigma^-\tilde{\rho}_{\text{Int}}\}\\ &+\frac{2g^2\Omega}{\hbar}\bar{N}(\Omega)\{2\sigma^+\tilde{\rho}_{\text{Int}}\sigma^- - \tilde{\rho}_{\text{Int}}\sigma^-\sigma^+ - \sigma^-\sigma^+\tilde{\rho}_{\text{Int}}\}\\ &-\frac{ig}{\hbar}\left[(\mathcal{E}\sigma^+ + \mathcal{E}^*\sigma^-),\tilde{\rho}_{\text{Int}}\right]\end{aligned} \tag{1.5.39}$$

where terms oscillating with frequencies $\pm 2\Omega$ have been omitted to obtain the last line. This equation cannot be reduced to diagonal and off-diagonal equations because the last term couples diagonal and off-diagonal terms. The simplest method of solution is to note that $\tilde{\rho}_{\text{Int}}$ is completely specified by the quantities

$$\left.\begin{aligned}\langle\sigma^\pm\rangle &\equiv \text{Tr}\left\{\tilde{\rho}_{\text{Int}}\sigma^\pm\right\}\\ \langle\sigma_z\rangle &\equiv \text{Tr}\left\{\tilde{\rho}_{\text{Int}}\sigma_z\right\}\end{aligned}\right\} \tag{1.5.40}$$

and to use the algebra of the matrices to derive

$$\frac{d\langle\sigma^+\rangle}{dt} = -\frac{2g^2\Omega}{\hbar}\left(2\bar{N}(\Omega)+1\right)\langle\sigma^+\rangle - \frac{ig}{\hbar}\mathcal{E}^*\langle\sigma_z\rangle \tag{1.5.41}$$

$$\frac{d\langle\sigma^-\rangle}{dt} = -\frac{2g^2\Omega}{\hbar}\left(2\bar{N}(\Omega)+1\right)\langle\sigma^-\rangle + \frac{ig}{\hbar}\mathcal{E}\langle\sigma_z\rangle \tag{1.5.42}$$

$$\frac{d\langle\sigma_z\rangle}{dt} = -\frac{4g^2\Omega}{\hbar}\left(2\bar{N}(\Omega)+1\right)\langle\sigma_z\rangle - \frac{4g^2\Omega}{\hbar} - \tfrac{1}{2}i\frac{g}{\hbar}\left(\mathcal{E}\langle\sigma^+\rangle - \mathcal{E}^*\langle\sigma^-\rangle\right). \tag{1.5.43}$$

We can solve this equation in the stationary situation, to obtain

$$\left.\begin{aligned}\langle\sigma_z\rangle_{\text{s}} &= \frac{-\left(2\bar{N}(\Omega)+1\right)}{\left(2\bar{N}(\Omega)+1\right)^2 + 2|\mathcal{E}|^2/g^2\Omega^2}\\ \langle\sigma^+\rangle_{\text{s}} = \langle\sigma^-\rangle_{\text{s}}^* &= \frac{2i\mathcal{E}/g\Omega}{\left(2\bar{N}(\Omega)+1\right)^2 + 2|\mathcal{E}|^2/g^2\Omega^2}\end{aligned}\right. \tag{1.5.44}$$

from which we can reconstruct $\tilde{\rho}^{\text{s}}_{\text{Int}}$ using

$$\tilde{\rho}^{\text{s}}_{\text{Int}} = \tfrac{1}{2} + \tfrac{1}{2}\langle\sigma_z\rangle_{\text{s}}\sigma_z + \langle\sigma^+\rangle_{\text{s}}\sigma^- + \langle\sigma^-\rangle_{\text{s}}\sigma^+. \tag{1.5.45}$$

From this it is clear that there are non-diagonal terms—these represent quantum mechanical coherence, and arise from the coherent driving field.

2. Quantum Statistics

The very concept of "quantum noise" implies some kind of a unification of statistics and quantum mechanics. It is therefore essential to set out the background to this unification in order to establish both notation and some less widely known matters relevant to the subject.

There are two rather separate subjects which are relevant to the subject "Quantum Statistics". The first of these is the aspect of statistics which arises in the very process of measurement, and this really amounts to the quantum theory of measurement. The usual treatments of the quantum theory of measurement found in texts on quantum mechanics concentrate on developing the formulation of a single measurement at a precise time on a system in a certain quantum state. Such a formulation is in principle sufficient for our needs, but in fact falls rather short of what is in practice done. A typical measurement in a noisy system of any kind (quantum mechanical or otherwise) will amount to the possibly continuous monitoring of a small number of simultaneously measurable quantities, which will probably be multiplied together with time delays to get correlation functions, or be Fourier analysed (either computationally by means of the fast Fourier transform, or physically with a spectrum analyser such as a Fabry-Pérot interferometer) to produce spectra.

Thus, in Sect. 2.2 of this chapter, a kind of quantum measurement theory is introduced which can generate the results of arbitrarily many measurements of the most general kind, and the correlation functions which ensue. The end result is that any time ordered correlation function can be generated by means of sequences of possible measurements.

The formalism of measurement theory is essential—but it must be combined with experience of the everyday world—which is that there is a great deal of randomness in the world as we see it, and this is well described by the quantum mechanical adaptation of the subject of statistical mechanics. Thus the final section of this chapter gives an outline of quantum statistical mechanics in a form suitable for the remainder of the book

2.1 The Density Operator

If a system has a state vector (in the Schrödinger picture) $|\psi, t\rangle$, then it is possible to define the *density operator* for the system by the outer product

$$\rho(t) = |\psi, t\rangle\langle\psi, t|. \tag{2.1.1}$$

2.1.2 Von Neumann's Equation

The equation of motion for the density operator in the form (2.1.4) is obtained directly from the Schrödinger equation for the time development of the state vector,

$$H|\psi_a, t\rangle = i\hbar \frac{d}{dt}|\psi_a, t\rangle \tag{2.1.23}$$

as

$$\dot{\rho}(t) = -\frac{\mathrm{i}}{\hbar}[H, \rho(t)] \tag{2.1.24}$$

where H is the Hamiltonian for the system. This equation has the formal solution

$$\rho(t) = \exp\left(-\frac{\mathrm{i}}{\hbar}Ht\right)\rho(0)\exp\left(\frac{\mathrm{i}}{\hbar}Ht\right). \tag{2.1.25}$$

In contrast to von Neumann's equation, Heisenberg's equation of motion for dynamical operators (in the Heisenberg picture) is

$$\dot{A}(t) = \frac{\mathrm{i}}{\hbar}[H, A(t)] \tag{2.1.26}$$

which differs in sign from (2.1.24).

2.2 Quantum Theory of Measurement

From the early days of quantum mechanics the contradiction between our experience that quantities can be measured and the quantum mechanical formulation of physics only in terms of probability amplitudes has given the field of quantum measurement theory a rather metaphysical character. There is a certain way in which the theory of quantum noise can clarify matters, as will be shown in Chap.7.

The original formulation of measurement theory is that of *von Neumann* [2.1], and in spite of its rather artificial appearance, it does seem to be in accord with reality. Thus, in this section, this basic and traditional formalism will be outlined and developed along the lines of *Caves* [2.2] to give us in the end a theory of repeated measurements on the same system. From this we can develop the theory of *correlation functions* as the basic measurable quantities of quantum theory.

This section will discuss only the basic and traditional formalism of quantum measurement theory. There are some mysteries in this formalism, which impinge directly on the field of quantum noise, and these will be discussed in Chap.7.

2.2.1 Precise Measurements

Von Neumann's theory of measurement is very simple in the case that the specification of the eigenvalue of one variable A specifies the state of the system—thus we can label the states as $|a\rangle$ where a is the eigenvalue of A. The measurement postulate states that if the system is originally in the state $|\psi, t\rangle$ and the value a of A is measured then

i) The value a is found with probability $|\langle a|\psi, t\rangle|^2$

ii) The state after the measurement is $|a\rangle$.

Thus the state vector is "instantaneously" reduced from $|\psi, t\rangle$ to $|a\rangle$. (A more modest interpretation is that the state transforms from $|\psi, t\rangle$ to $|a\rangle$ in the time taken to make the measurement.) This may be symbolized as

$$|\psi, t\rangle \rightarrow P_a|\psi, t\rangle \equiv \big(|a\rangle\langle a|\big)|\psi, t\rangle \tag{2.2.1}$$

Note that P_a is a *projector*, i.e. $P_a^2 = P_a$, so that the evolution implied by (2.2.1) is not unitary, as is obvious from the fact that the norm of the right hand side is $|\langle a|\psi, t\rangle|^2$, and not one.

The measurement postulate as it stands is a postulate on its own, which has been shown to be *compatible* with quantum mechanics, but has not yet been satisfactorily derived from the dynamical postulates of quantum mechanics. (See however [2.3] for an interesting attempt.) Most physicists are uncomfortable with the idea of measurement as being in some way in addition to dynamics—it is felt that we are all part of this physical world, and any conclusions we draw about it should arise from the same dynamical description that we apply to it.

2.2.2 Imprecise Measurements

It is not possible to build a measurement theory solely on the basis of precise measurements, for two reasons:

i) No perfect measuring apparatus has ever been built, so there is always some intrinsic error to any measurement in addition to that required by quantum mechanics.

ii) Usually a complete description of a system is effected most naturally by the use of more than one operator, e.g., the operators for x,y and z coordinates are necessary to specify the position of a particle.

We therefore have to work out a theory of measurement in which we do not measure sufficient variables to specify precisely what the final state is.

We will consider a system describable as above, by the eigenstates $|a\rangle$, and assume that these have a discrete spectrum. It is obviously possible to approximate any situation by this kind of description, since a discrete set can always be ordered, and hence described by a single operator whose eigenvalues represent the position in order.

It is experimentally true that a measurement of any kind appears to leave the system in a final state which is a *linear* function of the initial state, in the sense implied by (2.2.1), that the final state is given up to a normalization factor by a linear operator acting on it.

With an imprecise measurement there is some arbitrariness in defining what the measured value obtained by a measurement actually is. An imprecise measurement necessarily will sometimes give a wrong answer! A more balanced point of view is perhaps that the value we infer from a measurement of A has an error associated with it, and we can only say that the "true value" lies within some range of the measured value.

All of this can be compactly put together into a rather elegant formalism. Suppose $\bar{a}$ are the possible "measured values" of A which result as the end result of a measurement process. For each $\bar{a}$ we introduce the amplitudes $\Upsilon_{\bar{a}a}$ and the operator

$$\hat{\Upsilon}_{\bar{a}} \equiv \sum_a \Upsilon_{\bar{a}a}|a\rangle\langle a| = \sum_a \Upsilon_{\bar{a}a}P_a. \tag{2.2.2}$$

This definition of the operator $\hat{\Upsilon}_{\bar{a}}$ actually means that $\hat{\Upsilon}_{\bar{a}}$ is a function of the operator A which is being measured.

We assume that the amplitudes are normalized

$$\sum_{\bar{a}} |\Upsilon_{\bar{a}a}|^2 = 1. \tag{2.2.3}$$

This normalization property is related to the completeness of the possible measured values $\bar{a}$, and can be interpreted as meaning that the probability of obtaining one of all the possible values $\bar{a}$ is one—there are therefore no other possible values.

We can also introduce another operator, called an *effect*, $\hat{F}_{\bar{a}}$, given by

$$\hat{F}_{\bar{a}} = \hat{\Upsilon}_{\bar{a}}^\dagger \hat{\Upsilon}_{\bar{a}} = \sum_a |\Upsilon_{\bar{a}a}|^2 P_a. \tag{2.2.4}$$

Notice that there is a completeness relation

$$\sum_{\bar{a}} \hat{F}_{\bar{a}} = \sum_{a\bar{a}} |\Upsilon_{\bar{a}a}|^2 P_a = \sum_a P_a = 1, \tag{2.2.5}$$

which follows by use of (2.2.3).

We can now put all of this together to give a theory of imprecise measurement, via two postulates.

i) The probability of obtaining the value $\bar{a}$ as a result of this kind of measurement is

$$P(\bar{a}) = \langle\psi, t|\hat{F}_{\bar{a}}|\psi, t\rangle = \sum_a |\Upsilon_{\bar{a}a}|^2 |\langle a|\psi, t\rangle|^2. \tag{2.2.6}$$

ii) If the measured result is $\bar{a}$, then the state of the system immediately after the measurement is

$$|\psi_{\bar{a}}, t\rangle = \frac{\hat{\Upsilon}_{\bar{a}}|\psi, t\rangle}{\sqrt{P(\bar{a})}} = \sum_a |a\rangle \frac{\Upsilon_{\bar{a}a}\langle a|\psi, t\rangle}{\sqrt{P(\bar{a})}}. \tag{2.2.7}$$

Comments

i) Notice that this theory reduces to the precise measurement theory in the case that $\Upsilon_{\bar{a}a}$ is zero unless $\bar{a} = a$.

ii) A particular case of great interest occurs when $\Upsilon_{\bar{a}a}$ is either 1 or 0, so that

$$\hat{\Upsilon}_{\bar{a}}^2 = \hat{\Upsilon}_{\bar{a}} \tag{2.2.8}$$

and thus $\hat{\Upsilon}_{\bar{a}}$ is a projection operator. If we further consider the situation that each value of $\bar{a}$ corresponds to a range of values of a which does not overlap with that for any other value of $\bar{a}$, we have the orthogonality condition,

$$\hat{\Upsilon}_{\bar{a}}\hat{\Upsilon}_{\bar{b}} = \delta_{\bar{a}\bar{b}}\hat{\Upsilon}_{\bar{a}}. \tag{2.2.9}$$

In this case, the measurement of A is essentially equivalent to a precise measurement of the reduced operator $\tilde{A} = \sum_{\bar{a}} \hat{\Upsilon}_{\bar{a}}$ in which the unmeasured range is not affected. This happens in practice when the system can be described by two commuting operators, A_1 and A_2, and we make a measurement simply by applying the precise measurement theory to one of them, so that the relevant projection operator takes the form

$$\hat{\Upsilon}_{\bar{a}_1} = \sum_{a_2} |\bar{a}_1, a_2\rangle\langle\bar{a}_1, a_2|. \tag{2.2.10}$$

iii) In general, however, the imprecise measurement does not correspond to the precise measurement of any operator.

iv) The construction of $\hat{F}_{\bar{a}}$ shows that there are a large number of $\hat{\Upsilon}_{\bar{a}}$ which give the same effect, $\hat{F}_{\bar{a}}$, and from (2.2.6) it is clear that all the information about the measured values obtained from the measurement is contained in the effect $\hat{F}_{\bar{a}}$. The phases of the $\Upsilon_{\bar{a}a}$ are essential only for the elucidation of the final state of the system. However this information is very important if we wish to carry out a further measurement after the first.

v) In principle, any $\Upsilon_{\bar{a}a}$ are allowed for any possible basis set $|a\rangle$ in the quantum mechanical Hilbert space.

2.2.3 The Quantum Bayes Theorem

Caves [2.2] has pointed out that the measurement formulation has a very suggestive interpretation as a kind of quantum Bayes theorem. From (2.2.7) we can derive the result

$$|\langle a|\psi,t\rangle|^2|\Upsilon_{\bar{a}a}|^2 = P(\bar{a})|\langle a|\psi_{\bar{a}},t\rangle|^2 \tag{2.2.11}$$

$$\equiv P(a;\bar{a}). \tag{2.2.12}$$

We introduce the quantity $P(a;\bar{a})$ as a joint probability that the system had the eigenvalue a before measurement and the result $\bar{a}$ was obtained as a result of the measurement, for we can make the two interpretations, depending on whether we look at the LHS or the RHS of (2.2.11) :

LHS; Refers to the situation before measurement.

$P(a;\bar{a}) =$ *(probability that the system is in state a before measurement)* $\times$ *(probability of getting the value $\bar{a}$ given that it was in the state a before measurement)*

RHS; Refers to the situation after measurement.

$P(a;\bar{a}) =$ *(probability that the result of measurement was $\bar{a}$)* $\times$ *(probability that the system was in the state a after measurement)*

If we now consider two different reference states, $|a\rangle$ and $|b\rangle$, we can cancel out $P(\bar{a})$, to get

$$\frac{|\langle a|\psi_{\bar{a}},t\rangle|^2}{|\langle b|\psi_{\bar{a}},t\rangle|^2} = \frac{|\langle a|\psi,t\rangle|^2|\Upsilon_{\bar{a}a}|^2}{|\langle b|\psi,t\rangle|^2|\Upsilon_{\bar{a}b}|^2} \tag{2.2.13}$$

or, one sees that (2.2.13) can be written as

$$|\langle a|\psi_{\bar{a}},t\rangle|^2 \propto |\langle a|\psi,t\rangle|^2|\Upsilon_{\bar{a}a}|^2 \qquad \text{for fixed } \bar{a}, \tag{2.2.14}$$

which means, in words,

Probability the system is in the state $|a\rangle$, given that the result $\bar{a}$ has been measured $\propto$ Probability that the system was in state $|a\rangle$ before measurement

$\times$ probability that the result is $\bar{a}$ given that the state is $|a\rangle$.

This is in exact analogy to the Bayes theorem concepts of classical probability theory. The term $|\langle a|\psi,t\rangle|^2$ is the a priori probability of the value a. The term $|\Upsilon_{\bar{a}a}|^2$ is a conditional probability for getting the result $\bar{a}$, given that we start with $|a\rangle$—this can be seen directly from (2.2.6), where we only need to set $|\psi,t\rangle = |a\rangle$ to get

$$\{P(\bar{a}) \text{ given the initial state is } |a\rangle\} = |\Upsilon_{\bar{a}a}|^2. \tag{2.2.15}$$

Thus, like Bayes theorem, this result takes the conditional probabilities for a result, given an initial value, and the a priori probabilities of the initial values, and yields a result for the probability of a given value, given a particular result.

Finally we note that there is a more powerful result here, for probability amplitudes, following from (2.2.7). We see that

$$|\langle a|\psi_{\bar{a}},t\rangle|P(\bar{a})^{\frac{1}{2}} = \langle a|\psi,t\rangle\Upsilon_{\bar{a}a}. \tag{2.2.16}$$

Using different $|a\rangle, |b\rangle$, and cancelling $P(a)^{\frac{1}{2}}$ as above, we deduce that

$$\langle a|\psi_{\bar{a}},t\rangle \propto \langle a|\psi,t\rangle\Upsilon_{\bar{a}a} \qquad \text{for fixed } \quad \bar{a}, \tag{2.2.17}$$

which gives an amplitude interpretation of Bayes theorem. This is a much stronger result than the form (2.2.14), and fully embodies the quantum mechanical centrality of the concept of a probability amplitude. The formulation (2.2.17) is fully equivalent to the original postulates (2.2.6,7), but presents a much more natural way of looking at them.

2.2.4 More General Kinds of Measurements

We have to ask what kind of operations $\hat{\Upsilon}_{\bar{a}}$ are allowed by the theoretical development of Sect. 2.2.2. This is easily seen from (2.2.2), which implies that $\hat{\Upsilon}_{\bar{a}}$ is a function of the operator A which is itself Hermitian and an observable. This is of course quite a restrictive requirement; that there must exist some observable A of which $\hat{\Upsilon}_{\bar{a}}$ is a function, and indeed this is the observable which is being measured.

Is this assumption needed? The reason for it lies in our concept of what we would like to think a measurement is. This kind of measurement has the property that if the initial state is the eigenstate $|a_0\rangle$ of A, then the state after measurement is also $|a_0\rangle$. In this sense we can say that the system is not disturbed by the measurement when the initial value is precise. This arises basically by assumption, and as a consequence, $\hat{\Upsilon}_{\bar{a}}$ is a function of A. It is conventional to call this *a measurement of the first kind.*

There are physical observations which do not have this property. For example in photon counting we count photons by absorbing them, and no matter what state the system is initially in, the final state will have fewer photons. The measurement destroys the state being measured. We are all aware that this kind of measurement is quite important in everyday life. For example, one usually tests the quality of an apple by eating it. At the end of the experiment we have no apple, but consider the information gained to be relevant to similarly prepared apples, which others may choose to eat. Classical sampling theory does not normally take into account that sampling may destroy the sample, even though it is obvious that such destructive sampling is very common. It is conventional to say that these measurements are *measurements of the second kind.* In quantum mechanics we can conceptualize such a measurement very similarly to the way we have already done. The simplest thing to assume is that the operator $\hat{\Upsilon}_{\bar{a}}$, which is a function of the observable A, is replaced by a general operator $\hat{\Phi}(\bar{a})$, which can be represented as

$$\hat{\Phi}(\bar{a}) = \sum_{a,b} \Phi(\bar{a})_{ab}|a\rangle\langle b|. \tag{2.2.18}$$

We now define the effect corresponding to the operator $\hat{\Phi}(\bar{a})$ by

$$\hat{G}(\bar{a}) = \hat{\Phi}^\dagger(\bar{a})\hat{\Phi}(\bar{a}) \tag{2.2.19}$$

and the assumption of completeness is

$$\sum_{\bar{a}} \hat{G}(\bar{a}) = 1. \tag{2.2.20}$$

The measurement postulates are now framed from the point of view that $\bar{a}$ is not so much the value of an operator A as a label of a possible outcome of the measurement. We then postulate

i) The probability of obtaining the outcome $\bar{a}$ as a result of the measurement is

$$P(\bar{a}) = \langle\psi,t|\hat{G}(\bar{a})|\psi,t\rangle. \tag{2.2.21}$$

ii) If the outcome is $\bar{a}$, then the state of the system immediately after the measurement is

$$|\psi_{\bar{a}}, t\rangle = \frac{\hat{\Phi}(\bar{a})|\psi, t\rangle}{\sqrt{P(\bar{a})}} = \sum_{a,b} \frac{|a\rangle\Phi(\bar{a})_{a,b}\langle b|\psi, t\rangle}{\sqrt{P(\bar{a})}}. \tag{2.2.22}$$

The definition (2.2.18) of $\hat{\Phi}(\bar{a})$ means that it is a completely general linear operator; thus (2.2.22) will give the most general *linear* functional of the initial state which results from the process that we view as the measurement giving the result $\bar{a}$.

Notice that the completeness assumption (2.2.20) means that

$$\sum_{\bar{a}} P(\bar{a}) = 1 \tag{2.2.23}$$

as required for a probability

Example: Let us consider a two level atom, described by the two states

$$u(+) = \begin{pmatrix} 1 \\ 0 \end{pmatrix} \quad , \quad u(-) = \begin{pmatrix} 0 \\ 1 \end{pmatrix} \tag{2.2.24}$$

and assume that we wish to test whether the atom is in the excited state, $u(+)$, by absorbing the energy in it, and thus leaving it in the lower state. Let us also allow the method of measurement to be somewhat imperfect, so that even if the atom is in the upper state, we will not necessarily absorb the energy—however we do assume that if the energy is absorbed, then we do detect it.

The first outcome is that the energy is detected, and the atom left in the ground state; thus

$$\hat{\Phi}(1) = \begin{pmatrix} 0 & 0 \\ \lambda & 0 \end{pmatrix} \tag{2.2.25}$$

is a possible choice. The only possible other outcome is that the energy is not detected, in which case (because we assume that if the energy is released, it is detected) the atom, if it is in either the ground or the excited state, remains there. The only reasonable choice for the other operation is

$$\hat{\Phi}(2) = \begin{pmatrix} \mu & 0 \\ 0 & 1 \end{pmatrix} \tag{2.2.26}$$

where

$$|\mu|^2 = 1 - |\lambda|^2 \tag{2.2.27}$$

which ensures that

$$\hat{\Phi}(2)^\dagger\hat{\Phi}(2) + \hat{\Phi}(1)^\dagger\hat{\Phi}(1) = 1. \tag{2.2.28}$$

We can then see that on a general state $\begin{pmatrix} u \\ v \end{pmatrix}$ the final state is

$$\begin{pmatrix} 0 & 0 \\ \lambda & 0 \end{pmatrix} \begin{pmatrix} u \\ v \end{pmatrix} = \begin{pmatrix} 0 \\ \lambda u \end{pmatrix} \qquad \text{if the energy is detected,} \tag{2.2.29}$$

and

$$\begin{pmatrix} \mu & 0 \\ 0 & 1 \end{pmatrix} \begin{pmatrix} u \\ v \end{pmatrix} = \begin{pmatrix} \mu u \\ v \end{pmatrix} \qquad \text{if the energy is not detected.} \tag{2.2.30}$$

2.2.5 Measurements and the Density Operator

The general measurement theory can be quite simply expressed in terms of the density operator as follows. If the density operator is initially ρ, then the probability of the outcome $\bar{a}$ is

$$P(\bar{a}) = \mathrm{Tr}\left\{ \hat{\Phi}(\bar{a})\rho\, \hat{\Phi}^\dagger(\bar{a})\right\} \tag{2.2.31}$$

and after the measurement, the density operator has the form

$$\rho(\bar{a}) = \hat{\Phi}(\bar{a})\rho\hat{\Phi}^\dagger(\bar{a})/P(\bar{a}). \tag{2.2.32}$$

This expression in terms of the density operator is merely a transcription from the formulation in terms of state vectors. But there is a real generalization of the concept of a measurement which can only be expressed using the density operator.

Let us return to the example of Sect. 2.2.4, and consider the situation in which the energy may be absorbed from the atom, but yet not be detected, because the detector is not perfect. It is natural to try to remedy this by including in $\hat{\Phi}(2)$ an amplitude to de-excite the atom, thus we might modify the expression (2.2.26) for $\hat{\Phi}(2)$ to $\begin{pmatrix} \mu & 0 \\ \tau & 1 \end{pmatrix}$, where the term τ allows this possibility. However to preserve the completeness (2.2.28), we find we have to modify both $\hat{\Phi}(1)$ and $\hat{\Phi}(2)$ to

$$\hat{\Phi}(1) = \begin{pmatrix} 0 & 0 \\ \lambda & -\tau^*/\sqrt{1-|\mu|^2} \end{pmatrix} \tag{2.2.33}$$

$$\hat{\Phi}(2) = \begin{pmatrix} \mu & 0 \\ \tau & \lambda^*/\sqrt{1-|\mu|^2} \end{pmatrix}. \tag{2.2.34}$$

where $|\lambda|^2 + |\mu|^2 + |\tau|^2 = 1$. The new expression for $\hat{\Phi}(1)$ means that there is a probability of detecting the energy even when the atom is in the ground state. While this is not impossible, it is certainly not experienced in all detectors. It is a particular (an unlikely) way in which inefficiency can arise.

A more natural way is to introduce a probability of detection of the energy released; i.e. we separate the detection of the energy once it has been released, from the release of the energy by the atom. We can thus isolate three possibilities. If ϵ is the efficiency of detecting the energy once it is released, the operations are:

$\hat{\Phi}(1) \quad = \begin{pmatrix} 0 & 0 \\ \lambda & 0 \end{pmatrix}$ The atom is put into the ground state, and the energy is detected, with probability ϵ.

$\hat{\Phi}(2,\alpha) = \begin{pmatrix} \mu & 0 \\ 0 & 1 \end{pmatrix}$ The atom remains in the ground or the excited state, and no energy is detected.

$\hat{\Phi}(2,\beta) = \begin{pmatrix} 0 & 0 \\ \lambda & 0 \end{pmatrix}$ The atom is put into the ground state, and no energy is detected, with probability $1-\epsilon$.

The observable outcomes of the $\hat{\Phi}(2,\alpha)$ and $\hat{\Phi}(2,\beta)$ are the same, and the density operator after performing the detection, and observing outcome 2 (no energy detected) is the average of the two. Thus, the probability of outcome 2 is

$$P(2) \equiv \mathrm{Tr}\left\{\hat{\Phi}(2,\alpha)\,\rho\,\hat{\Phi}^\dagger(2,\alpha) + (1-\epsilon)\,\hat{\Phi}(2,\beta)\,\rho\,\hat{\Phi}^\dagger(2,\beta)\right\} \tag{2.2.35}$$

and the density operator after measurement is given by

$$\rho(2) = P(2)^{-1}\left\{\hat{\Phi}(2,\alpha)\,\rho\,\hat{\Phi}^\dagger(2,\alpha) + (1-\epsilon)\,\hat{\Phi}(2,\beta)\,\rho\,\hat{\Phi}^\dagger(2,\beta)\right\}. \tag{2.2.36}$$

On the other hand the probability of outcome 1 is

$$P(1) = \mathrm{Tr}\left\{\epsilon\,\hat{\Phi}(1)\,\rho\,\hat{\Phi}^\dagger(1)\right\} \tag{2.2.37}$$

and

$$\rho(1) = P(1)^{-1}\,\epsilon\,\hat{\Phi}(1)\,\rho\,\hat{\Phi}^\dagger(1). \tag{2.2.38}$$

Notice that the requirement $P(1)+P(2)=1$ is ensured by the completeness sum in the form

$$\epsilon\,\hat{\Phi}^\dagger(1)\,\hat{\Phi}(1) + (1-\epsilon)\,\hat{\Phi}^\dagger(2,\beta)\,\hat{\Phi}(2,\beta) + \hat{\Phi}^\dagger(2,\alpha)\,\hat{\Phi}(2,\alpha) = 1. \tag{2.2.39}$$

General Formulation: We can now formulate this kind of generalized detection as follows. We assume that the outcomes of a given detection process are labelled by the index $\bar{a}$, as before, and that each outcome can arise by a number of different operations $\hat{\Phi}(\bar{a},\alpha)$. Then the probability of measuring the outcome $\bar{a}$ is

$$P(\bar{a}) = \sum_\alpha \mathrm{Tr}\left\{\hat{\Phi}(\bar{a},\alpha)\rho\,\hat{\Phi}^\dagger(\bar{a},\alpha)\right\} \tag{2.2.40}$$

and the density operator after the measurement is

$$\rho(\bar{a}) = P(\bar{a})^{-1}\sum_\alpha \hat{\Phi}(\bar{a},\alpha)\rho\,\hat{\Phi}^\dagger(\bar{a},\alpha). \tag{2.2.41}$$

Notice that the efficiency ϵ of the example has been absorbed into the definition of the effect; thus the completeness takes the form

$$\sum_{\bar{a},\alpha} \hat{\Phi}^\dagger(\bar{a},\alpha)\hat{\Phi}(\bar{a},\alpha) = 1. \tag{2.2.42}$$

The type of mapping $\rho \to \rho(\bar{a})$ specified by (2.2.41) is known as a *completely positive map*, since if ρ is itself a positive definite operator, i.e. one such that $\langle u|\rho|u\rangle \geq 0$ for all vectors $|u\rangle$, then so is $\rho(\bar{a})$ positive definite.

This is the most general kind of formulation of a measurement which is usually made.

2.3 Multitime Measurements

We now want to extend the measurement theory developed for measurements at a single time to a theory of measurements at several times. The basic process is quite easy to formulate. A measurement at an initial time instantaneously transforms the state of the system, which then evolves according to a Hamiltonian time development until the next measurement which instantaneously transforms the state of the system. The Hamiltonian evolution continues until the next measurement and so on. Eventually we obtain the multitime joint probability for obtaining a sequence of measured values corresponding to the measurements carried out sequentially.

We will be able to show that by means of such a process we can, in principle, measure any time ordered correlation function of any quantum mechanical operators.

2.3.1 Sequences of Measurements

Let us, for simplicity, consider the process of carrying out a sequence of measurements on a *well defined initial quantum state* $|\psi, t_0\rangle$. We must first take account of the fact that during the times between successive measurements the system will evolve unitarily. Thus, provided no disturbance or measurement happens to the system between t and $t+\tau$,

$$|\psi, t+\tau\rangle = U(t+\tau, t)|\psi, t\rangle. \tag{2.3.1}$$

Suppose we will make measurements at times $t_q, q = 1, 2, 3, \ldots$, where $t_q > t_{q-1}$. We label the possible results of the qth measurement by $\bar{a}_q$. At some initial time $t_0 < t_1$ we assume the state of the system is $|\psi_0, t_0\rangle$.

The result can be developed by iteration. We assume that the results of the first $q-1$ samples are $\bar{a}_1, \bar{a}_2, \ldots \bar{a}_{q-1}$. The state of the system just after the $(q-1)$th measurement depends, in general, on the results of all previous measurements, so we will denote it by $|\psi_{\bar{a}_1,\bar{a}_2,\ldots\bar{a}_{q-1}}, t_{q-1}\rangle$. The state just before the qth measurement is given by the unitary time evolution from the state just after the $(q-1)$th measurement, namely

$$|\psi_{\bar{a}_1\bar{a}_2,\ldots\bar{a}_{q-1}}, t_q\rangle = U(t_q, t_{q-1})|\psi_{\bar{a}_1\bar{a}_2,\ldots\bar{a}_{q-1}}, t_{q-1}\rangle. \tag{2.3.2}$$

We now use (2.2.6) to compute that the probability of obtaining the result $\bar{a}_q$ at the qth measurement, given the results of all previous samplings, is

$$P(\bar{a}_q|\bar{a}_1, \bar{a}_2, \dots, \bar{a}_{q-1}) = \langle \psi_{\bar{a}_1}, \dots_{\bar{a}_{q-1}}, t_q|\hat{F}_{\bar{a}_q}|\psi_{\bar{a}_1,\dots\bar{a}_{q-1}}, t_q\rangle \tag{2.3.3}$$

and the wavefunction is

$$|\psi_{\bar{a}_1,\dots\bar{a}_q}, t_q\rangle = \frac{\hat{\Upsilon}_{\bar{a}_q}|\psi_{\bar{a}_1,\dots\bar{a}_{q-1}}, t_q\rangle}{\sqrt{P(\bar{a}_q|\bar{a}_1, \bar{a}_2, \dots, \bar{a}_{q-1})}}. \tag{2.3.4}$$

We can now iterate these equations to get

$$|\psi_{\bar{a}_1,\dots\bar{a}_q}, t_q\rangle = \frac{\left[\prod_{r=1}^{q} \Upsilon_{\bar{a}_r} U(t_r, t_{r-1})\right] |\psi_0, t_0\rangle}{\sqrt{P(\bar{a}_1; \dots; \bar{a}_q)}}, \tag{2.3.5}$$

where the product in the numerator is ordered so that the times increase to the left, and

$$P(\bar{a}_1; \dots; \bar{a}_q) = \prod_{r=1}^{q} P(\bar{a}_r|\bar{a}_1, \dots, \bar{a}_{r-1}). \tag{2.3.6}$$

Since the state is normalized, we deduce that

$$P(\bar{a}_1; \dots; \bar{a}_q) = \langle \psi_0, t_0| \left[\prod_{r=1}^{q} \hat{\Upsilon}_{\bar{a}_r} U(t_r, t_{r-1})\right]^\dagger \prod_{r=1}^{q} \hat{\Upsilon}_{\bar{a}_r} U(t_r, t_{r-1})|\psi_0, t_0\rangle. \tag{2.3.7}$$

2.3.2 Expression as a Correlation Function

If we define the Heisenberg picture operators corresponding to the $\hat{\Upsilon}_{\bar{a}}$ by

$$\hat{\Upsilon}_{\bar{a}}(t) = U^\dagger(t, t_0)\hat{\Upsilon}_{\bar{a}} U(t, t_0) \tag{2.3.8}$$

the formula (2.3.7) can be written

$$P(\bar{a}_1; \bar{a}_2; \dots \bar{a}_q) = \langle \psi_0, t_0|\hat{\Upsilon}^\dagger_{\bar{a}_1}(t_1) \dots \hat{\Upsilon}^\dagger_{\bar{a}_q}(t_q)\hat{\Upsilon}_{\bar{a}_q}(t_q) \dots \hat{\Upsilon}_{\bar{a}_1}(t_1)|\psi_0, t_0\rangle. \tag{2.3.9}$$

This is a kind of correlation function of the operators $\hat{\Upsilon}_{\bar{a}}(t)$ with each other.

2.3.3 General Correlation Functions

There is in principle no reason why the measurement cannot be a completely arbitrary operation $\hat{\Phi}(\bar{a})$, apart from the normalization necessary to make the completeness assumption (2.2.20) possible. This means that the arguments of Sect. 2.2.4 can be carried through to show that the joint probability for a sequence of outcomes $\bar{a}_1, \bar{a}_2, \dots$ can be related directly to the correlation functions of the $\hat{\Phi}(\bar{a}_i)$, which are essentially arbitrary; thus following (2.3.9)

$$P(\bar{a}_1; \bar{a}_2; \dots; \bar{a}_q) = \langle \psi_0, t_0|\hat{\Phi}^\dagger(\bar{a}_1, t_1) \dots \hat{\Phi}^\dagger(\bar{a}_q, t_q)\hat{\Phi}(\bar{a}_q, t_q) \dots \hat{\Phi}(\bar{a}_1, t_1)|\psi_0, t_0\rangle. \tag{2.3.10}$$

We can use the polarization identity

$$A^\dagger M B = \tfrac{1}{4}\Big\{(A+B)^\dagger M(A+B) - (A-B)^\dagger\, M(A-B) \\ -\mathrm{i}(A+\mathrm{i}B)^\dagger M(A+\mathrm{i}B) + \mathrm{i}(A-\mathrm{i}B)^\dagger M(A-\mathrm{i}B)\Big\} \tag{2.3.11}$$

to generate more general correlations by linear combinations of probabilities, and by assuming that the $\hat{\Phi}(\bar{a},t)$ are complete in the sense that any operator including the identity, can be expressed in terms of a linear combination of them, we can generate correlation functions of the kind

$$\langle \psi_0, t_0 | A_1(t_1) A_2(t_2) \dots A_n(t_n) B_m(s_m) \dots B_2(s_2) B_1(s_1) | \psi_0, t_0 \rangle \tag{2.3.12}$$

where the A_i and the B_i are arbitrary operators, and s_i and t_i are times such that

$$t_1 \le t_2 \le \dots \le t_n, \qquad s_1 \le s_2 \le \dots \le s_m. \tag{2.3.13}$$

Such correlation functions are generally called *time ordered correlation functions*, and are the only kind of correlation functions susceptible to direct measurement in this form.

2.4 Quantum Statistical Mechanics

The introduction of the density operator as a means of describing quantum mechanical states in the absence of complete information is the first step in formulating quantum statistics. While the technique is very flexible, there is still a lack of structure—we can conceive of large numbers of ways in which information may be incomplete, but these may not be very relevant to real life. The contact with reality comes from the introduction into quantum mechanics of the concepts of temperature, entropy, and statistical ensembles, which were so successful in the nineteenth century development of statistical mechanics. The formal method of doing this is via the concept of entropy maximization, subject to constraints.

2.4.1 Entropy

The concept of a macrostate, defined by the values of certain macroscopic variables, such as temperature, pressure, momentum, etc., of a system composed of a very large number of particles is fundamental to statistical mechanics. A macrostate is represented microscopically by myriads of possible configurations of the microscopic variables, (energy, momentum, etc., of the individual particles)—there is no unique microscopic configuration corresponding to a macrostate (other than in very exceptional cases).

The *entropy* of a macrostate is introduced as a measure of how many different microstates represent the same macrostate. In a quantum mechanical system, this gives a particularly simple formula for entropy. If we consider a density operator ρ which is diagonal in a certain basis, then ρ may be written as a matrix

$$\rho = \begin{pmatrix} p_1 & 0 & 0 & 0 & \dots \\ 0 & p_2 & 0 & 0 & \dots \\ 0 & 0 & p_3 & 0 & \dots \\ 0 & 0 & 0 & p_4 & \dots \\ \vdots & \vdots & \vdots & \vdots & \ddots \end{pmatrix}. \tag{2.4.1}$$

Suppose N different quantum states contribute to this matrix in equal proportions: then the logical measure of entropy is

$$S = k \log N \tag{2.4.2}$$

where k is Boltzmann's constant. In this case, we have also

$$\left.\begin{aligned} p_i &= \frac{1}{N} & \quad \mathrm{i} &\leq N \\ &= 0 & \quad \mathrm{i} &> N \end{aligned}\right\} \tag{2.4.3}$$

and we can use this to write the entropy in a form which does not depend on the particular basis, for it is clear that in this situation

$$\mathrm{Tr}\,\{\rho \log \rho\} = \sum p_i \log p_i = -\log N \tag{2.4.4}$$

where we have used $x \log x = 0$ for $x = 0$.

A change of basis is a unitary transformation, which leaves the trace invariant; hence as a general form of entropy S when (2.4.3) is satisfied,

$$S = -k\mathrm{Tr}\,\{\rho \log \rho\}\,. \tag{2.4.5}$$

Obviously the form of distribution (2.4.3) is not always satisfied: in this case one cannot easily say how many quantum states contribute to the density matrix, but (2.4.5) gives a measure of the *average* number of states which contribute to the density matrix.

2.4.2 Thermodynamic Equilibrium

The concept that thermodynamic equilibrium is obtained by maximizing entropy is introduced by many compelling and ingenious arguments, which I do not choose to repeat here. Excellent discussions are given by *Tolman* [2.4] and *Landau* and *Lifshitz* [2.5].

The maximization is carried out subject to constraints, which specify the macroscopic knowledge we have of the system. Fundamental to the idea of equilibrium is the idea that the system is stationary, i.e., the density operator has no time dependence. From von Neumann's equation (2.1.24), in the case that there is no explicit time dependence of the Hamiltonian, one sees immediately that in equilibrium the Hamiltonian and the density operator commute with each other. Maximization of entropy can be carried out on the assumption that ρ is diagonal in an energy representation, so that

$$\rho = \sum p_i |i\rangle\langle i| \tag{2.4.6}$$

$$H = \sum E_i |i\rangle\langle i| \tag{2.4.7}$$

$$S = -k \sum p_i \log p_i . \tag{2.4.8}$$

One is now in a position to specify exactly what is known about the system. The most common assumption is that the *mean energy content* (essentially the temperature) is known, i.e., that

$$\langle E\rangle = \sum p_i E_i \tag{2.4.9}$$

must be specified as a constraint when maximizing the entropy. Another constraint is, of course, that

$$\mathrm{Tr}\,\{\rho\} = \sum p_i = 1. \tag{2.4.10}$$

This means that we must maximize using Lagrange multipliers: namely, one requires

$$\delta \left\{\sum p_i \log p_i + \alpha \sum p_i + \beta \sum E_i p_i\right\} = 0 \tag{2.4.11}$$

for which the solution is

$$p_i = \exp(-\alpha - \beta E_i) \tag{2.4.12}$$

which can be written as

$$\rho = \exp(-\alpha - \beta H). \tag{2.4.13}$$

Using (2.4.10) we can solve for α to get the *canonical density operator*

$$\rho = Z(\beta)^{-1} \exp(-\beta H) \tag{2.4.14}$$

where $Z(\beta)$ is the *canonical partition function*, defined by

$$Z(\beta) = \mathrm{Tr}\,\{\exp(-\beta H)\} . \tag{2.4.15}$$

The most useful way of looking at β is to use it to define the temperature T through

$$\beta = 1/kT \tag{2.4.16}$$

where k is Boltzmann's constant.

Other ensembles can be useful. There is always the possibility that certain other constants of the motion may be specified; for the most general stationary solution of von Neumann's equation (2.1.24) is a density operator which is an arbitrary function of any constants of the motion, and these may be included as additional

In more general situations the heat bath need not be so explicitly constructed. For example, in a chemical reaction taking place in solution the "system" consists of the molecules of the particular species which react, while the "bath" consists of the non-reacting degrees of freedom of the solution as a whole. The "system" and the "bath" occupy the same space! Indeed, it may be possible, if the reaction does not liberate any significant amount of heat, to operate the reaction without a solution, e.g., a gas phase reaction, and in this case the "bath" consists of the kinetic degrees of freedom only.

2.5.1 Density Operators for "System" and "Heat Bath"

We shall therefore consider the situation in which the physics as a whole does admit a rather clear separation into "system" and "bath", but will not enquire exactly as to how this is achieved. We will therefore suppose that the states of the physical system as a whole can be written

$$|x,a\rangle \equiv |x\rangle|a\rangle \tag{2.5.1}$$

where
a represents the bath degrees of freedom,
x represents the system degrees of freedom.
For simplicity, let us assume the ranges of both a and x are discrete sets. We can consider now operators which act in the system only, and operators which act in the bath only. Of greatest interest are those of the system, since this is what is under study—the bath is merely an environment.

Suppose we are interested in the mean of a system operator, M. Since this is an operator in the system space only, we can write:

$$\langle x'a'|M|xa\rangle = \delta_{aa'}\langle x'|M|x\rangle. \tag{2.5.2}$$

If $|\psi\rangle$ is any quantum state, the mean of M is

$$\langle\psi|M|\psi\rangle = \sum_{xx'}\sum_{aa'}\langle\psi|xa\rangle\langle xa|M|x'a'\rangle\langle x'a'|\psi\rangle \tag{2.5.3}$$

and using (2.5.2)

$$= \sum_{xx'}\sum_{a}\langle\psi|xa\rangle\langle x'a|\psi\rangle\langle x|M|x'\rangle \tag{2.5.4}$$

$$= \mathrm{Tr}_{\mathrm{sys}}\left\{\rho_{\mathrm{sys}}M\right\} \tag{2.5.5}$$

where

$$\rho_{\mathrm{sys}} = \sum_{a}\sum_{xx'}|x'\rangle\langle x'a|\psi\rangle\langle\psi|xa\rangle\langle x| \tag{2.5.6}$$

is called the *reduced density operator* for the system. This expression makes it clear that

$$\rho_{\mathrm{sys}} = \mathrm{Tr}_{\mathrm{B}}\left\{|\psi\rangle\langle\psi|\right\}, \tag{2.5.7}$$

that is, it is the partial trace of the density operator for the whole physical system taken over the bath variables only. Alternatively, we can write

$$\rho_{\mathrm{sys}} = \sum_a P(a)|\psi_a\rangle\langle\psi_a| \tag{2.5.8}$$

where

$$P(a) = \sum_x |\langle xa|\psi\rangle|^2 \tag{2.5.9}$$

and $|\psi_a\rangle$ is a *system state*, given by

$$|\psi_a\rangle = \sum_x |x\rangle\langle xa|\psi\rangle/\sqrt{P(a)}\ . \tag{2.5.10}$$

The form (2.5.8) makes it clear that ρ_{sys} is a density operator such as was conceived in Sect. 2.1, and it is positive definite. It shows that a subsystem of a system described by a pure state density operator $|\psi\rangle\langle\psi|$ must be described by a mixed state density operator ρ_{sys}. Of course if the whole physical system is itself described by a density operator ρ, then the generalization of (2.5.8) is

$$\rho_{\mathrm{sys}} = \mathrm{Tr}_{\mathrm{B}}\left\{\rho\right\}. \tag{2.5.11}$$

2.5.2 Mutual Influence of "System" and "Bath"

The separation into "system" and "bath" is usually quite clear where it is appropriate, for example in quantum optics, the system could be the atom, and the heat bath the radiation field. We are often interested in the atoms only, and use the radiation field merely as a probe to investigate the atom, but there are situations in which the precise state of the radiation field which arises can be of interest.

The separation into the "system" and the "bath" then brings to mind the matter of their interaction with each other. The general supposition is that the bath is so large that the system can only have a negligible influence on it, and it is therefore possible to make approximations which depend on this assumption. This will be done in Chap.5. However, the influence of the system on the bath can be quite significant, for example, in the case of an atom in a radiation field, the atom may fluoresce, and the fluorescent light is itself the effect on the radiation field, without which the atom could not be detected.

The influence of the "bath" on the "system" is rather more obvious. The system and the bath can exchange energy, and this leads to *dissipation*, and to *fluctuations*. An excited atom will lose its energy by radiation into the bath, but it will also feel the random fluctuations of the electromagnetic field which pervades all of space—even at zero temperature there are still vacuum fluctuations. How to write and compute these effects is the subject of this book.

3. Quantum Langevin Equations

The mutual influence of a "system", with few degrees of freedom, and a "heat bath", with many degrees of freedom, on each other is the central concept in the physics of noise, both quantum and classical. The action of the many variables of the bath on the system is to modify the equations of motion of the system by the inclusion of apparently random terms. The archetypal version of this concept dates back to Langevin's introduction of the kind of equation which now bears his name: the equation for a Brownian particle moving in a viscous fluid under the influence of a potential in the form

$$m\ddot{x} = -V'(x) - \gamma\dot{x} + \sqrt{2\gamma kT}\ \xi(t).$$

The aim of this chapter is to derive analogous equations for quantum systems, and to develop some of the basic methods of using them in situations of physical interest.

No derivation can be carried out without some basic assumptions, and so far the only usable basis for a derivation has been the assumption that the heat bath consists of an assembly of harmonic oscillators. This is not a bad assumption, since the most useful form of quantum noise theory is that developed for quantum optics, in which the relevant heat bath is the electromagnetic field, which is exactly equivalent to such an assembly of oscillators. In other systems, arguments can be made that the bath can be so approximated, but there will always remain some doubts, based on the knowledge that the harmonic oscillator is special, so that generalizations to other systems can be misleading.

We will also find that the assembly of harmonic oscillators must have some rather special properties. In particular:

i) There must be a smooth dense spectrum of oscillator frequencies.

ii) The coupling of the system to the bath operators must be *linear* in the bath harmonic oscillator operators.

iii) The coupling constants of the system to the bath operators must be a smooth function of the frequency of the oscillators.

These properties are all provided rather naturally by a *quantum field*, of which the electromagnetic field is the foremost example. The oscillators which arise as the normal modes of a field in a large (or possibly infinite) volume do have a smooth dense frequency spectrum. Non-trivial couplings do arise which are linear in the field variables. Finally, the most natural coupling of a small system to a field is a *local* coupling, in which the coupling depends only on the value of the field at a single point. The fact that the normal mode variables of the field are essentially spatial Fourier transforms of the field variables means that such a local coupling corresponds to taking a coupling to the normal modes which is in fact independent of mode frequency.

In fact it will turn out that a field interpretation of the heat bath is almost always possible—indeed it may be possible in several ways. The most important conclusion to be drawn from this can be seen by considering the case of a single atom in the electromagnetic field. When viewed from the position of the atom there are two kinds of field modes, incoming and outgoing. The incoming modes influence the motion of the atom, whereas the outgoing modes are produced by the atom—they affect the atom because they carry away energy, and thus give rise to damping. The effect of the incoming modes is to produce random noise effects and to feed energy into the atom, which is carried away by the outgoing modes. Thus the kind of damping produced is *radiation* damping. However this view will always be true whenever a field interpretation of the bath is possible, and in my experience this is always possible.

We are then led to the concepts of noise *inputs* and *outputs*. Every system which can be modelled as being coupled to a bath of harmonic oscillators can be considered as being driven by a *noise input* and as radiating a *noise output*. In optical and transmission line systems the noise outputs may be directed into other systems for which they become the inputs. However, the inputs may consist of both noise and possibly a well defined coherent part which is perhaps best called a *signal*. Reverting to the case of an atom in an electromagnetic field, we can identify three parts in the input:

i) A *signal* which could arise from shining light from a light source such as a laser on the atom.

ii) *Noise* which arises from the fluctuations in the laser light, and is produced by random effects in the laser. Such noise could also arise from other sources of light, such as sunlight, which impinge on the atom.

iii) *Vacuum noise*, which arises even when there is no light source. The vacuum state of a quantum field is a state in which the mean square field is non-zero. Thus, these vacuum fluctuations will always be present, and must satisfy the equations of motion. Even in the vacuum—that is, in a completely dark room at absolute zero—the atom experiences a vacuum noise input, and radiates a vacuum noise output.

Mathematically, the effect of the vacuum noise is to ensure that the Langevin equations *always* have a driving term as well as a damping term. As explained in Chap. 1, without some such term the canonical commutation relations would decay to zero, violating quantum mechanics. Physically the effect is slightly alarming, since the effect of this vacuum noise can be viewed as rather small, or infinite, depending on one's point of view.

In one sense the noise is infinite, because, as mentioned in Chap. 1, the spectrum of quantum noise rises in proportion to the mode frequency, so that the total power is infinite. However, the net effect is to *renormalize* the constants involved. As a result the vacuum noise shifts the energy levels of an atom to give the Lamb shift, which is dealt with by the renormalization techniques of quantum electrodynamics. The net effect is very small. For lumped systems, such as the Josephson junction, the finite size translates into a high frequency cutoff, which makes the effect of the rising noise spectrum finite, and able to be calculated. However, this does not mean that it is easy to calculate in all cases, and considerable care must always be taken. For example, it is natural to assume that we can use an initial condition for the system-bath density operator which is the direct product of an operator for the system and one for the bath. In general this is true, but it may be a very exceptional kind of state, since the effect of the high frequency modes is very strong, and thus such a state will experience a very large transient as the bath and system come to equilibrium with each other. For example, a harmonic oscillator coupled to such a bath experiences a very large input of energy, which effectively raises the energies of all the energy levels. Nevertheless, measurable energy differences between the levels are only slightly affected.

All of this is rather daunting, but fortunately, we can get away with a drastic approximation which eliminates all these effects, known as the *rotating wave approximation*. This eliminates all the terms in the Hamiltonian which cause the undesired effects, while leaving the measurable effects much the same. It is only valid if the damping is very weak, but that is very often true. However, this does not eliminate vacuum noise—it merely eliminates the infinite effects, and is equivalent to a kind of renormalization. The remaining part of the vacuum noise is still there, and serves to preserve the commutation relations in this approximation.

3.1 The Harmonic Oscillator Heat Bath

The model of a heat bath as an assembly of harmonic oscillators has a Hamiltonian which can be written in the form

$$H_B = \sum_n \left\{ \frac{p_n^2}{2m_n} + \frac{k_n q_n^2}{2} \right\}. \tag{3.1.1}$$

The system Hamiltonian can be left arbitrary. There must, however, be a number of system variables which we can write as a vector $\boldsymbol{Z}$, with a finite number of elements Z_i. Thus the system Hamiltonian is called

$$H_{\text{sys}}(\boldsymbol{Z}). \tag{3.1.2}$$

This kind of heat bath could be a model of an elastic solid, or indeed of the electromagnetic field (since both of these can be viewed as assemblies of harmonic oscillators, one for each normal mode). The *system* could be an atom, as in quantum optics, or something like a macroscopic LC circuit. Examples will be treated later.

a) Coupling Between System and Bath: We can get a coupling between bath and system which yields an exact quantum Langevin equation, by writing

$$H = H_{\text{sys}}(\boldsymbol{Z}) + \sum_n \left\{ \frac{1}{2m_n} p_n^2 + \frac{k_n}{2}(q_n - X)^2 \right\}, \tag{3.1.3}$$

where X is a particular one of the system operators $\boldsymbol{Z}$. The coupling is physically very simple—it makes the potential energy depend on the deviation of X from all the q_n; in other words, it is as if each co-ordinate q_n is harmonically bound to a "position" X. Conventionally, the interaction term is often written as $-\sum_n k_n q_n X$ omitting the term $\frac{1}{2}\sum_n k_n X^2$. This second term is a function only of the system operators and therefore can be absorbed into the system Hamiltonian, so the two formulations are physically equivalent. However the formulation in terms of (3.1.3) turns out to be simpler, since the observed systematic motion is in fact given by $H_{\rm sys}$, without the need for any correction. The Hamiltonian (3.1.3) can be simplified considerably by the canonical transformation

$$\begin{aligned} q_n &\to p_n/\sqrt{k_n} \\ p_n &\to -q_n\sqrt{k_n} \\ k_n/m_n &\to \omega_n^2 \\ \sqrt{k_n} &\to \kappa_n \end{aligned} \tag{3.1.4}$$

which gives the Hamiltonian

$$H = H_{\rm sys}(Z) + \tfrac{1}{2}\sum_n \left\{(p_n - \kappa_n X)^2 + \omega_n^2 q_n^2\right\}. \tag{3.1.5}$$

For future reference, the equal times commutation relations implicit in this Hamiltonian are

$$\begin{aligned} [Z, p_n] &= [Z, q_n] = 0 \\ [p_n, p_m] &= [q_n, q_m] = 0 \\ [q_n, p_m] &= i\hbar\delta_{nm}. \end{aligned} \tag{3.1.6}$$

but no particular commutation relations are specified between the different components of Z—these will depend on the system under consideration.

b) Non-Linear Couplings: This kind of coupling to the bath is linear in the bath variables, and corresponds to energy being transferred to and from the bath by the absorption or emission of bath quanta. This is certainly what happens in the case of electromagnetic radiation, but in something like the theory of electrical resistance in a metal, a different formulation is appropriate, since it is known that electrical resistance arises from scattering of electrons by phonons and impurities. The scattering by phonons gives a temperature dependent resistance which arises because scattering depends on the average number of bath phonons, which of course depends on temperature. This cannot be reproduced by the linear coupling assumed. Non-linear couplings are not easy to create by the quantum Langevin equation, but can be dealt with by master equation methods, as will be shown in Chap.5. The residual scattering by impurities does not have this temperature dependence, but it is not really an incoherent effect, as was pointed out many years ago by *Landauer* [3.1]. The scattering randomly changes the phase of the electron wavefunction, but if the scatterer is fixed, this does not change the coherence, and interference between electron waves can be observed. However, this phenomenon is outside the scope of this book.

The most important quantum-mechanical aspect to be borne in mind is the matter of commutation relations. Thus, we can compute that

$$[\xi(t), \xi(t')] = \mathrm{i}\hbar \frac{d}{dt} f(t - t') \tag{3.1.29}$$

as a straightforward consequence of the definitions (3.1.14,15) and the creation and destruction operator commutation relations.

A more interesting commutation relation is that between the noise operator and the system variables. From (3.1.8) we can note that $a_n(t)$ is related to the canonical operators $p_n(t)$ and $q_n(t)$ at the time t, and therefore commutes with all the system operators at the same time. Hence we deduce the general result, for any system operator $Y(t)$, that

$$[Y(t), a_n(t_0)] = \kappa_n \sqrt{\frac{\omega_n}{2\hbar}} \int_{t_0}^{t} \mathrm{e}^{\mathrm{i}\omega_n(t'-t_0)} [Y(t), X(t')]\, dt' \tag{3.1.30}$$

and by using the definitions (3.1.14,15) of $\xi(t)$ and $f(t)$, we find

$$[Y(t), \xi(s)] = \int_{t_0}^{t} [Y(t), X(t')] \frac{d}{ds} f(s - t') dt' . \tag{3.1.31}$$

This relation will be of great utility in formulating the theory of *noise inputs* and *outputs* in Sect. 3.2. From its derivation it is clear that (3.1.31) must be consistent with the equation of motion in either of the forms (3.1.12,13), though this is not at all obvious at first glance.

a) Commutation Relations in the First Markov Approximation: Corresponding to the first Markov approximation, the commutation relations for the noise operators, (3.1.29,31) become (assuming $s, t > t_0$),

$$[\xi(t), \xi(t')] = 2\mathrm{i}\hbar\gamma \frac{d}{dt} \delta(t - t') \tag{3.1.32}$$

$$[Y(t), \xi(s)] = 2\gamma \frac{d}{ds} \{u(t - s)[Y(t), X(s)]\} \tag{3.1.33}$$

where

$$u(x) = \begin{cases} 1\,, & x > 0 \\ \frac{1}{2}\,, & x = 0 \\ 0\,, & x < 0. \end{cases} \tag{3.1.34}$$

Apart from the rather singular nature of the commutator at $t = s$, (3.1.33) has a very straightforward interpretation. Namely, for s in the future of t, this commutator vanishes, which means that one can specify the system variables independent of the *future* behaviour of the driving force. But the past behaviour of $\xi(t)$ does affect $Y(t)$, so this commutator does not vanish.

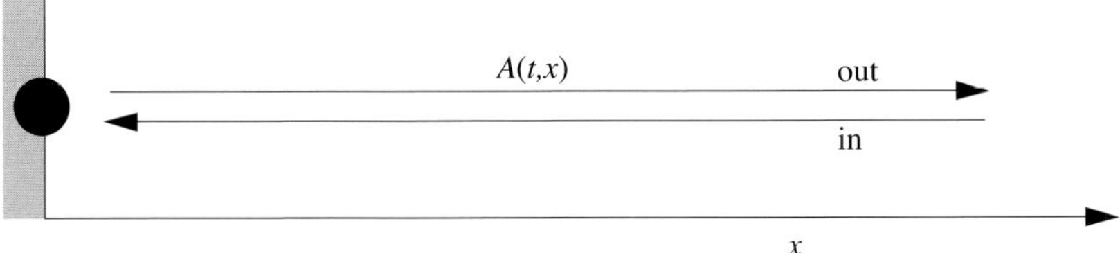

Fig. 3.1 The field interpretation of a system interacting with a heat bath. The system is a rather small localized object which interacts with a one sided field which streams in from the right, interacts with the system, and streams out to the right. Damping appears as *radiation* damping

3.2 The Field Interpretation—Noise Inputs and Outputs

In this section we will consider a particular implementation of the harmonic oscillator heat bath, in the form of a semi-infinite transmission line. Let us consider the system to be described by a Lagrangian $L_{\rm sys}(\boldsymbol{Z})$ interacting with a one dimensional electromagnetic field, $A(t, x)$, as in Fig.3.1. The full Lagrangian is then

$$L = L_{\rm sys}(\boldsymbol{Z}) + \tfrac{1}{2}\int_0^\infty dx\,\left\{\dot{A}(t,x)^2 - c^2[\partial_x A(t,x)]^2\right\} + X\int_0^\infty dx\,\kappa(x)\dot{A}(t,x). \tag{3.2.1}$$

Here the notation is as in Sect. 3.1; $\boldsymbol{Z}$ are the system operators, and X is the particular operator which interacts with the heat bath. The function $\kappa(x)$ determines the range of the interaction, and will be considered to be zero except for a small region in the vicinity of $x = 0$. Thus, the system can be viewed as being localized near $x = 0$, where it interacts only with the field $A(t, x)$ in that vicinity. The Hamiltonian corresponding to (3.2.1) is

$$H = H_{\rm sys}(\boldsymbol{Z}) + \tfrac{1}{2}\int_0^\infty dx\,\left\{[\pi(t,x) - X\kappa(x)]^2 + c^2[\partial_x A(t,x)]^2\right\} \tag{3.2.2}$$

where the canonical momentum $\pi(t, x)$ is given by

$$\pi(t,x) = \dot{A}(t,x) + X\kappa(x), \tag{3.2.3}$$

and $H_{\rm sys}(\boldsymbol{Z})$ is the Hamiltonian corresponding to the Lagrangian $L_{\rm sys}(\boldsymbol{Z})$, since the Lagrangian L contains no terms which depend on derivatives of $\boldsymbol{Z}$ other than those which already occur in $L_{\rm sys}(\boldsymbol{Z})$. We can define some Fourier transform variables by

$$A(t,x) = \sqrt{\frac{2}{\pi c}}\int_0^\infty q(\omega,t)\cos(\omega x/c)d\omega, \tag{3.2.4}$$

$$\pi(t,x) = \sqrt{\frac{2}{\pi c}}\int_0^\infty p(\omega,t)\cos(\omega x/c)d\omega, \tag{3.2.5}$$

and

$$\kappa(x) = \sqrt{\frac{2}{\pi c}} \int_0^\infty \tilde{\kappa}(\omega) \cos(\omega x/c) d\omega, \tag{3.2.6}$$

in terms of which the Hamiltonian takes the form

$$H = H_{\text{sys}}(Z) + \tfrac{1}{2} \int_0^\infty d\omega \left\{ [p(\omega, t) - \tilde{\kappa}(\omega) X]^2 + \omega^2 q(\omega)^2 \right\} \tag{3.2.7}$$

which is a continuum version of the Hamiltonian chosen in Sect. 3.1, and given in (3.1.5) in which $dn(\omega)/d\omega = 1$. We make the following comments.

a) Damping and Locality: The assumption made in Sect. 3.1.1c that gives rise to the first Markov approximation amounts to assuming that

$$\tilde{\kappa}(\omega) = \sqrt{2\gamma/\pi} \tag{3.2.8}$$

since $dn/d\omega = 1$ has been assumed. From the definition of $\tilde{\kappa}(\omega)$ in terms of $\kappa(x)$, (given in (3.2.6)), this corresponds to

$$\kappa(x) = 2\sqrt{\gamma c}\, \delta(x), \tag{3.2.9}$$

so that the assumption of constancy of $\tilde{\kappa}(\omega)$ is equivalent to the locality of the interaction in the x space. Although the three dimensional aspect of an atom is important, it is easy to see that this is the kind of interaction that an atom has with the radiation field, since the size of an atom is very small compared to the wavelength of the light with which it interacts. This is known in atomic physics as the electric dipole approximation.

b) The Process of Damping: Viewed from the perspective of a localized system interacting with a semi-infinite field, the process of damping is essentially that of *radiation*. The system may start with some energy, but this will induce outgoing waves which take the energy away. The ensuing wavetrain proceeds out to infinity, and as far as the system is concerned, the process is irreversible. However, at any finite time, it is possible to reflect the outgoing wave back and reverse the motion of the system.

c) Renormalization: The term $\frac{1}{2} \int_0^\infty d\omega\, \kappa(\omega)^2 X^2$ appears quite naturally as a result of the assumption of the straightforward Lagrangian (3.2.1), whereas in Sect. 3.1 it appeared almost arbitrary. The Lagrangian assumed here is a simplification of a full three dimensional electromagnetic Lagrangian, so that this "renormalization" of the system Hamiltonian can be considered quite natural.

3.2.1 Input and Output Fields

Using Lagrangian methods, the equation of motion for the field $A(t,x)$ is the wave equation in the presence of a source; namely

$$\ddot{A}(t,x) - c^2 \partial_x^2 A(t,x) = -\kappa(x)\dot{X}. \tag{3.2.10}$$

This equation can be solved by using the kind of solutions we have already derived in Sect. 3.1.2 (3.1.10), which in this case take the form

$$a(\omega,t) = \mathrm{e}^{-\mathrm{i}\omega(t-t_0)}a(\omega,t_0) - \kappa(\omega)\sqrt{\frac{\omega}{2\hbar}}\int_{t_0}^{t} dt'\mathrm{e}^{-\mathrm{i}\omega(t-t')}X(t') \tag{3.2.11}$$

in which $t_0 < t$ is an initial time, usually assumed to be in the remote past, and of course

$$a(\omega,t) = \frac{\omega q(\omega,t) + \mathrm{i}p(\omega,t)}{\sqrt{2\hbar\omega}}. \tag{3.2.12}$$

We can also consider the possibility of specifying a solution in terms of a *final condition* $t_1 > t$, which can usually be assumed to be in the remote future. Thus

$$a(\omega,t) = \mathrm{e}^{-\mathrm{i}\omega(t-t_1)}a(\omega,t_1) + \kappa(\omega)\sqrt{\frac{\omega}{2\hbar}}\int_{t}^{t_1} dt'\mathrm{e}^{-\mathrm{i}\omega(t-t')}X(t'). \tag{3.2.13}$$

We will now define *input* and *output* field operators in terms of these initial and final conditions, thus

$$A_{\mathrm{in}}(t) = \tfrac{1}{2}\int_0^\infty d\omega\sqrt{\frac{\hbar}{\pi\omega c}}\left\{a(\omega,t_0)\mathrm{e}^{-\mathrm{i}\omega(t-t_0)} + a^\dagger(\omega,t_0)\mathrm{e}^{\mathrm{i}\omega(t-t_0)}\right\} \tag{3.2.14}$$

$$A_{\mathrm{out}}(t) = \tfrac{1}{2}\int_0^\infty d\omega\sqrt{\frac{\hbar}{\pi\omega c}}\left\{a(\omega,t_1)\mathrm{e}^{-\mathrm{i}\omega(t-t_1)} + a^\dagger(\omega,t_1)\mathrm{e}^{\mathrm{i}\omega(t-t_1)}\right\}. \tag{3.2.15}$$

By substituting these solutions for $q(\omega,t)$ implicit in (3.2.13), into the definition of $A(t,x)$ in terms of $q(\omega,t)$, (3.2.4), we can obtain, after some careful manipulation

$$A(t,x) = A_{\mathrm{in}}\left(t+\frac{x}{c}\right) + A_{\mathrm{in}}\left(t-\frac{x}{c}\right) - \tfrac{1}{2}\int_{\frac{x}{c}-(t-t_0)}^{\frac{x}{c}+(t-t_0)} d\tau\,\kappa(c\tau)X\left(t-\left|\tau-\frac{x}{c}\right|\right) \tag{3.2.16}$$

$$= A_{\mathrm{out}}\left(t+\frac{x}{c}\right) + A_{\mathrm{out}}\left(t-\frac{x}{c}\right) + \tfrac{1}{2}\int_{\frac{x}{c}+(t-t_1)}^{\frac{x}{c}-(t-t_1)} d\tau\,\kappa(c\tau)X\left(t+\left|\tau-\frac{x}{c}\right|\right). \tag{3.2.17}$$

These two solutions are general solutions for arbitrary $\kappa(x)$. However, in all cases of interest $\kappa(x)$ is zero for x outside some finite range, even if it does not take the idealized delta function form (3.2.9). If we take x to be outside this finite range, then in the integrals in (3.2.16,17) τ is always less than x/c for nonzero $\kappa(c\tau)$, and we can write

$$|\tau - x/c| = x/c - \tau \tag{3.2.18}$$

in both cases. If t_0 is in the remote past, and t_1 in the remote future, we can also set the limits on the integrals to $\pm\infty$. It is then apparent that in both of these equations, the right hand sides can be expressed as a sum of a function of $t - x/c$ and a function of $t+x/c$. We can equate these functions in (3.2.16) with the corresponding functions in (3.2.17), and we find

$$A_{\text{in}}(t + x/c) = A_{\text{out}}(t + x/c) + \tfrac{1}{2}\int_{-\infty}^{\infty} d\tau\, \kappa(c\tau)X(t + x/c - \tau). \tag{3.2.19}$$

Since this is true for *any* t, even though x must be outside the range of $\kappa(x)$, we can see that

$$A_{\text{out}}(t) = A_{\text{in}}(t) - \tfrac{1}{2}\int_{-\infty}^{\infty} d\tau\, \kappa(c\tau)X(t - \tau) \tag{3.2.20}$$

is generally true for all finite t. (Equating the functions of $t - x/c$ gives the same result). Finally, using this result, we see that (3.2.16) gives, for x outside the range of $\kappa(x)$

$$A(t, x) = A_{\text{in}}(t + x/c) + A_{\text{out}}(t - x/c). \tag{3.2.21}$$

This means that outside the range of interaction, the total field is given by the sum of an incoming "in" field, and an outgoing "out" field. The expression of the "out" field by (3.2.20) shows that the "out" field is itself given by the sum of a *reflection* of the "in" field and a *radiated field*, $\frac{1}{2}\int d\tau\, \kappa(c\tau)X(t - \tau)$, whose behaviour is determined by the equation of motion for the system operator X.

This point of view makes very obvious sense in the field of quantum optics, where the input and output fields are very apparent as light beams. However, it can be seen here that even the vacuum is an input—one can view the zero temperature quantum noise as arising from the inward propagation of vacuum fluctuations, and it is these vacuum fluctuations which preserve the commutation relations from decay. Thus, there is *always* an input, even when the light beams are turned off.

3.2.2 Equations of Motion for System Operators

Although these have already been derived for the general case, it is interesting to see how they arise in the field context. Using the Hamiltonian (3.2.2) and the canonical commutation relations

$$[A(t, x), \pi(t, x')] = i\hbar\delta(x - x') \tag{3.2.22}$$

the equation of motion for any system operator is given by

$$\dot{Y} = \frac{\mathrm{i}}{\hbar}[H_{\text{sys}}, Y] - \frac{\mathrm{i}}{2\hbar}\int_{0}^{\infty} \kappa(x)dx[\dot{A}(t, x), [X, Y]]_{+}. \tag{3.2.23}$$

For simplicity, let us assume that $\kappa(x)$ has the delta function form (3.2.9). In this case it is possible to substitute for $A(t,x)$ using (3.2.16), to derive

$$A(t,x) = A_{\rm in}\left(t+\frac{x}{c}\right) + A_{\rm in}\left(t-\frac{x}{c}\right) - \sqrt{\frac{\gamma}{c}}\, X\left(t-\frac{x}{c}\right) \tag{3.2.24}$$

and thus

$$\dot{Y} = \frac{\mathrm{i}}{\hbar}[H_{\rm sys}, Y] + \frac{\mathrm{i}}{2\hbar}[\gamma\dot{X} - 2\sqrt{\gamma c}\,\dot{A}_{\rm in}(t), [X, Y]]_+. \tag{3.2.25}$$

This is of course exactly the same as the quantum Langevin equation (3.1.28) in the first Markov approximation, as derived in Sect. 3.1.1, with the substitution

$$\xi(t) \to 2\sqrt{\gamma c}\,\dot{A}_{\rm in}(t). \tag{3.2.26}$$

a) Noise Inputs and Outputs: Thus the view of the quantum Langevin equation provided by this model is of a system connected to an infinite transmission line. This transmission line can support both incoming and outgoing modes, and there is a boundary condition at the $x = 0$ end of this transmission line which takes the form, obtained by setting $x = 0$ and $\kappa(x) = 2\sqrt{\gamma c}\,\delta(x)$ into (3.2.19),

$$A_{\rm out}(t) = A_{\rm in}(t) - \sqrt{\frac{\gamma}{c}}\, X(t). \tag{3.2.27}$$

The "in" field can be considered to be a thermal noise *input* if the statistics of the modes $a(\omega, t_0), a^\dagger(\omega, t_0)$ are thermal. Since this specification is in terms of initial conditions in the remote past, we see that $A_{\rm in}(t)$ can be specified arbitrarily—it is in no way affected by the system operators. In contrast, $A_{\rm out}(t)$ is determined from the boundary condition (3.2.27), which requires the knowledge of $X(t)$, a system operator. Such operators are determined as solutions of the quantum Langevin equation (3.2.25). The damping, given by the term proportional to $\dot{X}$, is *radiation damping*—the quantum Langevin equation then displays two major terms, one which is the driving by the "in" field, and the other which represents the effect of loss of energy into the "out" field.

b) Causality: Finally, it is a straightforward matter to adapt the relationship (3.1.31) to yield in this case

$$[Y(t), A_{\rm in}(s)] = \sqrt{\frac{\gamma}{c}}\, u(t-s)[Y(t), X(s)]. \tag{3.2.28}$$

We can then use the boundary condition (3.2.27) to replace $A_{\rm in}(t)$ in terms of $A_{\rm out}(t)$ to obtain

$$[Y(t), A_{\rm out}(s)] = \sqrt{\frac{\gamma}{c}}\, u(s-t)[Y(t), X(s)]. \tag{3.2.29}$$

The causality interpretation is thus, that the system operators must be able to be specified independently of the *future* behaviour of the noise inputs, and hence $Y(t)$

commutes with $A_{\text{in}}(s)$ provided s is in the future of t. Conversely the output, once it leaves the system, is independent of $Y(t)$, i.e., $A_{\text{out}}(s)$ must be able to be specified independently of $Y(t)$ provided t is in the future of s. In fact, (3.2.28,29), follow simply from this causality condition and the boundary condition (3.2.27), provided we assume nothing singular happens at $s = t$.

c) Time Reversal: By substituting for $\dot{A}(t, x)$ in terms of $A_{\text{out}}(t)$, we can similarly derive a quantum Langevin equation in the form

$$\dot{Y} = \frac{i}{\hbar}[H_{\text{sys}}, Y] - \frac{i}{2\hbar}[\gamma \dot{X} + 2\sqrt{2\gamma c}\ \dot{A}_{\text{out}}(t), [X, Y]]_{+}. \tag{3.2.30}$$

Here the system is driven by the *output*, and the damping term has a reversed sign. Thus, we get a kind of "time reversed" quantum Langevin equation. In principle this need not be equivalent to time reversing the original equation, unless the Lagrangian assumed, (3.2.1), is itself time reversal invariant. If Θ is the antilinear time reversal operator, and we assume that

$$\Theta A(t, x)\Theta = A(-t, x) \tag{3.2.31}$$

$$\Theta L_{\text{sys}}[\boldsymbol{Z}(t)]\Theta = L_{\text{sys}}[\boldsymbol{Z}(-t)] \tag{3.2.32}$$

then time reversal of the Lagrangian (3.2.1) is assured if $\kappa(x)$ is real (which must be the case for a Hermitian Lagrangian) and

$$\Theta X(t)\Theta = -X(-t). \tag{3.2.33}$$

From (3.2.21) and (3.2.31), we note that

$$\begin{aligned} A(-t, x) &= \Theta A_{\text{in}}(t + x/c)\Theta + \Theta A_{\text{out}}(t - x/c)\Theta \\ &= A_{\text{in}}(-t + x/c) + A_{\text{out}}(-t - x/c) \end{aligned} \tag{3.2.34}$$

and equating separately the functions of $t + x/c, t - x/c$, we see that

$$\begin{aligned} \Theta A_{\text{in}}(t)\Theta &= A_{\text{out}}(-t) \\ \Theta A_{\text{out}}(t)\Theta &= A_{\text{in}}(-t). \end{aligned} \tag{3.2.35}$$

Using (3.2.31,35) it is then quite straightforward to show that the time reversed version of the quantum Langevin equation is indeed the "out" form (3.2.30).

d) Some Historical Comments: The model of a system in interaction with a bath of harmonic oscillators was first thoroughly investigated by *van Kampen* [3.3] in his doctoral thesis. His aim was to investigate the nature of quantum mechanical renormalization in a simplified model of quantum electrodynamics. The use of the model to generate quantum mechanical Langevin equations is due to *Ford et al.* [3.2], who showed that the same procedure could be used for the derivation of both the classical and the quantum mechanical Langevin equations.

The introduction of "inputs" and "outputs" into quantum mechanical systems has its origins in the LSZ formulation of quantum field theory [3.4], in which scattering theory is formulated on the basis of the unitary transformation between asymptotic non-interacting "in" and "out" fields. However in quantum field theory there is no "system" with a few degrees of freedom—only *fields* occur in the formalism, which is an exact formulation of the underlying theory, unlike the model treated here, which is a simplified representation of reality.

Caves [3.5] was the first to introduce the concept of "inputs" and "outputs" into the theory of quantum amplifiers, and from this formulation was able to derive general relations for noise performance of linear quantum amplifiers by treating them as unitary scattering operators which transform "in" fields to "out fields". This procedure was generalized by *Yurke* and *Denker* [3.6] to a "Quantum Network Theory", adapted to the treatment of quantum microwave systems. Their treatment was explicitly based on electrodynamics, and did not have the generality of the methods used here. However, all the essential properties of "inputs" and "outputs" were present, including the identification of the noise term in the quantum Langevin equation with the "in" field.

The formulation presented here is due to *Collett* [3.7], who was the first to use the boundary condition (3.2.27) to obtain properties of the outputs in terms of those of the system and the inputs, and hence to use the quantum Langevin equation as a practical computational tool.

3.3 The Noise Interpretation

The quantity $\xi(t)$ is an operator, and the equations so far derived cannot be interpreted straightforwardly as noise equations unless some assumption is made about the statistics of $\xi(t)$. Since $\xi(t)$ is determined directly in terms of heat bath operators at the time t_0, the nature of the state of the heat bath at this initial time is significant.

The physical picture which gives the clearest understanding is obtained as follows. We assume the system and the bath are initially independent and non-interacting, and introduce the interaction between them at the time t_0. In the Schrödinger picture this would mean that the density operator could be written at time t_0 as a direct product

$$\rho(t_0) = \rho_{\text{sys}}(t_0) \otimes \rho_B(t_0). \tag{3.3.1}$$

We are working in the Heisenberg picture, however, in which the density operator is equal to that at time t_0, hence we assume

$$\rho = \rho_{\text{sys}} \otimes \rho_B. \tag{3.3.2}$$

In the Heisenberg picture the density operator is a direct product at all times. This will mean that any quantum mechanical average of the bath variables is simply able to be written as, for example

$$\langle \xi(t)\xi(t')\rangle = \text{Tr}\left\{\rho_B \xi(t)\xi(t')\right\}. \tag{3.3.3}$$

The assumption of an initially factorizable density operator is intuitively appealing, since it has a simple classical interpretation. For example, if we place a Brownian particle in a fluid at time t_0, then initially the particle and the fluid are quite independent, and quantum mechanically this would mean a factorized density operator.

In a quantum mechanical situation this is not always so straightforward. If we consider the case of a charged particle in an electromagnetic field, it is clear that there is no way of actually disconnecting the particle from the electromagnetic field, so that there is no natural way of generating a factorized density operator. This is a problem familiar in quantum electrodynamics, and is overcome by the concept of renormalization. Unfortunately renormalization often involves infinite quantities—the coupled system is infinitely different from the factorized system. The same problem of infinities occurs in quantum noise theory, and manifests itself in an infinitely rising noise spectrum.

If a strictly local theory is used, we find that the density operator of the damped system is infinitely different from a factorized density operator. In practice the only way of dealing with this problem is to introduce a high frequency cutoff. It is then found that measurable quantities are usually independent of the cutoff.

In some situations the cutoff actually has a physical meaning, for example in a Josephson junction oscillator, the main damping is by electromagnetic radiation. The high frequency cutoff is simply an expression of the finite size of the junction, which no longer can be treated as a lumped circuit for frequencies such that the wavelength is smaller than the diameter of the junction.

3.3.1 Thermal Statistics

We may assume that the bath is initially *thermal*, so that in the Heisenberg picture

$$\rho_B = Z\exp(-H_B/kT) \tag{3.3.4}$$

and hence (using the notation of Sect. 3.1.1b)

$$\langle a_{nm}\rangle = \langle a^\dagger_{nm}\rangle = 0 \tag{3.3.5}$$

$$\langle a_{nm}a_{n'm'}\rangle = 0 \tag{3.3.6}$$

$$\langle a^\dagger_{nm}a_{n'm'}\rangle = \delta_{nn'}\delta_{mm'}\left\{\exp(-\hbar\omega_n/kT) - 1\right\}^{-1} \tag{3.3.7}$$

$$\equiv \delta_{nn'}\delta_{mm'}\bar{N}(\omega_n). \tag{3.3.8}$$

(Here, for brevity, we drop the t_0 dependence of $a, a^\dagger$.) We can straightforwardly compute that

$$\langle[\xi(t),\xi(t')]_+\rangle = 2\int_0^\infty d\omega\,\hbar\omega G(\omega)\frac{dn}{d\omega}\{\bar{N}(\omega)+\tfrac{1}{2}\}\cos\omega(t-t') \tag{3.3.9}$$

and in the case that $G(\omega)dn/d\omega$ is a constant, as in (3.1.24), we may write

$$\langle[\xi(t),\xi(t')]_+\rangle = \frac{2\gamma\hbar}{\pi}\int_0^\infty d\omega\,\omega\coth\left(\frac{\hbar\omega}{2kT}\right)\cos\omega(t-t') \tag{3.3.10}$$

and, of course $\langle\xi(t)\rangle = 0$. The high frequency behaviour of the spectral function $\omega\coth(\hbar\omega/2kT)$ is proportional to $|\omega|$, which is very badly behaved—classical noise theory usually considers only asymptotically constant noise spectra, which corresponds to white noise. Even white noise is not straightforward to handle, since it is merely continuous, and not differentiable. To my knowledge, there is no way of handling noise with a linearly rising noise spectrum other than by introducing a cutoff at high frequencies. The physical justification for this comes from the form (3.3.9), where the factor $G(\omega)dn/d\omega$ comes into play. If this factor drops off sufficiently rapidly at high frequencies, the rising noise spectrum can be made to drop off, and thus give a well behaved correlation function. The exact nature of $G(\omega)dn/d\omega$ is model dependent.

3.3.2 The Classical Limit

If we let $\hbar \to 0$, then $\xi(t)$ commutes with $\xi(t')$, and on using the limit

$$\lim_{\hbar\to 0} \hbar \coth\left(\frac{\hbar\omega}{2kT}\right) = \frac{2kT}{\omega} \tag{3.3.11}$$

(3.3.10) becomes

$$\langle \xi(t)\xi(t')\rangle = 2\gamma kT\delta(t-t'). \tag{3.3.12}$$

This limit is, of course, rather naive, since for any finite $\hbar$ the hyperbolic cotangent becomes infinite as $\omega \to \infty$. In practice the factor $G(\omega)dn/d\omega$ is not constant, but drops off at high frequencies. The limit is then

$$\langle \xi(t)\xi(t')\rangle = kTf(t-t') \tag{3.3.13}$$

and this approaches the result (3.3.12) as $G(\omega)dn/d\omega$ approaches a constant. Thus the wide bandwidth limit of the classical limit is a delta correlated noise, but the limits cannot be taken in the reverse order.

3.3.3 Behaviour of the Langevin Correlation Function as a Function of Time

The anticommutator correlation function (3.3.10) is strictly speaking divergent. This means that it can only be interpreted as a *generalized function* or *distribution*. Nevertheless, it is necessary to evaluate it to understand the kind of noise it implies. We can do this as follows. It is first best to separate out the divergent part explicitly, thus

$$\begin{aligned}\langle [\xi(t),\xi(t')]_+\rangle &= \frac{2\gamma\hbar}{\pi}\int_0^\infty d\omega \left[\omega\coth\left(\frac{\hbar\omega}{kT}\right) - \omega\right]\cos\omega(t-t') \\ &\quad + \frac{\gamma\hbar}{\pi}\int_{-\infty}^{\infty} d\omega\, |\omega| e^{i\omega(t-t')}. \end{aligned} \tag{3.3.14}$$

The first term is convergent for all $t-t'$. It can be evaluated by differentiating with respect to a the result [3.8]

$$\int_0^\infty \sin ax(\coth\beta x - 1)dx = \frac{\pi}{2\beta}\coth\frac{a\pi}{2\beta} - \frac{1}{a} \tag{3.3.15}$$

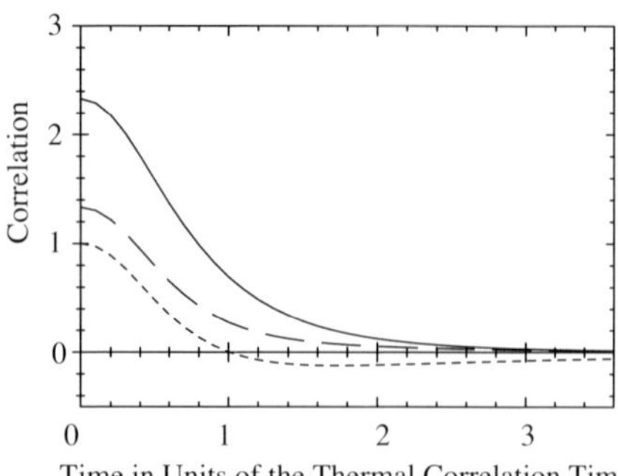

Fig. 3.2 Plot of the correlation function (3.3.14) against time, for the cutoff λ, equal to one thermal correlation time, and for temperatures 4.6pK (*solid line*), 2.3pK (*dashed line*), and 0.23 pK (*dotted line*)

to obtain

$$\frac{2\gamma\hbar}{\pi}\left\{-\left(\frac{\pi kT}{\hbar}\right)^2 \operatorname{cosech}^2\left(\frac{\pi kT}{\hbar}(t-t')\right)+\frac{1}{(t-t')^2}\right\}. \tag{3.3.16}$$

The other term is only able to be evaluated as a distribution or by introducing a cutoff. By introducing an exponential cutoff

$$\frac{\gamma\hbar}{\pi}\int_{-\infty}^{\infty} d\omega\,|\omega|e^{i\omega(t-t')}e^{-\lambda\omega} = \frac{2\gamma\hbar}{\pi}\frac{\lambda^2-(t-t')^2}{[\lambda^2+(t-t')^2]^2} \tag{3.3.17}$$

where λ is some very small time constant. The behaviour of (3.3.17) as $t \to t'$ is quite regular for any non-zero λ but diverges as $\lambda \to 0$. Formally, one can write

$$\langle[\xi(t),\xi(t')]_+\rangle = -\frac{2\gamma\pi k^2T^2}{\hbar}\operatorname{cosech}^2\left(\frac{\pi kT}{\hbar}(t-t')\right) \tag{3.3.18}$$

when $\lambda \to 0$, on the understanding that the behaviour near $t \to t'$ is given by adding (3.3.16) and (3.3.17).

This correlation displays a *thermal correlation time* which arises from the large $|t-t'|$ behaviour

$$\langle[\xi(t),\xi(t')]_+\rangle \approx -\frac{8\gamma\pi k^2T^2}{\hbar}\exp\left(-\frac{2\pi kT}{\hbar}|t-t'|\right) \quad \text{as } |t-t'|\to\infty \tag{3.3.19}$$

The thermal correlation time is given by

$$\tau_T = \hbar/2\pi kT. \tag{3.3.20}$$

This time becomes infinite when $T \to 0$; in fact the $T \to 0$ limit of the correlation function is no longer exponential, but inverse quadratic; i.e.,

$$\langle[\xi(t),\xi(t')]\rangle_+ \approx -2\gamma\hbar/\pi(t-t')^2 \quad \text{for } t \neq t'. \tag{3.3.21}$$

The slightly surprising thing about these results is the fact that the long time behaviour of the correlation function is *negative*—at long time differences there is an *anticorrelation*. To visualize what happens, it is helpful to graph the correlation function for some finite, but very small cutoff λ, and this is shown in Fig.3.2.

Notice that, from (3.3.10)

$$\int_{-\infty}^{\infty} dt \langle [\xi(t), \xi(t')]_+ \rangle = \pi \lim_{\omega \to 0} \left\{ \frac{2\gamma\hbar\omega}{\pi} \coth \frac{\hbar\omega}{2kT} \right\} = 4\gamma kT \tag{3.3.22}$$

so that the overall effect is for the cutoff dependent positive term to more than cancel the asymptotic negative term, except at $T = 0$, where the two effects cancel. This correlation function is a far more complicated object than the corresponding classical delta correlated result. The existence of the thermal correlation time τ_T does mean however that for any time scale significantly longer than this one can expect a description in terms of a kind of "quantum white noise", or a "quantum Markov process". The thermal correlation time is typically quite small; numerically $\tau_T = 1.27 \times 10^{-12}/T$ sec, so that at a temperature of 1K , τ_T is about a picosecond. There are time scales which are significantly faster than this; for example, the period of the light emitted in the $\mathrm{Na_D}$ transition is about 2×10^{-13}sec. However the time scale of an atomic decay is much slower, typically many thousands of periods. This means that we can expect a Markovian description of atomic decays, but not of the more detailed process of evolution over the time scale of the transition frequency.

3.3.4 Other Noise Statistics

The quantum Langevin equations derived in the preceding sections are not themselves specifically stochastic equations; they are valid independent of the statistics assumed for $\xi(t)$, and can be viewed simply as a convenient way of writing the Heisenberg equations of motion. Therefore other statistics of $\xi(t)$ are possible, such as *squeezed* statistics, which will be investigated in Chap.9. But the high frequency problems are not generally avoidable, since they arise from the commutation relations for the bath operators, which are always the same.

3.4 Examples and Applications

In this section a few of the more natural examples will be investigated, and in particular, the only soluble model, that is a free or harmonically bound particle, will be investigated.

3.4.1 A Particle Moving in a Potential

We have already considered this problem in Sect. 3.1.1a), and have found that the quantum Langevin equation corresponding to a particle with canonical co-ordinate and momentum q and p, and mass m, moving in potential $V(q)$ in which the operator X which couples to the heat bath is chosen to be q is, in the first Markov approximation,

$$m\ddot{q}(t) = -V'\big(q(t)\big) - \gamma\dot{q}(t) + \xi(t). \tag{3.4.1}$$

Apart from the cases of quadratic, linear, and constant potentials, this operator equation has not been solved.

Exercise. Suppose we make the replacement $X \to p$. Show the equation of motion is then very different.

The equation (3.4.1) is of exactly the same form as that studied in Sect. 1.4.1, and the commutator and mean anticommutator of $\xi(t)$ are as specified for a quantum Langevin equation. The requirement that the stationary distribution should be the Boltzmann distribution is not straightforward to prove, and will be left for Sect. 3.6.3

3.4.2 The Brownian Particle Langevin Equation

The case that $V(q) = 0$ represents a quantum analogue of the classical problem of Brownian motion, and is soluble by exactly the same method. The equation becomes

$$\dot{q} = p/m \tag{3.4.2}$$

$$\dot{p} = -\gamma p/m + \xi(t) \tag{3.4.3}$$

with the solution

$$p = \mathrm{e}^{-\gamma(t-t_0)/m} p_0 + \int_{t_0}^{t} dt' \mathrm{e}^{-\gamma(t-t')/m} \xi(t'). \tag{3.4.4}$$

a) Asymptotic Mean Energy: This is given by

$$\begin{aligned}\langle E(t)\rangle &= \frac{1}{2m}\langle p(t)^2\rangle \\ &= \frac{1}{2m}\mathrm{e}^{-2\gamma t/m}\langle p_0^2\rangle + \frac{1}{2m}\int_0^t\int_0^t dt' dt'' \mathrm{e}^{-\gamma(2t-t'-t'')/m}\langle \xi(t')\xi(t'')\rangle\end{aligned} \tag{3.4.5}$$

where the cross correlation term has been set to zero because of the assumed initial independence of p and the heat bath, and $t_0 = 0$ is chosen for convenience. Using the spectrum formula (3.3.10), we find, in the limit $t \to \infty$

$$\langle E(t)\rangle \to \frac{\gamma\hbar}{m\pi}\int_0^\infty d\omega \frac{\omega \bar{N}(\omega)}{(\gamma/m)^2+\omega^2} + \frac{\gamma\hbar}{2m\pi}\int_0^\infty d\omega \frac{\omega}{(\gamma/m)^2+\omega^2}. \tag{3.4.6}$$

In this formula, one sees a characteristic of all quantum noise problems; the infinite influence of the zero point fluctuations of the heat bath. The first term is finite, and can in principle be evaluated. In the limit of very weak coupling, $\gamma \to 0$ we note that

$$\lim_{\gamma\to 0}\frac{1}{\pi}\frac{\gamma/m}{(\gamma/m)^2+\omega^2} = \delta(\omega) \tag{3.4.7}$$

so that

$$\frac{\gamma\hbar}{m\pi}\int_0^\infty dw \frac{\omega \bar{N}(\omega)}{(\gamma/m)^2+\omega^2} = \tfrac{1}{2}\hbar \lim_{\omega\to 0} \omega\bar{N}(\omega) = \tfrac{1}{2}kT \tag{3.4.8}$$

which is exactly the thermal energy expected from both classical and quantum equipartition formulae for a free particle. However, the final term is actually infinite, but we can get some idea of its nature by using a frequency cutoff, which is equivalent to setting $\tilde{\kappa}(\omega)$ equal to zero above the cutoff, and constant below it. Replacing the limit ∞ by λ, the final term becomes

$$\frac{\gamma\hbar}{4m\pi}\ln(1+m^2\lambda^2/\gamma^2). \tag{3.4.9}$$

The most important thing about this term is its temperature independence. We cannot observe the actual energy of the particle, only energy changes, so that this added term does not affect the thermodynamics. For any reasonable cutoff, it is also small compared with $\frac{1}{2}kT$. For example, if $\gamma/m = 10^6$Hz, $\lambda = 10^{18}$Hz, and $T = 1$ deg K, then

$$\begin{aligned} \tfrac{1}{2}kT &= 7\times 10^{-24}\text{Joule}, \\ (3.4.9) &= 5\times 10^{-28}\text{Joule}. \end{aligned}$$

Exercise. Evaluate $\langle E(t)\rangle$ as a function of time, and show that a cutoff is needed for any $t > 0$.

b) Asymptotic Mean Square Displacement: We can now solve the position equation of motion (3.4.2), to get

$$q(t) - q_0 = \frac{p_0(1-\mathrm{e}^{-\gamma t/m})}{\gamma} + \frac{1}{m}\int_0^t dt' \int_0^{t'} dt''\mathrm{e}^{-\gamma(t'-t'')/m}\xi(t''). \tag{3.4.10}$$

As in classical Brownian motion, let us compute the mean square distance travelled in a time interval starting at the time t_1 and ending at the time t_2—the initial time can be chosen equal to t_0, the time at which the system and bath are assumed uncorrelated, but this is not necessary. Thus, we want to compute $\langle [q(t_2) - q(t_1)]^2\rangle$, and using (3.4.10) and substituting for the resulting correlation functions as previously, we find that

$$\begin{aligned} \langle [q(t_2)-q(t_1)]^2\rangle = \langle p_0^2\rangle &\frac{(\mathrm{e}^{-\gamma t_2/m} - \mathrm{e}^{-\gamma t_1/m})^2}{\gamma^2} \\ &+\frac{\gamma\hbar}{2\pi}\int_{-\infty}^{\infty} d\omega \frac{\omega\coth(\hbar\omega/2kT)}{\gamma^2+m^2\omega^2} \\ &\times\left[\frac{\mathrm{e}^{-\mathrm{i}\omega t_2}-\mathrm{e}^{-\mathrm{i}\omega t_1}}{-\mathrm{i}\omega} + \frac{\mathrm{e}^{-\gamma t_2/m}-\mathrm{e}^{-\gamma t_1/m}}{\gamma/m}\right] \\ &\times\left[\frac{\mathrm{e}^{\mathrm{i}\omega t_2}-\mathrm{e}^{\mathrm{i}\omega t_1}}{\mathrm{i}\omega} + \frac{\mathrm{e}^{-\gamma t_2/m}-\mathrm{e}^{-\gamma t_1/m}}{\gamma/m}\right]. \end{aligned} \tag{3.4.11}$$

This integral contains transient terms, which depend on the time t_1 and t_2, and not simply their difference. Further, by noting that the hyperbolic cotangent approaches 1 as $\omega \to \infty$, it is easy to see that the integral arising from the product of the last two terms in the brackets is divergent unless we introduce a cutoff. If we do introduce a cutoff, so the integrals are finite, and then consider that t_1 and t_2 are very large, though their difference is not, then all the transient terms vanish, and

$$\langle (q(t_2) - q(t_1))^2 \rangle = \frac{\gamma\hbar}{2\pi} \int_{-\infty}^{\infty} d\omega \frac{\omega \coth(\hbar\omega/2kT)}{\gamma^2 + m^2\omega^2} \frac{4\sin^2 \frac{\omega}{2}(t_2 - t_1)}{\omega^2}, \tag{3.4.12}$$

and the integral is convergent with or without a cutoff. We can now consider the situation where $|t_2 - t_1|$ is large compared with the values of ω^{-1} of significance in the integral, and use the standard formula used in time dependent perturbation theory

$$4\frac{\sin^2 \frac{\omega}{2}(t_2 - t_1)}{\omega^2} \to 2\pi |t_2 - t_1| \delta(\omega) \tag{3.4.13}$$

to get the asymptotic formula

$$\langle [q(t_2) - q(t_1)]^2 \rangle \approx \frac{2kT}{\gamma} |t_2 - t_1| \tag{3.4.14}$$

which is exactly the classical result. The interpretation of this result is thus as follows: The model assumes that at time t_0 (here equal to zero) the bath and the Brownian particle are uncorrelated. Immediately there is a large transient evolution, which appears to allow the particle to absorb a large amount of energy from the bath, and to travel a large mean square distance. The meaning of "large" in this context being that the quantities are dependent on the cutoff. Both of these effects can be traced to the linearly rising nature of $\omega \coth(\hbar\omega/2kT)$ as ω gets large, i.e., they are a result of the very large zero point energies available in the very high frequency oscillators of the heat bath. However, after this transient, for sufficiently large time differences the diffusion of the particle is given by exactly the same formula as in classical stochastics.

This transient behaviour is thus simply the evolution needed to destroy the rather artificial initial state with a density operator consisting of a direct product of a bath term and a system term.

c) The Case of Zero Temperature: In this case the replacement (3.4.13) is not valid, since at zero temperature

$$\omega \coth(\hbar\omega/2kT) \to |\omega| \tag{3.4.15}$$

and thus the argument of the integral is sharply varying at $\omega = 0$. The formula (3.4.12) yields instead

$$\langle [q(t_2) - q(t_1)]^2 \rangle = \frac{4\gamma\hbar}{\pi} \int_0^{\infty} \frac{d\omega \sin^2 \frac{\omega}{2}(t_2 - t_1)}{\omega(\gamma^2 + m^2\omega^2)} \tag{3.4.16}$$

which is still convergent. An asymptotic expression for large $|t_2 - t_1|$ can be developed.

By a setting $x = m\omega/\gamma$, and $a = \gamma(t_2 - t_1)/m$ we can reduce the evaluation of (3.4.16) to that of the real part of

$$\int_0^\infty dx \frac{1 - e^{iax}}{x(x^2+1)} = ia \int_0^\infty e^{iax} \ln\left(x/\sqrt{x^2+1}\right) dx \tag{3.4.17}$$

after a partial integration. This can then be written as

$$= \frac{d}{d\epsilon}\left\{ ia \int_0^\infty e^{iax} \left(x/\sqrt{x^2+1}\right)^\epsilon dx \right\}\Bigg|_{\epsilon=0} \tag{3.4.18}$$

and for large a, this can be written asymptotically as

$$\approx \frac{d}{d\epsilon}\left\{ ia\Gamma(1+\epsilon)(ia)^{-(1+\epsilon)}\right\}\Bigg|_{\epsilon=0} \tag{3.4.19}$$

by use of Watson's lemma [3.9]. Carrying out the differentiation, the result is

$$\approx \ln(ia) - \Gamma'(1) \tag{3.4.20}$$

and hence

$$\langle [q(t_2) - q(t_1)]^2 \rangle \approx \frac{2\hbar}{\pi\gamma} \ln\left(\frac{\gamma(t_2 - t_1)}{m}\right) \tag{3.4.21}$$

at $T = 0$. This is truly a *quantum* Brownian motion, since it is induced solely by the fluctuations in the zero point energy of the bath oscillators which is made clear by the factor $\hbar$ in (3.4.21). It is of course, much slower than the motion induced by thermal fluctuations, and does *not* follow the classical diffusion law. It also represents a *very* slow diffusion. If we take a proton, $m \approx 10^{-27}$kg, with $\gamma/m \approx 10^6$Hz, we find $\langle [q(t_2) - q(t_1)]^2 \rangle \approx 3 \times 10^{-14} \ln[10^6(t_2 - t_1)]\mathrm{m}^2$. If we wait perhaps for a day, $t_2 - t_1 \approx 10^5$s, then the mean square distance travelled is about $10^{-12}\mathrm{m}^2$. Thermal Brownian motion following the law (3.4.14) would mask this unless $T \ll 4 \times 10^{-16}$K, which is an almost impossibly low temperature.

3.4.3 The Harmonic Oscillator

For definiteness let us consider an electrical implementation of a damped harmonic oscillator, an LRC circuit. The system variables are the canonical variables P and Q, with a Hamiltonian

$$H_{\mathrm{sys}} = \frac{P^2}{2L} + \frac{Q^2}{2C}. \tag{3.4.22}$$

Using this Hamiltonian, and the choice of interaction $X \to Q$, we deduce the equations of motion from (3.1.13)

$$\dot{Q}(t) = P(t)/L \tag{3.4.23}$$

$$\dot{P}(t) = -Q(t)/C - \int_{t_0}^{t} f(t-t')\dot{Q}(t')\,dt' - f(t-t_0)Q(t_0) + \xi(t). \tag{3.4.24}$$

At this stage we shall not assume that $f(t-t')$ is a delta function, since we have already seen that this gives rise to divergences for the case of the free particle.

a) Fourier Transform Solution: The simplest method of solution is to take the Fourier transform, defined by

$$\begin{aligned} Q(t) &= \frac{1}{\sqrt{2\pi}} \int_{-\infty}^{\infty} d\omega\, \mathrm{e}^{\mathrm{i}\omega t}\tilde{Q}(\omega) \\ P(t) &= \frac{1}{\sqrt{2\pi}} \int_{-\infty}^{\infty} d\omega\, \mathrm{e}^{\mathrm{i}\omega t}\tilde{P}(\omega) \\ \xi(t) &= \frac{1}{\sqrt{2\pi}} \int_{-\infty}^{\infty} d\omega\, \mathrm{e}^{\mathrm{i}\omega t}\tilde{\xi}(\omega). \end{aligned} \tag{3.4.25}$$

We now must Fourier transform the equations of motion (3.4.23,24), and assume the initial condition is in the remote past, i.e., set $t_0 \to -\infty$. We use the convolution theorem for Fourier transforms

$$\frac{1}{\sqrt{2\pi}} \int_{-\infty}^{\infty} dt\, \mathrm{e}^{-\mathrm{i}\omega t} \left\{ \int_{-\infty}^{\infty} f^*(t-t')X(t')dt' \right\} = \sqrt{2\pi}\, \tilde{f}(\omega)^*\, \tilde{X}(\omega) \tag{3.4.26}$$

on the second term on the right hand side of (3.4.24). Since, however, the t' integral goes only up to t, we need the Fourier transform of $f(t)$ multiplied by a step function; this means that the Fourier transform of the equations will be

$$\mathrm{i}\omega\tilde{Q}(\omega) = \tilde{P}(\omega)/L \tag{3.4.27}$$

$$\mathrm{i}\omega\tilde{P}(\omega) = -\tilde{Q}(\omega)/C - \mathrm{i}\omega R(\omega)\tilde{Q}(\omega) + \tilde{\xi}(\omega) \tag{3.4.28}$$

in which

$$R(\omega) = \int_{0}^{\infty} dt\, \mathrm{e}^{-\mathrm{i}\omega t} f(t). \tag{3.4.29}$$

We use a quantity $r(\omega)$, defined by

$$f(t) = \frac{1}{\pi} \int_{-\infty}^{\infty} d\omega\, r(\omega)\mathrm{e}^{\mathrm{i}\omega t}. \tag{3.4.30}$$

Notice that the definition of $f(t)$ in (3.1.23) already specifies $r(\omega)$, so we can immediately see that

$$r(\omega) = r(-\omega) = \frac{\pi}{2} G(\omega) \frac{dn}{d\omega}. \tag{3.4.31}$$

Using the result

$$\int_0^\infty dt\, \mathrm{e}^{\mathrm{i}(\omega' - \omega)t} = \pi\delta(\omega' - \omega) + \mathrm{i}\frac{\mathrm{P}}{\omega' - \omega}, \tag{3.4.32}$$

where P means that the principal value is to be taken of any integral of the relevant part the RHS of (3.4.32), we find

$$R(\omega) = r(\omega) + \frac{\mathrm{i}}{\pi}\mathrm{P}\int \frac{d\omega' r(\omega')}{\omega' - \omega} \tag{3.4.33}$$

$$\equiv r(\omega) + \mathrm{i}s(\omega). \tag{3.4.34}$$

The function $R(\omega)$ is known in non-equilibrium statistical mechanics as a *complex impedance.* [3.10] Notice that, by definition $r(\omega) = r(-\omega)$ and hence $s(\omega) = -s(-\omega)$ so that $R(\omega)^* = R(-\omega)$.

The solutions of (3.4.27,28) are then

$$\tilde{Q}(\omega) = (-\omega^2 L + \mathrm{i}\omega R(\omega) + 1/C)^{-1}\tilde{\xi}(\omega) \tag{3.4.35}$$

$$\tilde{P}(\omega) = \mathrm{i}\omega L\tilde{Q}(\omega). \tag{3.4.36}$$

We note also that the commutation relations and mean anticommutator formula for $\xi(t)$ imply

$$[\tilde{\xi}(\omega), \tilde{\xi}(\omega')] = -2\hbar\omega r(\omega)\delta(\omega + \omega') \tag{3.4.37}$$

$$\langle[\tilde{\xi}(\omega), \tilde{\xi}(\omega')]_+\rangle = 4\hbar\omega r(\omega)(\bar{N}(\omega) + \tfrac{1}{2})\delta(\omega + \omega') \tag{3.4.38}$$

b) Preservation of Commutation Relations: In the special case that $r(\omega) = r$ = constant, we can show explicitly that the canonical commutation relations are preserved. Writing the canonical commutator in the form

$$[Q(t), P(t)] = \frac{1}{2\pi}\int\int d\omega d\omega' \mathrm{e}^{\mathrm{i}(\omega t + \omega' t)}[\tilde{Q}(\omega), \tilde{P}(\omega')] \tag{3.4.39}$$

and using (3.4.37) this can be simplified to

$$[Q(t), P(t)] = \frac{\mathrm{i}\hbar}{\pi}\int d\omega \frac{Lr\omega^2}{(\omega^2 L - 1/C)^2 + \omega^2 r^2} = \mathrm{i}\hbar. \tag{3.4.40}$$

c) Mean Energy: Statistical mechanics tells us that the mean energy of the Harmonic oscillator of a temperature T is given by

$$\langle H\rangle = \frac{\sum_n (n + \frac{1}{2})\hbar\Omega \mathrm{e}^{-n\hbar\Omega/kT}}{\sum_n \mathrm{e}^{-n\hbar\Omega/kT}} = \hbar\Omega(\bar{N}(\Omega) + \tfrac{1}{2}), \tag{3.4.41}$$

but, as noted in Sect. 1.4.1, this is an approximation, valid only in the limit of weak coupling. To verify that our methods agree with statistical mechanics, let us work out the mean energy directly from the solutions for $P(t)$ and $Q(t)$ implicit in (3.4.35,36). We find

$$\langle H\rangle = \left\langle \frac{P(t)^2}{2L} + \frac{Q(t)^2}{2C} \right\rangle \tag{3.4.42}$$

$$= \frac{1}{2\pi}\int d\omega d\omega' \left\{ \frac{\langle \tilde{P}(\omega)\tilde{P}(\omega')\rangle}{2L} + \frac{\langle \tilde{Q}(\omega)\tilde{Q}(\omega')\rangle}{2C} \right\}. \tag{3.4.43}$$

We can work out the averages by using the solutions (3.4.35,36). Because we use Fourier transforms, these will be stationary averages if the statistics of $\tilde{\xi}(\omega)$ are assumed stationary. We can then straightforwardly show from solving the Fourier transformed equations, that

$$\langle H_{\rm sys}\rangle = \frac{1}{4\pi}\int_{-\infty}^{\infty} d\omega \frac{R(\omega)\hbar\omega[2\bar{N}(\omega)+1]}{(\omega^2 L - 1/C)^2 + \omega^2|R(\omega)|^2}\left\{\frac{1}{C} + \omega^2 L\right\}. \tag{3.4.44}$$

This result does approximate the desired result. This can be seen as follows:

i) Assume $R(\omega) \to 0$ sufficiently rapidly as $\omega \to \pm\infty$ that the integral does not diverge; and that $|R(\omega)|^2$ is sufficiently small that the denominator has a sharp resonance behaviour at $\omega = 1/\sqrt{LC} \equiv \Omega$. Then we can replace most ω dependent factors by their values at $\omega = \Omega$, where the sharp resonance peak occurs, and use the integral in (3.4.40), to show that

$$\langle H\rangle \approx \hbar\Omega\left(\bar{N}(\Omega) + \tfrac{1}{2}\right) \tag{3.4.45}$$

as happens in statistical mechanics.

ii) However, as in the case of the Brownian particle (Sect. 3.4.2), if $R(\omega)$ is exactly constant, the mean energy actually diverges, for exactly the same reasons—the rising spectrum of the zero point contributions to the spectrum of quantum noise.

3.4.4 The Two Level Atom

We give a treatment here of the model of Sect. 1.5.1. Thus, in this case we assume that in the quantum Langevin equations (3.1.28), we set

$$H_{\rm sys} = \tfrac{1}{2}\hbar\Omega\sigma_z \tag{3.4.46}$$

$$X = \sqrt{\hbar/\Omega}\,\sigma_x \tag{3.4.47}$$

and in such atomic systems it is conventional to define γ by $\tilde{\kappa}(\omega) = \sqrt{\gamma/2\pi}$, in contrast to (3.2.8). Using the Pauli matrix commutation relations (these are summarized in Sect. 9.1.2), the quantum Langevin equations for each of the system operators, $\sigma_x, \sigma_y, \sigma_z$, are

$$\dot{\sigma}_x = -\Omega\sigma_y \tag{3.4.48}$$

$$\dot{\sigma}_y = \Omega\sigma_x + \frac{1}{\sqrt{\hbar\Omega}}[\xi(t), \sigma_z]_+ \tag{3.4.49}$$

$$\dot{\sigma}_z = -\gamma - \frac{1}{\sqrt{\hbar\Omega}}[\xi(t), \sigma_y]_+ . \tag{3.4.50}$$

In deriving the second two equations, $\dot{\sigma}_x$ has been replaced using the first equation. Unlike the case of the Brownian particle, or that of the harmonic oscillator, these equations are not exactly soluble. However, an approximate treatment can be readily given, which amounts to that of Sect. 1.5.1. The high frequency treatment of this model is also extensively investigated in Chap.9.

3.4.5 The Rotating Wave Approximation

Let us consider the harmonic oscillator of Sect. 3.4.3 in the situation in which the resonant frequency $\Omega = 1/\sqrt{LC}$ is very large, and the damping, determined by $f(t)$, is small. We want to show that the behaviour of this oscillator may be approximated by a rather simpler kind of expression as long as our only interest is in the behaviour of the oscillator within a relatively narrow bandwidth around the resonant frequency.

To do this we define the destruction operator

$$a(t) = \frac{1}{\sqrt{2\hbar}}\left\{\frac{L}{C}\right\}^{\frac{1}{4}} Q(t) + \frac{\mathrm{i}}{\sqrt{2\hbar}}\left\{\frac{L}{C}\right\}^{-\frac{1}{4}} P(t) \tag{3.4.51}$$

in terms of which the equation of motion becomes (setting $t_0 = -\infty$)

$$\dot{a}(t) = \mathrm{i}\Omega a(t) - \frac{1}{2L}\int_{-\infty}^{t} f(t-t')\,[a(t') - a^\dagger(t')]\,dt' + \frac{\mathrm{i}}{\sqrt{2\hbar}}\left\{\frac{L}{C}\right\}^{-\frac{1}{4}}\xi(t). \tag{3.4.52}$$

We now move to a rotating frame by setting

$$a(t) = \mathrm{e}^{-\mathrm{i}\Omega t}A(t). \tag{3.4.53}$$

Since the dominant term in (3.4.52) is the term $-\mathrm{i}\Omega a(t)$, we can expect that $A(t)$ changes only slowly with time—more precisely, $A(t)$ varies slowly compared with $f(t)$. We can then make an approximate equation by replacing $A(t')$ and $A^\dagger(t')$ with $A(t)$ and $A^\dagger(t)$, and using the definition (3.4.29) of $R(\omega)$. We get

$$\dot{A}(t) = \frac{\mathrm{i}}{\sqrt{2\hbar}}\left\{\frac{L}{C}\right\}^{-\frac{1}{4}} \mathrm{e}^{\mathrm{i}\Omega t}\xi(t) - A(t)\frac{R(\Omega)}{2L} - \mathrm{e}^{2\mathrm{i}\Omega t}A^\dagger(t)\frac{R^*(\Omega)}{2L}. \tag{3.4.54}$$

We now make some further approximations.

i) The last term oscillates very rapidly on the time scale which is of interest, and we can justifiably set it equal to its cycle averaged value, namely zero. This is the core of what is meant by the *rotating wave approximation,* because the solution is only valid when averaged over many cycles of the factor $e^{2i\Omega t}$.

ii) Associated with the first approximation is the realization that any solutions to these equations will automatically be valid only over a slow time scale for $A(t)$, so only the frequency components of $e^{i\Omega t}\xi(t)$ with frequency near zero will have any significant effect on the solutions.

Let us define noise operators $b(t)$, $b^\dagger(t)$ by

$$b(t) = ie^{i\Omega t}\xi(t)/\sqrt{2\hbar\Omega r(\Omega)} \tag{3.4.55}$$

$$b^\dagger(t) = -ie^{-i\Omega t}\xi(t)/\sqrt{2\hbar\Omega r(\Omega)}\ . \tag{3.4.56}$$

We can now compute the commutator of these two operators by inserting the Fourier decomposition of $\xi(t)$ given in (3.4.25); we get

$$[b(t), b^\dagger(t')] = \frac{1}{2\pi}\int_{-\infty}^{\infty} d\omega\, e^{i\omega(t-t')}\frac{(\Omega-\omega)r(\Omega-\omega)}{\Omega r(\Omega)} \tag{3.4.57}$$

and if we are only interested in $\omega \approx 0$ (that is, time scales slow compared to $1/\Omega$), we may replace the coefficient in this Fourier transform by its value at $\omega = 0$. We then obtain

$$[b(t), b^\dagger(t')] \approx \delta(t-t'). \tag{3.4.58}$$

Similar reasoning shows that we can also approximate the other commutators as

$$[b(t), b(t')] = [b^\dagger(t), b^\dagger(t')] = 0. \tag{3.4.59}$$

Furthermore, we can also similarly write the approximations for the expectations of anticommutators as

$$\langle [b(t), b(t')]_+\rangle = \langle [b^\dagger(t), b^\dagger(t')]_+\rangle = 0 \tag{3.4.60}$$

$$\langle [b(t), b^\dagger(t')]_+\rangle = \big(2\bar{N}(\Omega)+1\big)\delta(t-t'). \tag{3.4.61}$$

In terms of these variables (3.4.53) becomes

$$\dot{A}(t) = -\tfrac{1}{2}\gamma A(t) + \sqrt{\gamma}\ b(t). \tag{3.4.62}$$

in which $\gamma = R(\Omega)/L$.

iii) Finally let us note that $R(\Omega)$ differs from $r(\Omega)$ by an imaginary term only (defined by (3.4.34)), which effectively changes the oscillator frequency Ω to some slightly different frequency, which can be neglected in all practical applications. The final rotating wave approximation Langevin equation then becomes, after moving out of the rotating frame

$$\dot{a}(t) = -i\Omega a(t) - \frac{\gamma}{2}a(t) + \sqrt{\gamma}\ b(t) \tag{3.4.63}$$

In fact the last term should really be $e^{-i\Omega t}b(t)$, but since we make the white noise approximations (3.4.60,61), the exponential factor has no effect. However, it should be remembered that this equation will only give a good representation of the behaviour of $a(t)$ in a rather narrow bandwidth around Ω.

The rotating wave approximation has the major advantage of eliminating divergences. This arises because the counter-rotating terms have been dropped and because we approximate the noise spectrum by a flat spectrum around the frequency of interest. It is not easy to justify it rigorously in this context, but it will become clear when we derive the master equation in Sect.3.6.3.

Exercise. Solve (3.4.63), and show that it predicts that the equilibrium mean value of the energy is $\hbar\Omega[\bar{N}(\Omega) + \frac{1}{2}]$.

3.5 The Adjoint Equation

The nonlinear nature of the quantum Langevin equations in nearly all cases of interest makes them difficult to solve and hence to use. In Sect. 1.5.1 an alternative point of view was given, based upon a white noise approximation to quantum noise—this yielded the *master equation*, (1.5.31), for the density operator of a two level atomic system. This equation is very powerful in the situations where it is valid, for example in nearly all quantum optical situations. The master equation is an approximate equation for the reduced density operator of the *system*, which is obtained by tracing over the bath variables. The quantum Langevin equation is, in contrast, an equation for system *operators*. In order to connect the two formalisms it is useful to find an equation for the system density operator which is exact, and then to investigate approximations which lead to master equations of various kinds. The equation that makes this connection is the adjoint equation, which will now be derived.

3.5.1 Derivation of the Adjoint Equation

In the theory of classical noise the derivation of a Fokker-Planck equation equivalent to a Langevin equation can be done directly using the theory of stochastic differential equations, as is done in S.M. Sect. 3.4.4, and the central fact used is the delta-correlated nature of the noise source. However, we know that physically the noise is *not* delta-correlated, but merely has a short correlation time. This means that in the classical case we can regard Langevin equations like

$$\begin{aligned} \dot{q} &= p/m \\ \dot{p} &= -V'(q) - \gamma\dot{q} + \xi(t) \end{aligned} \tag{3.5.1}$$

as being equivalent to a Liouville equation for a corresponding probability density function $P(q,p,t)$; namely

$$\frac{\partial P}{\partial t} = -\frac{\partial}{\partial q}\left(\frac{p}{m}P\right) + \frac{\partial}{\partial p}\left([V'(q) + \gamma\frac{p}{m} - \xi(t)]P\right) \tag{3.5.2}$$

in which $\xi(t)$ is regarded as a given function of time—albeit a random function. Quantities of interest can be computed in principle by solving this equation for each realization of $\xi(t)$, and averaging over the various possible realizations. This kind of Liouville equation is known as the *stochastic Liouville equation* and was first introduced by *Kubo* [3.11]. Exact solution is of course rarely possible, and approximation techniques have to be used.

The adjoint equation is the quantum analogue of the stochastic Liouville equation. Let us consider firstly the case where the density operator in the Heisenberg picture factorizes into a direct product $\rho_{\text{sys}} \otimes \rho_B$. Let $Y(t)$ be an arbitrary operator in the Heisenberg picture, and Y a Schrödinger picture version of the same operator. Then we can consistently define a quantity $\mu(t)$ by

$$\text{Tr}_{\text{sys}}\{Y(t)\rho_{\text{sys}}\} = \text{Tr}_{\text{sys}}\{Y\mu(t)\} \tag{3.5.3}$$

if the equality is true for all system operators Y and $Y(t)$. In fact, this implicit definition of $\mu(t)$ can be made explicit as follows. Suppose e_i are a complete set of Schrödinger picture system operators, in the sense that any system operator can be expressed as a linear combination of them, and suppose that they are also orthogonal with respect to the trace :

$$\text{Tr}_{\text{sys}}\left\{\text{e}_i^\dagger \text{e}_j\right\} = \delta_{ij}. \tag{3.5.4}$$

From this and the assumed completeness it follows that any operator A can be written

$$A = \sum_i \text{Tr}_{\text{sys}}\{\text{e}_i A\}\, \text{e}_i^\dagger. \tag{3.5.5}$$

In particular, if G is any operator, then

$$G\text{e}_i = \sum_j G_{ij}\text{e}_j \tag{3.5.6}$$

where

$$G_{ij} = \text{Tr}_{\text{sys}}\left\{\text{e}_j^\dagger G \text{e}_i\right\}. \tag{3.5.7}$$

Let us denote the corresponding Heisenberg operators by $\text{e}_i(t)$. Since the time evolution is unitary, all algebraic relations will be preserved. In particular, if $G(t)$ is the Heisenberg operator corresponding to G, then the Heisenberg version of (3.5.6) is

$$G(t)\text{e}_i(t) = \sum_j G_{ij}\text{e}_j(t), \tag{3.5.8}$$

where the coefficients G_{ij} are exactly the same as in the Schrödinger version (3.5.6), i.e., are given by (3.5.7), in terms of the Schrödinger picture operators.

Let us now construct $\mu(t)$ explicitly. We can write

$$\mu(t) = \sum_i \text{Tr}_{\text{sys}}\{\text{e}_i(t)\rho_{\text{sys}}\}\, \text{e}_i^\dagger \tag{3.5.9}$$

since it is clear from (3.5.9) that

$$\mathrm{Tr}_{\mathrm{sys}}\left\{\mathrm{e}_i\mu(t)\right\} = \mathrm{Tr}_{\mathrm{sys}}\left\{\mathrm{e}_i(t)\rho_{\mathrm{sys}}\right\} \tag{3.5.10}$$

and the general form (3.5.3) follows by completeness; i.e. if

$$Y = \sum Y_i \mathrm{e}_i \qquad \text{then} \qquad Y(t) = \sum_i Y_i \mathrm{e}_i(t) \tag{3.5.11}$$

and thus $\mu(t)$ satisfies the implicit definition (3.5.3). Let us now take the quantum Langevin equation in the form (3.1.28). The equation of motion for $\dot{\mathrm{e}}_i(t)$ is then

$$\dot{\mathrm{e}}_i(t) = \frac{\mathrm{i}}{\hbar}\left[H_{\mathrm{sys}}, \mathrm{e}_i(t)\right] - \frac{\mathrm{i}}{2\hbar}\left[[X(t), \mathrm{e}_i(t)], \xi(t) - \gamma\dot{X}(t)\right]_+ . \tag{3.5.12}$$

To evaluate $\dot{\mu}(t)$, we need to evaluate $\mathrm{Tr}_{\mathrm{sys}}\left\{\dot{\mathrm{e}}_i(t)\rho_{\mathrm{sys}}\right\}$, and this will contain terms of the form (as well as similar terms with $\xi(t)$ on the left)

$$\mathrm{Tr}_{\mathrm{sys}}\left\{A(t)\mathrm{e}_i(t)B(t)\xi(t)\rho_{\mathrm{sys}}\right\} = \mathrm{Tr}_{\mathrm{sys}}\left\{\sum_j G_{ij}\mathrm{e}_j(t)\rho_{\mathrm{sys}}\right\}\xi(t) \tag{3.5.13}$$

where

$$G_{ij} = \mathrm{Tr}_{\mathrm{sys}}\left\{A\mathrm{e}_i B\mathrm{e}_j^\dagger\right\} = \mathrm{Tr}_{\mathrm{sys}}\left\{B\mathrm{e}_j^\dagger A\mathrm{e}_i\right\}, \tag{3.5.14}$$

and $A(t)$, $B(t)$ may be $X(t)$, $\mathrm{e}_i(t)$, $\dot{X}(t)$, etc. The corresponding term in $\dot{\mu}(t)$ is then

$$\sum_{i,j} \mathrm{Tr}_{\mathrm{sys}}\left\{\mathrm{e}_j(t)\rho_{\mathrm{sys}}\right\}\mathrm{Tr}_{\mathrm{sys}}\left\{B\mathrm{e}_j^\dagger A\mathrm{e}_i\right\}\mathrm{e}_i^\dagger\xi(t) \tag{3.5.15}$$

$$= \sum_j \mathrm{Tr}_{\mathrm{sys}}\left\{\mathrm{e}_j(t)\rho_{\mathrm{sys}}\right\} B\mathrm{e}_j^\dagger A\xi(t)$$

$$= B\mu(t)A\xi(t). \tag{3.5.16}$$

Using this kind of result, it is then straightforward to derive the *adjoint equation* for $\mu(t)$:

$$\boxed{\dot{\mu}(t) = -\frac{\mathrm{i}}{\hbar}[H_{\mathrm{sys}}, \mu(t)] + \frac{\mathrm{i}}{2\hbar}\left[[\gamma\dot{X} - \xi(t), \mu(t)]_+, X\right].} \tag{3.5.17}$$

In this equation, the operator $\dot{X}$ is given by

$$\dot{X} = \frac{\mathrm{i}}{\hbar}[H_{\mathrm{sys}}, X]. \tag{3.5.18}$$

3.5.2 Comments on the Adjoint Equation

a) Equivalence to the Langevin Equation: The adjoint equation is an exact consequence of the quantum Langevin equation, but is not really a version of the same thing in the Schrödinger picture. It is rather like the situation in which a quantum system is driven by a classical field, but in that case, $\gamma\dot{X}$ would be missing.

b) A Commutative Noise Representation: Rather miraculously, the operator nature of $\xi(t)$ can be almost completely eliminated. In the adjoint equation, $\xi(t)$ occurs only as an anticommutator. We can define an operator $\alpha(t)$ by

$$\alpha(t)\mu(t') = \tfrac{1}{2}[\xi(t), \mu(t')]_+ \qquad \text{for all } t \text{ and } t'. \tag{3.5.19}$$

Then in fact

$$\alpha(t)\alpha(t') = \alpha(t')\alpha(t). \tag{3.5.20}$$

The proof is quite simple. Using straightforward algebra it is easy to show that

$$[\alpha(t), \alpha(t')]\mu = \tfrac{1}{4}\left[[\xi(t), \xi(t')], \mu\right] \tag{3.5.21}$$

However the commutator $[\xi(t), \xi(t')]$ is a c-number, so that the commutator $[\alpha(t), \alpha(t')]$ is zero. By definition the multiplication of $\alpha(t)$ with $\alpha(t')$ is associative, so that $\alpha(t)$ is in fact equivalent to a c-number random function. This has enabled *Parkins* and *Gardiner* [3.12] to simulate the solutions of the adjoint equation by using an appropriate random sequence of numbers to represent $\alpha(t)$. It is also possible to use approximation techniques based on classical stochastic processes on the adjoint equation when it is represented in this form, as has been done by *Ritsch* and *Zoller* [3.13].

c) Correlation Functions Using the Adjoint Equation: Since the $\mathrm{e}_i(t)$ are a complete set of operators, it is sufficient to compute the correlation functions (using the notation of Sect. 3.5.1)

$$\mathrm{Tr}_{\mathrm{sys}}\left\{\mathrm{e}_i(t)\mathrm{e}_j(t')\rho_{\mathrm{sys}}\right\} \equiv \mu_{ij}(t,t') \tag{3.5.22}$$

from which

$$\langle \mathrm{e}_i(t)\mathrm{e}_j(t')\rangle = \mathrm{Tr}_{\mathrm{B}}\left\{\mu_{ij}(t,t')\rho_B\right\}. \tag{3.5.23}$$

What is the equation of motion for $\mu_{ij}(t,t')$? We consider the case $t \geq t'$, so that an initial condition at $t = t'$ can be obtained from the fact that

$$\mu_{ij}(t',t') = \mathrm{Tr}_{\mathrm{sys}}\left\{\mathrm{e}_i(t')\mathrm{e}_j(t')\rho_{\mathrm{sys}}\right\} = \mathrm{Tr}_{\mathrm{sys}}\left\{\mathrm{e}_i\mathrm{e}_j\mu(t')\right\}. \tag{3.5.24}$$

We use the quantum Langevin equation (3.5.12), and define

$$\mu_j(t,t') = \sum_i \mathrm{e}_i^\dagger \mu_{ij}(t,t') \tag{3.5.25}$$

and proceed as for the derivation of the adjoint equation. However, we meet a term corresponding to (3.5.13)

$$\mathrm{Tr}_{\mathrm{sys}}\left\{A(t)\mathrm{e}_i(t)B(t)\xi(t)\mathrm{e}_j(t')\rho_{\mathrm{sys}}\right\}. \tag{3.5.26}$$

In this term we can use our procedure to get a term in $\dot{\mu}_j(t,t')$

$$\sum_k \mathrm{Tr}_{\mathrm{sys}}\left\{\mathrm{e}_k(t)\xi(t)\mathrm{e}_j(t')\rho_s\right\} B\mathrm{e}_k^\dagger A \tag{3.5.27}$$

and the operator $\xi(t)$ is still in the middle of the trace. We can bring it fully to the right by using the commutation relation (3.1.33), by means of which (3.5.27) becomes

$$B\mu_j(t,t')A\xi(t)+\sum_k \mathrm{Tr}_{\mathrm{sys}}\left\{\mathrm{e}_k(t)\frac{d}{dt}\left[\gamma u(t'-t)[X(t),\mathrm{e}_j(t')]\right]\rho_{\mathrm{sys}}\right\} B\mathrm{e}_k^\dagger A. \tag{3.5.28}$$

Since we are considering only $t > t'$, we can write

$$= B\mu_j(t,t')A\xi(t) - \sum_k \mathrm{Tr}_{\mathrm{sys}}\left\{e_k(t')\gamma\delta(t-t')[X(t'),\mathrm{e}_j(t')]\rho_{\mathrm{sys}}\right\} B\mathrm{e}_k^\dagger A \tag{3.5.29}$$

$$= B\mu_j(t,t')A\xi(t) - \gamma\delta(t-t')\sum_k \mathrm{Tr}_{\mathrm{sys}}\{\mathrm{e}_k[X,\mathrm{e}_j]\mu(t')\}B\mathrm{e}_k^\dagger A. \tag{3.5.30}$$

The conclusion is that $\mu_j(t,t')$ obeys the adjoint equation in t for $t > t'$, but there is an initial delta function transient. This is equivalent to modifying the initial condition by an additional term

$$\frac{\mathrm{i}\gamma}{2\hbar}\left[X,[X,\mathrm{e}_j]\mu(t')\right], \tag{3.5.31}$$

which is often rather small, especially in quantum optical situations.

This transient arises because of the factorized form of density operator chosen, and is similar to those found in the exact solutions for quantum Brownian motion and the harmonic oscillator.

3.5.3 Summary of the Adjoint Equation

Using the quantity $\alpha(t)$, the adjoint equation takes the form

$$\dot{\mu}(t) = -\frac{\mathrm{i}}{\hbar}[H_{\mathrm{sys}},\mu(t)] + \frac{\mathrm{i}\gamma}{2\hbar}\left[[\dot{X},\mu(t)]_+,X\right] - \frac{\mathrm{i}}{\hbar}\alpha(t)[\mu(t),X] \tag{3.5.32}$$

in which $\alpha(t)$ can be regarded as a c-number quantity which, in the case of a thermal bath, is Gaussian and has the mean and correlation function given by (3.3.10). The transformation of the operator nature of the input noise into something essentially classical makes for a tremendous simplification in the use of the adjoint equation—in fact it enables us to carry out many calculations using the techniques of ordinary stochastic processes.

a) General Form of the Adjoint Equation: As written in (3.5.32), the adjoint equation corresponds only to the first Markov approximation of Sect. 3.1.2. The more general quantum Langevin equation (3.1.13) is of course similar except for the modification of the $\dot{X}$ term. The general equation contains the term

$$\int_{t_0}^{t} f(t-t')\dot{X}(t')dt'. \tag{3.5.33}$$

This cannot be expressed directly in terms of system operators at time t, but an approximate expression can be derived *in the case that the noise is small.* In this case $H_{\rm sys}$ is almost time independent and we can write approximately

$$X(t') = \exp\left(\frac{\mathrm{i}H_{\rm sys}(t'-t)}{\hbar}\right) X(t)\exp\left(\frac{-\mathrm{i}H_{\rm sys}(t'-t)}{\hbar}\right). \tag{3.5.34}$$

This does give an expression for $X(t')$ in terms of $X(t)$, so that we can *define* the Schrödinger operator $X(t')$ by

$$X(t') = \exp\left(\frac{\mathrm{i}H_{\rm sys}t'}{\hbar}\right) X\exp\left(\frac{-\mathrm{i}H_{\rm sys}t'}{\hbar}\right), \tag{3.5.35}$$

then the adjoint equation becomes

$$\begin{aligned}\dot{\mu}(t) = &-\frac{\mathrm{i}}{\hbar}[H_{\rm sys},\mu(t)]\\ &+\frac{\mathrm{i}}{2\hbar}\left[\int_{-\infty}^{t}\left[\dot{X}(t')f(t')dt',\mu(t)\right]_+, X\right] - \frac{\mathrm{i}}{\hbar}\alpha(t)[\mu(t),X]\end{aligned} \tag{3.5.36}$$

where t_0 has been set to $-\infty$, and it is assumed that $f(t_0) \to 0$ in this limit. This general form of the adjoint equation is not an exact consequence of the initial model, but will be valid in weak noise situations, even if the noise is not Markovian.

b) Applications: It will not be easy to get exact solutions for either of these forms of the adjoint equation, and a number of approximation techniques will be developed, most notably, those of the *Master Equation.* However there is also a possibility of showing that the quasiclassical Langevin equation mentioned in Sect. 1.2 derives its validity from the adjoint equation—this will be done in Sect. 2.4.2

3.6 The Master Equation

The adjoint equation gives a way of (in principle) evaluating $\mu(t)$ for any possible realization of $\alpha(t)$. This is far more than is usually required, since the detailed correlation between the noise, $\alpha(t)$, and the system is not usually of interest. However, average quantities are of interest, that is, averages over the possible values of $\alpha(t)$. Thus, we are interested in averaging $\mu(t)$ over $\alpha(t)$, using the known statistics of

$\alpha(t)$. This is in fact a well studied classical problem, since the adjoint equation can be written in the form

$$\frac{d\mu}{dt} = [A_0 + \alpha(t)A_1]\mu \tag{3.6.1}$$

in which A_0 and A_1 are linear operators given by

$$A_0\mu = -\frac{\mathrm{i}}{\hbar}[H_{\mathrm{sys}}, \mu] + \frac{\mathrm{i}}{2\hbar}\left[\int_{-\infty}^{0} \left[\dot{X}(t')f(t')dt', \mu(t)\right]_+, X\right] \tag{3.6.2}$$

$$A_1\mu = -\frac{\mathrm{i}}{\hbar}[\mu, X]. \tag{3.6.3}$$

Van Kampen [3.14] has studied this problem in some detail, and has shown how to obtain a reduced equation for $\langle\mu(t)\rangle$, the average of $\mu(t)$ over $\alpha(t)$. If we define

$$\rho(t) = \mathrm{Tr}_{\mathrm{B}}\left\{\mu(t)\rho_{\mathrm{B}}\right\} \equiv \langle\mu(t)\rangle \tag{3.6.4}$$

then all the averages of system variables follow from (3.5.3),

$$\left.\begin{aligned}\langle Y(t)\rangle &= \mathrm{Tr}_{\mathrm{sys}}\left\{\mathrm{Tr}_{\mathrm{B}}\left\{Y(t)\rho_{\mathrm{sys}}\otimes\rho_{\mathrm{B}}\right\}\right\} = \mathrm{Tr}_{\mathrm{B}}\left\{\mathrm{Tr}_{\mathrm{sys}}\left\{Y\mu(t)\right\}\rho_{\mathrm{B}}\right\}\\ &= \mathrm{Tr}_{\mathrm{sys}}\left\{Y\,\mathrm{Tr}_{\mathrm{B}}\left\{\mu(t)\rho_{\mathrm{B}}\right\}\right\}\\ &= \mathrm{Tr}_{\mathrm{sys}}\left\{Y\rho(t)\right\}.\end{aligned}\right\} \tag{3.6.5}$$

Thus, an equation of motion for $\rho(t)$ will enable us to deduce the time development of the averages of system operators. The mathematical result is an expansion in terms of a smallness parameter, which measures how small or how fast the fluctuations are. The result, to second order in this parameter, is

$$\dot{\rho}(t) = A_0\rho(t) + \int_0^\infty d\tau\langle\alpha(t)A_1 e^{A_0\tau}\alpha(t-\tau)A_1 e^{-A_0\tau}\rangle\rho(t). \tag{3.6.6}$$

The relevant smallness parameter is $||A_1||\;||\alpha||\;\tau_c$, where

$||A_1||$ is a measure of the magnitude of the operator A_1
$||\alpha||$ is the root mean square amplitude of $\alpha(t)$
τ_c is the correlation time of $\alpha(t)$

The method of derivation of (3.6.6) is known as van Kampen's cumulant expansion method.

a) Outline of van Kampen's Cumulant Expansion: Although it is not appropriate to derive (3.6.6) in full detail, an indication of the major steps in its derivation is appropriate, since the technique is fundamental to very many problems in quantum noise theory, and can be adapted to other situations where the existence of a

Putting these together gives a condition for the validity of (3.6.6)

$$\hbar\gamma/m \ll kT \tag{3.6.21}$$

which is equivalent to

$$\tau_T \ll \tau_D \tag{3.6.22}$$

where

τ_D is the damping constant of the system $= m/\gamma$

τ_T is the thermal correlation time of $E(t)$.

These are admittedly rather crude estimates, but should suffice for our purposes. The condition (3.6.22) is in the end rather reasonable—it simply requires that the correlation time of the noise be much less than the typical time scale of the damped motion of the system. Two cases can now be distinguished. In one of these, that of *quantum Brownian motion*, it is assumed that the correlation time of the noise is much shorter than the typical time scale of the systematic motion and the damping. The equation that arises is interesting, and has strong analogies with classical Brownian motion equations. However, it has not been widely applied for the most obvious of reasons—quantum Brownian motion is experimentally not a widely studied subject. In the second case, the assumption of rather slow systematic motion is not made, but in order to arrive at a reasonably tractable equation it is necessary to assume that the damping is very weak compared to the systematic motion. This is precisely what is needed in the study of quantum optics, where the systematic motion takes place at optical frequencies, but the damping (which is in this case radiation into the electromagnetic field) is very weak—the lifetime of an excited atom is typically many thousands of periods. The equation so derived is very widely applied, and is capable of describing most situations which arise in the field of quantum optics.

3.6.1 The Quantum Brownian Motion Master Equation

We now focus on the situation in which the correlation time of the noise operator $\alpha(t)$ is quite short. The term

$$\int_0^\infty d\tau \langle \alpha(t)\alpha(t-\tau)\rangle A_1 e^{A_0\tau} A_1 e^{-A_0\tau} \tag{3.6.23}$$

which arises in (3.6.6) can be considerably simplified if it is assumed that the factor $\exp(A_0\tau)$ changes very little in a correlation time; i.e., if we assume that

$$\exp(A_0\tau) \approx 1 \qquad \text{for } \tau \approx \tau_c. \tag{3.6.24}$$

From the expression for $\langle \alpha(t)\alpha(t-\tau)\rangle$ given in (3.3.10) we find that if the noise is assumed thermal

$$\int_0^\infty dt \langle \alpha(t)\alpha(t-\tau)\rangle = \frac{\pi kT}{2} G(\omega) \frac{dn}{d\omega}\bigg|_{\omega=0} \tag{3.6.25}$$

and in the case that we take the simpler form of the adjoint equation, arising from the assumption of a constant $G(\omega)dn/d\omega$, this reduces to

$$\int_0^\infty d\tau \langle \alpha(t)\alpha(t-\tau)\rangle = \gamma kT. \tag{3.6.26}$$

If we now substitute for A_1 using the definition (3.6.3), we find

$$\int_0^\infty dt \langle \alpha(t)\alpha(t-\tau)\rangle A_1 e^{A_0\tau} A_1 e^{-A_0\tau}\rho(t) \to -\frac{\gamma kT}{\hbar^2}\left[X,\left[X,\rho(t)\right]\right] \tag{3.6.27}$$

and from (3.6.6), we derive

$$\boxed{\frac{d\rho}{dt} = -\frac{\mathrm{i}}{\hbar}[H_{\mathrm{sys}},\rho] - \frac{\mathrm{i}\gamma}{2\hbar}\left[X,\left[\dot{X},\rho\right]_+\right] - \frac{\gamma kT}{\hbar^2}\left[X,\left[X,\rho\right]\right]} \tag{3.6.28}$$

which is known as the quantum Brownian motion master equation. We now treat some particular applications.

3.6.2 Quantum Brownian Motion of a Particle in a Potential

Suppose we take

$$\begin{aligned} H_{\mathrm{sys}} &\to \frac{p^2}{2m} + V(x) \\ X &\to x \end{aligned} \tag{3.6.29}$$

so that the quantum Langevin equation becomes

$$\begin{aligned} \dot{x} &= p/m \\ \dot{p} &= -V'(x) - \gamma p/m + \xi(t) \end{aligned} \tag{3.6.30}$$

and the master equation (3.6.28) becomes

$$\frac{d\rho}{dt} = -\frac{\mathrm{i}}{\hbar}[H_{\mathrm{sys}},\rho] - \frac{\mathrm{i}\gamma}{2m\hbar}\left[x,[p,\rho]_+\right] - \frac{\gamma kT}{\hbar^2}\left[x,[x,\rho]\right]. \tag{3.6.31}$$

As an operator equation, this looks rather formidable, but as an equation for the position space matrix elements $\langle x|\rho|y\rangle$, it can be seen to be more tractable. By using the explicit form (3.6.29) for H_{sys}, and the relation

$$p|x\rangle = \mathrm{i}\hbar\frac{\partial}{\partial x}|x\rangle \tag{3.6.32}$$

we can transform (3.6.31) to

$$\begin{aligned} \frac{d}{dt}\langle x|\rho|y\rangle = \Bigg\{&\frac{\mathrm{i}\hbar}{2m}\left(\frac{\partial^2}{\partial x^2} - \frac{\partial^2}{\partial y^2}\right) - \frac{\mathrm{i}}{\hbar}\big(V(x) - V(y)\big) \\ &- \frac{\gamma}{2m}(x-y)\left(\frac{\partial}{\partial x} - \frac{\partial}{\partial y}\right) - \frac{\gamma kT}{\hbar^2}(x-y)^2\Bigg\}\langle x|\rho|y\rangle \end{aligned} \tag{3.6.33}$$

which is a 2nd order partial differential equation—by no means a difficult thing to understand.

a) Stationary Solution (Approximate): The differential equation (3.6.33) is considerably simplified by a change of variables to

$$\begin{aligned} x &= u + \hbar z \\ y &= u - \hbar z \\ \langle x|\rho|y\rangle &= P(u, z) \end{aligned} \tag{3.6.34}$$

so that

$$\begin{aligned} \frac{\partial}{\partial x} &= \frac{1}{2}\left(\frac{\partial}{\partial u} + \frac{1}{\hbar}\frac{\partial}{\partial z}\right) \\ \frac{\partial}{\partial y} &= \frac{1}{2}\left(\frac{\partial}{\partial u} - \frac{1}{\hbar}\frac{\partial}{\partial z}\right). \end{aligned} \tag{3.6.35}$$

In these variables, (3.6.33) becomes

$$\frac{\partial P}{\partial t} = \left\{\frac{\mathrm{i}}{2m}\frac{\partial^2}{\partial u \partial z} + \frac{\big(V(u+\hbar z) - V(u-\hbar z)\big)}{\mathrm{i}\hbar} - \frac{\gamma}{m} z \frac{\partial}{\partial z} - 4\gamma k T z^2\right\} P. \tag{3.6.36}$$

In the limit that $\hbar$ is small, we can make the replacement

$$\frac{\mathrm{i}}{\hbar}\big(V(u+\hbar z) - V(u-\hbar z)\big) \to 2\mathrm{i} z V'(u) \tag{3.6.37}$$

and it is notable that Planck's constant no longer appears explicitly anywhere in the equation. By explicit substitution, one sees that the stationary solution is given by

$$P_s(u, z) = \mathcal{N} \exp\left\{-\frac{V(u)}{kT} - 2mkTz^2\right\} \tag{3.6.38}$$

where

$$\mathcal{N}^{-1} = \sqrt{\frac{\pi}{2mkT}} \int du \exp\left\{-\frac{V(u)}{kT}\right\}. \tag{3.6.39}$$

Of course Planck's constant does appear, since z involves $\hbar$: explicitly

$$\langle x|\rho|y\rangle = \mathcal{N} \exp\left\{-\frac{V\left(\frac{1}{2}(x+y)\right)}{kT} - \frac{mkT(x-y)^2}{2\hbar^2}\right\}. \tag{3.6.40}$$

It can be seen that in this basis the density operator ρ is very nearly diagonal, because of the exponential cutoff in $x - y$ over distances of order $\hbar/\sqrt{mkT}$.

Exercise. Show that symmetrized moments of x and p can be computed from (3.6.40), and that these are exactly the same as those given by classical physics.

b) Free Particle: We set $V(x) = 0$, and express P in terms of a Fourier transform with respect to the variable u

$$P(u, z) = \int dq e^{iqu} \tilde{P}(q, z). \tag{3.6.41}$$

This Fourier transform can also be regarded as the characteristic function, i.e.

$$\tilde{P}(q, z) = \langle \exp(-iq\hat{x} + 2i\hat{p}z) \rangle \tag{3.6.42}$$

where $\hat{x}, \hat{p}$ are the position and momentum operators. (Use the fact that $\hat{p}$ is a displacement operator, and factorize the exponential using the Baker-Hausdorff theorem.) The master equation now becomes

$$\frac{\partial \tilde{P}}{\partial t} + \left(\frac{q + 2\gamma z}{2m} \right) \frac{\partial \tilde{P}}{\partial z} = -4\gamma kTz^2 \tilde{P} \tag{3.6.43}$$

and this can be solved by the method of characteristics, as in Sect. 3.8.4 of S.M. The subsidiary equation is

$$\frac{dt}{1} = \frac{2mdz}{q + 2\gamma z} = -\frac{d\tilde{P}}{4\gamma kTz^2 \tilde{P}}. \tag{3.6.44}$$

Two particular integrals are obtained by integrating the first two and the second two of these equations:

$$u(z, t, q, \tilde{P}) = we^{-\gamma t/m} \tag{3.6.45}$$

$$v(z, t, q, \tilde{P}) = \tilde{P} \exp(2mkTw^2 - 4mqkTw/\gamma) w^{mkTq^2/\gamma^2} \tag{3.6.46}$$

where

$$w = z + q/2\gamma. \tag{3.6.47}$$

The general solution of the partial differential equation (3.6.43) can be written in the form $v = F(u)$, where F is an arbitrary function, and thus, the solution becomes

$$\tilde{P} = \exp(-2mkTw^2 + 4mqkTw/\gamma) w^{-mkTq^2/\gamma^2} F(we^{-\gamma t/m}). \tag{3.6.48}$$

Assuming that at $t = 0$, $\tilde{P} = f(q, w)$, we can deduce that

$$\tilde{P} = f(q, we^{-\gamma t/m}) \exp\left\{ -2mkTw^2 \left(1 - e^{-2\gamma t/m}\right) + \frac{4mqkTw}{\gamma} \left(1 - e^{-\gamma t/m}\right) \right\} e^{-(q^2 kT/\gamma)t}. \tag{3.6.49}$$

From this solution, we note that

i) there are two characteristic time constants

$$\tau_D = m/\gamma, \tag{3.6.50}$$

which is the time constant of frictional damping, and

$$\tau_q = \gamma/kTq^2, \tag{3.6.51}$$

which is a time constant with the same kind of q dependence as turns up in classical diffusion theory.

ii) There is always a range of q such that $\tau_D < \tau_q$, so we can characterize the time evolution of the $\tilde{P}$ by two periods. In a time of order τ_D, after substituting for w in terms of z

$$\tilde{P} \to f(q,0)\exp(-2mkTz^2)\exp\left[-\frac{kTq^2}{\gamma}\left(t+\frac{m}{2\gamma}\right)\right] \tag{3.6.52}$$

and at this stage , only q such that

$$q^2 \ll \frac{\gamma^2}{mkT} \tag{3.6.53}$$

are important.

c) Example—Damping of Quantum Coherence: *Savage* and *Walls* [3.15] considered the effect of noise and damping on the interference effects expected quantum mechanically from the superposition of two plane wave states. We consider a system initially in the pure state $|\varphi\rangle$, expressible as a sum of two plane waves, so that

$$\langle x|\varphi\rangle = \sqrt{\tfrac{1}{2}}\,\left(e^{ik_1x}+e^{ik_2x}\right). \tag{3.6.54}$$

Initially

$$\begin{aligned} P(u,z) &= \langle u+\hbar z|\varphi\rangle\langle\varphi|u-\hbar z\rangle \\ &= \tfrac{1}{2}\left\{e^{2ik_1\hbar z}+e^{2ik_2\hbar z}+e^{i\hbar z(k_1+k_2)}\left(e^{iu(k_1-k_2)}+e^{-iu(k_1-k_2)}\right)\right\} \end{aligned} \tag{3.6.55}$$

and

$$\begin{aligned} \tilde{P}(q,z) &= \tfrac{1}{2}\delta(q)\left(e^{2ik_1\hbar z}+e^{2ik_2\hbar z}\right) \\ &\quad +\tfrac{1}{2}e^{i\hbar z(k_1+k_2)}\left\{\delta(q-k_1+k_2)+\delta(q+k_1-k_2)\right\} \end{aligned} \tag{3.6.56}$$

Using (3.6.49), we can compute $f(q,w)$, and hence $\tilde{P}(q,w,t)$ which is found to be

$$\begin{aligned} \tilde{P}(q,w,t) = \tfrac{1}{2}\Big[&\delta(q)\left\{\exp\left(2ik_1\hbar w e^{-\gamma t/m}\right)+\exp\left(2ik_2\hbar w e^{-\gamma t/m}\right)\right\} \\ &+\left\{\delta(q+k_1-k_2)+\delta(q-k_1+k_2)\right\} \\ &\times\exp\left(i(k_1+k_2)\hbar(we^{-\gamma t/m}-q/2\gamma)\right)\Big] \\ &\times\exp\Big\{-2mkTw^2\left(1-e^{-2\gamma t/m}\right) \\ &\qquad+\frac{4mkTqw}{\gamma}\left(1-e^{-\gamma t/m}\right)-\frac{q^2kTt}{\gamma}\Big\}. \end{aligned} \tag{3.6.57}$$

This can be inverted to yield the density operator, but we shall only examine the diagonal matrix elements of the density operator here, since these refer to the measurable probabilities, which in the case of no damping, exhibit an interference pattern. Thus, from (3.6.55), initially

$$\langle x|\rho|x\rangle = 1 + \cos\left((k_1 - k_2)x\right), \tag{3.6.58}$$

which clearly shows an interference term. To evaluate the diagonal matrix element from (3.6.57) we set $z = 0$, i.e. $w = q/2\gamma$, and carry out the Fourier inversion on the result; we find

$$\langle x|\rho(t)|x\rangle = 1 + \mathrm{e}^{-\eta(t)} \cos\left((k_1 - k_2)x - \frac{\hbar}{2\gamma}(1 - \mathrm{e}^{-\gamma t/m})(k_1^2 - k_2^2)\right) \tag{3.6.59}$$

where

$$\eta(t) = (k_1 - k_2)^2 \frac{mkT}{\gamma^2}\left\{\frac{\gamma t}{m} - \frac{3}{2} + 2\mathrm{e}^{-\gamma t/m} - \tfrac{1}{2}\mathrm{e}^{-2\gamma t/m}\right\}. \tag{3.6.60}$$

Two effects are notable here

i) The position of the interference fringe moves, as a result of the extra term in the argument of the cosine in (3.6.59).

ii) The visibility of the fringe is damped out by the factor $\mathrm{e}^{-\eta(t)}$, and ultimately, there are no fringes.

3.6.3 The Quantum Optical Case

One of the most powerful instruments in the subject of quantum optics has been the quantum optical master equation. It has a rather long history, originating in the theory of nuclear magnetic resonance in the early fifties [3.16], and moving into quantum optics in the mid-sixties. In this section we will derive a rather general form of this equation using the quantum Langevin equation as a basis. This is not the only derivation, and it is important to realize that the rather particular nature of the heat bath Hamiltonian assumed in the derivation of the quantum Langevin equation is not necessary to obtain the same master equation. The more usual derivations are given in Sect. 5.1. In this derivation we resolve operators in (3.6.6) into eigenoperators of $H_{\rm sys}$, namely

$$X = \sum_m (X_m^+ + X_m^-) \tag{3.6.61}$$

where

$$[H_{\rm sys}, X_m^\pm] = \pm\hbar\omega_m X_m^\pm. \tag{3.6.62}$$

This is always possible if the eigenvectors of $H_{\rm sys}$ form a complete set. If we now consider the case when all the ω_m are much larger than γ (as is always the case in quantum optics), then we can omit the γ dependent term in A_0 (as defined in (3.6.2)) in the terms $e^{\pm A_0 \tau}$ in (3.6.6), and thus make the replacement

$$e^{\pm A_0 \tau} X_m^{\pm} \to e^{\pm i\omega_m \tau} X_m^{\pm} e^{\pm A_0 \tau}. \tag{3.6.63}$$

Using the correlation function (3.3.10), one finds, after some labour, the *quantum optical master equation*:

$$\begin{aligned}
\dot{\rho}_S(t) = &-\frac{i}{\hbar}[H_{sys}, \rho_S] \\
&-\sum_m \frac{\pi\omega_m}{2\hbar}\left(\bar{N}(\omega_m)+1\right)\kappa(\omega_m)^2[\rho_S X_m^+ - X_m^- \rho_S, X] \\
&-\sum_m \frac{\pi\omega_m}{2\hbar}\bar{N}(\omega_m)\kappa(\omega_m)^2[\rho_S X_m^- - X_m^+ \rho_S, X] \\
&+\frac{i}{2\hbar}\sum_m \mathrm{P}\int_{-\infty}^{\infty}\frac{d\omega\,\kappa(\omega)^2}{\omega_m-\omega}\omega\left(\bar{N}(\omega)+\tfrac{1}{2}\right)[[\rho_S, X_m^+ - X_m^-], X] \\
&+\frac{i}{4\hbar}\sum_m \mathrm{P}\int_{-\infty}^{\infty}\frac{d\omega\,\kappa(\omega)^2}{\omega_m-\omega}\omega_m[[\rho_S, X_m^+ + X_m^-]_+, X].
\end{aligned} \tag{3.6.64}$$

In deriving this equation, it is necessary :

i) To use the adjoint equation in the form implied by (3.6.1–3), where, in particular, the damping is taken with a memory function $f(t)$. We will see in Chap.5 that the second and third lines are the most significant, and account for damping and noise. The last two lines involve principal value integrals, and because of this are rather small. They give rise to rather small frequency shifts, known as the *Stark* shift, which is the term proportional to $\bar{N}(\omega)$ in the fourth line, and the *Lamb* shift, which is the sum of the term in the fifth line and the term proportional to $\frac{1}{2}$ in the fourth line. The Stark shift is thus the result of thermal noise, while the Lamb shift, which does not vanish at absolute zero, is the result of vacuum fluctuations. Notice that the integrals are both divergent unless $\kappa(\omega)$ drops off sufficiently rapidly at high frequencies. In the case of an atom, we can say that the actual size of the atom gives a cutoff. In a full quantum electrodynamics treatment (for which the original Hamiltonian would not apply) there is no physically justifiable cutoff, and in that case the full power of renormalization theory is required.

ii) To use the relationship

$$\begin{aligned}
\int_{-\infty}^{\infty} d\omega\, f(\omega) \int_0^{\infty} d\tau\, e^{-i\omega\tau} &= \int \left[\pi\delta(\omega) - \mathrm{P}\frac{i}{\omega}\right] f(\omega) d\omega \\
&= \pi f(0) - i\mathrm{P}\int_{-\infty}^{\infty} d\omega\, f(\omega)/\omega
\end{aligned} \tag{3.6.65}$$

In both of the equations (3.6.64,65) the symbol $\mathrm{P}\!\int$ means the Cauchy principal value of the integral.

a) The Rotating Wave Approximation: This kind of master equation is well attested to be capable of describing almost all kinds of phenomena which can be experienced in quantum optics, although it is not often written down in precisely the form we have given. As used in quantum optics, it is basically due to *Louisell* [3.17]. Usually the rotating wave approximation is also made, which involves the following.

i) Define an interaction picture system density operator by

$$\rho_{\mathrm{I}}(t) = \exp\left(\frac{\mathrm{i}}{\hbar}H_{\mathrm{sys}}t\right)\rho_{\mathrm{S}}(t)\exp\left(-\frac{\mathrm{i}}{\hbar}H_{\mathrm{sys}}t\right). \tag{3.6.66}$$

Note that the interaction picture master equation no longer has a term corresponding to the first line in (3.6.64), and use the relation (3.6.63) to commute $\exp(\pm\frac{\mathrm{i}}{\hbar}H_{\mathrm{sys}}t)$ with $X_m^{\pm}$. This generates certain terms with factors such as $\exp\left(\mathrm{i}(\omega_m+\omega_n)t\right)$, which is a very rapid time variation on the timescale of atomic decays, thus we may neglect them completely at optical frequencies.

ii) The Lamb and Stark shift terms are also dropped, since they are very small, leaving the interaction picture master equation in the form :

$$\begin{aligned}\dot{\rho}_{\mathrm{I}} = &\sum_m \frac{\pi\omega_m}{2\hbar}\left(\bar{N}(\omega_m)+1\right)\kappa(\omega_m)^2(2X_m^-\rho_{\mathrm{I}}X_m^+ - \rho_{\mathrm{I}}X_m^+X_m^- - X_m^+X_m^-\rho_{\mathrm{I}})\\ &+\sum_m \frac{\pi\omega_m}{2\hbar}\bar{N}(\omega_m)\kappa(\omega_m)^2(2X_m^+\rho_{\mathrm{I}}X_m^- - \rho_{\mathrm{I}}X_m^-X_m^+ - X_m^-X_m^+\rho_{\mathrm{I}}).\end{aligned} \tag{3.6.67}$$

In this form, the master equation describes transitions in an atomic system in a radiation field. Adaptations to include small additional nonlinearities and driving fields are commonly made by adding terms as follows.

b) Driving Fields—Inputs and Outputs: It is very common to consider a situation where a laser beam is incident on an atom. This means that the heat bath (in this case the electromagnetic field) is no longer characterized as having thermal statistics, but has as well some coherent excitation in a small range of modes. It is helpful to go back to the transmission line model of Sect. 3.2 to understand how to include an incident field. From this point of view, the inclusion of a coherent driving field is no problem, since we need only specify the "in" field. One simply makes the requirement

$$\langle A_{\mathrm{in}}(t)\rangle = a_{\mathrm{in}}(t) \tag{3.6.68}$$

which gives the mean time dependent excitation. The statistics of any fluctuations can be specified by setting

$$\xi(t) = 2\sqrt{\gamma c}\,\{\dot{A}_{\mathrm{in}}(t) - \dot{a}_{\mathrm{in}}(t)\} \tag{3.6.69}$$

and specifying the relevant correlation functions of $\xi(t)$. For example, a coherent driving field superimposed on a thermal background is obtained by taking the correlation function (3.3.9) for $\xi(t)$ as defined by (3.6.69). The quantum mechanical Langevin equation becomes

$$\dot{Y} = \frac{\mathrm{i}}{\hbar}[H_{\mathrm{sys}}, Y] - \frac{2\mathrm{i}\sqrt{\gamma c}}{\hbar}\dot{a}_{\mathrm{in}}(t)[X, Y] - \frac{\mathrm{i}}{2\hbar}\left[\xi(t) - \gamma\dot{X}, [X, Y]\right]_+ \tag{3.6.70}$$

and the master equations (3.6.64, 67) acquire an extra term

$$\frac{2\mathrm{i}\sqrt{\gamma c}}{\hbar}a_{\mathrm{in}}(t)[X, \rho], \tag{3.6.71}$$

and this corresponds to simply adding to the system Hamiltonian a corresponding driving term

$$H_{\mathrm{Driving}} = 2\sqrt{\gamma c}\ a_{\mathrm{in}}(t)X. \tag{3.6.72}$$

c) Small Anharmonicity: It is common in quantum optics to consider the system to consist of a single mode of the electromagnetic field inside a cavity, which communicates with the electromagnetic bath and driving modes through an almost perfect mirror. The system would then be a perfect harmonic oscillator—however one also introduces some kind of weak nonlinearity via a nonlinear medium within the cavity. It is thus possible to write

$$H_{\mathrm{sys}} = H_0 + H_{\mathrm{nl}} \tag{3.6.73}$$

where H_{nl} is very small compared to H_{sys}. How does this affect the analysis? Typically the effects of H_{nl} are of the same order of magnitude as the damping, so H_{nl} can be neglected in all the procedures leading to (3.6.64, 67). In particular this means that

i) ω_m are the transition frequencies of H_0, and *not* of H_{sys}, and since H_0 is harmonic all the ω_m are equal.

ii) X_m are eigenoperators of H_0, *not* of H_{sys} .

Thus the relevant interaction picture is defined in terms of H_0, so that (3.6.64) is modified simply by adding a term

$$-\frac{\mathrm{i}}{\hbar}[H_{\mathrm{nl}}, \rho_{\mathrm{I}}], \tag{3.6.74}$$

and we can replace ω_m by a single frequency, Ω, say. Notice that there is an interesting transition region between small nonlinearity, which has this effect, and large nonlinearity, which modifies the whole master equation by modifying the relevant energy levels.

d) Stationary Solution is the Boltzmann Distribution: If we neglect the Stark and Lamb shift terms in the master equation (3.6.64) the stationary solution is obviously the Boltzmann distribution, for the equations (3.6.64) necessarily imply that

$$\exp(-H_{\mathrm{sys}}/kT)X_m^{\pm} = \exp(\pm\hbar\omega_m/kT)X_m^{\pm}\exp(-H_{\mathrm{sys}}/kT) \tag{3.6.75}$$

from which, using the definition of $\bar{N}$ in (3.6.64), it is obvious that the corresponding terms in the two summations cancel. A general and correct inclusion of the effects of the Lamb and Stark shifts is more tricky, but cannot alter this conclusion as a lowest order result.

Let us now look at the result of Benguria and Kac (see Sect. 1.4.1) that the Boltzmann distribution solution for the stationary state requires quantum Gaussian statistics for $\xi(t)$ in the case (as here) that $[\xi(t_1), \xi(t_2)]$ is a c-number. This admittedly much less rigorous, but certainly far more physically transparent result does not seem to require Gaussian statistics, since only the correlation functions are involved. This is much the same as in the classical case, for there, the proof that there is a white noise limit of a non-white noise stochastic Liouville equation does not require a Gaussian physical noise. Nevertheless, the resultant Fokker-Planck equation is exactly equivalent to a white noise stochastic differential equation with Gaussian noise.

4. Phase Space Methods

Ever since the early days of Planck the harmonic oscillator has played a central rôle in quantum mechanics. This rôle arises from the fact that in practice very many forces are nearly harmonic as well as from the all pervading nature of electromagnetic fields, which, like all Bose fields, are exactly equivalent to assemblies of harmonic oscillators.

The harmonic oscillator has an infinite number of equally spaced energy levels, and thus represents one extreme case of systems with discrete energy levels. Systems with a finite number of energy levels do not exist, but in many situations it is possible to consider that interesting processes involve only a few energy levels of some system, and in these situations it is advantageous to consider an idealized Hamiltonian whose full range of energy levels comprises only those of interest. The most idealized extreme of this is known as the *two level atom*, in which we consider processes involving only two energy levels. Such an idealization is relevant to the theory of atomic decays, which involve only an upper level and a lower level, as well as the electromagnetic field to which the atom is coupled.

From the realm of quantum optics there have been developed a very rich profusion of techniques for dealing with both of these kinds of system—such methods are particularly necessary when dealing with systems composed of assemblies of either of the elementary harmonic oscillator or two level atom systems, for calculations involving such complexity can very rapidly become unwieldy without some clever techniques. Much of the thrust of these techniques lies in their ability to exploit classical analogues—most particularly analogues with classical noise theory. Using these techniques, such as the P-representation of *Glauber* [4.1] and *Sudarshan* [4.2] and the *Wigner* representation [4.3], purely harmonic systems can be reduced to non-operator systems. However the essentially quantum mechanical nature of the problem is present in terms of the *interpretation* of the apparently classical variables.

This chapter first develops these phase space techniques for harmonic oscillators. The phase space techniques are not actually applicable to their fullest advantage until quantum noise systems are dealt with in the next and subsequent chapters, but even in the case of non-noisy systems they bring a clarity to the problem of the transition from the quantum world to the classical world. The central idea which unifies all the harmonic oscillator techniques is the *coherent state*, first introduced by *Glauber* [4.1], which is the quantum state which most closely approaches the classical description of harmonic physics.

4.1 The Harmonic Oscillator in One Variable

The observation by Galileo that the period of a pendulum is independent of the amplitude of its motion introduced the two most important features of the harmonic oscillator into physics: namely, the amplitude independent period itself, and equally significantly, the fact that many physical systems, (such as a pendulum) are for small amplitudes good approximations to the harmonic oscillator.

The Hamiltonian for a particle moving in a potential $V(x)$ is given in both classical and quantum mechanics by

$$H = \frac{p^2}{2m} + V(x) \tag{4.1.1}$$

and here

$$\begin{aligned} p &= \text{momentum of particle} \\ x &= \text{position of the particle} \\ m &= \text{mass of the particle} \\ V(x) &= \tfrac{1}{2}kx^2, \text{ for the harmonic oscillator.} \end{aligned} \tag{4.1.2}$$

But there are other realizations of the harmonic oscillator; e.g.,the electrical example of the LC circuit, in which

$$\begin{aligned} x &= Q = \text{charge on a capacitor } C \\ m &= L = \text{inductance} \\ p &= \text{“magnetic momentum”} = L\dot{Q} \\ V(x) &= Q^2/2C. \end{aligned} \tag{4.1.3}$$

As written, the electric circuit is an exact realization of a harmonic oscillator, unlike the pendulum, and is the prototype of another large class of representations; namely a single mode of a free electromagnetic field behaves like a harmonic oscillator as will be shown in Sect. 2.5

We will use the language appropriate to the Hamiltonian (4.1.1) in most of our discussion, but it should always be borne in mind that we are talking about a general harmonic oscillator, which may be realized in many other ways.

4.1.1 Equations of Motion—Classical

Hamilton's equations of motion yield the well known equations

$$\begin{aligned} m\dot{x} &= p \\ \dot{p} &= -kx \end{aligned} \tag{4.1.4}$$

with the solutions

$$\begin{aligned} x(t) &= x(0)\cos\omega t \\ \omega &= \sqrt{k/m}\,. \end{aligned} \tag{4.1.5}$$

The quantity ω is of course the angular frequency of the oscillator. The motion of the particle is bounded as a consequence of the boundedness of the sine and cosine functions, or as a consequence of conservation of energy.

In the limit $\omega \to 0$ this is no longer the case, since

$$x(t) \to x(0) + \frac{p(0)}{m} t \tag{4.1.6}$$

unless m is infinite. This qualitative difference is very important in applications, since it is very frequently assumed that the amplitude of the oscillation is bounded and quite small—such approximations cannot be made for $\omega \to 0$, where (4.1.6) corresponds to the motion of a free particle.

4.1.2 Equations of Motion—Quantum

The canonical commutation relations

$$[x, p] = \mathrm{i}\hbar \tag{4.1.7}$$

form the basis of the quantization of the harmonic oscillator. The oscillator may be treated in either the Schrödinger picture, in which time independent operators act on time dependent quantum states $|\psi, t\rangle$ which obey the Schrödinger equation

$$H|\psi, t\rangle = \mathrm{i}\hbar \frac{\partial}{\partial t} |\psi, t\rangle, \tag{4.1.8}$$

or in the Heisenberg picture, in which time dependent operators act on time independent states; the operators $a(t)$ obey the Heisenberg equations of motion

$$[H, a(t)] = -\mathrm{i}\hbar \frac{da(t)}{dt}. \tag{4.1.9}$$

The two pictures are related by

$$\begin{array}{ll} \textit{Schrödinger} & \textit{Heisenberg} \\ |\psi, t\rangle & = \exp(-\mathrm{i}Ht/\hbar)|\psi\rangle \\ \exp(\mathrm{i}Ht/\hbar) a \exp(-\mathrm{i}Ht/\hbar) & = a(t) \end{array} \tag{4.1.10}$$

In our treatment of quantum noise, the Schrödinger picture will usually be the most convenient, but nevertheless the Heisenberg picture will play an important role, and will often give more easily interpretable equations. As a point of notation, we shall usually indicate which picture is being used by the explicit functional dependence on time t. Thus $|\psi\rangle$ is a Heisenberg state, and a is a Schrödinger operator, while $a(t)$ is a Heisenberg operator. However the notation will not be turned into a formal pedantry; where little confusion can arise, and brevity makes it advantageous, t may not be written explicitly.

4.1.3 The Schrödinger Picture: Energy Eigenvalues and Number States

The harmonic oscillator is best solved in the Schrödinger picture. We introduce the destruction operator a and its Hermitian conjugate $a^\dagger$ by

$$\begin{aligned} a &= \frac{\mathrm{i}p}{\sqrt{2\hbar\omega m}} + \sqrt{\frac{k}{2\hbar\omega}}\, x \\ a^\dagger &= \frac{-\mathrm{i}p}{\sqrt{2\hbar\omega m}} + \sqrt{\frac{k}{2\hbar\omega}}\, x . \end{aligned} \tag{4.1.11}$$

In terms of these operators

$$\begin{aligned} &H = \hbar\omega(a^\dagger a + \tfrac{1}{2}) \\ &[a, a^\dagger] = 1 . \end{aligned} \tag{4.1.12}$$

The solution of the Schrödinger equation is achieved if the eigenvalues and eigenvectors of $a^\dagger a$ can be determined. This is done straightforwardly. One first notes that

$$\begin{aligned} [H, a] &= -\hbar\omega a \\ [H, a^\dagger] &= \hbar\omega a^\dagger . \end{aligned} \tag{4.1.13}$$

From these we see that if

$$H|E\rangle = E|E\rangle \tag{4.1.14}$$

then

$$Ha^\dagger|E\rangle = (E + \hbar\omega)a^\dagger|E\rangle \tag{4.1.15}$$

$$Ha|E\rangle = (E - \hbar\omega)a|E\rangle . \tag{4.1.16}$$

If follows that there is a whole sequence of eigenvalues of the form $E + n\hbar\omega$, for any integral n either positive or negative. Because H is a positive operator it cannot have any negative eigenvalues. But if E_0 is the least non-negative eigenvalue, then the only way to terminate the decreasing sequence defined by (4.1.16) is to require

$$a|E_0\rangle = 0. \tag{4.1.17}$$

From this and (4.1.12) it follows that $E_0 = \frac{1}{2}\hbar\omega$, and that the eigenvalues of H are

$$E_n = (n + \tfrac{1}{2})\hbar\omega \qquad (n = 0, 1, 2, \cdots), \tag{4.1.18}$$

and we will denote the corresponding eigenstates by $|n\rangle$. The general solution of the Schrödinger equation is thus of the form

$$|\psi, t\rangle = \sum_n \exp\left[\mathrm{i}(n + \tfrac{1}{2})\omega t\right] A_n|n\rangle \tag{4.1.19}$$

where the A_n are certain coefficients and the states $|n\rangle$ are normalized and orthogonal

$$\langle m|n\rangle = \delta_{mn}. \tag{4.1.20}$$

then $\langle x(t)\rangle$ and $\langle p(t)\rangle$ are given in terms of initial values of these quantities by (4.1.37). Simple substitution gives the solutions for the Heisenberg picture creation and destruction operators (4.1.11) as

$$\begin{aligned} a(t) &= e^{-i\omega t}a(0) \\ a^\dagger(t) &= e^{i\omega t}a^\dagger(0). \end{aligned} \tag{4.1.39}$$

4.2 Coherent States and the Classical Limit

It is well known that the limit $\hbar \to 0$ corresponds to the approach of quantum mechanics to classical mechanics. The energy eigenfunctions have an asymptotic form in this case that is most easily determined by the WKB method of solution of the Schrödinger equation. For the harmonic oscillator, the eigenfunctions are asymptotically

$$\psi_n(x) \simeq \frac{1}{\sqrt{\pi}}\left[\frac{k}{2E - kx^2}\right]^{\frac{1}{4}} \begin{matrix}\cos\\ \sin\end{matrix}\left(\int \frac{\sqrt{m(2E - kx^2)}\ dx}{\hbar}\right) \tag{4.2.1}$$

where

$$E = (n + \tfrac{1}{2})\hbar\omega \tag{4.2.2}$$

and the choice of cos or sin corresponds to n even or odd.

This is valid in the region in which the particle is classically allowed to exist, i.e., in the region where $|x| < \sqrt{2E/k}$ and the kinetic energy is consequently positive. Outside that region the wavefunction vanishes faster than any power of $\hbar$.

The average probability distribution over the fast sinusoidal oscillations is obtained by squaring $\psi(x)$, and setting the $\cos^2$ or $\sin^2$ terms equal to $\frac{1}{2}$: it is

$$P(x) \simeq \frac{1}{2\pi}\left[\frac{k}{2E - kx^2}\right]^{\frac{1}{2}}. \tag{4.2.3}$$

This is recognizable as the classical distribution, in that the probability density to be at position x is inversely proportional to the speed at which the particle moves there.

A similar result obviously holds for the momentum space wavefunction. Both these results have the defect that is obvious from (4.2.3), namely momentum and position are not specified comparably, and neither approximates the classical situation in which momentum and position oscillate—we get simply averages over all phases of the classical motion.

To investigate what kind of states might correspond more closely with classical states, we must consider how one might construct states which specify x, p, and E with as little uncertainty as possible in all of them simultaneously. To this end we must investigate what limits are placed on such uncertainties in general, and this is best done via the Heisenberg picture.

4.2.1 Coherent States as Quasi-Classical States

The eigenfunctions of energy do not give states which approach the classical motion in the limit as $\hbar \to 0$, as we have just seen in (4.2.1). What one would like is a set of quantum states in which the uncertainties in $x(t)$ and $p(t)$ are related to their initial values in the same way as the classical values are, and in which the uncertainty in the variables is in some way distributed between the position and the momentum to much the same extent.

Suppose we define

$$\begin{aligned} A(t) &= a(t) - \langle a(t) \rangle \\ A^\dagger(t) &= a^\dagger(t) - \langle a^\dagger(t) \rangle \end{aligned} \tag{4.2.4}$$

then these operators have the same commutation relations as $a(t)$ and $a^\dagger(t)$. Using (4.1.37), we see that the variances of $x(t)$ and $p(t)$, and the correlation between $x(t)$ and $p(t)$ are given by

$$\begin{aligned} &\langle \Delta x(t)^2 \rangle \\ &= \frac{\hbar}{2\sqrt{km}} \{ \langle A(0)^2 \rangle e^{-2i\omega t} + \langle A^\dagger(0)^2 \rangle e^{2i\omega t} + \langle A^\dagger(0)A(0) + A(0)A^\dagger(0) \rangle \} \end{aligned} \tag{4.2.5}$$

$$\begin{aligned} &\langle \Delta p(t)^2 \rangle \\ &= -\frac{\hbar\sqrt{km}}{2} \{ \langle A(0)^2 \rangle e^{-2i\omega t} + \langle A^\dagger(0)^2 \rangle e^{2i\omega t} - \langle A^\dagger(0)A(0) + A(0)A^\dagger(0) \rangle \} \end{aligned} \tag{4.2.6}$$

$$\begin{aligned} &\langle \tfrac{1}{2} \{ \Delta p(t) \Delta x(t) + \Delta x(t) \Delta p(t) \} \rangle \\ &= -\frac{i\hbar}{2} \{ \langle A(0)^2 \rangle e^{-2i\omega t} - \langle A^\dagger(0)^2 \rangle e^{2i\omega t} \}. \end{aligned} \tag{4.2.7}$$

These variances are time independent if (and only if) both $\langle A(0)^2 \rangle$ and $\langle A(0)^{\dagger 2} \rangle$ are zero. This corresponds to what we would like to see the system moving in a wavepacket with well defined *constant* uncertainties in the canonical variables.

The minimum uncertainty arises by minimizing $\langle A^\dagger(0)A(0) + A(0)A^\dagger(0) \rangle$. But we have just noted that $A(t)$ and $A^\dagger(t)$ satisfy the same commutation relations as a and $a^\dagger$, so that the eigenvalues of $A^\dagger(0)A(0)$ will be the same as those of $a^\dagger a$, since the derivation in Sect. 4.1.3 depends only on the commutation relations. Thus the eigenvalues of $A^\dagger(0)A(0) + A(0)A^\dagger(0)$ are of the form $m + \frac{1}{2}$, where m is a non-negative integer, and clearly the minimum mean is obtained in the state $m = 0$. We call this state $|\alpha\rangle$, and in analogy to (4.1.17), $|\alpha\rangle$ satisfies

$$A(0)|\alpha\rangle = 0. \tag{4.2.8}$$

Furthermore, we define the quantity α by

$$\alpha = \langle a \rangle. \tag{4.2.9}$$

Thus, the minimum uncertainty state is $|\alpha\rangle$, which satisfies

$$\boxed{a|\alpha\rangle = \alpha|\alpha\rangle,} \tag{4.2.10}$$

and this is known as a *coherent state*. It was first introduced by *Glauber* [4.1].

This equation can be considered as an alternative definition of a coherent state; namely, the coherent state $|\alpha\rangle$ is the eigenvector of the destruction operator. Since the destruction operator is not Hermitian, the coherent states need not be (and are not) orthogonal; nor is the eigenvalue real.

From (4.2.5) we see that

$$\begin{aligned} \langle\alpha|\Delta x(t)^2|\alpha\rangle &= \hbar/2\sqrt{km} \\ \langle\alpha|\Delta p(t)^2|\alpha\rangle &= \hbar\sqrt{km}\,/2 \\ \langle\alpha|\{\Delta p(t)\Delta x(t)+\Delta x(t)\Delta p(t)\}|\alpha\rangle &= 0. \end{aligned} \tag{4.2.11}$$

By comparing with (1.1.6), one can see that the minimum uncertainty theoretically possible is indeed achieved in the coherent state.

Exercise. Since a is not Hermitian, an eigenvector need not exist, although in fact it does exist. Show that there is no eigenstate of the creation operator $a^\dagger$.

4.2.2 Coherent State Solution for the Harmonic Oscillator

The solutions (4.1.39) are equivalent to

$$\mathrm{e}^{\mathrm{i}Ht/\hbar}a\mathrm{e}^{-\mathrm{i}Ht/\hbar} = \mathrm{e}^{-\mathrm{i}\omega t}a. \tag{4.2.12}$$

The solution of the Schrödinger equation with an initial coherent state $|\alpha\rangle$ is

$$|\alpha, t\rangle = \mathrm{e}^{-\mathrm{i}Ht/\hbar}|\alpha\rangle \tag{4.2.13}$$

and using (4.2.12) and (4.2.10) we see that

$$a|\alpha, t\rangle = \mathrm{e}^{-\mathrm{i}Ht/\hbar}\left(\mathrm{e}^{\mathrm{i}Ht/\hbar}a\mathrm{e}^{-\mathrm{i}Ht/\hbar}\right)|\alpha\rangle \tag{4.2.14}$$

$$= \mathrm{e}^{-\mathrm{i}Ht/\hbar}\mathrm{e}^{-\mathrm{i}\omega t}\alpha|\alpha\rangle \tag{4.2.15}$$

$$= \alpha\mathrm{e}^{-\mathrm{i}\omega t}|\alpha, t\rangle. \tag{4.2.16}$$

Thus, from (4.2.10) we see that $|\alpha, t\rangle$ must itself be proportional to the coherent state $|\alpha\mathrm{e}^{-\mathrm{i}\omega t}\rangle$. This means that the coherent states form a set of solutions of the Schrödinger equation which are determined completely by the classical solutions, which are given by the real and imaginary parts of $\alpha\mathrm{e}^{-\mathrm{i}\omega t}$.

The precise constant of proportionality cannot be determined without a determination of the phase of $|\alpha\rangle$. This is a matter of definition, which is made precise in the next section. The full solution is given in Sect. 4.3.2(g).

The formulae for the variances (4.2.11) show that they are proportional to Planck's constant, so that in the classical limit they become negligible. Thus the coherent state solution does represent exactly what was wanted—a set of solutions which directly approach the classical solutions.

4.3 Coherent States

This section is largely of the nature of a compendium of properties of coherent states which are often needed. The reader need not study it deeply, but merely use it for reference.

4.3.1 Properties of the Coherent States

a) Expression in Terms of Number States: By using the relations (4.1.23) it is easy to show that the only solution to the defining equation (4.2.10) which also satisfies $\langle\alpha|\alpha\rangle = 1$ is (up to a phase)

$$|\alpha\rangle = \exp\left(-\tfrac{1}{2}|\alpha|^2\right)\sum_{n=0}^{\infty}\frac{\alpha^n}{\sqrt{n!}}|n\rangle. \tag{4.3.1}$$

b) Unitary Transformation of the Vacuum: We can also show that

$$|\alpha\rangle = \exp\left(\alpha a^\dagger - \alpha^* a\right)|0\rangle. \tag{4.3.2}$$

This involves the use of the Baker-Hausdorff formula: for any two operators A and B such that $[A, B]$ commutes with both of them one can write

$$\begin{aligned}\exp(A + B) &= \exp(A)\exp(B)\exp\left(-\tfrac{1}{2}[A, B]\right)\\ &= \exp(B)\exp(A)\exp\left(\tfrac{1}{2}[A, B]\right)\end{aligned} \tag{4.3.3}$$

and this is proved in Appendix A. Using this identity, we see that (4.3.2) gives

$$|\alpha\rangle = \exp\left(-\tfrac{1}{2}|\alpha|^2\right)\exp\left(\alpha a^\dagger\right)\exp\left(-\alpha^* a\right)|0\rangle \tag{4.3.4}$$

and noting that $a|0\rangle = 0$, we see

$$|\alpha\rangle = \exp\left(-\tfrac{1}{2}|\alpha|^2\right)\sum\frac{\alpha^n (a^\dagger)^n}{n!}|0\rangle, \tag{4.3.5}$$

which yields (4.3.1) when we use the expression (4.1.24) for $|n\rangle$.

c) Scalar Product:

$$\langle\alpha|\beta\rangle = \exp\left(\alpha^*\beta - \tfrac{1}{2}|\alpha|^2 - \tfrac{1}{2}|\beta|^2\right) \tag{4.3.6}$$

$$|\langle\alpha|\beta\rangle|^2 = \exp\left(-|\alpha - \beta|^2\right). \tag{4.3.7}$$

Notice that no two coherent states are actually orthogonal to each other. However, if α and β are significantly different from each other, the two states are almost orthogonal.

d) Completeness Formula:

$$1 = \frac{1}{\pi}\int d^2\alpha\,|\alpha\rangle\langle\alpha|. \tag{4.3.8}$$

Here,

$$\begin{aligned} \alpha &= \alpha_x + \mathrm{i}\alpha_y, \\ d^2\alpha &= d\alpha_x d\alpha_y \end{aligned} \tag{4.3.9}$$

and the integral is over the whole complex plane. We prove this as follows : If $|A\rangle$ is an arbitrary vector, then write

$$|A\rangle = \sum_n A_n|n\rangle \tag{4.3.10}$$

so that

$$\frac{1}{\pi}\int d^2\alpha\,|\alpha\rangle\langle\alpha|A\rangle = \frac{1}{\pi}\sum_n\int A_n|\alpha\rangle\langle\alpha|n\rangle d^2\alpha. \tag{4.3.11}$$

Substitute (4.3.1) and change to polar coordinates by

$$\alpha = re^{\mathrm{i}\theta} \tag{4.3.12}$$

$$d^2\alpha = r\,dr\,d\theta. \tag{4.3.13}$$

Hence,

$$\begin{aligned} (4.3.11) &= \frac{1}{\pi}\sum_{m,n}\int A_n \mathrm{e}^{-r^2} r^{n+m} \mathrm{e}^{\mathrm{i}(m-n)\theta}(n!m!)^{-\frac{1}{2}}|m\rangle r\,dr\,d\theta \\ &= 2\sum_n\int A_n \mathrm{e}^{-r^2} r^{2n+1}(n!)^{-1}|n\rangle\,dr \end{aligned} \tag{4.3.14}$$

where we have used

$$2\delta_{n,m} = \frac{1}{\pi}\int d\theta\,\mathrm{e}^{\mathrm{i}(m-n)\theta}. \tag{4.3.15}$$

Now noting

$$\int_0^\infty dr\,\mathrm{e}^{-r^2} r^{2n+1} = n!/2, \tag{4.3.16}$$

we find

$$(4.3.11) = \sum_n A_n|n\rangle = |A\rangle. \tag{4.3.17}$$

Formulae (4.3.6,7) together indicate *that the coherent states are not orthogonal* for different α and β, and that since there is a factor $1/\pi$ in front of the integral (4.3.8) *the coherent states are overcomplete.* In fact, for any r, (4.3.1) shows that we can write

$$|n\rangle = \frac{1}{2\pi}\exp\left(\tfrac{1}{2}r^2\right) r^{-n}\sqrt{n!}\int d\theta\,\mathrm{e}^{-in\theta}|\alpha\rangle \tag{4.3.18}$$

which indicates that the states for any fixed $r = |\alpha|$ are complete. This overcompleteness is, however, not a disadvantage because of the very simple connection between coherent states and the physics of classical states, and because of the fact that the *Bargmann states*, defined by

$$||\alpha\rangle = \exp\left(\tfrac{1}{2}|\alpha|^2\right)|\alpha\rangle = \sum_{n=0}^{\infty} \frac{\alpha^n}{\sqrt{n!}}|n\rangle, \tag{4.3.19}$$

are analytic functions of α. This property is very important in what follows.

e) Expansion of Arbitrary States in Terms of Coherent States: Consider an arbitrary state $|f\rangle$. Then using the completeness relation (4.3.8) we have

$$|f\rangle = \frac{1}{\pi}\int d^2\alpha|\alpha\rangle f(\alpha^*)\exp\left(-\tfrac{1}{2}|\alpha|^2\right), \tag{4.3.20}$$

where

$$f(\alpha^*) = \langle\alpha|f\rangle \exp\left(\tfrac{1}{2}|\alpha|^2\right) = \langle\alpha||f\rangle \tag{4.3.21}$$

is *an analytic function of* α^*. With this proviso, the expansion (4.3.20) is unique. If functions of both α^* and α are permitted, the expansion is no longer unique, as *Glauber* [4.1] shows.

The scalar product of two states $|f\rangle$ and $|g\rangle$ is straightforwardly shown to be

$$\langle g|f\rangle = \frac{1}{\pi}\int d^2\alpha[g(\alpha^*)]^* f(\alpha^*)e^{-|\alpha|^2} \tag{4.3.22}$$

which is obviously a Hilbert space of analytic functions. It provides, in fact, the soundest mathematical starting point for the study of the harmonic oscillator, the creation destruction operators and all the formalism of this chapter.

f) Expansion of an Operator in Coherent States: Consider any operator T in the quantum Hilbert space. Using the identity resolution twice,

$$\begin{aligned} T &= 1\cdot T\cdot 1 = \frac{1}{\pi^2}\int d^2\alpha d^2\beta|\alpha\rangle\langle\alpha|T|\beta\rangle\langle\beta| \\ &= \frac{1}{\pi^2}\int d^2\alpha d^2\beta|\alpha\rangle\langle\beta|T(\alpha^*,\beta)\exp(-\tfrac{1}{2}|\alpha|^2 - \tfrac{1}{2}|\beta|^2) \end{aligned} \tag{4.3.23}$$

where

$$\begin{aligned} T(\alpha^*,\beta) &= \exp\left(\tfrac{1}{2}|\alpha|^2 + \tfrac{1}{2}|\beta|^2\right)\langle\alpha|T|\beta\rangle \\ &= \langle\alpha||T||\beta\rangle \end{aligned} \tag{4.3.24}$$

and from the analyticity of the states $||\alpha\rangle$, $||\beta\rangle$, we see that $T(\alpha^*,\beta)$ *is an analytic function of* α^* *and* β*, and, with this proviso, is unique*. Notice, for example, that if

$$T = (a^\dagger)^m(a^n), \tag{4.3.25}$$

then

$$\begin{aligned} T(\alpha^*, \beta) &= \langle\alpha|(a^\dagger)^m(a^n)|\beta\rangle \exp\left(\tfrac{1}{2}|\alpha|^2 + \tfrac{1}{2}|\beta|^2\right) \\ &= (\alpha^*)^m(\beta)^n\langle\alpha|\beta\rangle \exp\left(\tfrac{1}{2}|\alpha|^2 + \tfrac{1}{2}|\beta|^2\right) \\ &= (\alpha^*)^m(\beta)^n \exp(\alpha^*\beta). \end{aligned} \quad (4.3.26)$$

g) Any Operator T is Determined by Its Expectation in all Coherent States: For if T is any operator,

$$\langle\alpha|T|\alpha\rangle = \sum_{n,m}\langle n|T|m\rangle \mathrm{e}^{-|\alpha|^2}(\alpha^*)^n(\alpha)^m/\sqrt{n!m!} \quad (4.3.27)$$

so that

$$\langle n|T|m\rangle = \frac{1}{\sqrt{n!m!}}\frac{\partial^n}{\partial\alpha^{*n}}\frac{\partial^m}{\partial\alpha^m}\left(\mathrm{e}^{\alpha\alpha^*}\langle\alpha|T|\alpha\rangle\right)\Big|_{\alpha=0}. \quad (4.3.28)$$

The derivatives are formal derivatives and, as in analytic function theory, are to be interpreted as

$$\begin{aligned} \alpha &= x + \mathrm{i}y \\ \frac{\partial}{\partial\alpha} &= \frac{1}{2}\left(\frac{\partial}{\partial x} - \mathrm{i}\frac{\partial}{\partial y}\right) \\ \frac{\partial}{\partial\alpha^*} &= \frac{1}{2}\left(\frac{\partial}{\partial x} + \mathrm{i}\frac{\partial}{\partial y}\right) \end{aligned} \quad (4.3.29)$$

in which case, $\langle n|T|m\rangle$ are the combinations of coefficients of power series in real variables x and y.

h) Coherent States are Eigenstates of a: Namely,

$$a|\alpha\rangle = \alpha|\alpha\rangle$$

and (4.3.30)

$$\langle\alpha|a^\dagger = \langle\alpha|\alpha^*$$

which are proved directly from the definition and were the original basis for investigating them.

i) Normal Products: In evaluating matrix elements, *normal products* of operators in which all destruction operators stand to the right of creation operators, are useful. Thus,

$$\begin{aligned} \langle\alpha|a^\dagger a a^\dagger|\beta\rangle &= \langle\alpha|a^\dagger a^\dagger a + a^\dagger[a, a^\dagger]|\beta\rangle \\ &= \langle\alpha|a^\dagger a^\dagger a + a^\dagger|\beta\rangle \\ &= (\alpha^{*2}\beta + \alpha^*)\langle\alpha|\beta\rangle. \end{aligned} \quad (4.3.31)$$

The symbol : : around an expression means that it is to be considered a normal product: thus,

$$:(a + a^\dagger)(a + a^\dagger): = a^{\dagger 2} + a^2 + 2a^\dagger a. \quad (4.3.32)$$

From (4.3.30) it follows that the matrix element between coherent states $\langle\alpha|$ and $|\beta\rangle$ of any normally ordered function $F(a^\dagger, a)$ of creation and destruction operators is given by $F(\alpha^*, \beta)$. Thus, for example,

$$\langle\alpha| : (a + a^\dagger)^3 : |\beta\rangle = (\beta + \alpha^*)^3. \tag{4.3.33}$$

j) Poissonian Number Distribution of Coherent States: The state $|n\rangle$ is known as an n *quantum state* since its energy is $n\hbar\omega$ more than that of the vacuum $|0\rangle$, the zero quantum state. In quantum mechanics, therefore, the probability of observing n quanta in a coherent state $|\alpha\rangle$ is

$$P_\alpha(n) = |\langle n|\alpha\rangle|^2 = \left|\exp(-\tfrac{1}{2}\alpha^2)\frac{\alpha^n}{\sqrt{n\;!}}\right|^2 \tag{4.3.34}$$

$$= \frac{\mathrm{e}^{-|\alpha|^2}|\alpha|^{2n}}{n!} \tag{4.3.35}$$

which is a Poisson distribution with mean $|\alpha|^2$. Since the number n corresponds to the eigenvalue of the number operator N, we have

$$\left.\begin{aligned}\langle N\rangle &= \langle\alpha|N|\alpha\rangle = \sum_n nP(n) = |\alpha|^2\\ \langle N^2\rangle &= \langle\alpha|a^\dagger a a^\dagger a|\alpha\rangle\\ &= \langle\alpha|a^\dagger a^\dagger a a + a^\dagger[a, a^\dagger]a|\alpha\rangle\\ &= |\alpha|^4 + |\alpha|^2.\end{aligned}\right\} \tag{4.3.36}$$

Hence,

$$\langle N(N-1)\rangle = |\alpha|^4 = \langle N\rangle^2 \tag{4.3.37}$$

as required for a Poisson.

k) Wavefunction for a Coherent State: Using the definition of a and $a^\dagger$ in terms of p and x (4.1.11), and the usual $p \to -i\hbar d/dx$, the coherent state definition (4.2.10)

$$a|\alpha\rangle = \alpha|\alpha\rangle \to \left(\sqrt{\frac{1}{2\eta}}\frac{d}{dx} + \sqrt{\frac{\eta}{2}}\,x\right)\langle x|\alpha\rangle = \alpha\langle x|\alpha\rangle \tag{4.3.38}$$

for which the normalized solution is

$$\langle x|\alpha\rangle = \left(\frac{\eta}{\pi}\right)^{\frac{1}{4}}\exp\left[-\Big(\mathrm{Im}(\alpha)\Big)^2 - \frac{\eta}{2}\left(x - \alpha\sqrt{\frac{2}{\eta}}\right)^2\right]. \tag{4.3.39}$$

Defining

$$\alpha = \frac{i\bar{p}}{\hbar}\frac{1}{\sqrt{2\eta}} + \bar{x}\sqrt{\frac{\eta}{2}} \tag{4.3.40}$$

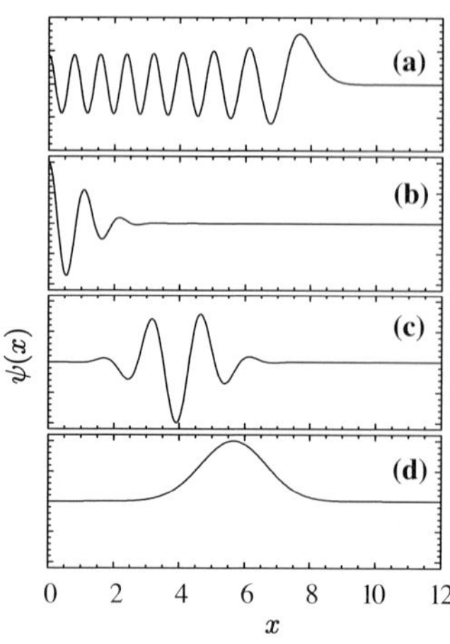

Fig. 4.1 Plot of coherent state wavefunctions, compared with the energy eigenfunction with the same mean energy: **(a)** Energy eigenfunction from (4.1.32); **(b)**, **(c)**, **(d)**, Coherent state wavefunctions for $\omega t = 0, \pi/4, \pi/2$ from the Gaussian wavepacket (4.3.41)

we can write

$$\langle x|\alpha\rangle = \left(\frac{\eta}{\pi}\right)^{\frac{1}{4}} \exp\left[-\frac{\eta}{2}(x-\bar{x})^2 + \frac{\mathrm{i}\bar{p}x}{\hbar}\right]. \tag{4.3.41}$$

This corresponds to a Gaussian wavepacket centred on $\bar{x}$ modulated by a plane wave $\exp(\mathrm{i}\bar{p}x/\hbar)$, which corresponds to a mean momentum $\bar{p}$, i.e. a minimum uncertainty Gaussian wavepacket.

4.3.2 Coherent States are Driven Oscillator Wavefunctions

The driven harmonic oscillator is described by adding a term $xE(t)$ to the Hamiltonian for the harmonic oscillator. If $x(t)$ is a position co-ordinate of an oscillator, then $E(t)$ could be proportional to some classical external field of force, for example gravitation, or an electric field. The Schrödinger picture Hamiltonian expressed in terms of a and $a^\dagger$ can be written (with $F(t) = \left(\hbar\sqrt{2\eta}\right)^{-1} E(t)$)

$$H = \hbar\omega(a^\dagger a + \tfrac{1}{2}) + \hbar F(t)(a + a^\dagger). \tag{4.3.42}$$

We can solve the Schrödinger equation by use of the interaction picture as follows. Write

$$\begin{aligned} H_0 &= \hbar\omega\left(a^\dagger a + \tfrac{1}{2}\right), \\ H_I(t) &= \hbar F(t)(a + a^\dagger). \end{aligned} \tag{4.3.43}$$

Then the Schrödinger equation is

$$\frac{d|\psi(t)\rangle}{dt} = -\frac{\mathrm{i}}{\hbar}[H_0 + H_I(t)]|\psi(t)\rangle. \tag{4.3.44}$$

If we let

$$|\phi(t)\rangle = \exp\left(\frac{\mathrm{i}H_0 t}{\hbar}\right)|\psi(t)\rangle \tag{4.3.45}$$

then

$$\frac{d|\phi(t)\rangle}{dt} = -\frac{\mathrm{i}}{\hbar}\exp\left(\frac{\mathrm{i}H_0 t}{\hbar}\right) H_I(t)\exp\left(\frac{-\mathrm{i}H_0 t}{\hbar}\right)|\phi(t)\rangle. \tag{4.3.46}$$

Notice that the solutions (4.2.10) imply

$$\begin{aligned} \exp\left(\frac{\mathrm{i}H_0 t}{\hbar}\right) a \exp\left(\frac{-\mathrm{i}H_0 t}{\hbar}\right) &= \mathrm{e}^{-\mathrm{i}\omega t} a \\ \exp\left(\frac{\mathrm{i}H_0 t}{\hbar}\right) a^\dagger \exp\left(\frac{-\mathrm{i}H_0 t}{\hbar}\right) &= \mathrm{e}^{\mathrm{i}\omega t} a^\dagger \end{aligned} \tag{4.3.47}$$

so that we find

$$\frac{d|\phi(t)\rangle}{dt} = -\mathrm{i}F(t)\left\{a^\dagger \mathrm{e}^{\mathrm{i}\omega t} + a\mathrm{e}^{-\mathrm{i}\omega t}\right\}|\phi(t)\rangle. \tag{4.3.48}$$

The two terms satisfy the conditions of the Baker-Hausdorff generalization as given in Appendix 4A. If we define

$$\alpha(t) = -\mathrm{i}\int_0^t dt'\, \mathrm{e}^{\mathrm{i}\omega(t-t')}F(t'), \tag{4.3.49}$$

the solution can be written

$$|\phi(t)\rangle = \exp\left\{a^\dagger \mathrm{e}^{-\mathrm{i}\omega t}\alpha(t) - a\mathrm{e}^{\mathrm{i}\omega t}\alpha^*(t) + \mathrm{i}\lambda(t)\right\}|\phi(0)\rangle \tag{4.3.50}$$

where $\lambda(t)$ is a phase given by

$$\lambda(t) = -\tfrac{1}{2}\mathrm{i}\int_0^t ds\int_0^t ds'\, \epsilon(s-s')F(s)F(s')\mathrm{e}^{\mathrm{i}\omega(s-s')}, \tag{4.3.51}$$

and the antisymmetric step function $\epsilon(x)$ is defined in (4A.17). Hence

$$|\psi(t)\rangle = \exp\left\{a^\dagger\alpha(t) - a\alpha^*(t) + \mathrm{i}\lambda(t)\right\}\exp\left\{-\frac{\mathrm{i}H_0 t}{\hbar}\right\}|\psi(0)\rangle. \tag{4.3.52}$$

a) The Initial State is the Vacuum State: In this case we get a coherent state solution:

$$|\psi(t)\rangle = \exp\left\{\mathrm{i}\left(\lambda(t) - \tfrac{1}{2}\omega t\right)\right\}|\alpha(t)\rangle. \tag{4.3.53}$$

Apart from the phase factors, the solution is a coherent state whose parameter $\alpha(t)$ is given by the solution of the classical equation of motion.

b) The Initial State is a Coherent State : Let the initial state be $|\alpha_0\rangle$. Using the representation of the coherent state (4.3.2), and following a similar procedure, we find that

$$|\psi(t)\rangle = \mathrm{e}^{-\mathrm{i}\omega t/2}\exp\left\{\mathrm{i}\lambda(t) + \tfrac{1}{2}\left(\alpha(t)\alpha_0^*\mathrm{e}^{\mathrm{i}\omega t} - \alpha^*(t)\alpha_0\mathrm{e}^{-\mathrm{i}\omega t}\right)\right\}|\alpha(t) + \alpha_0\mathrm{e}^{-\mathrm{i}\omega t}\rangle. \tag{4.3.54}$$

We see that if $F(t)$ vanishes the solution is simply given by

$$|\psi(t)\rangle = e^{-i\omega t/2}|\alpha_0 e^{-i\omega t}\rangle. \tag{4.3.55}$$

This is the minimum uncertainty state of Sect. 4.2.1.

c) Other Initial Conditions: A wide variety of initial conditions is available, but we have seen that only in the case of initial coherent states (including the vacuum as a special case) do we find that the state vector evolves so as to remain a coherent state. Thus, although we have found a large class of states which behave very similarly to classical states, we have not explained why they should be preferred to any other kind of state. That can only be explained in the context of damping theory, as is discussed in Sect. 7.3.1 where it is shown that the coupling of a quantum system to an environment consisting of a large number of degrees of freedom can select out the coherent state as that which will be observed in practice.

4.4 Phase Space Representations of the Harmonic Oscillator Density Operator

The classical limit of quantum mechanics is most naturally given by a coherent state description. For example, in Sect. 4.2.1 we saw that the coherent states were the unique states which give a time independent wavefunction in which both $x(t)$ and $p(t)$ have a minimum uncertainty, and in Sect. 2.2.2 it was shown that a classical driving field would automatically drive any coherent state into another coherent state—the parameter $\alpha(t)$ being the solution of the corresponding classical equation.

In all practical situations which arise in noisy systems the description of the system is best carried out in terms of the density operator for the system. In this section we will give a number of methods of expressing the density operator in terms of a c-number function of a coherent state variable α. These methods lead to considerable simplification of the quantum equations of motion, and are therefore of considerable practical use.

4.4.1 The Q-Representation

We have seen already in Sect. 4.3 that any operator can be expanded in terms of coherent states, but in fact this does not yield a very useful way of expressing the physics of the density operator in terms of coherent states. A more useful and interesting result is that of Sect. 4.3.1g, that any operator T is determined by its diagonal coherent state matrix elements $\langle\alpha|T|\alpha\rangle$. This result has embodied in it the overcompleteness and non-orthogonality of the coherent states, for such a result is not possible for a complete orthonormal set: the non-diagonal matrix elements are also required.

The connection can be made in a variety of ways, the most common of which are known as the Q- and P-representations, and the Wigner function representation.

They arise from the consideration of three different aims. In the Q-representation we attempt to generate a probability density for the system by using the diagonal matrix elements of the density operator for this purpose. The P-representation arises from trying to represent any density operator as an ensemble of coherent states. Finally Wigner invented the function which bears his name as a result of his desire to find a joint distribution for canonically conjugate variables p and q which would be something like a classical probability distribution.

All of these phase space representations have their advantages and disadvantages—nevertheless they are all very useful.

a) Definition: We can define the Q-Function corresponding to a density operator ρ by

$$Q(\alpha, \alpha^*) = \frac{1}{\pi}\langle\alpha|\rho|\alpha\rangle. \tag{4.4.1}$$

We note that the normalization of the density operator gives a normalization of $Q(\alpha, \alpha^*)$, for

$$1 = \mathrm{Tr}\,\{\rho\} = \mathrm{Tr}\left\{\frac{1}{\pi}\int d^2\alpha|\alpha\rangle\langle\alpha|\rho\right\} \tag{4.4.2}$$

$$= \frac{1}{\pi}\int d^2\alpha\langle\alpha|\rho|\alpha\rangle \tag{4.4.3}$$

so that

$$\int d^2\alpha\, Q(\alpha, \alpha^*) = 1. \tag{4.4.4}$$

A similar procedure shows that the averages of *antinormally* ordered products of creation and destruction operators is

$$\langle a^r (a^\dagger)^s\rangle = \mathrm{Tr}\left\{a^r (a^\dagger)^s \rho\right\} = \int d^2\alpha\, \alpha^r (\alpha^*)^s Q(\alpha, \alpha^*). \tag{4.4.5}$$

Thus, the Q-Function can be viewed as a kind of probability: $Q(\alpha, \alpha^*)$ is positive, normalized, and the (antinormally ordered) quantum moments can be determined in terms of simple moments of $Q(\alpha, \alpha^*)$.

b) Examples of Q-Functions:

i) Coherent State: The density operator is that of a coherent state

$$\rho = |\alpha_0\rangle\langle\alpha_0| \tag{4.4.6}$$

and using (4.3.7)

$$Q(\alpha, \alpha^*) = \frac{1}{\pi}|\langle\alpha|\alpha_0\rangle|^2 = \frac{1}{\pi}\exp(-|\alpha - \alpha_0|^2). \tag{4.4.7}$$

The exponential form of this expression shows that $Q(\alpha, \alpha^*)$ is only significantly different from zero if $\alpha \approx \alpha_0$.

In terms of the macroscopic momentum and position variables,

$$\begin{aligned}\alpha &= \bar{x}\sqrt{\frac{\eta}{2}} + \frac{i}{\hbar}\bar{p}\frac{1}{\sqrt{2\eta}} \\ \alpha_0 &= \bar{x}_0\sqrt{\frac{\eta}{2}} + \frac{i}{\hbar}\bar{p}_0\frac{1}{\sqrt{2\eta}},\end{aligned} \tag{4.4.8}$$

one finds that the Q-function is only significantly different from zero if

$$\begin{aligned}\bar{x} - \bar{x}_0 &\le (2\hbar)^{\frac{1}{2}}(km)^{-\frac{1}{4}} \\ \bar{p} - \bar{p}_0 &\le (2\hbar)^{\frac{1}{2}}(km)^{\frac{1}{4}}.\end{aligned} \tag{4.4.9}$$

Essentially, the Q-function is in this case a kind of probability distribution concentrated in the minimum uncertainty region around the position $\bar{x}_0, \bar{p}_0$.

ii) A Number State: In this case the density operator is

$$\rho = |n\rangle\langle n| \tag{4.4.10}$$

and

$$\begin{aligned}Q_n(\alpha,\alpha^*) &= \frac{1}{\pi}|\langle\alpha|n\rangle|^2 \\ &= \frac{1}{\pi}\exp(-|\alpha|^2)(|\alpha|^2)^n/n!.\end{aligned} \tag{4.4.11}$$

The Q-function is independent of the phase of α, and has a maximum at $|\alpha|^2 = n$. The mean and variance of $|\alpha|^2$ are straightforwardly computed as

$$\begin{aligned}\langle|\alpha|^2\rangle &= n+1 \\ \mathrm{var}\{|\alpha|^2\} &= n+1\end{aligned} \tag{4.4.12}$$

so that for sufficiently large n the distribution is quite sharply peaked around $|\alpha|^2 = n+1$

iii) A Thermal State: The density operator for a thermal state is

$$\rho_T = \left(1 - e^{-\hbar\omega/kT}\right)\sum_n |n\rangle\langle n| e^{-n\hbar\omega/kT} \tag{4.4.13}$$

and the Q-function is

$$Q(\alpha,\alpha^*) = \frac{\left(1 - e^{-\hbar\omega/kT}\right)}{\pi}\sum_n e^{-n\hbar\omega/kT}\exp(-|\alpha|^2)|\alpha|^{2n}/n! \tag{4.4.14}$$

so that

$$\boxed{Q_T(\alpha,\alpha^*) = \frac{\left(1 - e^{-\hbar\omega/kT}\right)}{\pi}\exp\left\{-|\alpha|^2\left(1 - e^{-\hbar\omega/kT}\right)\right\}.} \tag{4.4.15}$$

This is a bivariate Gaussian in α and α^*, with

$$\begin{aligned}\langle\alpha\rangle &= \langle\alpha^*\rangle = 0\\ \langle\alpha^2\rangle &= \langle\alpha^{*2}\rangle = 0\\ \langle|\alpha|^2\rangle &= \left(1 - \mathrm{e}^{-\hbar\omega/kT}\right)^{-1}.\end{aligned} \tag{4.4.16}$$

Notice that as $kT \to 0$, $\langle|\alpha|^2\rangle \to 1$. This is correct, since $\langle|\alpha|^2\rangle = \langle aa^\dagger\rangle$.

At sufficiently large temperatures, we can approximate the $\exp(-\hbar\omega/kT)$ factor, to get

$$Q_T(\alpha,\alpha^*) \to \frac{\hbar\omega}{\pi kT}\exp\left(-\frac{\hbar\omega|\alpha|^2}{kT}\right) \qquad \text{as } T \to \infty \tag{4.4.17}$$

which looks like a Boltzmann distribution.

c) The Q-Function as a Quasiprobability: The Q-function reduces the evaluation of means to that of ordinary averages over the two variables α, α^*, and behaves very like a probability distribution. The main problem of the Q-function is that not all positive normalizable Q-functions correspond to positive definite normalizable density operators. For example, we know that

$$\langle aa^\dagger\rangle = \sum_n (n+1)\langle n|\rho|n\rangle \geq 1 \tag{4.4.18}$$

for any density operator. Hence

$$\langle|\alpha|^2\rangle \geq 1 \tag{4.4.19}$$

for any Q-function. This means that $\delta^2(\alpha)$ is not a permissible Q-function, since it would give $\langle|\alpha|^2\rangle = 0$. Similarly, no delta function like $\delta^2(\alpha - \alpha_0)$ is permissible. The physical effect is quite expected: from (4.2.5) we can see that a delta function Q-function would mean that $\langle\Delta x(t)^2\rangle$ and $\langle\Delta p(t)^2\rangle$ were negative.

d) The Density Operator in Terms of the Q-Function: A final question has to be answered: Namely, given the Q-function, how does one construct the density operator? The answer is simple, but the derivation is slightly roundabout. We first show that $Q(\alpha,\alpha^*)$ can be expressed as an absolutely convergent power series in α, and α^*. To do this we need to prove that for any density operator ρ, the number state matrix elements satisfy

$$|\langle n|\rho|m\rangle| \leq 1. \tag{4.4.20}$$

We know that ρ can be expressed as

$$\rho = \sum P_A|A\rangle\langle A| \tag{4.4.21}$$

so that

$$|\langle n|\rho|m\rangle| \leq \sum P_A|\langle n|A\rangle|\,|\langle m|A\rangle|. \tag{4.4.22}$$

We note that for any x and y

$$2xy \le x^2 + y^2 \tag{4.4.23}$$

so that

$$|\langle n|\rho|m\rangle| \le \sum \tfrac{1}{2} P_A \left\{ |\langle n|A\rangle|^2 + |\langle m|A\rangle|^2 \right\}. \tag{4.4.24}$$

However, since

$$\sum |\langle n|A\rangle|^2 = \langle A|A\rangle = 1 \tag{4.4.25}$$

we can deduce

$$|\langle n|A\rangle|^2 \le 1 \qquad \text{for all } n, \tag{4.4.26}$$

hence

$$|\langle n|\rho|m\rangle| \le \sum P_A = 1. \tag{4.4.27}$$

Using the expression for coherent states (4.3.1), we see that

$$Q(\alpha,\alpha^*) = \mathrm{e}^{-\alpha\alpha^*} \sum_{n,m} \frac{\langle n|\rho|m\rangle}{\pi\sqrt{n!m!}} \alpha^m \alpha^{*n}. \tag{4.4.28}$$

The double power series is clearly absolutely convergent, because of the bound (4.4.20), and since $\mathrm{e}^{-\alpha\alpha^*}$ has an absolutely convergent power series, the product of these two series is also absolutely convergent.

Suppose this power series is

$$Q(\alpha,\alpha^*) = \sum_{n,m} Q_{n,m} \alpha^m (\alpha^*)^n. \tag{4.4.29}$$

Then, we can straightforwardly compute

$$\langle n|\rho|m\rangle = \pi\sqrt{n!m!} \sum_r Q_{n-r,m-r}. \tag{4.4.30}$$

As an alternative expression we can write ρ as a *normally ordered* power series in a and $a^\dagger$;

$$\rho = \pi \sum Q_{n,m} (a^\dagger)^n a^m. \tag{4.4.31}$$

If such a power series exists, then the definition $a|\alpha\rangle = \alpha|\alpha\rangle$ immediately shows that $\langle\alpha|\rho|\alpha\rangle/\pi$ is identical with (4.4.29): thus the density operator is obtained by replacing $\alpha^m(\alpha^*)^n$ in the power series for $Q(\alpha,\alpha^*)$ by the corresponding *normally ordered* product of creation and destruction operators.

4.4.2 The Quantum Characteristic Function

Since the Q-function determines the density operator, the Fourier transform of the Q-function will also have this property. The (normally ordered) *quantum characteristic function* is defined by

$$\chi(\lambda,\lambda^*) = \mathrm{Tr}\left\{\rho \exp(\lambda a^\dagger)\exp(-\lambda^* a)\right\}. \tag{4.4.32}$$

The quantum characteristic function has the property, analogous to that of the classical characteristic function, that the *normally ordered* moments are given by its derivatives at $\lambda = 0$; precisely

$$\langle (a^\dagger)^r a^s\rangle = (-)^s \frac{\partial^{r+s}}{\partial^r\lambda \partial^s\lambda^*}\chi(\lambda,\lambda^*)\bigg|_{\lambda=0}. \tag{4.4.33}$$

Using the Baker-Hausdorff theorem (Sect. 4.3.1b) to reverse the order of the exponentials, and the completeness formula (4.3.8), we find

$$\chi(\lambda,\lambda^*) = \mathrm{Tr}\left\{\frac{1}{\pi}\int d^2\alpha\, \rho \exp(|\lambda|^2)\exp(-\lambda^* a)|\alpha\rangle\langle\alpha|\exp(\lambda a^\dagger)\right\} \tag{4.4.34}$$

so that

$$\chi(\lambda,\lambda^*) = \exp(|\lambda|^2)\int d^2\alpha\, \exp(\lambda\alpha^* - \alpha\lambda^*)Q(\alpha,\alpha^*). \tag{4.4.35}$$

This is a Fourier transform in two real variables; $\mathrm{Re}(\alpha)$ and $\mathrm{Im}(\alpha)$. Since $Q(\alpha,\alpha^*)$ is positive and is integrable, the Fourier transform exists, and is bounded as $|\lambda| \to \infty$

a) Some Examples of Quantum Characteristic Functions:

i) Coherent State:

$$\rho = |\alpha\rangle\langle\alpha|; \qquad \chi(\lambda,\lambda^*) = \exp(\lambda\alpha^* - \alpha\lambda^*). \tag{4.4.36}$$

ii) Number State:

$$\rho = |n\rangle\langle n|; \qquad \chi(\lambda,\lambda^*) = \sum_{m=0}^{n} \frac{n!}{(m!)^2(n-m)!}(-|\lambda|^2)^m. \tag{4.4.37}$$

iii) Thermal State: This is computed from the Fourier transform of (4.4.15) using (4.4.35)

$$\chi(\lambda,\lambda^*) = \exp\left\{-\frac{|\lambda|^2}{e^{\hbar\omega/kT}-1}\right\}. \tag{4.4.38}$$

4.4.3 The P-Representation

The density operator corresponding to a coherent state $|\alpha\rangle$ is of course given by $|\alpha\rangle\langle\alpha|$. It might happen that the variable α is not well defined, and we may wish to consider an ensemble of coherent states; that is, we may consider a density operator of the form

$$\rho = \int d^2\alpha\, P(\alpha, \alpha^*)|\alpha\rangle\langle\alpha|. \tag{4.4.39}$$

Glauber [4.1] and *Sudarshan* [4.2] both showed that such a representation of the density operator did exist for a large class of ρ, and indeed *Klauder et al.* [4.5] showed that, provided $P(\alpha, \alpha^*)$ is permitted to be a sufficiently singular generalized function, such a representation exists for any density operator ρ.

Notice that

$$\begin{aligned} 1 = \mathrm{Tr}\,\{\rho\} &= \int d^2\alpha\, P(\alpha, \alpha^*) \sum_n \langle n|\alpha\rangle\langle\alpha|n\rangle \\ &= \int d^2\alpha\, P(\alpha, \alpha^*)\langle\alpha|\alpha\rangle \end{aligned} \tag{4.4.40}$$

so that

$$\int d^2\alpha P(\alpha, \alpha^*) = 1. \tag{4.4.41}$$

Further, for any normal product (Sect. 4.3.1i) of creation and destruction operators (that is, a product in which all creation operators are to the left of all destruction operators) we can write

$$\langle (a^\dagger)^r a^s \rangle = \mathrm{Tr}\left\{(a^\dagger)^r a^s \rho\right\} = \mathrm{Tr}\left\{a^s \rho (a^\dagger)^r\right\} \tag{4.4.42}$$

$$= \int d^2\alpha\, P(\alpha, \alpha^*) \sum_n \langle n|a^s|\alpha\rangle\langle\alpha|(a^\dagger)^r|n\rangle \tag{4.4.43}$$

and using the property of the coherent state (4.3.30), we deduce

$$\langle (a^\dagger)^r a^s \rangle = \int d^2\alpha\, (\alpha^*)^r \alpha^s P(\alpha, \alpha^*). \tag{4.4.44}$$

Thus the quantity $P(\alpha, \alpha^*)$ plays the role of a kind of probability density for the variables α and α^*, in that the means of *normally ordered* products of quantum operators are simple moments of $P(\alpha, \alpha^*)$.

a) Relationship Between the P- and Q-Functions: The P- and Q-functions play equivalent roles for normal and antinormal products respectively, and their relationship is best exhibited by use of the quantum characteristic function. The *normally* *ordered* quantum characteristic function defined by (4.4.32) is easily shown to be given by

$$\chi(\lambda, \lambda^*) = \int d^2\alpha \, \exp(\lambda\alpha^* - \lambda^*\alpha) P(\alpha, \alpha^*). \tag{4.4.45}$$

The relationships (4.4.45) and (4.4.35) thus provide the connection between the P- and Q-functions. Looking at the examples of characteristic functions, we immediately discover that the P-function may correspond to a characteristic function which does not have an ordinary Fourier transform. For example, from (4.4.37) we see that $\chi(\lambda, \lambda^*)$ can be a polynomial, which means that $P(\alpha, \alpha^*)$ would be a finite sum of delta functions and finite order derivatives of a delta function. In fact it is easy to see that a quantum characteristic functions could be any convergent infinite series in λ and λ^*, and this would give P-function consisting of an infinite sum of derivatives of delta functions of arbitrarily high order. Formal manipulations can be justified for these [4.5], and in the sense of these rather singular distributions, it can be shown that a P-function always exists. However the usefulness of such a P-function in practice is limited.

The relations (4.4.45) and (4.4.35) can be Fourier transformed to give the more direct connection

$$Q(\alpha, \alpha^*) = \frac{1}{\pi} \int d^2\beta \, \exp(-|\alpha - \beta|^2) P(\beta, \beta^*). \tag{4.4.46}$$

b) Examples of P-Functions: The usefulness of the P-function becomes apparent when one sees that for many common situations, it does exist. For example

i) Coherent State:

$$\rho = |\alpha_0\rangle\langle\alpha_0|, \qquad P(\alpha, \alpha^*) = \delta^2(\alpha - \alpha_0). \tag{4.4.47}$$

ii) Thermal State: Computed from the Fourier transform of (4.4.38)

$$P(\alpha, \alpha^*) = \frac{1}{\pi} \left(\mathrm{e}^{\hbar\omega/kT} - 1 \right) \exp\left\{ -|\alpha|^2 (\mathrm{e}^{\hbar\omega/kT} - 1) \right\}. \tag{4.4.48}$$

In fact the P-function is naturally adapted to the study of thermal or quasithermal situations. For as $T \to 0$, the thermal Gaussian P-function (4.4.48) becomes infinitely sharp—thus we can say that all the broadening of this P-function is associated with the existence of thermal fluctuations. On the other hand at high temperatures,

$$P(\alpha, \alpha^*) \to \frac{\hbar\omega}{\pi kT} \exp\left(-\frac{\hbar\omega|\alpha|^2}{kT} \right) \qquad \text{as } T \to \infty \tag{4.4.49}$$

which is the same as the high temperature limit of the corresponding Q-function, (4.4.17), and essentially amounts to the classical Boltzmann distribution.

iii) A Number State: No functional form exists for a P-function for a number state, but a solution can be given in terms of derivatives of delta functions.

and we can insert the unitary transformation into this to get

$$\chi_W(\lambda,\lambda^*) = \mathrm{Tr}\left\{U(\theta)\rho\, U^{-1}(\theta)U(\theta)\exp(\lambda a^\dagger - \lambda^* a)U^{-1}(\theta)\right\} \tag{4.4.102}$$

$$\begin{aligned} &= \mathrm{Tr}\Big\{\mathcal{N}\exp(-\sqrt{n^2-m^2}\; a^\dagger a) \\ &\times \exp\big(-(\lambda^*\cosh\theta - \lambda\sinh\theta)a + (\lambda\cosh\theta - \lambda^*\sinh\theta)a^\dagger\big)\Big\}. \end{aligned} \tag{4.4.103}$$

But the density operator is now of the same form as that of a thermal state, so we can use the result for its Wigner characteristic function to deduce

$$\chi_W(\lambda,\lambda^*) = \exp\left\{-\tfrac{1}{2}|\lambda^*\cosh\theta - \lambda\sinh\theta|^2 \coth\left(\tfrac{1}{2}\sqrt{n^2-m^2}\,\right)\right\} \tag{4.4.104}$$

and substituting for $\cosh\theta$ and $\sinh\theta$ by (4.4.98), we find the quantum characteristic function is given by

$$\chi(\lambda,\lambda^*) = \exp(\tfrac{1}{2}|\lambda|^2)\chi_W(\lambda,\lambda^*) \tag{4.4.105}$$

$$= \exp\left\{\tfrac{1}{2}|\lambda|^2 - \tfrac{1}{2}\left[\frac{m(\lambda^2+\lambda^{*2})}{2\sqrt{n^2-m^2}} + \frac{|\lambda|^2 n}{\sqrt{n^2-m^2}}\right]\coth\left[\tfrac{1}{2}\sqrt{n^2-m^2}\,\right]\right\}. \tag{4.4.106}$$

c) P-, Q- and Wigner Functions: Let us define quantities

$$N = -\frac{1}{2} + \frac{n/2}{\sqrt{n^2-m^2}}\coth\left[\tfrac{1}{2}\sqrt{n^2-m^2}\,\right] \tag{4.4.107}$$

$$M = M^* = -\frac{m/2}{\sqrt{n^2-m^2}}\coth\left[\tfrac{1}{2}\sqrt{n^2-m^2}\,\right] \tag{4.4.108}$$

so that the characteristic function can be written

$$\chi(\lambda,\lambda^*) = \exp\left(-N|\lambda|^2 + \tfrac{1}{2}M^*\lambda^2 + \tfrac{1}{2}M\lambda^{*2}\right). \tag{4.4.109}$$

From this characteristic function it is straightforward to show that the P-, Q- and Wigner functions are

$$\begin{aligned} P(\alpha,\alpha^*) &= \frac{1}{\pi}\left(N^2 - |M|^2\right)^{-\frac{1}{2}} \\ &\times \exp\left\{\frac{-N|\alpha|^2 + \frac{1}{2}M^*\alpha^2 + \frac{1}{2}M\alpha^{*2}}{N^2-|M|^2}\right\} \end{aligned} \tag{4.4.110}$$

$$\begin{aligned} Q(\alpha,\alpha^*) &= \frac{1}{\pi}\left((N+1)^2 - |M|^2\right)^{-\frac{1}{2}} \\ &\times \exp\left\{\frac{-(N+1)|\alpha|^2 + \frac{1}{2}M^*\alpha^2 + \frac{1}{2}M\alpha^{*2}}{(N+1)^2-|M|^2}\right\} \end{aligned} \tag{4.4.111}$$

$$W(\alpha,\alpha^*) = \frac{1}{\pi}\left((N+\tfrac{1}{2})^2 - |M|^2\right)^{-\frac{1}{2}} \times \exp\left\{\frac{-(N+\frac{1}{2})|\alpha|^2 + \frac{1}{2}M^*\alpha^2 + \frac{1}{2}M\alpha^{*2}}{(N+\frac{1}{2})^2 - |M|^2}\right\}. \tag{4.4.112}$$

These three functions are obviously very similar. For all of them we find

$$\langle\alpha^2\rangle = \langle\alpha^{*2}\rangle = M \tag{4.4.113}$$

but

$$\langle|\alpha|^2\rangle_Q = N+1 \tag{4.4.114}$$

$$\langle|\alpha|^2\rangle_P = N \tag{4.4.115}$$

$$\langle|\alpha|^2\rangle_W = N+\tfrac{1}{2}. \tag{4.4.116}$$

However, the conditions for these functions to exist as normalizable Gaussians are different from the conditions for them to correspond to physical density operators. There is a physical requirement

$$N(N+1) \geq |M|^2 \tag{4.4.117}$$

which arises from the requirement that

$$\langle(a^\dagger + za)(a + z^*a^\dagger)\rangle \geq 0 \qquad \text{for all } z. \tag{4.4.118}$$

Notice that this means that there are density operators which satisfy this condition, but do not yield normalizable P-functions, since the P-function is normalizable only if $N^2 \geq |M|^2$, which is weaker than (4.4.116). Conversely, the conditions for normalisability of the Q-function and the Wigner function, respectively $(N+1)^2 \geq |M|^2$ and $N(N+1) \geq |M|^2 - 1/4$, are both weaker than the condition for a physical density operator, so that there exist normalizable Q- and Wigner functions which do not correspond to a positive density operator.

d) Higher Moments for Gaussian Density Operators: Since a Gaussian density operator implies a Gaussian P-, Q- and Wigner function, the relations for the moments of classical Gaussian variables can be used to obtain corresponding relations in the quantum case. Thus, according to S.M. Sect. 2.8.1, if X_i are Gaussian variables with $\langle X_i\rangle = 0$, then

$$\langle X_i X_j X_k \cdots\rangle = \frac{(2N)!}{N!2^N}\{\sigma_{ij}\sigma_{kl}\sigma_{mn}\cdots\}_{\text{sym}} \tag{4.4.119}$$

where $\sigma_{ij} = \langle X_i X_j\rangle$, the order of the moment is $2N$, and the subscript "sym" means the symmetrized form of the product of σ's.

Because a Gaussian density operator gives Gaussian P-, Q-, or Wigner functions, we will be able to deduce from (4.4.119) relations for normal, antinormal and symmetric ordering. Thus, for example

$$\begin{aligned}\langle a^{\dagger 4}\rangle &= 3\langle a^{\dagger 2}\rangle\langle a^{\dagger 2}\rangle \\ \langle a^{\dagger 3}a\rangle &= 3\langle a^{\dagger 2}\rangle\langle a^{\dagger}a\rangle \\ \langle a^{\dagger 2}a^2\rangle &= \langle a^{\dagger 2}\rangle\langle a^2\rangle + 2\langle a^{\dagger}a\rangle^2\end{aligned} \tag{4.4.120}$$

and similar equations will be true for antinormal and symmetric ordering.

In the study of optical correlations and the Hanbury-Brown Twiss effect, (treated in Sect. 8.3.4) we will find formulae involving fourth order moments of two different Gaussian modes. A result which is then applied is the particular case of this general Gaussian result:

$$\langle a_1^{\dagger}a_2^{\dagger}a_2a_1\rangle = \langle a_1^{\dagger}a_2^{\dagger}\rangle\langle a_2a_1\rangle + \langle a_1^{\dagger}a_2\rangle\langle a_2^{\dagger}a_1\rangle + \langle a_1^{\dagger}a_1\rangle\langle a_2^{\dagger}a_2\rangle. \tag{4.4.121}$$

In the common case that $\langle a_1^{\dagger}a_2^{\dagger}\rangle = \langle a_2a_1\rangle = 0$, this reduces to

$$g^{(2)} = 1 + |g^{(1)}|^2 \tag{4.4.122}$$

where

$$g^{(1)} = \frac{\langle a_1^{\dagger}a_2\rangle}{\sqrt{\langle a_1^{\dagger}a_1\rangle\langle a_2^{\dagger}a_2\rangle}} \tag{4.4.123}$$

$$g^{(2)} = \frac{\langle a_1^{\dagger}a_2^{\dagger}a_2a_1\rangle}{\langle a_1^{\dagger}a_1\rangle\langle a_2^{\dagger}a_2\rangle}. \tag{4.4.124}$$

These results will be true when a_1, a_2 are any linear combinations of destruction operators, which is what turns up in the theory of photon counting.

4.5 Operator Correspondences and Equations of Motion

The action of an operator on a density operator ρ is mirrored by the action of a corresponding differential operator on the P-, Q- or Wigner function. This can make the solution of the equations of motion quite simple in certain cases. We already know the basic relations

$$a|\alpha\rangle = \alpha|\alpha\rangle \tag{4.5.1}$$

$$\langle\alpha|a^{\dagger} = \alpha^*\langle\alpha|. \tag{4.5.2}$$

For the other possible ways of acting with the $a, a^\dagger$, it is convenient to use the Bargmann states, defined in Sect. 4.3.1d, $\| \alpha\rangle$, so that

$$a^\dagger \| \alpha\rangle = \sum_n \frac{\alpha^n}{\sqrt{n!}}\sqrt{n+1}\,|n+1\rangle \tag{4.5.3}$$

$$= \frac{\partial}{\partial\alpha} \| \alpha\rangle. \tag{4.5.4}$$

Similarly,

$$\langle\alpha \| a = \frac{\partial}{\partial\alpha^*}\langle\alpha \| . \tag{4.5.5}$$

In terms of the Bargmann states, we can write the P-representation in the form

$$\rho = \int d^2\alpha \| \alpha\rangle\langle\alpha \| \, e^{-\alpha\alpha^*} P(\alpha, \alpha^*). \tag{4.5.6}$$

Thus, we have

$$a^\dagger\rho = \int d^2\alpha \frac{\partial}{\partial\alpha}(\| \alpha\rangle)\langle\alpha \| \, e^{-\alpha\alpha^*} P(\alpha, \alpha^*). \tag{4.5.7}$$

and integrating by parts

$$= \int d^2\alpha \| \alpha\rangle\langle\alpha \| \, e^{-\alpha\alpha^*} \left(\alpha^* - \frac{\partial}{\partial\alpha}\right) P(\alpha, \alpha^*). \tag{4.5.8}$$

We thus can make an *operator correspondence* between $a^\dagger$ and $\alpha^* - \partial/\partial\alpha$. A similar formula holds for a. Summarizing, with the obvious correspondences arising from (4.5.1,2), we have

$$\begin{aligned}
a\rho &\leftrightarrow \alpha P(\alpha, \alpha^*) \\
a^\dagger\rho &\leftrightarrow \left(\alpha^* - \frac{\partial}{\partial\alpha}\right) P(\alpha, \alpha^*) \\
\rho a &\leftrightarrow \left(\alpha - \frac{\partial}{\partial\alpha^*}\right) P(\alpha, \alpha^*) \\
\rho a^\dagger &\leftrightarrow \alpha^* P(\alpha, \alpha^*).
\end{aligned} \tag{4.5.9}$$

For the Q- and Wigner functions, a similar set of relations holds. These are

$$\begin{aligned}
a\rho &\leftrightarrow \left(\alpha + \frac{\partial}{\partial\alpha^*}\right) Q(\alpha, \alpha^*) \\
a^\dagger\rho &\leftrightarrow \alpha^* Q(\alpha, \alpha^*) \\
\rho a &\leftrightarrow \alpha Q(\alpha, \alpha^*) \\
\rho a^\dagger &\leftrightarrow \left(\alpha^* + \frac{\partial}{\partial\alpha}\right) Q(\alpha, \alpha^*).
\end{aligned} \tag{4.5.10}$$

For the Wigner function it is simplest to use (4.4.76) to show that

$$\begin{aligned}
x\rho &\leftrightarrow \left(x + \frac{i\hbar}{2}\frac{\partial}{\partial p}\right)\bar{W}(x,p)\\
\rho x &\leftrightarrow \left(x - \frac{i\hbar}{2}\frac{\partial}{\partial p}\right)\bar{W}(x,p)\\
p\rho &\leftrightarrow \left(p - \frac{i\hbar}{2}\frac{\partial}{\partial x}\right)\bar{W}(x,p)\\
\rho p &\leftrightarrow \left(p + \frac{i\hbar}{2}\frac{\partial}{\partial x}\right)\bar{W}(x,p).
\end{aligned} \tag{4.5.11}$$

Using the relation (4.4.72) between α, $\bar{x}$, and $\bar{p}$, it is straightforward that

$$\begin{aligned}
a\rho &\leftrightarrow \left(\alpha + \frac{1}{2}\frac{\partial}{\partial\alpha^*}\right)W(\alpha,\alpha^*)\\
a^\dagger\rho &\leftrightarrow \left(\alpha^* - \frac{1}{2}\frac{\partial}{\partial\alpha}\right)W(\alpha,\alpha^*)\\
\rho a &\leftrightarrow \left(\alpha - \frac{1}{2}\frac{\partial}{\partial\alpha^*}\right)W(\alpha,\alpha^*)\\
\rho a^\dagger &\leftrightarrow \left(\alpha^* + \frac{1}{2}\frac{\partial}{\partial\alpha}\right)W(\alpha,\alpha^*).
\end{aligned} \tag{4.5.12}$$

4.5.1 Application to the Driven Harmonic Oscillator

We consider the Hamiltonian

$$H = \hbar\omega(a^\dagger a + \tfrac{1}{2}) + (\lambda a^\dagger + \lambda^* a) \tag{4.5.13}$$

for which the von Neumann's equation is

$$i\hbar\frac{\partial\rho}{\partial t} = \hbar\omega[a^\dagger a, \rho] + \lambda[a^\dagger, \rho] + \lambda^*[a, \rho]. \tag{4.5.14}$$

We now turn this into an equation for $P(\alpha, \alpha^*)$ by using the operator correspondences (4.5.9). Thus

$$a^\dagger a\rho \to \left(\alpha^* - \frac{\partial}{\partial\alpha}\right)\alpha P \tag{4.5.15}$$

$$-\rho a^\dagger a \to -\left(\alpha - \frac{\partial}{\partial\alpha^*}\right)\alpha^* P. \tag{4.5.16}$$

Notice that the order of the operators in (4.5.16) reverses, since acting on ρ they operate from the right, whereas on P, they operate from the left. Also,

$$[a^\dagger, \rho] \to \left[\left(\alpha^* - \frac{\partial}{\partial \alpha}\right) - \alpha^*\right] P = -\frac{\partial}{\partial \alpha} P \tag{4.5.17}$$

and similarly,

$$[a, \rho] \to \frac{\partial}{\partial \alpha^*} P \tag{4.5.18}$$

so that we find

$$\frac{\partial P}{\partial t} = \mathrm{i}\left(-\omega \frac{\partial}{\partial \alpha}\alpha + \omega \frac{\partial}{\partial \alpha^*}\alpha^* - \frac{\lambda}{\hbar}\frac{\partial}{\partial \alpha} + \frac{\lambda^*}{\hbar}\frac{\partial}{\partial \alpha^*}\right) P. \tag{4.5.19}$$

This corresponds to a Liouville equation for the P- function. A word of caution. It is tempting to treat α and α^* as independent variables which is not strictly true, and in writing all the above correspondences, one should really write

$$\begin{aligned} \alpha &= x + \mathrm{i}y \\ \alpha^* &= x - \mathrm{i}y \\ \frac{\partial}{\partial \alpha} &= \frac{1}{2}\left(\frac{\partial}{\partial x} - \mathrm{i}\frac{\partial}{\partial y}\right) \\ \frac{\partial}{\partial \alpha^*} &= \frac{1}{2}\left(\frac{\partial}{\partial x} + \mathrm{i}\frac{\partial}{\partial y}\right) \end{aligned} \tag{4.5.20}$$

and $\lambda = \mu + \mathrm{i}\nu$. In terms of these real variables, (4.5.19) becomes

$$\frac{\partial P}{\partial t} = \left[\frac{\partial}{\partial x}(\omega y + \nu/\hbar) - \frac{\partial}{\partial y}(\omega x + \mu/\hbar)\right] P \tag{4.5.21}$$

which is a Liouville equation, equivalent to the differential equations

$$\begin{aligned} \frac{dx}{dt} &= -\omega y - \nu/\hbar \\ \frac{dy}{dt} &= \omega x + \mu/\hbar \end{aligned} \tag{4.5.22}$$

which are equivalent to one complex differential equation

$$\frac{d\alpha}{dt} = \mathrm{i}(\omega\alpha + \lambda/\hbar) \tag{4.5.23}$$

with the solution

$$\alpha = -\lambda/\hbar\omega + \xi \mathrm{e}^{\mathrm{i}\omega t}. \tag{4.5.24}$$

The solution for $P(\alpha, \alpha^*)$ is, assuming a deterministic initial condition,

$$P(\alpha, \alpha^*, t) = \delta\left(x - \mathrm{Re}\left\{-\frac{\lambda}{\hbar\omega} + \xi \mathrm{e}^{\mathrm{i}\omega t}\right\}\right)\delta\left(y - \mathrm{Im}\left\{-\frac{\lambda}{\hbar\omega} + \xi \mathrm{e}^{\mathrm{i}\omega t}\right\}\right) \tag{4.5.25}$$

$$= \delta^2(\alpha + \lambda/\hbar\omega - \xi \mathrm{e}^{\mathrm{i}\omega t}) \tag{4.5.26}$$

where the delta function of a complex variable is defined by these equations. Notice that λ may depend on time, in which case (4.5.23) becomes

$$\frac{d\alpha}{dt} = \mathrm{i}[\omega\alpha + \lambda(t)/\hbar] \tag{4.5.27}$$

whose solution is

$$\alpha(t) = \alpha(0)\mathrm{e}^{\mathrm{i}\omega t} + \mathrm{i}\int_0^t dt' \mathrm{e}^{\mathrm{i}\omega(t-t')}\lambda(t')/\hbar \tag{4.5.28}$$

and the corresponding P is

$$P(\alpha, \alpha^*, t) = \delta^2[\alpha - \alpha(t)]. \tag{4.5.29}$$

Exercise. Show that exactly the same equation (4.5.19) arises for the Q-function and the Wigner function.

From the results above, it can be seen that, if one views any one of P, Q, or W as a quasiprobability, one gets an equation of motion which is the same as that of a classical system. This is the basis of the idea of quantum-classical correspondence. It is a formal expression of the fact that in harmonic systems the classical and quantum systems both possess only one frequency.

Exercise. Suppose we have a simplified anharmonic oscillator Hamiltonian

$$H = \tfrac{1}{2}\hbar\omega(a^\dagger a + \tfrac{1}{2}) + g(a^\dagger a)^2. \tag{4.5.30}$$

What is the P-representation equation of motion in this case?

4.5.2 The Wigner Function and the Quasiclassical Langevin Equation

We will now show that it is possible to find a rather simple stochastic differential equation which is in some cases exactly equivalent to the adjoint equation. To do this it is most advantageous to use the Wigner function which was introduced in Sect. 4.4.4. The equation of motion for the Wigner function, in the case that H_{sys} is the Hamiltonian for a particle moving in a potential, i.e.,

$$H_{\mathrm{sys}} = \frac{p^2}{2m} + V(x), \tag{4.5.31}$$

is derived by the substitution of the operator correspondences for the Wigner function, (4.5.11) into the adjoint equation, (3.5.32). The result is

$$\begin{aligned}\frac{\partial W}{\partial t} &= \left\{-\frac{\partial}{\partial x}\frac{p}{m} + \frac{\partial}{\partial p}\big(V'(x) + \gamma p/m - \alpha(t)\big)\right\} W \\ &\quad + \left\{\sum_{n=1}^{\infty}\frac{(\mathrm{i}\hbar/2)^{2n}}{(2n+1)!}\frac{\partial^{2n+1}}{\partial p^{2n+1}}V^{2n+1}(x)\right\} W.\end{aligned} \tag{4.5.32}$$

There are a number of situations in which the second line vanishes or is negligible:

i) If $V(x) = Ax + Bx^2$, i.e., we are dealing with a harmonic oscillator, a linear potential, or a free particle.

ii) If $\hbar \to 0$.

iii) If $\gamma \to \infty$: the large friction case. For in this case, $\alpha(t)$ also becomes larger, and this represents the dominant noise term. The last line in (4.5.32) is a kind of noise, which is independent of γ, and is then presumably negligible in this limit. (This argument certainly lacks rigour, but can probably be made rigorous.)

In all these cases, we are left with a conventional stochastic Liouville equation for the Wigner function, equivalent to the classical Langevin equation

$$\begin{aligned}\dot{x} &= p/m \\ \dot{p} &= -V'(x) - \gamma p/m + \alpha(t)\end{aligned} \tag{4.5.33}$$

in which, however, the noise spectrum is given by the right hand side of (3.3.10).

Schmid [4.6] has called this equation the *quasiclassical Langevin equation*. It is a mixture of classical stochastic theory and quantum mechanics, in which the variables are non-operator quantities, while the spectrum of the noise is given by the expression given by Callen and Welton, as noted in Sect. 1.2. Its validity is limited, however, according to the above derivation as well as that of Schmid, to situations in which the friction coefficient γ is large. How large γ must be can be determined by the following argument. By using the Wigner function in the adjoint equation we get an exact representation of the quantum mechanical operators. Thus, the terms on the right hand side of (4.5.32) which do not involve $\alpha(t)$ or γ faithfully represent the operator H_{sys}. The characteristic frequencies which turn up will then be the correct quantized frequencies, i.e., they will correspond to the atomic energy levels, and if we apply van Kampen's cumulant expansion to (4.5.32) we will get the master equation, in the same way as was done in Sect. 3.6.1, Sect. 3.6.3. By dropping the terms with higher derivatives, one finds that the characteristic frequencies of the Liouville operator are those of the classical orbits in the potential $V(x)$, which are of course quite different from the quantum frequencies, though they do approach the classical frequencies in the large scale limit. For these to be compatible with each other the friction must be sufficiently large to broaden the quantum energy levels to the extent that the motion is no longer distinguishable from the classical motion. At the very least, the broadening should be so large that the levels overlap. This will happen most easily when levels are very close together

The quasiclassical Langevin equation been used by *Koch et al.* [4.7] to analyse the experiments on quantum noise mentioned in Sect. 1.2 in which the spectrum of quantum noise was measured. In this analysis, a high friction limit was assumed, which means that the analysis is within the regime of validity of the quasiclassical Langevin equation.

In fact there does exist a body of knowledge called "stochastic electrodynamics" in which various authors [4.8] have built up a theory in which a classical particle interacts with a random electromagnetic field, whose statistics are chosen to give a Planck spectrum. A generalization to a full three dimensional formulation of inputs and outputs, in which the appropriate wave equation corresponded to electrodynamics, would yield a quasiclassical Langevin equation of exactly this form, which would have the same range of validity. Exact agreement between stochastic electrodynamics and quantum theory is found for assemblies of harmonic oscillators, and for free particles. From this point of view this is not surprising—but the terms in the second line of (4.5.32) will make their presence felt in all other cases, and stochastic electrodynamics cannot be a valid representation of reality for general situations.

Appendix 4A. The Baker-Hausdorff Formula

In the most general form, consider two time dependent operators, $A(t)$, $B(t)$, such that for all t, t'

$$[A(t), A(t')] = [B(t), B(t')] = 0 \tag{4A.1}$$

$$[A(t), B(t')] = f(t, t') \tag{4A.2}$$

and $f(t, t')$ commutes with $A(t'')$ and $B(t'')$ for all t, t', t''.

Suppose that $V(t)$ is an operator such that

$$\frac{dV(t)}{dt} = \{A(t) + B(t)\}V(t), \qquad V(0) = 1. \tag{4A.3}$$

This equation cannot be solved simply since $[A(t) + B(t), A(t') + B(t')] \neq 0$. However, the equation

$$\frac{dv(t)}{dt} = B(t)v(t), \qquad v(0) = 1. \tag{4A.4}$$

does have the simple solution $v(t) = \exp(\int_0^t B(t')\,dt')$, because $[B(t), B(t')] = 0$, from (4A.1). This motivates us to write

$$U(t) = \exp\left\{-\int_0^t B(t')\,dt'\right\} V(t) \tag{4A.5}$$

so that, using (4A.1),

$$\frac{dU(t)}{dt} = \exp\left\{-\int_0^t B(t')\,dt'\right\} A(t)V(t) \tag{4A.6}$$

$$= \left\{A(t) + \int_0^t f(t, t')\,dt'\right\} U(t) \tag{4A.7}$$

where, in going from (4A.6) to (4A.7), we have used the formula

$$\exp(-F)G\exp(F) = \sum_n \frac{(-1)^n}{n!}[F, G]_n \tag{4A.8}$$

where the iterated commutator is defined by

$$[F,G]_n = [F,[F,G]_{n-1}] \tag{4A.9}$$

and

$$[F,G]_1 = [F,G], \tag{4A.10}$$

and have used the fact that the first commutator $f(t,t')$ commutes with $B(t'')$ for all t'', so that only the first two terms in the expansion (4A.8) survive. Now using the fact that $f(t,t')$ commutes with all $A(t'')$, we can solve (4A.7), and resubstitute into (4A.5) to get $V(t)$ as

$$V(t) = \exp\left\{\int_0^t B(s)\,ds\right\}\exp\left\{\int_0^t A(s)\,ds\right\}\exp\left\{\int_0^t ds\int_0^s ds' f(s,s')\right\}. \tag{4A.11}$$

4A.1 Corollaries

a) Baker-Hausdorff Formula: If $A(t)$, $B(t)$ are constants, A, B, then clearly (4A.3) can be solved by direct integration to give $V(t)$:

$$V(t) = \exp\big((A+B)t\big). \tag{4A.12}$$

Setting $t = 1$ in both this equation and and in (4A.11), we deduce

$$V(1) = \exp(A+B) = \exp(B)\exp(A)\exp(\tfrac{1}{2}[A,B]) \tag{4A.13}$$

$$= \exp(A)\exp(B)\exp(-\tfrac{1}{2}[A,B]) \tag{4A.14}$$

which is the Baker-Hausdorff formula.

b) Generalization of the Baker-Hausdorff Formula: Now notice that

$$\bar{A} \equiv \int_0^t ds\,A(s), \qquad \bar{B} \equiv \int_0^t ds\,B(s), \tag{4A.15}$$

satisfy the necessary conditions for the Baker-Hausdorff formula (4A.13). We can therefore use (4A.13) to rewrite the formula (4A.11) in the form

$$V(t) = \exp\left\{\int_0^t ds\{A(s)+B(s)\}\right\}\exp\left\{\tfrac{1}{2}\int_0^t ds\int_0^t ds'\epsilon(s-s')f(s,s')\right\}. \tag{4A.16}$$

in which $\epsilon(x)$ is the antisymmetric step function

$$\epsilon(x) = \begin{cases} +1 & \text{for } x>0, \\ -1 & \text{for } x<0. \end{cases} \tag{4A.17}$$

5. Quantum Markov Processes

The concept of a Markov process is a very powerful and surprisingly general tool in the study of classical irreversible phenomena. Classically, the essence of the Markov idea is that we need only know the probability distribution at time t in order to predict it for all future times. At the deepest level in classical physics this is exactly true—Hamilton's equations of motion are first order in time, so that the knowledge of the canonical co-ordinates and momenta at one time determines them for all future times.

At a coarser level, even classically the Markov property must be regarded as an approximation. For, in both classical and quantum systems, damping of a macroscopic system comes from the transfer of energy to microscopic degrees of freedom, which, in the macroscopic description, give rise to noise terms in the macroscopic equations of motion. If the motion of the noise degrees of freedom is very fast, then on any slower time scale they will appear to be almost delta correlated, and the equation of motion will appear like a white noise stochastic differential equation—that is, a Markov process. It is also very fortunate that it is quite easy to embed a non-Markov process in a larger Markov process, basically by considering non-Markov noise to be itself given as the solution of a Markovian equation of motion.

The theory of Markov processes has two main analytic descriptions, by means of stochastic differential equations, or by Fokker-Planck equations, or the corresponding discrete versions of these, and the interplay between one method and the other yields a powerful and flexible way of understanding and computing physically interesting quantities. As well as this there is the dual description of the stochastic differential equation by the Ito and the Stratonovich techniques, a fact which has caused much confusion and even evangelical zeal [5.1]. It is unfortunately the case that both of these techniques have their uses, and it is not possible to avoid the use of either definition. Indeed, there is considerable insight to be gained from knowing and understanding both techniques.

Until quite recently, there has been no description of quantum Markov processes which possesses the completeness of the classical version; that is, a duality between stochastic differential equations and the master equation (the quantum version of the Fokker-Planck equation), nor had the concepts of Ito and Stratonovich quantum stochastic differential equations been formulated. The reason for this lies in the rather rigid requirements of quantum mechanics, the most important of which is to ensure that the canonical commutation relations are preserved.

Nevertheless, quantum Langevin equations and the master equation have quite a long history. Quantum Langevin equations were first actually used in a quantum optical setting by *Senitzky* [5.2]. The first real application was to laser theory, which was done almost simultaneously by *Lax* [5.3] and *Haken* [5.4]. The master equation approach was developed in a quantum optical context by *Louisell* (see [5.5] for a summary or his book [5.6] for a more detailed presentation.) In all of these approaches it was necessary to develop rather careful methods which did not obscure the quantum mechanics underlying the system under study. An approach based on the use of the Pauli master equation, which does not consider off diagonal density operator matrix elements, would simply not have been adequate.

This chapter will develop all quantum analogues of the classical Markovian techniques. It starts with two further derivations of the master equation, one of them a traditional "physical" derivation, and the other based on the method of adiabatic elimination of fast variables, using projection operator techniques to assist in the algebra. This method gives a rather clear understanding of the limits involved. These derivations, unlike those in Chap.3, do not depend on any particular model of a heat bath, and therefore demonstrate that the concept of a master equation is really much more general than one would have supposed from the work of that chapter.

The next section is devoted to the formulation of the full multitime structure of a quantum Markov process, giving abstract formulae for all multitime correlation functions, and shows how close the structures of classical and quantum Markov processes are to each other.

The next section develops the theory of quantum stochastic differential equations as a Markovian version of the noise input and output theory of Chap.3. The quantum stochastic differential equation (QSDE) is a much more rigidly prescribed object than its classical counterpart, and methods are given for writing down both Ito and Stratonovich versions corresponding to a given coupling to a heat bath. Here the heat bath is of harmonic oscillators. The final section develops the equivalence between the master equation and the QSDE, and shows how we may relate the correlation structure of the outputs to those of the inputs and the operators whose equations of motion are given by the QSDE.

5.1 The Physical Basis of the Master Equation

In the previous chapter a derivation of the master equation was given. This section concentrates on two further ways of deriving the master equation. The first is firmly grounded in physical intuition, and gives the clearest physical idea of the physics behind the master equation. An alternative derivation based on the idea of projection operators is presented next, for this gives a particularly compact way of deriving the master equation at the cost of introducing a certain degree of abstraction.

5.1.1 Derivation of the Quantum Optical Master Equation

We consider a situation described by a Hamiltonian of the form

$$H = H_{\text{sys}} + H_B + H_{\text{Int}} \tag{5.1.1}$$

where the constituent Hamiltonians describe respectively the system, the heat bath, and the interaction. The total density operator for the system and reservoir, $\rho_{\text{tot}}(t)$, in the Schrödinger picture, satisfies

$$\dot{\rho}_{\text{tot}} = -\frac{\mathrm{i}}{\hbar}[H_{\text{sys}} + H_B + H_{\text{Int}}, \rho_{\text{tot}}]. \tag{5.1.2}$$

We want an equation for a quantity which can give us the means of *system* operators, and this is, of course, given by the *reduced density operator*

$$\hat{\rho}(t) = \text{Tr}_{\text{B}}\left\{\rho_{\text{tot}}(t)\right\}. \tag{5.1.3}$$

a) Transformation to the Interaction Picture: It is most convenient to transform the equation of motion into the interaction picture, defined by setting

$$\rho_I(t) = \exp\left[\frac{\mathrm{i}}{\hbar}(H_{\text{sys}} + H_B)t\right] \rho_{\text{tot}}(t) \exp\left[-\frac{\mathrm{i}}{\hbar}(H_{\text{sys}} + H_B)t\right]. \tag{5.1.4}$$

It is straightforward to show that $\rho_I(t)$ obeys the interaction picture equation of motion

$$\dot{\rho}_I(t) = -\frac{\mathrm{i}}{\hbar}\left[H_{\text{Int}}(t), \rho_I(t)\right] \tag{5.1.5}$$

where

$$H_{\text{Int}}(t) = \exp\left[\frac{\mathrm{i}}{\hbar}(H_{\text{sys}} + H_B)t\right] H_{\text{Int}} \exp\left[-\frac{\mathrm{i}}{\hbar}(H_{\text{sys}} + H_B)t\right]. \tag{5.1.6}$$

Notice that, from (5.1.3,4),

$$\hat{\rho}(t) = \text{Tr}_{\text{B}}\left\{\exp\left[-\frac{\mathrm{i}}{\hbar}(H_{\text{sys}} + H_B)t\right] \rho_I(t) \exp\left[\frac{\mathrm{i}}{\hbar}(H_{\text{sys}} + H_B)t\right]\right\} \tag{5.1.7}$$

and since H_B is a function only of bath variables, we can use the cyclic property of the trace to cancel the factors involving H_B, to get

$$\hat{\rho}(t) = \exp\left[-\frac{\mathrm{i}}{\hbar}H_{\text{sys}}t\right] \rho(t) \exp\left[\frac{\mathrm{i}}{\hbar}H_{\text{sys}}t\right], \tag{5.1.8}$$

where

$$\rho(t) \equiv \text{Tr}_{\text{B}}\left\{\rho_I(t)\right\} \tag{5.1.9}$$

is the interaction picture reduced density operator.

b) Initial Conditions: We assume that the system and the bath are initially independent, so that the total density operator factorizes into a direct product

$$\rho_{\text{tot}}(0) = \hat{\rho}(0) \otimes \rho_B. \tag{5.1.10}$$

In addition, we assume that *the reservoir is so large that its statistical properties are unaffected by the weak coupling to the system.* This assumption is essential for the derivation.

c) Integration of the Equations of Motion: We can return to (5.1.5) and integrate from 0 to t subject to the initial condition (5.1.10). This gives, after two iterations

$$\rho_I(t) = \rho_I(0) - \frac{\mathrm{i}}{\hbar}\int_0^t dt' \left[H_{\mathrm{Int}}(t'), \rho_I(0)\right] - \frac{1}{\hbar^2}\int_0^t dt' \int_0^{t'} dt'' \left[H_{\mathrm{Int}}(t'), [H_{\mathrm{Int}}(t''), \rho_I(t'')]\right] . \tag{5.1.11}$$

One could proceed to iterate this way, and would obtain a power series in the perturbation $H_{\mathrm{Int}}(t)$. However, this would not be very useful—to obtain an exponential decay law, for example, clearly requires an infinite number of iterations. One therefore differentiates (5.1.11) with respect to t, to get an integro-differential equation for $\rho_I(t)$

$$\dot{\rho}_I(t) = -\frac{\mathrm{i}}{\hbar}\left[H_{\mathrm{Int}}(t), \rho_I(0)\right] - \frac{1}{\hbar^2}\int_0^t dt' \left[H_{\mathrm{Int}}(t), [H_{\mathrm{Int}}(t'), \rho_I(t')]\right] . \tag{5.1.12}$$

d) Trace over the Bath Variables: If we trace both sides over the bath variables, and use (5.1.9), we obtain

$$\dot{\rho}_I(t) = -\frac{1}{\hbar^2}\int_0^t dt' \, \mathrm{Tr_B}\left\{\left[H_{\mathrm{Int}}(t), [H_{\mathrm{Int}}(t'), \rho_I(t')]\right]\right\} \tag{5.1.13}$$

where we have assumed

$$\mathrm{Tr_B}\left\{H_{\mathrm{Int}}(t)\rho_I(0)\right\} = 0 \tag{5.1.14}$$

and

$$\rho_I(0) = \rho_{\mathrm{tot}}(0) = \hat{\rho}(0) \otimes \rho_B \tag{5.1.15}$$

in (5.1.10). This means that we assume that the interaction has no diagonal elements in the representation in which H_B is diagonal. In practice this is not a great restriction, and we can always redefine H_{sys} and H_{Int} to include any such diagonal elements in H_{sys}.

e) Weak Coupling Assumption: At this point we assume that H_{Int} is very much less than either H_B or H_{sys}, and that the reservoir density operator is not significantly affected by the interaction. If this is the case, we might be justified in replacing $\rho_I(t')$ in (5.1.13) by a factorized approximation

$$\rho_I(t') \approx \rho(t') \otimes \rho_B \tag{5.1.16}$$

in which we have made the following assumptions :

i) The bath density operator is not significantly changed by the interaction.

ii) However we have allowed the system density operator to change significantly, on the ground that the system is much smaller than the bath, and hence the fractional effect of the interaction is much bigger on the system.

iii) We have assumed that the density operator may be written approximately as a direct product. In fact, a much weaker assumption is normally sufficient, since in practice the interaction can be written as a sum of terms like

$$X_{\text{sys}} A_B \tag{5.1.17}$$

where both of these are reasonably simple operators which act respectively only in the system and the bath spaces. All we really need then, is that approximations like

$$\mathrm{Tr_B}\left\{\left[A_B(t),[A_B(t'),\rho_I(t')]\right]\right\} \approx \rho(t')\otimes \mathrm{Tr_B}\left\{\left[A_B(t),[A_B(t'),\rho_B]\right]\right\} \tag{5.1.18}$$

are valid. Essentially, this means that bath correlation functions are not significantly changed by the interaction.

f) Markov Approximation: Using the weak coupling assumption, we find that (5.1.13) can be reduced to an integro-differential equation for $\rho(t)$; namely

$$\dot\rho(t) = -\frac{1}{\hbar^2}\int_0^t dt'\,\mathrm{Tr_B}\left\{\left[H_{\text{Int}}(t),[H_{\text{Int}}(t'),\rho(t')\otimes\rho_B]\right]\right\}. \tag{5.1.19}$$

The final approximation is now to turn this into a differential equation, and this is done by making an assumption about the bath correlation functions, such as (5.1.18), which turn up in this equation. Since the interaction is assumed weak, the rate of change of the interaction picture system density operator will be quite slow compared to that of the bath operators, such as $A_B(t)$, $A_B(t')$, which will vary on a time scale determined from (5.1.6) by H_B. In fact, the bath correlation functions will generally be determined by a *thermal* choice of ρ_B, and in this case the correlation times will be like the thermal correlation time of Sect. 3.3.3, which are much shorter than the typical time constant expected from $\rho(t)$. If this is the case, we can say that the factor $\rho(t')$ changes insignificantly over the time taken for the correlation functions in (5.1.19) to vanish. In this case we can

i) Set $\rho(t')\to\rho(t)$.

ii) For $t \gg$ thermal correlation time, we can let the lower limit of the time integral go to $-\infty$.

In this case, we deduce the master equation :

$$\dot\rho(t) = -\frac{1}{\hbar^2}\int_0^\infty d\tau\,\mathrm{Tr_B}\left\{\left[H_{\text{Int}}(t),[H_{\text{Int}}(t-\tau),\rho(t)\otimes\rho_B]\right]\right\}. \tag{5.1.20}$$

In this form, by inserting appropriate forms for H_{Int} and H_{sys}, we can derive a number of useful master equations. The approximation, $\rho(t') \to \rho(t)$ is known as the

Markov approximation since it yields a first order (operator) differential equation for $\rho(t)$—this means that the knowledge of $\rho(t)$ at one point in time $t = t_0$ is sufficient to determine $\rho(t)$ for all $t > t_0$. This is exactly analogous to the Markov property in classical stochastic processes, where a similar property is true. The Markov property is a highly desirable property from a mathematical point of view, because a whole structure of measurement theory can be built around it in a compact and self-contained way. The elegance of this structure leads to the formulation in the abstract of the concept of the quantum Markov process as a branch of mathematics. Nevertheless, it is important to remember that it is an assumption, based on the existence of short correlation times in the heat bath, and the use of perturbation theory.

5.1.2 A Derivation Based on Projection Operators

We start from the equation of motion (5.1.2) in the form

$$\dot{\rho}_{\text{tot}} = (\mathcal{L}_{\text{sys}} + \mathcal{L}_B + \mathcal{L}_{\text{Int}})\rho_{\text{tot}} \tag{5.1.21}$$

where the *Liouvillians* $\mathcal{L}$ are a shorthand notation for the commutator terms. One is interested in an equation for

$$\hat{\rho}(t) \equiv \text{Tr}_{\text{B}}\left\{\rho_{\text{tot}}(t)\right\}, \tag{5.1.22}$$

and this can be viewed as a projection of the operator $\rho_{\text{tot}}(t)$ into a subspace. One can formalize this by writing a projection operator, $\mathcal{P}$, given by

$$\mathcal{P}\rho_{\text{tot}}(t) = \text{Tr}_{\text{B}}\left\{\rho_{\text{tot}}(t)\right\} \otimes \rho_B \tag{5.1.23}$$

where ρ_B is a density operator for the bath. One does not need to specify ρ_B at this stage, but it will turn out to be most convenient to choose

$$\mathcal{L}_B\rho_B = 0 \tag{5.1.24}$$

which is a condition satisfied by a thermal equilibrium bath density operator; that is by $\exp(-H_B/kT)$. We also define

$$\mathcal{Q} = 1 - \mathcal{P} \tag{5.1.25}$$

and from (5.1.21), can derive two equations. First write

$$\rho_{\text{tot}} = \mathcal{P}\rho_{\text{tot}} + \mathcal{Q}\rho_{\text{tot}} \equiv v(t) + w(t). \tag{5.1.26}$$

We then use the following properties

i) $$\mathcal{P}\mathcal{L}_{\text{sys}} = \mathcal{L}_{\text{sys}}\mathcal{P}, \tag{5.1.27}$$

which is true because $\mathcal{P}$ and $\mathcal{L}_{\text{sys}}$ operate in different spaces and

ii) $$\mathcal{P}\mathcal{L}_B = \mathcal{L}_B\mathcal{P} = 0. \tag{5.1.28}$$

The first equality is true because the equation of motion given by the bath Hamiltonian must conserve probability; the second is true because of the definition (5.1.23) and the choice (5.1.24).

iii) $$\mathcal{P}\mathcal{L}_{\text{Int}}\mathcal{P} = 0 \tag{5.1.29}$$

This is equivalent to the condition (5.1.14), and is normally true

iv) $$\mathcal{P}^2 = \mathcal{P}, \qquad \mathcal{Q}^2 = \mathcal{Q} \tag{5.1.30}$$

which are true by definition (5.1.23).

The most easily understood analysis comes from using the Laplace transform of the equations of motion. Defining

$$\begin{aligned} \tilde{v}(s) &= \int_0^\infty \mathrm{e}^{-st} v(t)dt \\ \tilde{w}(s) &= \int_0^\infty \mathrm{e}^{-st} w(t)dt \end{aligned} \tag{5.1.31}$$

we can soon derive

$$s\tilde{v}(s) - v(0) = \mathcal{L}_{\text{sys}}\tilde{v}(s) + \mathcal{P}\mathcal{L}_{\text{Int}}\tilde{w}(s) \tag{5.1.32}$$

$$s\tilde{w}(s) - w(0) = (\mathcal{L}_{\text{sys}} + \mathcal{L}_B + \mathcal{Q}\mathcal{L}_{\text{Int}})\tilde{w}(s) + \mathcal{Q}\mathcal{L}_{\text{Int}}\tilde{v}(s), \tag{5.1.33}$$

and if we solve for $\tilde{w}(s)$, and substitute for $\tilde{w}(s)$ in (5.1.32), we get

$$\begin{aligned} &s\tilde{v}(s) - \left\{v(0) + \mathcal{P}\mathcal{L}_{\text{Int}}[s - \mathcal{L}_{\text{sys}} - \mathcal{L}_B - \mathcal{Q}\mathcal{L}_{\text{Int}}]^{-1} w(0)\right\} \\ &\quad = \left\{\mathcal{L}_{\text{sys}} + \mathcal{P}\mathcal{L}_{\text{Int}}[s - \mathcal{L}_{\text{sys}} - \mathcal{L}_B - \mathcal{Q}\mathcal{L}_{\text{Int}}]^{-1}\mathcal{Q}\mathcal{L}_{\text{Int}}\right\}\tilde{v}(s). \end{aligned} \tag{5.1.34}$$

Two limits can now be distinguished.

a) The Weak Coupling Limit: Suppose the interaction is proportional to a small parameter introduced by setting

$$\mathcal{L}_{\text{Int}} \to \epsilon\mathcal{L}_{\text{Int}}. \tag{5.1.35}$$

Then making this substitution, and preserving only lowest order terms in the homogeneous and inhomogeneous terms of (5.1.34) separately

$$\boxed{\begin{aligned} &s\tilde{v}(s) - \left\{v(0) + \epsilon\mathcal{P}\mathcal{L}_{\text{Int}}[s - \mathcal{L}_{\text{sys}} - \mathcal{L}_B]^{-1} w(0)\right\} \\ &\quad = \left\{\mathcal{L}_{\text{sys}} + \epsilon^2\mathcal{P}\mathcal{L}_{\text{Int}}[s - \mathcal{L}_{\text{sys}} - \mathcal{L}_B]^{-1}\mathcal{Q}\mathcal{L}_{\text{Int}}\right\}\tilde{v}(s). \end{aligned}} \tag{5.1.36}$$

The term of order ϵ can be neglected as a small change to the initial condition whose influence does not accumulate with time. However, the second order term on the right hand side can be significant if (as is the case) it gives rise to a kind of behaviour totally different from that of $\mathcal{L}_{\rm sys}$—in any case the solution of (5.1.36) will yield an accumulation of this term. Neglecting the correction to the initial condition, and inverting the Laplace transform we find

$$\dot{v}(t) = \mathcal{L}_{\rm sys}v(t) + \epsilon^2\mathcal{P}\mathcal{L}_{\rm Int}\int_0^\infty d\tau \exp\{(\mathcal{L}_{\rm sys}+\mathcal{L}_B)\tau\}\,\mathcal{Q}\mathcal{L}_{\rm Int}v(t-\tau), \tag{5.1.37}$$

and inserting the definitions of the various quantities involved, it is not hard to see that this equation is equivalent to the master equation (5.1.19). The Markov approximation is then easily justified, since the time scale of $\rho(t)$ is of order ϵ^2, which is much slower than all other time scales.

b) The Quantum Brownian Motion Limit: Suppose we have a parametric dependence of the form

$$\gamma^2\mathcal{L}_B + \gamma\mathcal{L}_{\rm Int} + \mathcal{L}_{\rm sys} \tag{5.1.38}$$

and that γ is very large, corresponding to fast, large fluctuations. This is similar to the adiabatic elimination problem of classical stochastics. A similar analysis shows that the equation of motion is, in the limit of $\gamma \to \infty$,

$$\dot{v}(t) = \mathcal{L}_{\rm sys}v(t) - \mathcal{P}\mathcal{L}_{\rm Int}\mathcal{L}_B^{-1}\mathcal{L}_{\rm Int}v(t) \tag{5.1.39}$$

and it is not difficult to show that this corresponds to the quantum Brownian motion master equation of Sect. 3.6.1

5.1.3 Relationship to the Quantum Optical Master Equation

The master equation in the form (5.1.20) can eventually be reduced to exactly the same form as that in Sect. 3.6.3. We note that in that case we can write

$$H_B = \tfrac{1}{2}\sum_n \{p_n^2 + \omega_n^2 q_n^2\} \tag{5.1.40}$$

$$H_{\rm Int} = -\sum_n \kappa_n p_n X \tag{5.1.41}$$

$$H_{\rm sys} = H_{\rm sys}(Z) + \tfrac{1}{2}\sum_n \kappa_n^2 X^2 \tag{5.1.42}$$

and in the weak coupling situation, the κ_n are implicitly understood to be proportional to the smallness parameter ϵ. Thus, the Liouvillian can be written in the form

$$\mathcal{L} = \mathcal{L}_{\rm sys} + \mathcal{L}_B + \epsilon^2\mathcal{L}'_{\rm sys} + \epsilon\mathcal{L}_{\rm Int}. \tag{5.1.43}$$

The terms in this equation are straightforward. It is possible now to show that (5.1.37) (which is itself the same as (5.1.20)) is identical with the master equation of Sect. 3.6.3 derived by van Kampen's cumulant expansion. The essence of the demonstration relies on the following.

i) Note that

$$\exp\{(\mathcal{L}_{\text{sys}} + \mathcal{L}_B)\tau\}\, g = \mathrm{e}^{-\mathrm{i}H_B\tau/\hbar}\mathrm{e}^{-\mathrm{i}H_{\text{sys}}\tau/\hbar} g \mathrm{e}^{\mathrm{i}H_B\tau/\hbar}\mathrm{e}^{\mathrm{i}H_{\text{sys}}\tau/\hbar}. \tag{5.1.44}$$

ii) Note also that, since $\mathcal{P}\mathcal{L}_{\text{Int}}\mathcal{P} = 0$ and $v = \mathcal{P}\rho_{\text{tot}}$,

$$\mathcal{Q}\mathcal{L}_{\text{Int}}v = \mathcal{L}_{\text{Int}}v. \tag{5.1.45}$$

iii) Make the Markov approximation, $v(t-\tau) \to v(t)$.

iv) Remember that $\mathcal{L}_B\mathcal{P} = 0$. (5.1.46)

v) Then the second term in (5.1.37) becomes

$$-\frac{\epsilon^2}{\hbar^2}\mathcal{P}\left[\sum \kappa_n p_n(0)X(0), \left[\int_0^\infty d\tau \sum \kappa_n p_n(-\tau)X(-\tau),\, \rho_B \otimes \hat{\rho}(t)\right]\right] \tag{5.1.47}$$

where

$$\begin{aligned} p_n(t) &= \mathrm{e}^{\mathrm{i}H_Bt/\hbar} p_n \mathrm{e}^{-\mathrm{i}H_Bt/\hbar} \\ X(t) &= \mathrm{e}^{\mathrm{i}H_{\text{sys}}t/\hbar} X \mathrm{e}^{-\mathrm{i}H_{\text{sys}}t/\hbar} \end{aligned} \tag{5.1.48}$$

and $\hat{\rho}(t)$ is as in (5.1.3,8). Now let us define

$$\xi(t) = \sum \kappa_n p_n(t) \tag{5.1.49}$$

which is similar to the definition (3.1.14), but differs in the fact that here $p_n(t)$ are, because of the definition (5.1.48), like free field canonical operators. Thus the definition (5.1.49) is essentially of identical form to that in (3.1.14). Using the definition of $\mathcal{P}$, we find that (5.1.47) is

$$\begin{aligned} -\frac{\epsilon^2}{\hbar^2}\rho_B \otimes \int_0^\infty d\tau\, \Big\{ &\langle \xi(0)\xi(-\tau)\rangle \left[X(0), X(-\tau)\hat{\rho}(t)\right] \\ &+ \langle \xi(-\tau)\xi(0)\rangle \left[\hat{\rho}(t)X(-\tau), X(0)\right]\Big\} \end{aligned} \tag{5.1.50}$$

where

$$\langle \xi(0)\xi(-\tau)\rangle = \mathrm{Tr}_{\mathrm{B}}\left\{\xi(0)\xi(-\tau)\rho_B\right\}. \tag{5.1.51}$$

vi) Now note that

$$\tfrac{1}{2}\langle[\xi(0), \xi(-\tau)]_+\rangle = \langle \alpha(\tau)\alpha(0)\rangle \tag{5.1.52}$$

where $\alpha(t)$ is as defined in (3.5.19), and

$$\tfrac{1}{2}\langle[\xi(0), \xi(-\tau)]\rangle = \tfrac{1}{2}\mathrm{i}\hbar\frac{d}{dt}f(t). \tag{5.1.53}$$

Here $f(t)$ is defined in (3.1.15) and this latter relation follows from (3.1.29). Then expressing the bath correlation functions in terms of the commutator and anticommutators, we get two kinds of term. Those involving the anticommutator correspond exactly to the integral term in (3.6.6). The other term is integrated by parts, and a boundary term cancels with the term arising from $\epsilon^2 \mathcal{L}'_{\text{sys}}$, to yield the $A_0\rho(t)$ term, evaluated in the situation where $f(t)$is not necessarily a delta function, as in (3.6.2).

5.1.4 Quantum Optical Master Equation with Arbitrary Bath

The master equation in the form (5.1.20) makes no restriction on the precise kind of bath or indeed on the bath state. We can consider the bath Hamiltonian to be arbitrary, and attempt to derive a master equation for a quite general H_{Int}. Let us suppose that the Schrödinger picture H_{Int} can be written

$$H_{\text{Int}} = \hbar \sum_m (X_m^+ \Gamma_m + X_m^- \Gamma_m^\dagger) \tag{5.1.54}$$

where the $X_m^\pm$ are eigenoperators of the system Hamiltonian satisfying

$$[H_{\text{sys}}, X_m^\pm] = \pm\hbar\omega_m X_m^\pm . \tag{5.1.55}$$

This form is quite general, since any system operator can be decomposed into eigenoperators of H_{sys}. If we substitute (5.1.54) into the master equation, there are a number of different terms. To illustrate the method of derivation and demonstrate the approximations, consider the term

$$-\int_0^t dt' \sum_{m,n} X_m^+ e^{i\omega_m t} X_n^- e^{-i\omega_n t'} \rho(t') \text{Tr}_\text{B} \left\{ \Gamma_m(t) \Gamma_n^\dagger(t') \rho_B \right\} . \tag{5.1.56}$$

i) We assume that the bath state is stationary, so that $\text{Tr}_\text{B} \left\{ \Gamma_m(t) \Gamma_n^\dagger(t') \rho_B \right\}$ is a function of $t - t'$ only. This means that terms with $\omega_m \neq \omega_n$ are rapidly varying functions of t, and will make many oscillations during the typical time scale over which $\rho(t)$ alters significantly. This is the *rotating wave approximation* or RWA, first noted in Sect. 3.6.3a.

ii) We make the Markov approximation as in the previous section, thus replacing $\rho(t')$ by $\rho(t)$.

iii) After doing this, the term takes the form

$$-\sum_m X_m^+ X_m^- \rho(t) \int_0^t d\tau e^{i\omega_m \tau} \text{Tr}_\text{B} \left\{ \Gamma_m(\tau) \Gamma_m^\dagger(0) \rho_B \right\} . \tag{5.1.57}$$

The other terms take similar forms. The time t can be considered to be as large as we please compared with the correlation time of the trace, and can be considered infinite.

iv) Thus, everything can be written in terms of

$$\begin{aligned}
\int_0^\infty d\tau\, e^{i\omega_m\tau}\mathrm{Tr_B}\left\{\Gamma_m(\tau)\Gamma_m^\dagger(0)\rho_B\right\} &\equiv \tfrac{1}{2}K_m + i\delta_m \\
\int_0^\infty d\tau\, e^{-i\omega_m\tau}\mathrm{Tr_B}\left\{\Gamma_m(0)\Gamma_m^\dagger(\tau)\rho_B\right\} &\equiv \tfrac{1}{2}K_m - i\delta_m \\
\int_0^\infty d\tau\, e^{i\omega_m\tau}\mathrm{Tr_B}\left\{\Gamma_m^\dagger(\tau)\Gamma_m(0)\rho_B\right\} &\equiv \tfrac{1}{2}G_m + i\epsilon_m \\
\int_0^\infty d\tau\, e^{-i\omega_m\tau}\mathrm{Tr_B}\left\{\Gamma_m^\dagger(0)\Gamma_m(\tau)\rho_B\right\} &\equiv \tfrac{1}{2}G_m - i\epsilon_m
\end{aligned} \tag{5.1.58}$$

(The correlations involving terms like $\Gamma\Gamma$ and $\Gamma^\dagger\Gamma^\dagger$ are usually assumed to be zero if the bath is in a thermal state. However, if the bath is *squeezed*—see Chap.10—this need not be the case.) The master equation finally takes the form (in the interaction picture)

$$\begin{aligned}
\dot\rho(t) = &-i\sum_m[\delta_m X_m^+X_m^- + \epsilon_m X_m^-X_m^+, \rho] \\
&+\tfrac{1}{2}\sum_m K_m\left(2X_m^-\rho X_m^+ - X_m^+X_m^-\rho - \rho X_m^+X_m^-\right) \\
&+\tfrac{1}{2}\sum_m G_m\left(2X_m^+\rho X_m^- - X_m^-X_m^+\rho - \rho X_m^-X_m^+\right)
\end{aligned} \tag{5.1.59}$$

The effect of the ϵ_m and δ_m terms is to add a small perturbing Hamiltonian term—these are *Lamb* and *Stark shift* terms (see Sect. 3.6.3), and are usually neglected. The final terms are just those of the RWA master equation previously derived in (3.6.67).

v) Notice from (5.1.57) , assuming that the Γ_m are indeed destruction operators of some kind, that G_m should vanish at $T = 0$, while K_m should not. Thus the terms in (5.1.59) which involve K_m are present at absolute zero, and must represent transitions from upper to lower energy levels. The terms involving G_m represent transitions to higher energy levels.

Exercise. By considering the diagonal matrix elements of (5.1.59), show that the identification of upward and downward transitions is indeed correct.

vi) In conclusion, we see that the master equation form is expected to have a wide range of validity. However, the only characteristics of the bath that have any relevance are the correlation functions as in (5.1.57)

Exercise—A Bath of Two Level Atoms. Consider a heat bath which consists of a large number of two level atoms, so that the bath Hamiltonian is

$$H_B = \sum \tfrac{1}{2}\hbar\Omega_a\sigma_z^a. \tag{5.1.60}$$

Suppose the coupling to the bath is given by

$$\Gamma_m = \sum_a g_a^m \sigma_a^-, \qquad \Gamma_m^\dagger = \sum_a g_a^m \sigma_a^+ \tag{5.1.61}$$

where the g_a^m are real numbers. Suppose also that the density operator ρ_B is the direct product of individual density matrices

$$\rho_a = \begin{pmatrix} N_a^+ & 0 \\ 0 & N_a^- \end{pmatrix}, \qquad \text{with} \qquad N_a^+ + N_a^- = 1. \tag{5.1.62}$$

Show that, by making the approximation

$$\begin{aligned} &\sum_a \to \int d\Omega\, n(\Omega) \\ &g_a^m \to g^m(\Omega) \qquad N_a^\pm \to N^\pm(\Omega) \end{aligned} \tag{5.1.63}$$

where $n(\Omega)$ is a density of states with energy $\hbar\Omega$, that in this case

$$\begin{aligned} K_m &= 2\pi \left\{g_m(\omega_m)\right\}^2 n(\omega_m) N^-(\omega_m) \\ G_m &= 2\pi \left\{g_m(\omega_m)\right\}^2 n(\omega_m) N^+(\omega_m) \\ \delta_m &= \mathrm{P}\int \frac{d\Omega\, n(\Omega) g(\Omega)^2 N^-(\Omega)}{\omega_m - \Omega} \\ \epsilon_m &= \mathrm{P}\int \frac{d\Omega\, n(\Omega) g(\Omega)^2 N^+(\Omega)}{\omega_m - \Omega}. \end{aligned} \tag{5.1.64}$$

Note that there is no restriction on N_a^+ and N_a^-—this means that $N_a^+ > N_a^-$ is permissible. Physically such a population inversion could only be maintained by some kind of pumping, as indeed happens in a laser (see Chap.9).

5.1.5 Relationship to the Quantum Brownian Motion Master Equation

If we take the quantum Brownian motion limit in the case that the various Hamiltonians are as defined in (5.1.40–42, we find the quantum Brownian motion master equation in exactly the same form (3.6.28) as derived in Sect. 3.6.1.

5.1.6 Notational Matters

Abstractly viewed, the master equation of either kind can be written in the form

$$\frac{d\rho(t)}{dt} = L(t)\rho(t) \tag{5.1.65}$$

where $L(t)$ is a linear operator whose form depends on the particular master equation. The operator $L(t)$ may or may not have an explicit time dependence, depending on a number of factors. Notice that the master equations as displayed in Chap.3 are originally written in the Schrödinger picture, though to make the rotating wave approximation in (3.6.67) it is necessary to go to the interaction picture. The derivation of Sect. 5.1.1 gives an interaction picture master equation, while that of Sect. 5.1.2 does not use the interaction picture. If none of the terms in the Hamiltonian $H = H_{\rm sys} + H_{\rm Int} + H_B$ has any explicit time dependence, then neither will there be any explicit time dependence in the resulting operator $L(t)$ if this is the operator relevant for evolution in the Schrödinger picture. Thus, we can say that for the quantum Brownian motion case

$$L\rho = -\frac{\mathrm{i}}{\hbar}[H_{\rm sys}, \rho] - \frac{\mathrm{i}\gamma}{2\hbar}\left[X, [\dot{X}, \rho]_+\right] - \frac{\gamma kT}{\hbar^2}\left[X, [X, \rho]\right] \tag{5.1.66}$$

and in the quantum optical case

$$\begin{aligned} L\rho = &-\frac{\mathrm{i}}{\hbar}[H_{\rm sys}, \rho] \\ &-\sum_m \frac{\pi\omega_m}{2\hbar}\left(\bar{N}(\omega_m) + 1\right)\kappa(\omega_m)^2[\rho X_m^+ - X_m^-\rho, X] \\ &-\sum_m \frac{\pi\omega_m}{2\hbar}\bar{N}(\omega_m)\kappa(\omega_m)^2[\rho X_m^- - X_m^+\rho, X] \\ &+\frac{\mathrm{i}}{2\hbar}\sum_m \mathrm{P}\int_{-\infty}^{\infty}\frac{d\omega\,\omega\kappa(\omega)^2}{\omega_m - \omega}\left(\bar{N}(\omega) + \tfrac{1}{2}\right)[[\rho, X_m^+ - X_m^-], X] \\ &+\frac{\mathrm{i}}{4\hbar}\sum_m \mathrm{P}\int_{-\infty}^{\infty}\frac{d\omega\,\omega_m\kappa(\omega)^2}{\omega_m - \omega}[[\rho, X_m^+ + X_m^-]_+, X]. \end{aligned} \tag{5.1.67}$$

In the interaction picture an unavoidable time dependence will arise in the quantum Brownian motion case, since the operator X does not commute with $H_{\rm sys}$. However, in the quantum optical case the resulting time dependence can be eliminated by the procedure used to derive the rotating-wave master equation, (3.6.64). This is possible because in this case the time evolution arising from $H_{\rm sys}$ is very much faster than that of the damping, which means that all the resulting time dependence can be averaged over many cycles—leaving only the time independent terms. Explicit time dependence of $L(t)$ will arise in practice only from explicit time dependence in $H_{\rm sys}$ or $H_{\rm Int}$ (or even in H_B). In this case the Laplace transform methods of Sect. 5.1.2 do not work, and must be adapted accordingly. It can be seen therefore, that the Schrödinger picture gives the simplest general way of looking at the master equation, since explicit time dependences then only arise from some explicit time dependence in the physics of the situation.

A general notation, that will be used in the remainder of this book will be of the form

$$\dot{\rho}(t) = L(t)\rho(t) \tag{5.1.68}$$

where $L(t)$ will be written L when no explicit time dependence is present. The solution of an equation like (5.1.68) is given in terms of the time-ordered product

$$\rho(t) = \mathbf{T}\exp\left\{\int_{t_0}^{t} dt'\,L(t')\right\}\rho(t_0). \tag{5.1.69}$$

This is a formula of little use for computing answers, but it is of use for analytical purposes. Of particular use is the alternative method of writing the solution as a product limit

$$\rho(t) = \lim_{N\to\infty} \prod_{n=0}^{N} [1 + L(t_n)\Delta t_n]\rho(t_0) \tag{5.1.70}$$

where

$$t_0 < t_1 < t_2 \cdots < t_N < t \tag{5.1.71}$$

and

$$\Delta t_n = t_{n+1} - t_n . \tag{5.1.72}$$

5.2 Multitime Structure of Quantum Markov Processes

The solutions (5.1.69) or (5.1.70) can be formally written

$$\rho(t) = V(t, t_0)\rho(t_0), \tag{5.2.1}$$

where $V(t, t_0)$ is the evolution operator, and satisfies the same equation as $\rho(t)$,

$$\dot{V}(t, t_0) = L(t)V(t, t_0), \tag{5.2.2}$$

but with the initial condition

$$V(t_0, t_0) = 1. \tag{5.2.3}$$

The notation in terms of $V(t, t_0)$ will be of particular use when we formulate the concept of a quantum Markov process. Since $V(t, t_0)$ obeys the differential equation (5.2.2), it follows that

$$V(t, t_1)V(t_1, t_0) = V(t, t_0). \tag{5.2.4}$$

This is known as the *semigroup property* of the evolution operator [5.7].

5.2.1 Computation of Multitime Averages

From the reduced density operator it is possible to compute any average of any product of system operators, provided these are all evaluated at the same time. This is obvious, since the product of any number of system operators is again a system operator. Unfortunately this is not sufficient to satisfy the curiosity of a physicist. It is very often necessary to evaluate time correlation functions of system operators, since these have a direct connection with measurable quantities, such as spectra, etc. For example, we may wish to evaluate a correlation function such as

$$\langle A(t+\tau)B(t)\rangle = \mathrm{Tr}_{\mathrm{sys}}\left\{\mathrm{Tr}_{\mathrm{B}}\left\{A(t+\tau)B(t)\rho_s \otimes \rho_B\right\}\right\} \tag{5.2.5}$$

where $A(t+\tau)$ and $B(t)$ are Heisenberg picture system operators. An exact answer for (5.2.5) is of course given by

$$\langle A(t+\tau)B(t)\rangle = \mathrm{Tr}_{\mathrm{sys}}\left\{\mathrm{Tr}_{\mathrm{B}}\left\{\mathrm{e}^{\mathrm{i}H(t+\tau)/\hbar}A\mathrm{e}^{-\mathrm{i}H(t+\tau)/\hbar}\mathrm{e}^{\mathrm{i}Ht/\hbar}B\mathrm{e}^{-\mathrm{i}Ht/\hbar}\rho_s\otimes\rho_B\right\}\right\}. \tag{5.2.6}$$

However, this exact equation is not really what we would like, which would be an equation involving only the system operators, and the reduced dynamics on the system, as given by the master equation. A transformation of (5.2.6) to a form which does involve only the reduced dynamics is quite simple to give. We can rewrite (5.2.6) as

$$\langle A(t+\tau)B(t)\rangle = \mathrm{Tr}_{\mathrm{sys}}\left\{A\,\mathrm{Tr}_{\mathrm{B}}\left\{\mathrm{e}^{-\mathrm{i}H\tau/\hbar}B\rho_{\mathrm{tot}}(t)\mathrm{e}^{iH\tau/\hbar}\right\}\right\}, \tag{5.2.7}$$

since A is proportional to the identity in the bath space. The equation of motion for the term

$$X(\tau,t)\equiv\mathrm{e}^{-\mathrm{i}H\tau/\hbar}B\rho_{\mathrm{tot}}(t)\mathrm{e}^{\mathrm{i}H\tau/\hbar} \tag{5.2.8}$$

in terms of τ is

$$\mathrm{i}\hbar\frac{\partial}{\partial\tau}X(\tau,t) = [H, X(\tau,t)], \tag{5.2.9}$$

and in exactly the same way as we derived the master equations in Sect. 5.1.1, Sect. 5.1.2, we can show that $\mathrm{Tr}_{\mathrm{B}}\{X(\tau,t)\}$ also obeys the master equation as a function of τ. We now use the evolution operator to write

$$\begin{aligned}\mathrm{Tr}_{\mathrm{B}}\{X(\tau,t)\} &= V(t+\tau,t)\mathrm{Tr}_{\mathrm{B}}\{X(0,t)\}\\ &= V(t+\tau,t)\{B\rho(t)\}\end{aligned} \tag{5.2.10}$$

where in this case $\rho(t)$ is the Schrödinger picture reduced density operator, as in Sect. 5.1.6. Putting these together, we find that the correlation function is given by

$$\boxed{\langle A(t+\tau)B(t)\rangle = \mathrm{Tr}_{\mathrm{sys}}\{AV(t+\tau,t)\{B\rho(t)\}\}.} \tag{5.2.11}$$

a) Definition of Operator Ordering: It is important to realize that when we write an equation like (5.2.11) that $L(t)$ and $V(t+\tau,t)$ are *two sided* operators (as is clear from the definitions (5.1.66,67) while the operators A, B and the density operator $\rho(t)$ are ordinary quantum mechanical operators. This means that there is an element of non-associativity. For example

$$L\{A\rho(t)\}\neq\{LA\}\rho(t), \tag{5.2.12}$$

as is clear from the definitions (5.1.66,67). A similar property pertains to the evolution operator. However, in practice, the explicit specification of the ordering by brackets can be cumbersome, and the convention is that any evolution operator operates on everything to its right. Thus, conventionally, we would write (5.2.11) as

$$\langle A(t+\tau)B(t)\rangle = \mathrm{Tr}_{\mathrm{sys}}\left\{AV(t+\tau)B\rho(t)\right\}. \tag{5.2.13}$$

b) General Formulae for Time-Ordered Correlation Functions: It is quite straightforward to show that time ordered correlation functions can always be written in a form analogous to (5.2.14). Thus, we consider a correlation function of the form

$$\langle A_0(s_0)A_1(s_1)\dots A_m(s_m)B_n(t_n)B_{n-1}(t_{n-1})\dots B_0(t_0)\rangle \tag{5.2.14}$$

where the terms are ordered

$$\begin{aligned} &t_n \geq t_{n-1} \geq \dots \geq t_0, \\ &s_m \geq s_{m-1} \geq \dots \geq s_0. \end{aligned} \tag{5.2.15}$$

The relative order of the s and the t is not specified. This kind of time ordered correlation function looks a little arbitrary at first glance, but, as we have seen in Sect. 2.3.2 this is the class of correlation function which can be obtained by repeated measurements on the system. To develop the formula for (5.2.14), assume that s_m is the latest time and that t_n is the next latest. Then

$$\begin{aligned} &\langle A_0(s_0)A_1(s_1)\dots A_m(s_m)B_n(t_n)B_{n-1}(t_{n-1})\dots B_0(t_0)\rangle = \\ &\quad \mathrm{Tr}_{\mathrm{s}}\left\{\mathrm{Tr}_{\mathrm{B}}\left\{A_m(s_m)B_n(t_n)\dots B_0(t_0)\rho_s\otimes\rho_B A_0(s_0)\dots A_{m-1}(s_{m-1})\right\}\right\} \end{aligned} \tag{5.2.16}$$

and using the same argument as for (5.2.11),

$$\begin{aligned} &= \mathrm{Tr}_s\{A_m V(s_m,t_n)B_n \\ &\qquad \times\mathrm{Tr}_{\mathrm{B}}\left\{B_{n-1}(t_{n-1})\dots B_0(t_0)\rho_s\otimes\rho_B A_0(s_0)\dots A_{m-1}(s_{m-1})\right\}\} \end{aligned} \tag{5.2.17}$$

It will be noted that the particular interlacing of the times t with the times s is significant. The precise procedure can be described as follows:
(1) Order the times t and s in sequence, and rename them τ_i, so that

$$\tau_0 \leq \tau_1 \leq \tau_2 \leq \dots \leq \tau_{r-1} \leq \tau_r \qquad (r = n+m+1). \tag{5.2.18}$$

(2) Let F_i be the Schrödinger picture operator corresponding to τ_i, and define the operator f_i by

$$\begin{aligned} f_i\rho &= F_i\rho \qquad \text{if } F_i \text{ is one of the } B'\text{s} \\ f_i\rho &= \rho F_i \qquad \text{if } F_i \text{ is one of the } A'\text{s}. \end{aligned} \tag{5.2.19}$$

(3) Then

$$\begin{aligned} &\langle A_0(s_0)A_1(s_1)\dots A_m(s_m)B_n(t_n)B_{n-1}(t_{n-1})\dots B_0(t_0)\rangle \\ &\quad = \mathrm{Tr}_{\mathrm{s}}\left\{f_r V(\tau_r,\tau_{r-1})f_{r-1}V(\tau_{r-1},\tau_{r-2})\dots V(\tau_1,\tau_0)f_0\rho(t_0)\right\}. \end{aligned} \tag{5.2.20}$$

Exercise. Show that (5.2.20) gives the following

$$\langle A(t_0)B(t_1)C(t_1)D(t_0)\rangle = \mathrm{Tr}_{\mathrm{sys}}\{BCV(t_1,t_0)D\rho(t_0)A\} \quad (5.2.21)$$

$$\langle A(t)B(t+\tau)\rangle = \mathrm{Tr}_{\mathrm{sys}}\{BV(t+\tau,t)\rho(t)A\} \quad (5.2.22)$$

$$\langle B(t+\tau)A(t)\rangle = \mathrm{Tr}_{\mathrm{sys}}\{BV(t+\tau,t)A\rho(t)\}\,. \quad (5.2.23)$$

5.2.2 The Markov Interpretation

The time development equation for the density operator ρ, together with the formula (5.2.20) for the time-ordered correlation functions provide the ingredients for the definition of a quantum Markov process. The essential formula in classical Markov theory is the Chapman-Kolmogorov equation for the conditional probability, $p(x_1,t_1|x_0,t_0)$; namely,

$$\int dy\, p(x_1,t_1|y,t)p(y,t|x_0,t_0) = p(x_1,t_1|x_0,t_0). \quad (5.2.24)$$

This has the same formal structure as the equation for the evolution operator

$$V(t_1,t)V(t,t_0) = V(t_1,t_0). \quad (5.2.25)$$

To show this we simply define an operator $W(t_1,t_0)$ by

$$\{W(t_1,t_0)f\}(x_1) = \int dx_0 p(x_1,t_1|x_0,t_0)f(x_0). \quad (5.2.26)$$

Similarly, the correlation function equation

$$\begin{aligned}\langle f_r(x_r,\tau_r)f_{r-1}(x_{r-1},\tau_{r-1})\dots f_0(x_0,\tau_0)\rangle \\ = \int dx_r dx_{r-1}\dots dx_0 f_r(x_r)f_{r-1}(x_{r-1})\dots f_0(x_0) \\ \times p(x_r\tau_r|x_{r-1},\tau_{r-1})p(x_{r-1},\tau_{r-1}|x_{r-2},\tau_{r-2})\dots \\ \times p(x_1,\tau_1|x_0,\tau_0)p(x_0,\tau_0)\end{aligned} \quad (5.2.27)$$

is equivalent to

$$\begin{aligned}\langle f_r(x_r,\tau_r)f_{r-1}(x_{r-1},\tau_{r-1})\dots f_0(x_0,\tau_0)\rangle \\ = \int dx\Big\{f_r(x)\{W(\tau_r,\tau_{r-1})f_{r-1}W(\tau_{r-1},\tau_{r-2})f_{r-2}\dots \\ \times W(\tau_1,\tau_0)f_0p(\tau_0)\}(x)\Big\},\end{aligned} \quad (5.2.28)$$

which is in exact analogy to (5.2.20)—the differences being that here there is no time-ordering required since the functions commute, and the quantum mechanical trace is replaced by the integral over x.

The Lindblad Form: The physical essence of the Markov property is the existence of an evolution equation as a first order differential equation in time for the density operator or the conditional probability. However, not every such equation will produce positive probabilities or a positive semidefinite density operator. The relevant conditions in classical stochastics give rise to the differential Chapman-Kolmogorov equation (S.M. Chap.3). In the case of a quantum Markov process case, it can be shown [5.7] that L must be of the *Lindblad* form

$$L\rho = -\frac{\mathrm{i}}{\hbar}[H,\rho] + \sum_J [2A_J\rho A_J^\dagger - \rho A_J^\dagger A_J - A_J^\dagger A_J\rho]. \tag{5.2.29}$$

Here H is some Hermitian operator, and the A_J are arbitrary. (The rigorous proofs actually require bounded operators for the A_J.) The quantum optical master equation is of this form, so it can be concluded that the positivity of the density operator is guaranteed for it.

However the quantum Brownian motion master equation cannot be written in the Lindblad form, and therefore has solutions which are unphysical [5.8]. Nevertheless, the unphysical solutions normally appear only in rapid initial transients, where one expects the Markov approximation itself to be invalid.

5.2.3 Quantum Regression Theorem

Linear systems are always of particular interest in physics, because of their simplicity and the fact that they are exactly soluble. It is often the case that the equations of motion for the means of certain operators are linear. In that case, it is possible to show that the corresponding time correlation functions obey exactly the same equations. This result was first derived by *Lax* [5.3] and is known as the *quantum regression theorem.* Suppose for certain set of operators Y_i, the master equation can be shown to yield for any initial $\hat{\rho}$,

$$\partial_t\langle Y_i(t)\rangle = \sum_j G_{ij}(t)\langle Y_j(t)\rangle, \tag{5.2.30}$$

then we assert that

$$\partial_t\langle Y_i(t+\tau)Y_l(t)\rangle = \sum_j G_{ij}(\tau)\langle Y_j(t+\tau)Y_l(t)\rangle. \tag{5.2.31}$$

For

$$\langle Y_i(t+\tau)Y_l(t)\rangle = \mathrm{Tr_s}\left\{Y_iV(t+\tau,t)Y_l\rho(t)\right\} \tag{5.2.32}$$

and the right hand side is an average of Y_i at time $t+\tau$ with the choice of initial density operator

$$\rho \to Y_l\rho(t). \tag{5.2.33}$$

Since by hypothesis any initial ρ is permitted, and (5.2.30) is linear, we may in fact generate any initial condition whatsoever. Hence choosing ρ as given by (5.2.33), the hypothesis (5.2.30) yields the *quantum regression theorem*, (5.2.31).

The form (5.2.33) is in fact quite general provided one accepts that there may be an infinite number of operators Y_i, because the operators A, B, V in (5.2.11) are all linear operators. This means that they may be expressed in terms of any linearly independent basis, and that (5.2.33) can be viewed as a re-expression of (5.2.11) in the basis of the Y_i. It is a common terminology to define the quantum regression theorem as the general equation for time correlation functions, (5.2.20), with the definition of the evolution operators given by (5.2.2).

5.3 Inputs, Outputs and Quantum Stochastic Differential Equations

The quantum optical master equation forms the definition of a quantum Markov process, but unlike the classical Markov process there is as yet in this book no quantum stochastic differential equation in a corresponding analogy to the classical stochastic differential equation. In this section we will show how appropriate "quantum optical" approximations may be made in the input-output formalism of Sect. 3.2 to yield an idealized *white noise* quantum stochastic differential equation formalism which is in direct analogy to the classical stochastic differential equation. The work in this section is largely adapted from the work of *Collett* and *Gardiner* [5.9]

5.3.1 Idealized Hamiltonian

The simplifications necessary are made by introducing the idealized Hamiltonian

$$H = H_{\rm sys} + H_B + H_{\rm int} \tag{5.3.1}$$

$$H_B = \hbar \int_{-\infty}^{\infty} d\omega\, \omega b^\dagger(\omega) b(\omega) \tag{5.3.2}$$

$$H_{\rm int} = \mathrm{i}\hbar \int_{-\infty}^{\infty} d\omega \kappa(\omega) \left[b^\dagger(\omega) c - c^\dagger b(\omega) \right] \tag{5.3.3}$$

where the $b(\omega)$ are boson annihilation operators for the bath, with

$$\left[b(\omega), b^\dagger(\omega') \right] = \delta(\omega - \omega') \tag{5.3.4}$$

and c is one of the several possible system operators. We do not specify either $H_{\rm sys}$ or the kind of system operators or their commutation relations. There are two principal idealizations in the form of the Hamiltonian—the rotating wave approximation, and the fact that the range of the ω integration is $(-\infty, \infty)$ rather than $(0, \infty)$. These are intimately connected, as follows.

i) Without the rotating wave approximation the form of H_{int} would be something like

$$i\hbar \int_0^\infty d\omega\, \kappa(\omega) \left[b^\dagger(\omega) + b(\omega)\right] \left[c - c^\dagger\right] \tag{5.3.5}$$

which has counter-rotating terms like $b^\dagger(\omega)c^\dagger$. The arguments for dropping these are based on the smallness of H_{Int}, so that the motion of the operators c and $c^\dagger$ will be given essentially by the free Hamiltonian H_{sys}. In quantum optical systems, only high frequencies are involved. Thus the time dependence of $c^\dagger$ is perhaps like $e^{i\Omega t}$ and terms like $b^\dagger(\omega)c^\dagger$ have a time dependence like $e^{i(\omega+\Omega)t}$, which is always rapidly oscillating, while $b^\dagger(\omega)c$ has a time dependence like $e^{i(\omega-\Omega)t}$ which is almost constant near $\omega \approx \Omega$, i.e., near resonance.

ii) Essentially, one argues that only terms which are almost resonant are important. This allows one to extend the lower limit of the ω integrals to $-\infty$. Though this is formally absurd, the added terms are all non-resonant, and contribute very little. Notice that the extension of the lower limit of integration to $-\infty$ only makes sense if the rotating wave approximation has been made, since otherwise the terms like $e^{i(\omega+\Omega)t}$ would be resonant at $\omega = -\Omega$, and would produce spurious effects.

iii) These simplifications, however, do generate a very simple formalism which is essentially a formulation of *quantum white noise*. As in classical white noise, the required delta function correlations can only be produced if the full range of positive and negative frequencies is available—the extension of the range of integration to $(-\infty, \infty)$ is essential for this.

Thus although the Hamiltonians (5.3.2,3) do look physically absurd, within the rotating wave approximation they are very little different from their more realistic counterparts, but have the merit of producing a mathematical formulation which is very simple and, moreover, will be shown to be *Markovian*.

5.3.2 Derivation of the Langevin Equations

We follow a procedure very similar to that used in Chap.3. From (5.3.1–3) we derive the Heisenberg equations of motion for $b(\omega)$, and an arbitrary system operator a. They are

$$\dot{b}(\omega) = -i\omega b(\omega) + \kappa(\omega)c, \tag{5.3.6}$$

$$\dot{a} = -\frac{i}{\hbar}\left[a, H_{\text{sys}}\right] + \int d\omega\, \kappa(\omega) \left\{b^\dagger(\omega)\,[a, c] - \left[a, c^\dagger\right] b(\omega)\right\} \tag{5.3.7}$$

and we solve (5.3.6) to obtain

$$b(\omega) = e^{-i\omega(t-t_0)} b_0(\omega) + \kappa(\omega) \int_{t_0}^t e^{-i\omega(t-t')} c(t')dt'. \tag{5.3.8}$$

Here $b_0(\omega)$ is the value of $b(\omega)$ at $t = t_0$ and has the same commutation relations as $b(\omega)$. We substitute in (5.3.7) to obtain

$$\begin{aligned}\dot{a} = &-\frac{\mathrm{i}}{\hbar}\left[a, H_{\mathrm{sys}}\right] \\ &+\int d\omega\, \kappa(\omega)\left\{\mathrm{e}^{\mathrm{i}\omega(t-t_0)} b_0^\dagger(\omega)\,[a, c] - \left[a, c^\dagger\right] \mathrm{e}^{-\mathrm{i}\omega(t-t_0)} b_0(\omega)\right\} \\ &+\int d\omega\, [\kappa(\omega)]^2 \int_{t_0}^{t} dt' \left\{\mathrm{e}^{\mathrm{i}\omega(t-t')} c^\dagger(t')\,[a, c] - \left[a, c^\dagger\right] \mathrm{e}^{-\mathrm{i}\omega(t-t')} c(t')\right\}. \end{aligned} \tag{5.3.9}$$

For notational convenience in (5.3.9) we omit the time argument on the system operators when it is t but write it explicitly otherwise. (Thus $a \equiv a(t)$.)

The equations are exact so far. In the same way as in Chap.3 we introduce what we shall call the first Markov approximation, that the coupling constant is independent of frequency.

First Markov Approximation: This takes a simpler form from that of Sect. 3.1.1; namely

$$\kappa(\omega) = \sqrt{\gamma/2\pi}\ . \tag{5.3.10}$$

This approximation can be used to put (5.3.9) into the form of a damping equation. We use the properties

$$\int_{-\infty}^{\infty} d\omega \mathrm{e}^{-\mathrm{i}\omega(t-t')} = 2\pi\delta(t-t') \tag{5.3.11}$$

and

$$\int_{t_0}^{t} c(t')\delta(t-t')dt' = \tfrac{1}{2}c(t). \tag{5.3.12}$$

(The second result has the factor of $\frac{1}{2}$ because the peak of the delta function is at the end of the interval of integration.) In the same way as in Chap.3 we define an "in" field by

$$b_{\mathrm{in}}(t) = \frac{1}{\sqrt{2\pi}}\int d\omega \mathrm{e}^{-\mathrm{i}\omega(t-t_0)} b_0(\omega) \tag{5.3.13}$$

which satisfies in this case the commutation relation

$$[b_{\mathrm{in}}(t), b_{\mathrm{in}}^\dagger(t')] = \delta(t-t'). \tag{5.3.14}$$

Using (5.3.11–13) we readily derive the *quantum Langevin equation*

$$\dot{a} = -\frac{\mathrm{i}}{\hbar}\left[a, H_{\mathrm{sys}}\right] - \left[a, c^\dagger\right]\left[\frac{\gamma}{2}c + \sqrt{\gamma}\, b_{\mathrm{in}}(t)\right] + \left[\frac{\gamma}{2}c^\dagger + \sqrt{\gamma}\, b_{\mathrm{in}}^\dagger(t)\right][a, c]. \tag{5.3.15}$$

This equation is clearly analogous to that given in (3.1.28), though a slight difference in definitions of the first Markov approximation means that no time derivative terms turn up. We will see that this is a narrow bandwidth approximation to the more exact quantum Langevin equation. The commutation rule (5.3.14) can be similarly justified as having the same effect as the commutator (3.1.32) as long as only a narrow bandwidth is being considered.

Some points to note are:

a) Noise Terms: The terms depending on $b_{\text{in}}(t), b_{\text{in}}^{\dagger}(t)$ are to be taken as noise terms. The definition of $b_{\text{in}}(t)$ in terms of the values $b_0(\omega)$ of $b(\omega)$ at time $t = t_0$ ensures that these operators may be freely specified on the same basis as initial conditions. Similarly, we may freely specify the state of the system to be such that, at $t = t_0$, the system and bath density operators both factorize. It is *not* necessary (nor indeed is it possible) to make any further "independence" assumption. In fact it is not necessary to make any particular assumption about the initial state of the system at all for the Langevin equation to be valid in the form (5.3.15). However, the terms $b_{\text{in}}(t), b_{\text{in}}^{\dagger}(t)$ can only be reasonably interpreted as noise when the state of the system is initially factorized and the state of $b_{\text{in}}(t)$ is incoherent (e.g., a thermal state). In the case, that, for example $b_{\text{in}}(t)$ is in a coherent state, we would have a classical driving field being applied to the system, and other intermediate situations can easily be envisaged.

b) There Are Alternative Forms: These arise from the fact that the terms $\frac{1}{2}\gamma c(t)+\sqrt{\gamma}\, b_{\text{in}}(t)$ and $\frac{1}{2}\gamma c^{\dagger}(t)+\sqrt{\gamma}\, b_{\text{in}}^{\dagger}$ commute with all system operators. For, from (5.3.4–10)

$$\frac{1}{\sqrt{2\pi}}\int d\omega b(\omega) = b_{\text{in}}(t) + \frac{\sqrt{\gamma}}{2}c(t). \tag{5.3.16}$$

Since the $b(\omega)$ are bath operators, and commute with all system operators at the same time, this proves the result.

c) Consistency with Calculus: The basic rule of ordinary calculus for noncommuting operators is that of the product—if a_1 and a_2 are two operators

$$\frac{d}{dt}(a_1 a_2) = \dot{a}_1 a_2 + a_1 \dot{a}_2. \tag{5.3.17}$$

This rule is also valid for any commutator, i.e., for any A

$$[A, a_1 a_2] = [A, a_1]\, a_2 + a_1\, [A, a_2]\,. \tag{5.3.18}$$

Only the first term on the right of (5.3.15) is in the explicit form of a commutator, but the parts involving γ are of the form of commutators multiplied by terms like (5.3.16) which commute with all system operators. Thus the time derivative of any product will be correctly given by (5.3.17) and by the substitution $a \to a_1 a_2$ in (5.3.15). We find that there are no Ito-like terms, even if the input is noise. We come back to this in Sect. 5.3.3.

d) "Out" Fields: If we consider $t_1 > t$, we can write, analogously to (5.3.8),

$$b(\omega) = \mathrm{e}^{-\mathrm{i}\omega(t-t_1)}b_1(\omega) - \kappa(\omega)\int_t^{t_1} \mathrm{e}^{-\mathrm{i}\omega(t-t')}c(t')\,dt' \tag{5.3.19}$$

and similarly define

$$b_{\text{out}}(t) = \frac{1}{\sqrt{2\pi}} \int d\omega e^{-i\omega(t-t_1)} b_1(\omega). \tag{5.3.20}$$

Carrying out the same procedure, we arrive at a version of the Langevin equation written in terms of the "out" operators;

$$\dot{a} = -\frac{i}{\hbar}\left[a, H_{\text{sys}}\right] + \left[a, c^\dagger\right]\left[\frac{\gamma}{2}c - \sqrt{\gamma}\, b_{\text{out}}(t)\right] - \left[\frac{\gamma}{2}c^\dagger - \sqrt{\gamma}\, b_{\text{out}}(t)\right][a, c] \tag{5.3.21}$$

in which we see that

$$\begin{aligned} \sqrt{\gamma}\, b_{\text{in}}(t) &\to \sqrt{\gamma}\, b_{\text{out}}(t), \\ \tfrac{1}{2}\gamma c &\to -\tfrac{1}{2}\gamma c, \end{aligned} \tag{5.3.22}$$

and furthermore

$$\frac{1}{\sqrt{2\pi}} \int d\omega b(\omega) = b_{\text{out}}(t) - \frac{\sqrt{\gamma}}{2} c(t) \tag{5.3.23}$$

is derived similarly to (5.3.16). From this follows the identity that

$$\boxed{b_{\text{out}}(t) - b_{\text{in}}(t) = \sqrt{\gamma}\, c(t)} \tag{5.3.24}$$

which also can be used to transform between the forward Langevin equation (5.3.15) and the time-reversed Langevin equation (5.3.21). This is the analogue of the results of Sect. 3.2.2c.

Exercise—Damped Harmonic Oscillator. Set

$$\begin{aligned} &H_{\text{sys}} \to \hbar\Omega a^\dagger a \\ &c \to a \\ &[a, a^\dagger] = 1. \end{aligned} \tag{5.3.25}$$

Show that the quantum Langevin equation becomes

$$\dot{a} = -\left(i\Omega + \frac{\gamma}{2}\right) a - \sqrt{\gamma}\, b_{\text{in}}(t). \tag{5.3.26}$$

Notice that this represents a *damped* motion for $\langle a \rangle$ independent of the statistics of $b_{\text{in}}(t)$, because of the linearity of the equation in the variables a and $b_{\text{in}}(t)$.

Exercise—Two Level Atom. Set

$$H_{\text{sys}} \to \tfrac{1}{2}\hbar\Omega\sigma_z, \qquad c \to \sigma^-. \tag{5.3.27}$$

Show that the quantum Langevin equations become

$$\begin{aligned} \dot{\sigma}^- &= -(i\Omega + \frac{\gamma}{2})\sigma^- + \sqrt{\gamma}\, \sigma_z b_{\text{in}}(t) \\ \dot{\sigma}^+ &= (i\Omega - \frac{\gamma}{2})\sigma^+ + \sqrt{\gamma}\, \sigma_z b_{\text{in}}^\dagger(t) \\ \dot{\sigma}_z &= -\gamma(1 + \sigma_z) - 2\sqrt{\gamma}\, \{\sigma^+ b_{\text{in}}(t) + b_{\text{in}}^\dagger(t)\sigma^-\}. \end{aligned} \tag{5.3.28}$$

These equations are non-linear in $\boldsymbol{\sigma}$ and $b_{\text{in}}(t)$—hence we cannot express, for example, $\langle \sigma^+ b_{\text{in}}(t) \rangle$ as a function of the $\langle \boldsymbol{\sigma} \rangle$. By making white noise assumptions (as in Sect. 5.3.5) these averages can be computed. Notice that we cannot conclude that the damping is independent of the statistics of the bath, as in the case of the harmonic oscillator.

5.3.3 Inputs and Outputs, and Causality

The quantities $b_{\text{in}}(t)$ and $b_{\text{out}}(t)$ will be interpreted as inputs and outputs to the system. The condition (5.3.24) can be viewed as a boundary condition, relating input, output, and internal modes, and is the analogue of the boundary condition of Chap.3.

If (5.3.15) is solved to give values of the system operators in terms of their past values and those of $b_{\text{in}}(t)$, then it is clear that $a(t)$ is independent of $b_{\text{in}}(t')$ for $t' > t$; that is, the system variables do not depend on the values of the input in the future. Hence we deduce

$$\left[a(t), b_{\text{in}}(t')\right] = 0, \quad t' > t \tag{5.3.29}$$

and using similar reasoning, we deduce from (5.3.24)

$$\left[a(t), b_{\text{out}}(t')\right] = 0, \quad t' < t. \tag{5.3.30}$$

Defining the step function $u(t)$ as

$$u(t) = \begin{cases} 1, & t > 0 \\ \frac{1}{2}, & t = 0 \\ 0, & t < 0, \end{cases} \tag{5.3.31}$$

and using (5.3.24) we obtain the specific results

$$\left[a(t), b_{\text{in}}(t')\right] = -u(t - t')\sqrt{\gamma}\,\left[a(t), c(t')\right] \tag{5.3.32}$$

$$\left[a(t), b_{\text{out}}(t')\right] = u(t' - t)\sqrt{\gamma}\,\left[a(t), c(t')\right] \tag{5.3.33}$$

which are analogous to the result (3.1.33). We now have, in principle, a complete specification of a system with input and output. We *specify* the input $b_{\text{in}}(t)$, and solve (5.3.15) for $a(t)$. We then compute the output from the known $a(t)$ and $b_{\text{in}}(t)$ by use of the boundary condition, (5.3.24).

The commutators (5.3.32 ,33) are an expression of quantum causality—that only the future motion of the system is affected by the present input, and that only the future value of the output is affected by the present values of the system operators.

5.3.4 Several Inputs and Outputs

It is quite straightforward to generalize the formalism to a situation in which the system is coupled to several heat baths, so that

$$H_B = \sum_l \hbar \int_{-\infty}^{\infty} d\omega\, b^\dagger(l,\omega) b(l,\omega) \tag{5.3.34}$$

$$H_{\text{int}} = \mathrm{i}\hbar \sum_l \int_{-\infty}^{\infty} d\omega \kappa(l,\omega)[b^\dagger(l,\omega)c_l - c_l^\dagger b(l,\omega)] \tag{5.3.35}$$

and in this case the quantum Langevin equation becomes

$$\dot{a} = -\frac{\mathrm{i}}{\hbar}[a, H_{\mathrm{sys}}] - \sum_l [a, c_l^\dagger]\{\tfrac{1}{2}\gamma_l c_l + \sqrt{\gamma_l}\, b_{\mathrm{in}}(l,t)\} + \sum_l \{\tfrac{1}{2}\gamma_l c_l^\dagger + \sqrt{\gamma_l}\, b_{\mathrm{in}}^\dagger(l,t)\}[a, c_l]. \tag{5.3.36}$$

The relations between the inputs and outputs are of course then

$$b_{\mathrm{out}}(l,t) = b_{\mathrm{in}}(l,t) + \sqrt{\gamma_l}\, c_l(t). \tag{5.3.37}$$

5.3.5 Formulation of Quantum Stochastic Differential Equations

The results of the previous section, while relevant to the study of a system being driven by a noisy input from a heat bath, are not genuinely stochastic results, since no assumptions have been made concerning the density operator of the bath. We will now develop the formalism by defining *white noise* inputs as the idealized noisy input, similar to classical white noise.

a) The Quantum Wiener Process: The fields $b_{\mathrm{in}}(t)$ defined in Sect. 5.3.2–Sect. 5.3.3 will provide the input to the system described by H_{sys}. The particular quantum state or ensemble of quantum states of the "in" operators determines the nature of the input. There will always be some quantum noise arising from the zero-point fluctuations of the input, and depending on the input ensemble, there may be additional noise, such as thermal noise.

The input ensemble which corresponds most closely to a classical white noise input is not a thermal ensemble, but one in which the input density operator ρ_{in} is such that

$$\begin{aligned} \mathrm{Tr}\left\{\rho_{\mathrm{in}} b_{\mathrm{in}}^\dagger(t) b_{\mathrm{in}}(t')\right\} &\equiv \langle b_{\mathrm{in}}^\dagger(t) b_{\mathrm{in}}(t')\rangle = \bar{N}\delta(t-t'). \\ \mathrm{Tr}\left\{\rho_{\mathrm{in}} b_{\mathrm{in}}(t) b_{\mathrm{in}}^\dagger(t')\right\} &\equiv \langle b_{\mathrm{in}}(t) b_{\mathrm{in}}^\dagger(t')\rangle = (\bar{N}+1)\delta(t-t'), \end{aligned} \tag{5.3.38}$$

which corresponds to

$$\begin{aligned} \langle b_0^\dagger(\omega) b_0(\omega')\rangle &= \bar{N}\delta(\omega-\omega'), \\ \langle b_0(\omega) b_0^\dagger(\omega')\rangle &= (\bar{N}+1)\delta(\omega-\omega'). \end{aligned} \tag{5.3.39}$$

This corresponds to an ensemble in which the number of quanta per unit bandwidth is constant, and this is not the case in a thermal ensemble, in which in (5.3.39) $\bar{N}$ would be replaced by $\bar{N}(\omega)$, given by

$$\bar{N}(\omega) = 1/\left[\exp(\hbar\omega/kT) - 1\right]. \tag{5.3.40}$$

Thus, to an even larger extent than in classical stochastics, quantum white noise is an idealization, not actually attained in any real system.

To define quantum stochastic integration, we define the *quantum Wiener process* by

$$B(t, t_0) = \int_{t_0}^{t} b_{\text{in}}(t')dt' \tag{5.3.41}$$

in which we find that

$$\langle B^\dagger(t, t_0)B(t, t_0)\rangle = \bar{N}(t - t_0), \tag{5.3.42}$$

$$\langle B(t, t_0)B^\dagger(t, t_0)\rangle = (\bar{N} + 1)(t - t_0), \tag{5.3.43}$$

$$\left[B(t, t_0), B^\dagger(t, t_0)\right] = t - t_0. \tag{5.3.44}$$

In addition, we specify that the distribution of $B(t, t_0), B^\dagger(t, t_0)$ is quantum Gaussian (see Sect. 4.4.5), by which we mean that the density operator is

$$\rho(t, t_0) = (1 - e^{-\kappa}) \exp\left[-\frac{\kappa B^\dagger(t, t_0)B(t, t_0)}{t - t_0}\right] \tag{5.3.45}$$

in which

$$\bar{N} = 1/(e^\kappa - 1). \tag{5.3.46}$$

It is clear that any normal-ordered moment of order n in $B(t, t_0)$ and $B^\dagger(t, t_0)$ will be a constant times $(t - t_0)^{n/2}$.

The moments of order n with any ordering of $B(t, t_0), B^\dagger(t, t_0)$ will, as a consequence of the commutation relation (5.3.44), also always be proportional to $(t - t_0)^{n/2}$—a factor of importance in manipulating stochastic differentials.

5.3.6 Quantum Ito Stochastic Integration

In ordinary stochastic integration, there is a choice of the Ito or the Stratonovich definition. The Ito form has some mathematical advantages, which arise from the increment being independent of the integration variable. However, the rules of calculus are not those of ordinary calculus: only in the Stratonovich form is this the case.

In the quantum situation we have the added complication that variables do not commute. We can define both Ito and Stratonovich quantum stochastic integration, and can show that only the Stratonovich form preserves the rules of (noncommuting) calculus. But this can only be proved via a route involving the quantum Ito calculus, so it will be necessary to define both kinds of integration even to use the apparently simpler Stratonovich form.

If $g(t)$ is any *system operator* we define the *quantum Ito integral* by

$$(\mathbf{I}) \int_{t_0}^{t} g(t')dB(t') = \lim_{n\to\infty} \sum g(t_i)\left[B(t_{i+1}, t_0) - B(t_i, t_0)\right] \tag{5.3.47}$$

where $t_0 < t_1 < t_2 < \cdots < t_n = t$, and the limit is a mean-square limit in terms of the density operator (5.3.45). A similar definition can be used for $\int_{t_0}^{t} g(t')dB^\dagger(t')$.

The Ito increments $dB(t)$ and $dB^\dagger(t)$ commute with $g(t)$, which follows from (5.3.29) and the definition (5.3.47), since

$$\left[g(t_i), B(t_{i+1}, t_0) - B(t_i, t_0)\right] = -\sqrt{\gamma} \int_{t_i}^{t_{i+1}} dt'\, u(t_i - t') \left[g(t_i), c(t')\right] = 0. \tag{5.3.48}$$

Hence

$$(\mathbf{I}) \int_{t_0}^{t} g(t')dB(t') = (\mathbf{I}) \int_{t_0}^{t} dB(t')g(t') \tag{5.3.49}$$

and similarly for integrals with respect to $dB^\dagger(t)$.

5.3.7 Ito Quantum Stochastic Differential Equation

If $\bar{N}$ is defined as in (5.3.46), we define an Ito QSDE in the form

$$\begin{aligned}(\mathbf{I})\, da = &-\frac{\mathrm{i}}{\hbar}\left[a, H_{\mathrm{sys}}\right] dt + \frac{\gamma}{2}(\bar{N}+1)(2c^\dagger a c - a c^\dagger c - c^\dagger c a)dt \\ &+ \frac{\gamma}{2}\bar{N}(2cac^\dagger - acc^\dagger - cc^\dagger a)dt \\ &- \sqrt{\gamma}\left[a, c^\dagger\right] dB(t) + \sqrt{\gamma}\, dB^\dagger(t)\left[a, c\right].\end{aligned} \tag{5.3.50}$$

We will show that this equation is equivalent to the quantum Langevin equation (5.3.15); but first, show that the second order calculus rule appropriate to Ito integration,

$$d(ab) = a\, db + b\, da + da\, db, \tag{5.3.51}$$

is true. To do this we need the identities

$$\begin{aligned}\left[dB(t)\right]^2 = \left[dB^\dagger(t)\right]^2 &= 0, \\ dB(t)dB^\dagger(t) &= (\bar{N}+1)dt, \\ dB^\dagger(t)dB(t) &= \bar{N}\, dt.\end{aligned} \tag{5.3.52}$$

All other products, including $dt\, dB$, $dt\, dB^\dagger(t)$, and higher orders are set equal to zero. These will not be derived here, but are easy to derive in exactly the same way as in the non-quantum case.

The proof is then straightforward. We compute da and db using (5.3.50), substitute these into (5.3.51) using (5.3.52), and the result is that (after some rearrangement) $d(ab)$ as derived from the substitution $a \to ab$ in (5.3.50) is the same as the form (5.3.51). From (5.3.51) we can derive the rules of calculus for any polynomial. In the case that z is a variable which commutes with dz, we obtain

$$df(z) = f'(z)dz + \tfrac{1}{2}f''(z)dz^2 \tag{5.3.53}$$

which leads to Ito rules. Where z does not commute with dz, (5.3.53) can be interpreted in the sense that all products in a power-series expansion of $f'(z)$ and $f''(z)$ are completely symmetrized in terms of z and dz.

5.3.8 The Quantum Stratonovich Integral

The quantum Stratonovich integral is defined by

$$(\mathbf{S})\int_{t_0}^{t} g(t')dB(t') = \lim_{n\to\infty}\sum \frac{g(t_i)+g(t_{i+1})}{2}\left[B(t_{i+1},t_0) - B(t_i,t_0)\right]. \tag{5.3.54}$$

We notice that the Stratonovich increment does *not* commute with $g(t)$, and in fact, using (5.3.33), it is straightforward to show that we must take

$$(\mathbf{S})\int_{t_0}^{t} g(t')dB(t') - (\mathbf{S})\int_{t_0}^{t} dB(t')g(t') = \frac{\sqrt{\gamma}}{2}\int_{t_0}^{t} dt'\left[g(t'),c(t')\right]. \tag{5.3.55}$$

We can show this more rigorously by deriving the connection between the two kinds of stochastic integral.

5.3.9 Connection between the Ito and Stratonovich Integral

Let us assume that all operators obey the quantum Ito equation (5.3.50). Let us define $\bar{t}_i = \frac{1}{2}(t_i + t_{i+1})$, so that we can rewrite (5.3.54) as

$$\begin{aligned}(\mathbf{S})&\int_{t_0}^{t} g(t')dB(t')\\ &= \lim_{n\to\infty}\left[\sum g(\bar{t}_i)\left[B(t_{i+1}) - B(\bar{t}_i)\right] + \sum g(\bar{t}_i)\left[B(\bar{t}_i) - B(t_i)\right]\right].\end{aligned} \tag{5.3.56}$$

We then write

$$g(\bar{t}_i) = g(t_i) + dg(t_i), \tag{5.3.57}$$

where $dg(t_i)$ is obtained from (5.3.50), with

$$\begin{aligned} dt_i &= \bar{t}_i - t_i,\\ dB(t_i) &= B(\bar{t}_i) - B(t_i),\end{aligned} \tag{5.3.58}$$

which will be valid to lowest order. We then find that

$$
\begin{aligned}
(\mathbf{S}) \int_{t_0}^{t} & g(t')dB(t') \\
&= \lim_{n\to\infty} \Big\{ \sum g(\bar{t}_i)\big(B(t_{i+1}) - B(\bar{t}_i)\big) + \sum g(t_i)\big(B(\bar{t}_i) - B(t_i)\big) \\
&\quad + \sqrt{\gamma} \sum [g(t_i), c^\dagger(t_i)]\big(B(\bar{t}_i) - B(t_i)\big)\big(B(\bar{t}_i) - B(t_i)\big) \\
&\quad + \sqrt{\gamma} \sum [g(t_i), c(t_i)]\big(B^\dagger(\bar{t}_i) - B^\dagger(t_i)\big)\big(B(\bar{t}_i) - B(t_i)\big) \Big\} . \qquad (5.3.59)
\end{aligned}
$$

We now use (5.3.52) in the last part, and combine the first two terms into the Ito integral, to get

$$
(\mathbf{S}) \int_{t_0}^{t} g(t')dB(t') = (\mathbf{I}) \int_{t_0}^{t} g(t')dB(t') + \tfrac{1}{2}\sqrt{\gamma}\, \bar{N} \int_{t_0}^{t} [g(t'), c(t')]dt' \qquad (5.3.60)
$$

and similarly

$$
(\mathbf{S}) \int_{t_0}^{t} dB(t')g(t') = (\mathbf{I}) \int_{t_0}^{t} g(t')dB(t') + \tfrac{1}{2}\sqrt{\gamma}\, (\bar{N}+1) \int_{t_0}^{t} [g(t'), c(t')]dt' \qquad (5.3.61)
$$

$$
(\mathbf{S}) \int_{t_0}^{t} g(t')dB^\dagger(t') = (\mathbf{I}) \int_{t_0}^{t} g(t')dB^\dagger(t') - \tfrac{1}{2}\sqrt{\gamma}\, (\bar{N}+1) \int_{t_0}^{t} [g(t'), c^\dagger(t')]dt' \qquad (5.3.62)
$$

$$
(\mathbf{S}) \int_{t_0}^{t} dB^\dagger(t')g(t') = (\mathbf{I}) \int_{t_0}^{t} g(t')dB^\dagger(t') - \tfrac{1}{2}\sqrt{\gamma}\, \bar{N} \int_{t_0}^{t} [g(t'), c^\dagger(t')]dt' . \qquad (5.3.63)
$$

5.3.10 Stratonovich Quantum Stochastic Differential Equation

Substituting for the Ito integral implicit in (5.3.50), we find the equivalent quantum Stratonovich equations

$$
\begin{aligned}
(\mathbf{S})\ da = & -\frac{\mathrm{i}}{\hbar}[a, H_{\mathrm{sys}}]dt - \frac{\gamma}{2}\big([a, c^\dagger]c - c^\dagger[a, c]\big)\, dt \\
& - \sqrt{\gamma}\, [a, c^\dagger]dB(t) + \sqrt{\gamma}\, dB^\dagger(t)[a, c]. \qquad (5.3.64)
\end{aligned}
$$

Thus the quantum Ito equation (5.3.50) is equivalent to the quantum Stratonovich equation (5.3.64). This is exactly of the same form as the quantum Langevin equation(5.3.15), which thus justifies the definition of the Ito equation. Notice also that the commutation relations (5.3.32,33) follow from (5.3.60–63).

Finally, we can compute (**S**) $d(ab)$ from the corresponding Ito form, and readily verify that

$$(\mathbf{S})\, d(ab) = a\, db + da\, b \tag{5.3.65}$$

so that ordinary (non-commuting) calculus is valid, as expected for a Stratonovich differential. Thus the stochastic calculus for the quantum Langevin equation is that originally assumed.

5.3.11 Comparison of the Two Forms of QSDE

Let us now summarize the differences between the two forms of QSDE.

a) Stratonovich:

i) The Stratonovich form is the "natural" physical choice, since it is what arises directly from the physical considerations in Sect. 5.3.2.

ii) However, we note that the increment neither commutes with system operators, nor is it stochastically independent of them.

iii) A direct proof that ordinary calculus is true is difficult to present.

iv) Because the commutator of the increment and a system variable depends on the precise form of the QSDE, it is not possible to define the quantum Stratonovich integrals without a knowledge of the QSDE.

v) The QSDE in Stratonovich form should also be valid for nonwhite noise and is the same for any $\bar{N}$ if the noise is white. The only assumption necessary to obtain the Stratonovich QSDE is that of a constant $\kappa(\omega)$, (5.3.10).

b) Ito:

i) Not a natural physical choice.

ii) Increment commutes with and is statistically independent of system operators at the same time.

iii) Ordinary calculus is not true, but the appropriate Ito calculus is easy to derive.

iv) Because $\bar{N}$ appears in the QSDE, it is not possible to define the QSDE without knowledge of $\bar{N}$ and bath statistics, and the Ito QSDE (5.3.50) is exact only in the case of quantum white noise.

v) The Ito QSDE is thus explicitly a white noise theory, and is essentially as valid as the quantum optical master equation, to which we will shortly show it is exactly equivalent.

c) Small Noise Expansions: In Sect. 8.3.1 in the development of laser theory, we will see how macroscopic QSDE's can arise for macroscopic variables which are obtained as the sum of many microscopic variables, all obeying the same QSDE, but with independent noises. An example is the atoms in a gas laser—in this case the sum of all the noises does not increase as rapidly with the number of atoms as do the systematic terms, but only in the Ito formulation. This means that the Ito damping terms correspond to the observable macroscopic properties.

Exercise. Compare the Ito and Stratonovich formulations of the quantum stochastic differential equations for a damped two level atom. Note that if the noise terms are dropped, the stationary value of σ_z according to Ito is the expected statistical mechanics value, while Stratonovich gives $\sigma_z = -1$. Of course this does not mean Stratonovich is wrong—rather, it means that some approximations are more robust in the Ito formulation than in the Stratonovich.

5.3.12 Noise Sources of Several Frequencies

Let us go back to (5.3.15), and move to an interaction picture in which all system operators are transformed by the substitution

$$a(t) = \mathrm{e}^{\mathrm{i}H_{\mathrm{sys}}t/\hbar}\hat{a}(t)\mathrm{e}^{-\mathrm{i}H_{\mathrm{sys}}t/\hbar}. \tag{5.3.66}$$

This obviously applies to the operators $c, c^\dagger$. We now write, as in (3.6.61), an expansion of $c, c^\dagger$ in eigenoperators of H_{sys},

$$c(t) = \sum_m X_m^-, \quad c^\dagger(t) = \sum_m X_m^+ \tag{5.3.67}$$

where

$$\left[H_{\mathrm{sys}}, X_m^\pm\right] = \pm\hbar\omega_m X_m^\pm. \tag{5.3.68}$$

Thus,

$$\hat{c}(t) = \sum_m X_m^- \mathrm{e}^{-\mathrm{i}\omega_m t} \tag{5.3.69}$$

and if we substitute this in the quantum Langevin equation (5.3.15), we get

$$\begin{aligned}\frac{d\hat{a}}{dt} = &-\left[\hat{a}, \sum_m X_m^+ \mathrm{e}^{\mathrm{i}\omega_m t}\right]\left(\frac{\gamma}{2}\sum_n X_n^- \mathrm{e}^{-\mathrm{i}\omega_n t} + \sqrt{\gamma}\, b_{\mathrm{in}}(t)\right) \\ &+\left(\frac{\gamma}{2}\sum_n X_n^+ \mathrm{e}^{\mathrm{i}\omega_n t} + \sqrt{\gamma}\, b_{\mathrm{in}}^\dagger\right)\left[\hat{a}, \sum_m X_m^- \mathrm{e}^{-\mathrm{i}\omega_m t}\right].\end{aligned} \tag{5.3.70}$$

We can now make a rotating wave approximation in two forms.

i) In the terms involving products of terms like $X_m^\pm X_n^\mp$, keep only the terms with $m = n$, since other terms oscillate rapidly, and will have negligible effect.

ii) This will leave terms involving $\mathrm{e}^{\mathrm{i}\omega_m t}b_{\mathrm{in}}(t)$ or its Hermitian conjugate. Only the non-rapidly rotating parts of this can significantly affect the motion of $\hat{a}(t)$, and we can thus assume that this will behave like a white noise term corresponding to this frequency.

Thus we set

$$\mathrm{e}^{\mathrm{i}\omega_m t}b_{\mathrm{in}}(t) \to b_{\mathrm{in}}(m, t) \tag{5.3.71}$$

and

$$\left[b_{\mathrm{in}}^\dagger(m, t), b_{\mathrm{in}}(n, t)\right] = \delta_{mn}\delta(t - t') \tag{5.3.72}$$

$$\left\langle b_{\mathrm{in}}^\dagger(m, t) b_{\mathrm{in}}(n, t)\right\rangle = \bar{N}(\omega_m)\delta_{mn}\delta(t - t') \tag{5.3.73}$$

where $\bar{N}(\omega_m)$ is the Planck factor,

$$\bar{N}(\omega_m) = \{\exp(\hbar\omega_m/kT) - 1\}^{-1}. \tag{5.3.74}$$

Following the same reasoning as before, we can then identify (5.3.70) with the Ito QSDE in the Heisenberg picture

$$\begin{aligned} da = & -\frac{i}{\hbar}[a, H_{\text{sys}}] \\ & +\sum_m \frac{\gamma_m}{2}\left(\bar{N}(\omega_m)+1\right)\left(2X_m^+ a X_m^- - aX_m^+X_m^- - X_m^+X_m^- a\right) \\ & +\sum_m \frac{\gamma_m}{2}\bar{N}(\omega_m)\left(2X_m^- a X_m^+ - aX_m^-X_m^+ - X_m^-X_m^+ a\right) \\ & -\sum_m \sqrt{\gamma_m}\,[a, X_m^+]dB(m,t) + \sum_m \sqrt{\gamma_m}\,dB^\dagger(m,t)[a, X_m^-]. \end{aligned} \tag{5.3.75}$$

This form is written for generality—the derivation actually yields $\gamma_m = \gamma$ for all m. The derivation of the master equation just given then carries over, and we find a master equation equivalent to (3.6.67), except that we must set

$$\frac{\pi\omega_m}{\hbar}\kappa(\omega_m)^2 = \gamma. \tag{5.3.76}$$

Thus the full generality of the quantum optical master equation can only be obtained by allowing γ to depend on frequency—essentially this means dropping the first Markov approximation. This is in fact not too difficult to do, so that we may actually allow γ_m in (5.3.75) to be different from each other.

The resulting equation will be valid when the coupling to the bath is weak and each possible atomic transition will be driven by a noise source tuned to its frequency, and independent of the sources corresponding to other transitions.

5.4 The Master Equation

In this section we will show that the quantum optical master equation can be derived directly from the QSDE. Unlike the case of the more exact quantum Langevin equations of Chap.3, the equivalence is (in the white noise case) exact. This means that the short correlation time approximation has already been made by assuming quantum white noise.

5.4.1 Description of the Density Operator

We consider uncorrelated initial conditions, that is, we assume that the density matrix can be written as a direct product

$$\rho = \rho_s(t_0) \otimes \rho_B(t_0). \tag{5.4.1}$$

where x_{mn} are certain complex numbers, and that the equations of motion for the matrix elements of the density operator are

$$\frac{d}{dt}\langle s|\rho|r\rangle = \Bigg\{ i\omega_{rs} - \sum_{n<r} \frac{\gamma_{rn}}{2}[\bar{N}(\omega_{rn})+1]|x_{rn}|^2 - \sum_{n<s} \frac{\gamma_{sn}}{2}[\bar{N}(\omega_{sn})+1]|x_{sn}|^2$$
$$- \sum_{m>r} \frac{\gamma_{mr}}{2}\bar{N}(\omega_{mr})|x_{mr}|^2 - \sum_{m>s} \frac{\gamma_{ms}}{2}\bar{N}(\omega_{ms})|x_{ms}|^2 \Bigg\} \langle s|\rho|r\rangle$$
$$+\delta_{rs}\Bigg\{ \sum_{m>r} \gamma_{mr}[\bar{N}(\omega_{mr})+1]|x_{mr}|^2\langle m|\rho|m\rangle + \sum_{n<r} \gamma_{rn}\bar{N}(\omega_{rn})|x_{mr}|^2\langle n|\rho|n\rangle \Bigg\}. \tag{5.4.16}$$

Notice that the off-diagonal matrix elements obey damped uncoupled equations of motion. The diagonal matrix elements are occupation probabilities, $p_r = \langle r|\rho|r\rangle$. For these probabilities the equation becomes

$$\dot{p}_r = \sum_{m>r} \gamma_{mr}|x_{mr}|^2\{p_m[\bar{N}(\omega_{mr})+1] - p_r\bar{N}(\omega_{mr})\}$$
$$+\sum_{n<r} \gamma_{nr}|x_{nr}|^2\{p_n\bar{N}(\omega_{rn}) - p_r[\bar{N}(\omega_{mr})+1]\}. \tag{5.4.17}$$

This is an ordinary stochastic master equation—essentially the Pauli master equation of Sect. 1.5—whose solutions approach a stationary state given by

$$p_r(\mathrm{ss}) \propto \exp(-\hbar\omega_r/kT). \tag{5.4.18}$$

5.4.5 Equivalence of QSDE and Master Equation

For every master equation we can write an equivalent QSDE, and conversely. However the master equation can be derived in situations where the heat bath is *not* composed of harmonic oscillators, as was demonstrated in Sect. 5.1.4. Nevertheless, in this case the *effect* of the heat bath can be represented via a QSDE, which implicitly defines an equivalent harmonic oscillator heat bath. This equivalent form is very useful when discussing inputs and outputs to a field coupled to a system which is also damped by a coupling to a more general bath. By formulating the heat bath as an equivalent QSDE with appropriate inputs and outputs, all sources of damping and noise acquire an equivalent status—all of this in the Markovian limit, of course.

5.4.6 Correlation Functions of Inputs, System, and Outputs

In the case of an optical system the input fields $b_{\mathrm{in}}(t)$ are identified as being proportional to appropriate modes of the electromagnetic field impingent on some reasonably well localized system, such as an atom. Thus it is a very reasonable question in such a situation to ask how specified input statistics are transformed into the statistics of the output field. This is a question which has only recently become important to answer, with the advent of *squeezing*, and it was for this purpose that the input-output formalism was derived.

We will normally specify the correlation functions of the input, and will wish to compute those of the system and those of the output. From this will arise two principal problems.

First, given the correlation functions of the input, how do we compute those of the system? In certain situations we know the answer. If the QSDE for the system is linear, it can be explicitly solved, and correlations directly calculated. It does not matter for linear systems what kind of statistics the input has, a direct solution for internal modes and output is possible.

In nonlinear situations we can get a master equation exactly equivalent to the internal QSDE, *provided* the input field consists of ordinary white quantum noise, or squeezed white noise. In practice this cannot be exactly true, either because the noise is not white, but has a nonflat spectrum, or because the particular model we have formulated is not exactly valid. However, there *are* many cases where the white-noise approximation is reasonably valid, and the internal modes of these can be treated using the master equation and the numerous techniques available for that. Thus the correlation functions of the internal modes can be calculated in very many cases.

The second consideration is how to compute the correlation functions of the output. We will show that these can be related directly to those of the internal modes, but are not identical in all cases.

a) Output Correlation Functions: The output correlation functions of most interest are the normally ordered correlation functions of the form

$$\langle b_{\text{out}}^{\dagger}(t_1)b_{\text{out}}^{\dagger}(t_2)\dots b_{\text{out}}^{\dagger}(t_n)b_{\text{out}}(t_{n+1})\dots b_{\text{out}}(t_m)\rangle . \tag{5.4.19}$$

It does not in fact matter what time ordering is chosen in (5.4.19) since all the $b_{\text{out}}(t_r)$ commute with each other. Products which are not normally ordered can be related to normally ordered products by use of the commutation relations for the output field, which we have shown are the same as those for the input field.

Thus we may consider with complete generality that the correlation function is of the form (5.4.19) and that the $b_{\text{out}}^{\dagger}(t_r)$ are time *antiordered* (i.e., $t_1 < t_2 < \cdots < t_n$) and the $b_{\text{out}}(t_r)$ are *time ordered* (i.e., $t_{n+1} > t_{n+2} > \cdots > t_m$).

We may now substitute for $b_{\text{out}}(t_r)$ by the relation (5.3.24)

$$b_{\text{out}}(t_r) = \sqrt{\gamma}\, c(t_r) + b_{\text{in}}(t_r). \tag{5.4.20}$$

We note that from (5.3.32,33) $c^{\dagger}(t_r)$ will commute with all $b_{\text{in}}^{\dagger}(t_s)$ which occur to the right of $c^{\dagger}(t_r)$, since we have shown $t_r < t_s$. Similarly, $c(t_{r'})$ will commute with all $b_{\text{in}}(t_{s'})$ which occur to the left of $c(t_{r'})$, since we have chosen $t_{r'} > t_{s'}$. Thus we may write the correlation function in a form in which we make the substitution (5.4.20) in (5.4.19), and reorder the terms so that

i) the $c^{\dagger}(t_r)$ are time *antiordered*, and are to the left of all $c(t_r)$;

ii) the $c(t_{r'})$ are *time ordered*;

iii) the $b_{\text{in}}^{\dagger}(t_s)$ all stand to the left of all other operators; and

iv) the $b_{\text{in}}(t_{s'})$ all stand to the right of all other operators.

We now consider various cases.

b) Input in the Vacuum State: In this case, $b_{\text{in}}(t)\rho_{\text{in}} = \rho_{\text{in}}b^\dagger_{\text{in}}(t) = 0$, and we derive

$$\langle b^\dagger_{\text{out}}(t_1)\, b^\dagger_{\text{out}}(t_2) \dots b^\dagger_{\text{out}}(t_n) b_{\text{out}}(t_{n+1}) \dots b_{\text{out}}(t_m)\rangle =$$

$$\gamma^{(m+n)/2} \left\langle \tilde{T}\left[c^\dagger(t_1)c^\dagger(t_2)\dots c^\dagger(t_n)\right] T\left[c(t_{n+1})\dots c(t_m)\right]\right\rangle , \qquad (5.4.21)$$

where $\tilde{T}$ is the time-antiordered product and T the time-ordered product.
c) Input is in a Coherent State: In this case

$$\begin{aligned} b_{\text{in}}(t)\rho_B &= \beta(t)\rho_B, \\ \rho_B b^\dagger_{\text{in}}(t) &= \beta^*(t)\rho_B, \end{aligned} \qquad (5.4.22)$$

where $\beta(t)$ is the coherent field amplitude; a c-number. In this case we may replace $b, b^\dagger$ by β, β^*, respectively, and since β and β^* are mere numbers, reorder them so that

$$\begin{aligned} &\langle b^\dagger_{\text{out}}(t_1)\, b^\dagger_{\text{out}}(t_2) \dots b^\dagger_{\text{out}}(t_n) b_{\text{out}}(t_{n+1}) \dots b_{\text{out}}(t_m)\rangle \\ &\quad = \Big\langle \tilde{T}\{[\sqrt{\gamma}\, c^\dagger(t_1) + \beta^*(t_1)] \dots\} T\{[\sqrt{\gamma}\, c(t_{n+1}) + \beta(t_{n+1})] \\ &\qquad\quad \dots [\sqrt{\gamma}\, c(t_m) + \beta(t_m)]\}\Big\rangle . \end{aligned} \qquad (5.4.23)$$

d) The Input is Quantum White Noise: By carrying out this procedure we are left with the problem of evaluating terms like $\langle c^\dagger(t_1)c^\dagger(t_2)b_{\text{in}}(s_1)b_{\text{in}}(s_2)\rangle$ when the density matrix is not coherent or vacuum, but represents a more general state. The solution of this problem for arbitrary statistics of the input is rather difficult, but we can give a way of doing this calculation in the case that the input is quantum white noise (clearly, this will also allow us to carry out the computation when the input also has an additional coherent part). This requires us, more generally, to calculate terms like $\langle a_1(t)a_2(t')dB(s)dB(s')\rangle$ where $a_1(t)$ and $a_2(t)$ obey a single QSDE of the form (Ito)(5.3.50) which we can write in an abbreviated form

$$d\boldsymbol{a} = A\big(\boldsymbol{a}(t)\big)\, dt + G^\dagger\big(\boldsymbol{a}(t)\big)\, dB(t) + G\big(\boldsymbol{a}(t)\big)\, dB^\dagger(t). \qquad (5.4.24)$$

(We shall only treat the case where the simple master equation (5.4.12) is relevant. The more general situation of (3.6.67) is essentially a many-input version of the same procedure; one input and one output being available for each frequency band around the ω_m). For simplicity, let us consider first the evaluation of $\langle a(t')dB(t)\rangle$. If $t > t'$, then $dB(t)$ is independent of $a(t')$, and so we deduce

$$\langle a(t')dB(t)\rangle = 0, \quad t > t'. \qquad (5.4.25)$$

For $t < t'$, we argue as follows. We discretize time and define

$$\begin{aligned} a_0 &= a(t), & a_n &= a(t_n), \\ A_n &= A(\boldsymbol{a}_n), & G_n &= G(\boldsymbol{a}_n), \end{aligned} \qquad (5.4.26)$$

so that (because we use the Ito form of the QSDE)

$$a_n = a_{n-1} + A_{n-1}\Delta t_n + G^\dagger_{n-1}\Delta B_n + G_{n-1}\Delta B^\dagger_n . \qquad (5.4.27)$$

To solve the QSDE, we repeatedly use (5.4.27) starting with the initial condition a_0, and we can eventually write

$$a_n = \sum_{r,s} a_n^{r,s} \{\Delta B_1^\dagger\}^r \{\Delta B_1\}^s, \tag{5.4.28}$$

where the coefficients depend on a_0 and $\Delta B_m^\dagger, \Delta B_{m'}$, for $1 \leq m, m' \leq n$. The discretized form of $\langle a(t')dB(t)\rangle$ is (to lowest order in Δt)

$$\langle a_n \Delta B_1 \rangle = \langle a_n^{1,0} \rangle \langle \Delta B_1^\dagger \Delta B_1 \rangle = \bar{N} \Delta t_1 \langle a_n^{1,0} \rangle. \tag{5.4.29}$$

We now need an expression for $\langle a_n^{1,0} \rangle$. Suppose we modify the QSDE (5.4.24) by the substitution

$$\begin{aligned} dB(t) &\to dB(t) + \epsilon(t)dt, \\ dB^\dagger(t) &\to dB^\dagger(t) + \epsilon^*(t)dt, \end{aligned} \tag{5.4.30}$$

where $\epsilon(t)$ is a classical c-number driving field. Then discretizing again, it is clear that

$$\langle a_n^{1,0} \rangle = \frac{1}{\Delta t_1} \left\langle \frac{\partial}{\partial \epsilon_1^*} a_n^\epsilon \right\rangle_{\epsilon=0}, \tag{5.4.31}$$

and in a continuum notation this becomes

$$\langle a(t')dB(t)\rangle = \left[\bar{N} \frac{\delta}{\delta \epsilon^*(t)} \langle a^\epsilon(t') \rangle \right]_{\epsilon=0}. \tag{5.4.32}$$

We now evaluate this expression. We note that from the master equation (5.4.12)

$$\begin{aligned} \langle a^\epsilon(t') \rangle &= \langle a \rho^\epsilon(t') \rangle \\ &= \left\langle aT \left[\exp\left(\int_t^{t'} \hat{L}^\epsilon(s) ds \right) \right] \rho(t) \right\rangle \end{aligned} \tag{5.4.33}$$

and, discretizing, this can be written

$$\langle a_n^\epsilon \rangle = \langle a(1 + \hat{L}_n^\epsilon \Delta t_n)(1 + \hat{L}_{n-1}^\epsilon \Delta t_{n-1}) \dots (1 + \hat{L}_1^\epsilon \Delta t_1)\rho(t) \rangle. \tag{5.4.34}$$

But only the term involving L_1^ϵ depends on ϵ_1^*, and, in fact, the coefficient of ϵ_1^* in $L_1^\epsilon \rho$ is $\sqrt{\gamma}\ [c, \rho]$, so that

$$\frac{\partial}{\partial \epsilon_1^*} \langle a_n^\epsilon \rangle = \langle a(1 + L_n^\epsilon \Delta t_n) \dots (1 + L_2^\epsilon \Delta t_2) \times \sqrt{\gamma}\ [c, \rho(t)] \rangle \tag{5.4.35}$$

and we then find

$$\begin{aligned} \left[\frac{\delta}{\delta \epsilon^*(t)} \langle a^\epsilon(t') \rangle \right]_{\epsilon=0} &= \sqrt{\gamma}\ \langle a \exp[\hat{L}(t' - t)][c, \rho(t)] \rangle \\ &= \sqrt{\gamma}\ \langle [a(t'), c(t)] \rangle. \end{aligned} \tag{5.4.36}$$

Thus, we find the general expression

$$\langle a(t')dB(t)\rangle = \sqrt{\gamma}\,\bar{N}\,dt\,u(t'-t)\langle[a(t'),c(t)]\rangle. \tag{5.4.37}$$

We therefore derive in general by similar methods

$$\left.\begin{aligned}\langle a(t')b_{\text{in}}(t)\rangle &= \sqrt{\gamma}\,\bar{N}u(t'-t)\langle[a(t'),c(t)]\rangle\\ \langle b_{\text{in}}^{\dagger}(t)a(t')\rangle &= \sqrt{\gamma}\,\bar{N}u(t'-t)\langle[c^{\dagger}(t),a(t')]\rangle,\\ \langle a(t')b_{\text{in}}^{\dagger}(t)\rangle &= \sqrt{\gamma}\,(\bar{N}+1)u(t'-t)\langle[c^{\dagger}(t),a(t')]\rangle,\\ \langle b_{\text{in}}(t)a(t')\rangle &= \sqrt{\gamma}\,(\bar{N}+1)u(t'-t)\langle[a(t'),c(t)]\rangle\end{aligned}\right\} \tag{5.4.38}$$

and we note that these are consistent with the commutation relations (5.3.32,33). We can now use the results of Sect. 5.3.3 to derive

$$\langle b_{\text{out}}(t)b_{\text{out}}(t')\rangle = \gamma(\bar{N}+1)\langle T[c(t)c(t')]\rangle - \gamma\bar{N}\langle\tilde{T}[c(t)c(t')]\rangle, \tag{5.4.39}$$

$$\langle b_{\text{out}}^{\dagger}(t)b_{\text{out}}(t')\rangle = \gamma(\bar{N}+1)\langle c^{\dagger}(t)c(t')\rangle - \gamma\bar{N}\langle c(t')c^{\dagger}(t)\rangle + \bar{N}\delta(t-t'), \tag{5.4.40}$$

$$\langle b_{\text{out}}^{\dagger}(t)b_{\text{out}}^{\dagger}(t')\rangle = \gamma(\bar{N}+1)\langle\tilde{T}[c^{\dagger}(t)c^{\dagger}(t')]\rangle - \gamma\bar{N}\langle T[c^{\dagger}(t)c^{\dagger}(t')]\rangle. \tag{5.4.41}$$

Comments

i) It is very interesting to see that the only correlation functions of system operators which arise in this procedure are the kind of time ordered correlation functions which arise out of a consideration of the most general kind of measurement as given in (2.3.12). *Barchielli* [5.11] has shown how the process of measuring the correlation function can be viewed as a kind of quantum measurement. Essentially, the quantum field $b(t)$ can be regarded as a kind of physical probe, which gives information about the system.

ii) Indeed, even if we do not use Barchielli's viewpoint, it is certainly true in practice that the usual way of getting information about atoms is to shine light, microwaves, or perhaps sound on it, and to measure a scattered or fluorescent field of some kind. Thus, this *is* a measurement theory, and it is pleasing to see that it is consistent with that of Chap.2.

6. Applying the Master Equation

The previous chapter can be regarded as the statement of the mathematics of quantum Markov processes, and this chapter will show how the theory can be applied to a number of situations. It starts with an analysis of solutions of the master equation using number state matrix elements, and then moves into the very powerful methods based on the phase space representations introduced in Chap.4. Using these solutions, one is able to show how the quantum theory of damping and noise gives some very reasonable explanations of how classical physics arises as a macroscopic limit of quantum physics.

It finally introduces the ideas of generalized P-representations, introduced by the *Gardiner* and *Drummond*, by which many quantum Markov processes may be represented as classical diffusion processes. These have formed a very powerful tool in the practical solution of practical master equations.

6.1 Using the Number State Basis

For some simple systems, solution of the master equation in a number state basis is the most practical method. This section gives a few simple examples.

6.1.1 The Damped Harmonic Oscillator—Quantum Optical Case

We take the master equation in the form (3.6.67). In the case of the harmonic oscillator, we can write (using the notation of (4.1.32))

$$X \to x = \frac{(a + a^\dagger)}{\sqrt{2\eta}} \tag{6.1.1}$$

and it is clear that there is only one X_m^+ and one X_m^-; thus

$$\begin{aligned} X_m^- &= a/\sqrt{2\eta} \\ X_m^+ &= a^\dagger/\sqrt{2\eta} \end{aligned} \tag{6.1.2}$$

and the master equation takes the form (in the Schrödinger picture)

$$\begin{aligned} \dot{\rho} = &-\mathrm{i}\omega[a^\dagger a, \rho] \\ &+\tfrac{1}{2}K(\bar{N}+1)(2a\rho a^\dagger - a^\dagger a\rho - \rho a^\dagger a) \\ &+\tfrac{1}{2}K\bar{N}(2a^\dagger\rho a - aa^\dagger\rho - \rho aa^\dagger) \end{aligned} \tag{6.1.3}$$

in which

$$\begin{aligned} &\bar{N}(\omega) \to \bar{N} \\ &\pi\kappa(\omega)^2/2\hbar\eta \to K. \end{aligned} \tag{6.1.4}$$

Diagonal Matrix Elements:

i) The diagonal matrix element

$$\langle n|\rho|n\rangle \equiv P(n) \tag{6.1.5}$$

represents the probability of there being n quanta in the system. We easily check that (using the properties of $a^\dagger$ and a defined in Sect. 4.1.3

$$\begin{aligned} \partial_t P(n) = K(\bar{N}+1)[(n+1)P(n+1) - nP(n)] \\ +K\bar{N}[nP(n-1) - (n+1)P(n)]. \end{aligned} \tag{6.1.6}$$

Exercise. Show that the equations for the coherences $\langle n|\rho|m\rangle$ decouple, so that elements of the form $\langle n|\rho|n+r\rangle$ for a fixed r have equations independent of those for a different r.

ii) The fact that (6.1.6) is a version of a classical Markov process (a birth-death process) indicates that the connection between the classical and quantum version of a Markov process is more than mere formalism. In fact (6.1.6) is also a version of the Pauli master equation, (1.5.1).

iii) Notice that the transition probabilities of this classical master equation can be written in the form (in the notation of S.M. Chap.7)

$$\begin{aligned} t^+(n) &= K\bar{N}(n+1) \\ t^-(n) &= K(\bar{N}+1)n \end{aligned} \tag{6.1.7}$$

so that the probability of creating a quantum has a part proportional to $n+1$. A chemical reaction of the form

$$\begin{aligned} &B \rightleftharpoons X \\ &A + X \to 2X \end{aligned} \tag{6.1.8}$$

would have a similar master equation.

iv) The solution in the stationary state is

$$P_s(n) = \left(\frac{\bar{N}}{1+\bar{N}}\right)^n \frac{1}{1+\bar{N}}. \tag{6.1.9}$$

This is the usual Boltzmann distribution, in which one can identify

$$\frac{\bar{N}}{1+\bar{N}} = \exp(-\hbar\omega/kT) \tag{6.1.10}$$

which means

$$\bar{N} = \frac{1}{\exp(\hbar\omega/kT) - 1} \tag{6.1.11}$$

which determines $\bar{N}$ in terms of T , or conversely. We note that

$$\begin{aligned} \langle n\rangle_s &= \bar{N} \\ \mathrm{var}\{n\}_s &= \bar{N}^2. \end{aligned} \tag{6.1.12}$$

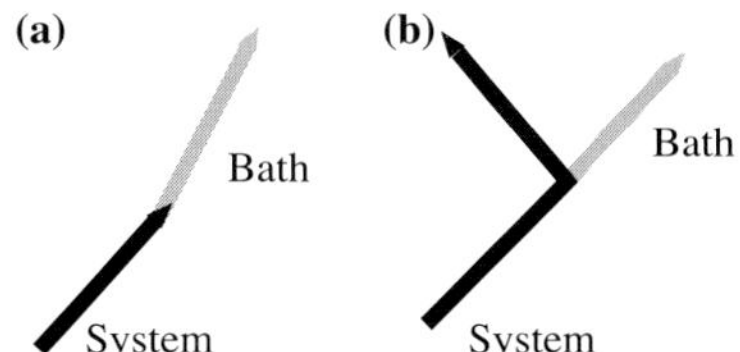

Fig. 6.1 Comparison of standard damping **(a)**, in which a system quantum is destroyed while producing a bath quantum (and conversely), and **(b)** phase damping, in which a system quantum is scattered when it absorbs or emits a bath quantum

6.1.2 The Phase Damped Oscillator

The damped harmonic oscillator of the previous section has a basic system-bath Hamiltonian of the form

$$H_{\text{Int}} = \sum_i \left\{ \Gamma_i a^\dagger + \Gamma_i^\dagger a \right\}, \tag{6.1.13}$$

which implies that an oscillator quantum can lose energy by creating a bath quantum, and conversely. This kind of damping transfers energy between the bath and the system.

It is possible to have an interaction which can be viewed as a *scattering* process, with an interaction Hamiltonian

$$H_{\text{Int}} = \sum_i \left\{ \Gamma_i a^\dagger a + \Gamma_i^\dagger a^\dagger a \right\} \tag{6.1.14}$$

in which a bath quantum can be absorbed or emitted, but in this case the number of oscillator quanta remains unchanged. The situation is illustrated diagrammatically in Fig.6.1. Such an interaction commutes with the system Hamiltonian, so no energy is transferred—only the phase of the system state is changed. The master equation is quite simple

$$\dot{\rho} = -i\omega[a^\dagger a, \rho] + \tfrac{1}{2}K(2\bar{N}+1)\{2a^\dagger a \rho a^\dagger a - (a^\dagger a)^2\rho - \rho(a^\dagger a)^2\} \tag{6.1.15}$$

and the equations of motion for the matrix elements $\langle n|\rho|m\rangle$ are

$$\frac{d\langle n|\rho|m\rangle}{dt} = \{-i\omega(n-m) - \tfrac{1}{2}K(2\bar{N}+1)(n-m)^2\}\langle n|\rho|m\rangle \tag{6.1.16}$$

with the solution

$$\langle n|\rho(t)|m\rangle = e^{-i\omega(n-m)t} e^{-(2\bar{N}+1)K(n-m)^2 t/2}\langle n|\rho(0)|m\rangle. \tag{6.1.17}$$

Comments:

i) Notice that the off-diagonal matrix elements decay away to zero, leaving a stationary diagonal solution, which in fact is

$$\langle n|\rho_s|m\rangle = \delta_{nm}\langle n|\rho(0)|m\rangle. \tag{6.1.18}$$

Thus this interaction produces no thermalization—its effect is to preserve the occupation probabilities, while erasing all coherences.

ii) The off-diagonal matrix elements decay at a rate proportional to $(n - m)^2$—that is, the further from the diagonal, the faster the damping. If the system is *macroscopic* $n - m$ will be enormous, so this decay will be very fast. Thus it would be almost impossible to observe the coherences in such a case, because of the very rapid decay.

6.2 Quantum Classical Correspondence

An alternative method of developing the idea of a quantum Markov process is to use the phase space methods of Chap.4 to derive phase space equations of motion which are very often of the classical Fokker-Planck form. The interpretation of these Fokker-Planck equations is not always straightforward, but the method is a very powerful technique for finding solutions to many quantum Markov problems. Since the phase space equations are classical non-operator equations, a wide variety of standard analytical techniques can be employed for the solution.

This section will concentrate on a number of specific examples, starting with the simplest, the harmonic oscillator.

6.2.1 Use of the P-Representation

We have seen in Chap.3 that a density operator can be expressed in terms of quasiprobability $P(\alpha)$, which behaves rather like a classical probability, though it need not always be positive or even exist as an ordinary function. If we now express the density matrix under study here as a P-representation

$$\rho = \int d^2\alpha |\alpha\rangle\langle\alpha| P(\alpha), \tag{6.2.1}$$

then using the operator correspondences of (4.5.9) it is easily shown that the quantum master equation (6.1.3) leads to the Fokker-Planck equation (remember that operator products written on the right of ρ are reversed in order when the correspondence is made)

$$\frac{\partial P}{\partial t} = \left\{ \tfrac{1}{2}K \left(\frac{\partial}{\partial\alpha}\alpha + \frac{\partial}{\partial\alpha^*}\alpha^* \right) - i\omega \left(\frac{\partial}{\partial\alpha}\alpha - \frac{\partial}{\partial\alpha^*}\alpha^* \right) + K\bar{N}\frac{\partial^2}{\partial\alpha\partial\alpha^*} \right\} P. \tag{6.2.2}$$

a) Interpretation as a Classical Fokker-Planck Equation: This is a form of complex Ornstein-Uhlenbeck process. To show this we write

$$\alpha = x + iy \tag{6.2.3}$$

and

$$\frac{\partial}{\partial\alpha} = \frac{1}{2}\left(\frac{\partial}{\partial x} - i\frac{\partial}{\partial y} \right) \tag{6.2.4}$$

and get

$$\frac{\partial P}{\partial t} = \left\{\omega\left(\frac{\partial}{\partial x}y - \frac{\partial}{\partial y}x\right) + \frac{K}{2}\left(\frac{\partial}{\partial x}x + \frac{\partial}{\partial y}y\right) + \frac{K\bar{N}}{2}\left(\frac{\partial^2}{\partial x^2} + \frac{\partial^2}{\partial y^2}\right)\right\} P. \tag{6.2.5}$$

b) Stochastic Differential Equations: The classical stochastic differential equations equivalent to (6.2.5) are

$$\begin{aligned} dx &= -(\tfrac{1}{2}Kx + \omega y)dt + \sqrt{\tfrac{1}{2}K\bar{N}}\; dW_1(t) \\ dy &= -(\tfrac{1}{2}Ky - \omega x)dt + \sqrt{\tfrac{1}{2}K\bar{N}}\; dW_2(t). \end{aligned} \tag{6.2.6}$$

These represent the equations for a damped oscillator, very like those which would occur classically. They are often written as one complex Langevin equation:

$$d\alpha = -(\tfrac{1}{2}K - \mathrm{i}\omega)\alpha\, dt + \sqrt{K\bar{N}}\;\, d\eta(t), \tag{6.2.7}$$

where $d\eta(t)$ is the increment of a complex Wiener process satisfying

$$\begin{aligned} &\langle d\eta(t)\rangle = \langle d\eta^*(t)\rangle = \langle d\eta(t)d\eta(t')\rangle = \langle d\eta^*(t)d\eta^*(t')\rangle = 0 \\ &\langle d\eta(t)d\eta^*(t)\rangle = dt \end{aligned} \tag{6.2.8}$$

and explicitly given by

$$d\eta(t) = [dW_1(t) + \mathrm{i}dW_2(t)]/\sqrt{2}\,. \tag{6.2.9}$$

The solutions of the Ornstein-Uhlenbeck process are given in S.M. Sect. 4.4.6. We find

$$\begin{aligned} &\langle\alpha(t)\rangle - \alpha(0)\exp[-(K/2 - \mathrm{i}\omega)t] \\ &\langle\alpha^*(t)\alpha(t)\rangle = \langle\alpha^*(0)\alpha(0)\rangle \mathrm{e}^{-Kt} + \bar{N}(1 - \mathrm{e}^{-Kt}) \end{aligned} \tag{6.2.10}$$

and the stationary variances are

$$\begin{aligned} &\langle\alpha^2\rangle_s = \langle\alpha^{*2}\rangle_s = 0 \\ &\langle\alpha\alpha^*\rangle_s = \langle a^\dagger a\rangle_s = \bar{N}. \end{aligned} \tag{6.2.11}$$

Only when $\bar{N} = 0$ do we find that $\langle a^\dagger a\rangle_s$ vanishes, i.e., at zero temperature.

c) Initial Coherent State: If the oscillator is initially in a coherent state $|\alpha(0)\rangle$, the P-function is initially of the form $\delta^2(\alpha - \alpha(0))$. This corresponds to the initial condition for the SDE (6.2.7) having zero variance. Since the equation of motion (6.2.7) has zero noise at zero temperature, $\alpha(t)$ will have zero variance for all time. Thus the time dependent solution for the P-function will be $\delta^2(\alpha - \alpha(t))$, and hence the time dependent solution for the state will be the time dependent coherent state $|\alpha(t)\rangle$.

This result is generalizable to multivariable harmonic systems with damping which arises from zero temperature absorption of single quanta—in such situations, an initial coherent state remains coherent. For example, it will be shown in Chap.8 that a coherent light beam passing through a beam splitter remains coherent, and that detection of photons also gives this picture of damping.

d) Limitations on the Classical Interpretation: At first glance, this appears to yield a complete equivalence between a classical and a quantum system, but this is not so. The most obvious difference is the fact that the probability density which is the solution of the FPE is not required to be positive. Thus, there are initial conditions which cannot be realized classically. The more subtle point is a question of the exact relationship between the conditional probability, for which the Fokker-Planck equation is the evolution equation, and the Markovian nature of the quantum process. Put at its simplest—from the FPE we can calculate $\langle\alpha(t+\tau)\alpha(t)\rangle$ and similar averages—how do these relate to the measurable quantum quantities?

6.2.2 Time Correlation Functions in the P-Representation

As a first example, consider the correlation functions

$$\langle a^\dagger(t+\tau)a(t)\rangle = \mathrm{Tr}\left\{a^\dagger V(t+\tau,t)\{a\rho(t)\}\right\} \tag{6.2.12}$$

where we have used (5.2.11) to write the correlation function in terms of the evolution operator of the quantum Markov process. Using the P-representation for $\rho(t)$, we can write:

$$\langle a^\dagger(t+\tau)a(t)\rangle = \mathrm{Tr}\left\{a^\dagger V(t+\tau,t)\int d^2\alpha\,\alpha P(\alpha,t)|\alpha\rangle\langle\alpha|\right\} \tag{6.2.13}$$

where the operator correspondence (4.5.9) has been used. The term

$$V(t+\tau,t)\int d^2\alpha\,\alpha P(\alpha,t)|\alpha\rangle\langle\alpha| \tag{6.2.14}$$

is a solution of the master equation, for which the initial condition at time t is the operator whose P-function is $\alpha P(\alpha,t)$. However, using the FPE (6.2.2), this can be written

$$\int d^2\alpha'\,\alpha' P(\alpha,t+\tau|\alpha',t)P(\alpha',t) \tag{6.2.15}$$

where $P(\alpha,t+\tau|\alpha',t)$ is the conditional probability which is a solution of the FPE—i.e., the solution of the FPE with the initial condition $\delta^2(\alpha-\alpha')$. Substituting into (6.2.13), and using the cyclic property of the trace, we find

$$\langle a^\dagger(t+\tau)a(t)\rangle = \mathrm{Tr}\{a^\dagger\int d^2\alpha\int d^2\alpha'\,\alpha' P(\alpha,t+\tau|\alpha',t)P(\alpha',t)|\alpha\rangle\langle\alpha|\} \tag{6.2.16}$$

$$= \int d^2\alpha\int d^2\alpha'\alpha^*\alpha' P(\alpha,t+\tau\,;\alpha',t) \tag{6.2.17}$$

and hence

$$\langle a^\dagger(t+\tau)a(t)\rangle = \langle \alpha^*(t+\tau)\alpha(t)\rangle . \tag{6.2.18}$$

This result is very simple and reinforces the feeling that there is a direct mapping from the quantum process to a classical process. However, on the left hand side of (6.2.18), the operators do not commute, while the c-numbers $\alpha(t+\tau)$ and $\alpha(t)$ on the right hand side do commute. So let us now calculate

$$\langle a(t)a^\dagger(t+\tau)\rangle = \mathrm{Tr}\{\{V(t+\tau,t)\rho a\}a^\dagger\} \tag{6.2.19}$$

where we have followed the prescription of Sect. 5.2.1b. Using the same procedure as above,

$$(6.2.19) = \mathrm{Tr}\left\{\left(V(t+\tau,t)\int d^2\alpha\left[\left(\alpha - \frac{\partial}{\partial\alpha^*}\right)P(\alpha)\right]|\alpha\rangle\langle\alpha|\right)a^\dagger\right\} \tag{6.2.20}$$

and finally

$$\int d^2\alpha' \int d^2\alpha \left\{\left(\alpha^* - \frac{\partial}{\partial\alpha}\right)P(\alpha,t+\tau|\alpha',t)\right\}\left(\alpha' - \frac{\partial}{\partial\alpha'^*}\right)P(\alpha',t). \tag{6.2.21}$$

In a certain formal sense, this can be written

$$\langle a(t)a^\dagger(t+\tau)\rangle = \left\langle \left[\alpha^* - \frac{\partial}{\partial\alpha}\right](t+\tau)\left[\alpha - \frac{\partial}{\partial\alpha^*}\right](t)\right\rangle \tag{6.2.22}$$

but it must be noted that the notation in (6.2.22) must be interpreted as merely a shorthand for (6.2.21). In particular, the fact that $\alpha' - \partial/\partial\alpha'^*$ acts only on $P(\alpha',t)$ and *not* on $P(\alpha,t+\tau|\alpha',t)$ is not clear from (6.2.22). If we introduce the notation

$$\langle \alpha^*, t+\tau|[\alpha',t]\rangle = \int d^2\alpha\, \alpha^* P(\alpha,t+\tau|\alpha',t), \tag{6.2.23}$$

which is the mean of α^* at time $t+\tau$ conditional on it having the value α' at time t, then after integrating by parts and discarding surface terms, (6.2.21) can be written

$$\langle a(t)a^\dagger(t+\tau)\rangle = \langle \alpha^*(t+\tau)\alpha(t)\rangle + \int d^2\alpha' \frac{\partial}{\partial\alpha'^*}\langle \alpha^*, t+\tau|[\alpha',t]\rangle P(\alpha',t). \tag{6.2.24}$$

6.2.3 Application to the Damped Harmonic Oscillator

Using standard methods, from the equations (6.2.2) or (6.2.7) we find, in the stationary situation,

$$\langle \alpha^*(t+\tau)\alpha(t)\rangle = \bar{N}\mathrm{e}^{-(K/2+\mathrm{i}\omega)\tau} \tag{6.2.25}$$

$$\langle \alpha^*, t+\tau|[\alpha',t]\rangle = \alpha'^*\mathrm{e}^{-(K/2+\mathrm{i}\omega)\tau} \tag{6.2.26}$$

so that

$$\begin{aligned} \langle a^\dagger(t+\tau)a(t)\rangle &= \bar{N}\mathrm{e}^{-(K/2+\mathrm{i}\omega)\tau} \\ \langle a(t)a^\dagger(t+\tau)\rangle &= (\bar{N}+1)\mathrm{e}^{-(K/2+\mathrm{i}\omega)\tau}. \end{aligned} \tag{6.2.27}$$

6.2.4 General Form for Time Correlation Functions in the P-Representation

The general result follows from (5.2.20). If $W(t_1,t_2)$ is the evolution operator of the Fokker-Planck equation, then we can make the correspondence:

$$\begin{aligned} &\langle A_0(s_0)A_1(s_1)\dots A_m(s_m)B_n(t_n)B_{n-1}(t_{n-1})\dots B_0(t_0)\rangle \\ &= \int d^2\alpha f_r(\alpha)W(\tau_r,\tau_{r-1})f_{r-1}(\alpha)W(\tau_{r-1},\tau_{r-2})\dots \\ &\qquad \dots\times W(\tau_1,\tau_0)f_0(\alpha)P(\alpha,\tau_0) \end{aligned} \tag{6.2.28}$$

$$\text{where}\quad f_r(\alpha) = \begin{cases} \alpha & \text{if } F_r = a \text{ and } F_r \text{ is one of the } B\text{'s}, \\ \alpha - \dfrac{\partial}{\partial\alpha^*} & \text{if } F_r = a \text{ and } F_r \text{ is one of the } A\text{'s}, \\ \alpha^* & \text{if } F_r = a^\dagger \text{ and } F_r \text{ is one of the } A\text{'s}, \\ \alpha^* - \dfrac{\partial}{\partial\alpha} & \text{if } F_r = a^\dagger \text{ and } F_r \text{ is one of the } B\text{'s}. \end{cases}$$

Here, of course, the operator W has the property for an arbitrary function $f(\alpha)$

$$W(\tau_r,\tau_{r-1})f(\alpha) = \int d^2\alpha' P(\alpha,\tau_r|\alpha',\tau_{r-1})f(\alpha'). \tag{6.2.29}$$

Exercise. Verify the result (6.2.28), and show that it gives the two previous results.

Exercise. Show that

$$\langle a^\dagger(t)a^\dagger(t+\tau)a(t+\tau)a(t)\rangle = \langle|\alpha(t+\tau)|^2|\alpha(t)|^2\rangle \tag{6.2.30}$$

and that this result can be generalized to any correlation function which can be expressed as a product of two terms, one consisting only of $a^\dagger$'s, ordered with times increasing to the right, and the other of a's, with times decreasing to the right.

6.3 Some Amplifier Models

In this section we consider some amplifiers by using the various possible phase space methods, but not by input-output methods, as in Sect. 5.3. We consider simply the growth of an initial value of a field mode in the amplifying medium from which the signal cannot escape. For these simple models the QSDE methods are much more transparent, but they are also excellent examples of the use of phase space methods.

6.3.1 A Simple Amplifier

If we consider a system consisting of a harmonic oscillator coupled to a bath of two level atoms, we may use the master equation (5.1.59) in the form (omitting frequency shifts, and working in the interaction picture)

$$\dot{\rho}(t) = \tfrac{1}{2}\kappa(2a^\dagger \rho a - aa^\dagger \rho - \rho aa^\dagger) + \tfrac{1}{2}\gamma(2a\rho a^\dagger - a^\dagger a\rho - \rho a^\dagger)a \tag{6.3.1}$$

and we note that from (5.1.64),

$$\begin{aligned} &\kappa > \gamma && \text{if there is a net inversion,} \\ &\kappa < \gamma && \text{if there is no net inversion .} \end{aligned}$$

In the same way as we derived the Langevin equation (6.2.7) we can derive the equation of motion

$$d\alpha = \tfrac{1}{2}(\kappa - \gamma)\alpha dt + \sqrt{\kappa}\, d\eta(t). \tag{6.3.2}$$

From this equation it can be seen that if there is a net inversion, the variable $\alpha(t)$ will grow—this arises from energy being fed into the system by the bath. This is the reverse of the damping situation, where energy is dissipated into the bath. However, the amplification is noisy, since the stochastic term is present for all non-zero κ. If the initial density operator corresponds to a pure coherent state $|\alpha_0\rangle$, the time dependent density operator can be deduced from the P-function. This will be a Gaussian, whose means and variances can be found by solving (6.3.2). Explicitly

$$\alpha(t) = \alpha_0 \exp[\tfrac{1}{2}(\kappa - \gamma)t] + \sqrt{\kappa} \int_0^t d\eta(t') \exp\left[\tfrac{1}{2}(\kappa - \gamma)(t - t')\right] \tag{6.3.3}$$

$$\langle \alpha(t) \rangle = \langle \alpha^*(t) \rangle^* = \alpha_0 \exp\left[\tfrac{1}{2}(\kappa - \gamma)t\right] \tag{6.3.4}$$

$$\begin{aligned} \text{var}[\alpha(t)] &= \langle \alpha(t)^2 \rangle - \langle \alpha(t) \rangle^2 = 0 \\ \text{var}[\alpha^*(t)] &= \langle \alpha^*(t)^2 \rangle - \langle \alpha^*(t) \rangle^2 = 0 \\ \langle \alpha^*(t), \alpha(t) \rangle &= \langle \alpha^*(t)\alpha(t) \rangle - \langle \alpha^*(t) \rangle \langle \alpha(t) \rangle \\ &= \frac{\kappa}{\kappa - \gamma} \{\exp[(\kappa - \gamma)t] - 1\}\,. \end{aligned} \tag{6.3.5}$$

In fact, because of the complex nature of α, α^*, the quantity $\langle \alpha^*(t), \alpha(t) \rangle$ actually expresses the variance. From these, we can say that $P(\alpha, \alpha^*)$ is a Gaussian with these parameters, and that the density operator is a quantum Gaussian, with the appropriate parameters.

i) From (6.3.3) we see that the initial field is continuously amplified with time, but there is an added noise term which also grows at a similar rate. The signal to noise ratio is given by

$$
\begin{aligned}
S(t) &= \langle\alpha(t)\rangle/\sqrt{\langle\alpha(t),\alpha^*(t)\rangle} \\
&= \alpha_0\sqrt{\frac{(\kappa-\gamma)}{\kappa\left\{1-\exp\left[-(\kappa-\gamma)t\right]\right\}}} \qquad (6.3.6)\\
&\to \alpha_0\sqrt{(\kappa-\gamma)/\kappa} \quad \text{as } t\to\infty \quad \text{provided } \kappa>\gamma. \qquad (6.3.7)
\end{aligned}
$$

The best signal to noise ratio arises from the highest gain situation, $\gamma = 0$. The noise term arises from the constant κ, which shows that if there is gain ($\kappa > \gamma$) then there will be noise.

ii) Suppose the actual operators corresponding to the EM field represented by $a, a^\dagger$ are (in the Schrödinger picture, not the interaction picture)

$$
\begin{aligned}
x &= (a + a^\dagger)/2 \\
y &= (a - a^\dagger)/2\mathrm{i}.
\end{aligned} \qquad (6.3.8)
$$

Then

$$
\begin{aligned}
\langle x(t)\rangle &= \exp[\tfrac{1}{2}(\kappa-\gamma)t]\{x_0\cos\omega t + y_0\sin\omega t\} \\
\langle y(t)\rangle &= \exp[\tfrac{1}{2}(\kappa-\gamma)t]\{-x_0\sin\omega t + y_0\cos\omega t\}
\end{aligned} \qquad (6.3.9)
$$

and the variances and correlation are

$$
\begin{aligned}
\mathrm{var}[x(t)] &= \tfrac{1}{4} + \tfrac{1}{2}\langle a^\dagger(t), a(t)\rangle \\
&= \tfrac{1}{4} + \frac{\kappa}{2(\kappa-\gamma)}\left\{\exp\left[(\kappa-\gamma)t\right]-1\right\} \qquad (6.3.10)\\
\mathrm{var}[y(t)] &= \tfrac{1}{4} + \frac{\kappa}{2(\kappa-\gamma)}\left\{\exp\left[(\kappa-\gamma)t\right]-1\right\} \qquad (6.3.11)
\end{aligned}
$$

and

$$
\langle x(t), y(t)\rangle + \langle y(t), x(t)\rangle = 0. \qquad (6.3.12)
$$

Thus, in the physical operators, x, y, we see three terms:

—The term $1/4$ in the variance: this arises from zero point fluctuations.

—The remainder of the variance: this is noise arising from the amplification process.

—The amplified signal.

iii) A physical picture of what happens is as follows. The bath of atoms is in an optical cavity, and most of the atoms are excited. The process of stimulated emission amplifies any electromagnetic field already present, but this is a noisy process. Further, the process of spontaneous emission adds yet more noise. As the signal grows, the bath of excited atoms will eventually deplete, and this model will break down, unless there is some pumping mechanism which maintains the atoms in their excited states.

6.3.2 Comparison of P-, Q- and Wigner Function Methods

When they were first introduced in Chap.4, of all the quasiprobabilities the Q-function looked the most attractive. It is positive, and always exists. The Wigner function always exists, but need not be positive, and the P-function need neither exist, nor be positive. In this section, we will compare the various methods on the same example, and show that in fact there are advantages to all of these methods, but that the P-function is most easily interpreted as a quasiprobability because of one major advantage—namely, the P-function for a coherent state $|\alpha_0\rangle\langle\alpha_0|$ is simply $\delta^2(\alpha - \alpha_0)$, while for the Q- and Wigner functions, a delta function is not admissible.

6.3.3 The Degenerate Parametric Amplifier

This can be made optically by putting a crystal with a non-linear dielectric constant in a cavity, and driving it with a light field whose frequency is twice that of the cavity resonance.

This leads to the Hamiltonian (in the Schrödinger picture)

$$H_{\text{sys}} = \hbar\omega a^\dagger a + \frac{\mathrm{i}\hbar\epsilon}{2}\left(a^{\dagger 2} e^{-2\mathrm{i}\omega t} - a^2 \mathrm{e}^{2\mathrm{i}\omega t}\right). \tag{6.3.13}$$

We assume that ϵ is quite small, so that it is correct to use the same damping term as in the last two sections, and simply add on the nonlinearity (see Sect. 3.6.3).

a) Use of P-function Methods: The equations of motion for the P-function are then straightforwardly computed, in the interaction picture, to be

$$\begin{aligned}\frac{\partial P(\alpha)}{\partial t} = \Bigg\{ &\frac{\kappa}{2}\left(\frac{\partial}{\partial\alpha}\alpha + \frac{\partial}{\partial\alpha^*}\alpha^*\right) - \epsilon\left(\frac{\partial}{\partial\alpha^*}\alpha + \frac{\partial}{\partial\alpha}\alpha^*\right) \\ &+\kappa\bar{N}\frac{\partial^2}{\partial\alpha\partial\alpha^*} + \frac{\epsilon}{2}\left(\frac{\partial^2}{\partial\alpha^2} + \frac{\partial^2}{\partial\alpha^{*2}}\right)\Bigg\} P.\end{aligned} \tag{6.3.14}$$

Notice that this equation is a Fokker-Planck equation of exactly the kind which turns up in classical noise processes, apart from two points:

i) The noise terms are of two types. The term proportional to $\bar{N}$ represents thermal noise, as it did in the previous examples. The term proportional to ϵ arises from the non-linearity of the medium, and must be regarded as a purely quantum phenomenon.

ii) The overall diffusion matrix need not be positive. In fact the condition for positivity is

$$\kappa\bar{N} \geq |\epsilon|. \tag{6.3.15}$$

This means that, provided the temperature (since $\bar{N} = \left[\exp(\hbar\omega/\kappa T) - 1\right]^{-1}$ in the case of a thermal bath) is sufficiently high, this is a genuine Fokker-Planck equation, but otherwise, although (6.3.14) is still valid, the evolution is not the same as that of an ordinary diffusion process. (However, generalized P-representations can deal with this problem—see Sect. 6.4).

The stochastic differential equations are

$$\begin{aligned} d\alpha &= \left(-\frac{\kappa}{2}\alpha + \epsilon\alpha^*\right) dt + d\eta(t) \\ d\alpha^* &= \left(-\frac{\kappa}{2}\alpha^* + \epsilon\alpha\right) dt + d\eta^*(t) \end{aligned} \tag{6.3.16}$$

where

$$\begin{aligned} d\eta(t)^2 &= \epsilon dt, \\ d\eta^*(t)^2 &= \epsilon dt \\ d\eta^*(t)d\eta(t) &= \kappa\bar{N}dt. \end{aligned} \tag{6.3.17}$$

The equations are best simplified in terms of the *quadrature phases* (see Chap.10 for more about these) :

$$u = \frac{\alpha + \alpha^*}{2}, \quad v = \frac{\alpha - \alpha^*}{2\mathrm{i}}, \tag{6.3.18}$$

in terms of which they are

$$\begin{aligned} du &= \left(-\frac{\kappa}{2} + \epsilon\right) u dt + dW_u(t) \\ dv &= \left(-\frac{\kappa}{2} - \epsilon\right) v dt + dW_v(t) \end{aligned} \tag{6.3.19}$$

where

$$\begin{aligned} dW_u(t)^2 &= \tfrac{1}{2}(\kappa\bar{N} + \epsilon)dt \\ dW_v(t)^2 &= \tfrac{1}{2}(\kappa\bar{N} - \epsilon)dt \\ dW_u(t)dW_v(t) &= 0. \end{aligned} \tag{6.3.20}$$

The solutions for the quadrature phase equations (6.3.19) are of course quite simple

$$u(t) = \exp\left[(-\tfrac{1}{2}\kappa + \epsilon)t\right] u(0) + \int_0^t dW_u(t') \exp\left[(-\tfrac{1}{2}\kappa + \epsilon)(t - t')\right] \tag{6.3.21}$$

$$v(t) = \exp\left[(-\tfrac{1}{2}\kappa - \epsilon)t\right] v(0) + \int_0^t dW_v(t') \exp\left[-(\tfrac{1}{2}\kappa + \epsilon)(t - t')\right]. \tag{6.3.22}$$

Provided $\epsilon > \frac{1}{2}\kappa$, the variable u displays gain, and provided also $\epsilon < \kappa\bar{N}$, the simple SDE viewpoint is valid. However, at the same time the variable $v(t)$ is attenuated. The variances are also easily computed to be

$$\begin{aligned} \mathrm{var}[u(t)] &= \mathrm{e}^{(2\epsilon-\kappa)t}\mathrm{var}[u(0)] + \frac{\kappa\bar{N} + \epsilon}{2(2\epsilon - \kappa)}\left(\mathrm{e}^{(2\epsilon-\kappa)t} - 1\right) \\ \mathrm{var}[v(t)] &= \mathrm{e}^{-(2\epsilon+\kappa)t}\mathrm{var}[v(0)] + \frac{\kappa\bar{N} - \epsilon}{2(2\epsilon + \kappa)}\left(1 - \mathrm{e}^{-(2\epsilon+\kappa)t}\right). \end{aligned} \tag{6.3.23}$$

Notice that the variances in the actual measurable variables are given by adding 1/4 to each of these, as noted in (6.3.10,11).

Comments:

i) What is most interesting in this system, is the appearance of the terms proportional to ϵ in the diffusion matrix, since these are noise terms which arise from the Hamiltonian (and therefore reversible) motion of the system. However, because of the restriction that $\kappa\bar{N} \geq |\epsilon|$ for positivity of the diffusion matrix, it is in principle not possible to use this method to study the quantum noise in the absence of classical noise.

ii) However, in the absence of classical noise the interaction picture Heisenberg equations of motion arising from (6.3.13) are

$$\begin{aligned} \frac{da}{dt} &= \epsilon a^\dagger \\ \frac{da^\dagger}{dt} &= \epsilon a. \end{aligned} \tag{6.3.24}$$

and the solutions are easily found to be computable in terms of

$$x = \frac{1}{2}(a + a^\dagger),\ y = \frac{\mathrm{i}}{2}(a^\dagger - a) \tag{6.3.25}$$

as

$$\begin{aligned} x(t) &= \mathrm{e}^{\epsilon t} x(0) \\ y(t) &= \mathrm{e}^{-\epsilon t} y(0), \end{aligned} \tag{6.3.26}$$

and hence

$$\begin{aligned} \mathrm{var}[x(t)] &= \mathrm{e}^{2\epsilon t}\mathrm{var}[x(0)] \\ \mathrm{var}[y(t)] &= \mathrm{e}^{-2\epsilon t}\mathrm{var}[y(0)]. \end{aligned} \tag{6.3.27}$$

Setting $\kappa = 0$ in (6.3.23) and adding 1/4, it can be seen that the solutions (6.3.23) are still valid in the case of no classical noise, even though the method is not. The reason for this will be made clear in Sect. 6.4 on generalized P-representations.

iii) We can now see however the difference between the classical noise and the quantum noise. In the case of pure quantum noise, we see that

$$\mathrm{var}[x(t)]\mathrm{var}[y(t)] = \mathrm{var}[x(0)]\mathrm{var}[y(0)]. \tag{6.3.28}$$

and, among other things, this means that the Hamiltonian motion transforms a minimum uncertainty state into another minimum uncertainty state. Essentially, we can say that only the noise required by Heisenberg's uncertainty principle is present.

b) Use of the Q-Function: The Q-function always exists, so it is of interest to see how it might be used in this situation. Using the operator correspondences given in

Sect. 4.5 an equation of motion for the Q-function is

$$\frac{\partial Q(\alpha)}{\partial t} = \left\{ \frac{\kappa}{2}\left(\frac{\partial}{\partial\alpha}\alpha + \frac{\partial}{\partial\alpha^*}\alpha^*\right) - \epsilon\left(\frac{\partial}{\partial\alpha^*}\alpha + \frac{\partial}{\partial\alpha}\alpha^*\right) + \kappa(\bar{N}+1)\frac{\partial^2}{\partial\alpha\partial\alpha^*} - \frac{\epsilon}{2}\left(\frac{\partial^2}{\partial\alpha^2} + \frac{\partial^2}{\partial\alpha^{*2}}\right)\right\} Q(\alpha). \tag{6.3.29}$$

Comments:

i) The equation does not look very different from the P-function equation. The only changes are in fact

$$\bar{N} \to \bar{N}+1$$

and $$\frac{\epsilon}{2}\left(\frac{\partial^2}{\partial\alpha^2} + \frac{\partial^2}{\partial\alpha^{*2}}\right) \to -\frac{\epsilon}{2}\left(\frac{\partial^2}{\partial\alpha^2} + \frac{\partial^2}{\partial\alpha^{*2}}\right). \tag{6.3.30}$$

Thus the Hamiltonian noise is changed in sign, and the thermal term is augmented by a term that remains even at zero temperature. The condition for the diffusion matrix to be positive is now

$$\kappa(\bar{N}+1) \geq |\epsilon|, \tag{6.3.31}$$

and this is not always satisfied.

ii) At first glance, this is paradoxical—the Q-function is always an acceptable probability function, but the corresponding equation of motion does not have a positive diffusion matrix, and hence does not always have solutions which are positive. The answer to this paradox lies in the restricted range of initial conditions. From (4.4.19), we see that in the case of the Q-function $\langle|\alpha|^2\rangle \geq 1$, so we can only admit as initial values Q-functions which satisfy this condition. If the condition (6.3.31) is not satisfied, this will mean that an initial state with $\langle|\alpha|^2\rangle \geq 1$ will evolve in such a way as to reduce the variance in one quadrature phase, but never below a value corresponding to a positive Q-function.

Exercise. Show that the variances of the quadrature phases with respect to the Q-function evolve in time as

$$\mathrm{var}[u(t)] = \mathrm{e}^{(2\epsilon-\kappa)t}\mathrm{var}[u(0)] + \frac{\kappa(\bar{N}+1)-\epsilon}{2(2\epsilon-\kappa)}\left\{\mathrm{e}^{(2\epsilon-\kappa)t} - 1\right\} \tag{6.3.32}$$

$$\mathrm{var}[v(t)] = \mathrm{e}^{-(2\epsilon-\kappa)t}\mathrm{var}[v(0)] + \frac{\kappa(\bar{N}+1)+\epsilon}{2(2\epsilon+\kappa)}\left\{1 - \mathrm{e}^{-(2\epsilon+\kappa)t}\right\} \tag{6.3.33}$$

and that for no allowable initial conditions can these ever be negative.

c) Use of the Wigner Function: We know that the Wigner function is also guaranteed to exist, and this time a real advantage ensues; the equation for the Wigner function is found to be (by using the correspondences (4.5.12))

$$\frac{\partial W}{\partial t} = \left\{ \frac{\kappa}{2}\left(\frac{\partial}{\partial\alpha}\alpha + \frac{\partial}{\partial\alpha^*}\alpha^*\right) - \epsilon\left(\frac{\partial}{\partial\alpha^*}\alpha + \frac{\partial}{\partial\alpha}\alpha^*\right) + \kappa(\bar{N}+\tfrac{1}{2})\frac{\partial^2}{\partial\alpha\partial\alpha^*}\right\} W \tag{6.3.34}$$

and in this case there is *no* Hamiltonian noise. A representation in terms of SDEs is possible for all ϵ. Using the same procedures for the P-function shows that the variances (6.3.23) are valid for all ϵ, $\bar{N}$ and κ.

d) Comments:

i) The fact that there is no Hamiltonian noise in this case, and that the Hamiltonian noise changes sign between P- and Q-functions indicates that the identification of these terms as noise in any physical sense is subject to some difficulties.

ii) The other noise term is proportional to $\bar{N} + \frac{1}{2}$, which can be interpreted as thermal noise plus vacuum fluctuations. Because the Wigner function corresponds to symmetrically ordered moments, we find that the variances of the operator quadrature phases correspond to those of the Wigner function quadrature phases. These variances satisfy

$$\mathrm{var}[x(t)] \equiv \mathrm{var}[u(t)] = \mathrm{var}[x(0)]\mathrm{e}^{(2\epsilon-\kappa)t} + \frac{\kappa(\bar{N}+\frac{1}{2})}{2(2\epsilon-\kappa)}\left\{\mathrm{e}^{(2\epsilon-\kappa)t} - 1\right\} \tag{6.3.35}$$

$$\mathrm{var}[y(t)] \equiv \mathrm{var}[v(t)] = \mathrm{var}[y(0)]\mathrm{e}^{-(2\epsilon+\kappa)t} + \frac{\kappa(\bar{N}+\frac{1}{2})}{2(2\epsilon+\kappa)}\left\{1 - \mathrm{e}^{-(2\epsilon+\kappa)t}\right\}. \tag{6.3.36}$$

In both cases the original variance is amplified (or attenuated as the case may be) and augmented by an amplified (or attenuated) contribution from the vacuum fluctuations, proportional to $(\bar{N} + \frac{1}{2})$.

Exercise—Other Amplifiers. Develop similar treatments of the other kinds of amplifier considered in Sect. 7.2

Exercise—Quantum Brownian Motion Master Equation. Use the Wigner function and the correspondences (4.5.11) to show that the quantum Brownian motion master equation (3.6.28) is equivalent in Wigner function representation to

$$\dot{x} = p/m \tag{6.3.37}$$

$$\dot{p} = -V'(x) - \gamma p/m + \sqrt{\frac{\gamma kT}{2}}\,\xi(t) \tag{6.3.38}$$

where

$$\langle \xi(t)\xi(t')\rangle = \delta(t-t') \tag{6.3.39}$$

provided $V(x)$ is a harmonic or linear potential. This is of course Kramers' equation, which describes classical Brownian motion. If the potential is non-harmonic, extra terms like those in (4.5.32), as used in the derivation of the quasiclassical Langevin equation, will also appear.

6.4 Generalized P-Representations

We have already seen that the analogy between the P-, Q- or Wigner functions, and a genuine probability distribution is imperfect. The P-function does not always exist, the Q-function always exists and is positive, but such a simple Q-function as $\delta^2(\alpha - \alpha_0)$ does not correspond to any physical state. Finally, although the Wigner function always exists, it need not be positive, and like the Q-function, a delta function does not correspond to any physical state. We have seen how these problems arise in the study of the degenerate parametric amplifier in Sect. 6.3.2 where both the P- and Q-functions gave a Fokker-Planck equation whose diffusion matrix was not always positive definite, though the Wigner function did not have this disadvantage. Nevertheless, the solutions involving all of these methods are the same (of course) where valid, and even where the P- and Q-function methods cannot be justified, application of simple Langevin equation methods gives the correct answer. It was this fact that motivated *Drummond* and *Gardiner* [6.1] to introduce a class of generalized P-representations, by expanding in non-diagonal coherent state projection operators. The generalized P-representations are defined as follows. We set

$$\rho = \int_{\mathcal{D}} \Lambda(\alpha, \beta) P(\alpha, \beta) d\mu(\alpha, \beta), \tag{6.4.1}$$

where

$$\Lambda(\alpha, \beta) = \frac{|\alpha\rangle\langle\beta^*|}{\langle\beta^*|\alpha\rangle}, \tag{6.4.2}$$

$d\mu(\alpha, \beta)$ is the integration measure which may be chosen to define different classes of possible representations and $\mathcal{D}$ is the domain of integration. The projection operator $\Lambda(\alpha, \beta)$ is analytic in (α, β).Useful choices of the integration measure are:

a) Glauber-Sudarshan P-Representation:

$$d\mu(\alpha, \beta) = \delta^2(\alpha^* - \beta) d^2\alpha\, d^2\beta. \tag{6.4.3}$$

This measure corresponds to the Glauber-Sudarshan P-representation.

b) Complex P-Representation:

$$d\mu(\alpha, \beta) = d\alpha\, d\beta. \tag{6.4.4}$$

Here (α, β) are treated as complex variables which are to be integrated on individual contours C, C'. The existence of this representation under certain circumstances is demonstrated in the next section. In particular, this representation exists for an operator expanded in a finite basis of number states. This is a characteristic situation where nonclassical photon statistics (photon antibunching) may arise, and where the Glauber-Sudarshan P-representation would be singular. This representation is called the complex P-representation since complex values of $P(\alpha, \beta)$ occur. It gives rise to a $P(\alpha, \beta)$ which can be shown to satisfy a FPE obtained by replacing α, α^* with α, β in the usual P-function FPE. Under certain circumstances, exact solutions to Fokker-Planck equations occur which cannot be normalized as P-functions. These can be handled with the present representation by choosing C, C' (paths of integration) in the complex phase space of (α, β).

c) Positive P-Representation:

$$d\mu(\alpha, \beta) = d^2\alpha d^2\beta. \tag{6.4.5}$$

This representation allows (α, β) to vary independently over the whole complex plane. In the next section we will show that $P(\alpha, \beta)$ always exists for a physical density operator and can always be chosen positive, and hence we call it the positive P-representation. This means that $P(\alpha, \beta)$ has all the properties of a genuine probability. It will also be shown that provided any FPE exists for time development in the Glauber-Sudarshan representation, a corresponding FPE exists with a positive semidefinite diffusion coefficient for the positive P-representation. This enables stochastic differential equations, and a correspondence between the quantum Markov process and ordinary diffusion processes to be derived. In all representations, it is, of course, true that observable moments are given by

$$\langle (a^\dagger)^m a^n \rangle = \int_{\mathcal{D}} d\mu(\alpha, \beta) \beta^m \alpha^n P(\alpha, \beta). \tag{6.4.6}$$

6.4.1 The R-Representation

Glauber [6.2] defined an *R-representation* by means of the operator expansion given in Sect. 4.3.1f:

$$\rho = \frac{1}{\pi^2} \int \int d^2\alpha \, d^2\beta \, |\alpha\rangle R(\alpha^*, \beta) \langle \beta| \exp[-\tfrac{1}{2}(|\alpha|^2 + |\beta|^2)] \tag{6.4.7}$$

in which

$$R(\alpha^*, \beta) = \langle \alpha|\rho|\beta\rangle \exp[\tfrac{1}{2}(|\alpha|^2 + |\beta|^2)]. \tag{6.4.8}$$

While the representation is analytic in α^*, β (and therefore nonsingular), it is also by definition non-positive, and has a normalization that includes a Gaussian weight factor. For this reason, it cannot have a Fokker-Planck equation or any direct interpretation as a quasiprobability. Nevertheless, the existence of this representation does demonstrate that a calculation of normally ordered observables for any ρ is possible with a nonsingular representation.

6.4.2 Existence Theorems

We will show in this section that the generalized P-representations have quite strong existence properties. We do this with a number of theorems. NB: for brevity, we shall use the notation $\boldsymbol{\alpha} = (\alpha, \beta)$.

Theorem 1: A complex P-representation exists for an operator with an expansion in a finite number of number states.
***Proof*:** Let

$$\rho = \sum_{n,m} C_{nm} (a^\dagger)^m |0\rangle\langle 0| a^n. \tag{6.4.9}$$

Then, by Cauchy's theorem,

$$\rho = \oint_C \oint_{C'} \Lambda(\boldsymbol{\alpha}) P(\boldsymbol{\alpha}) d\mu(\boldsymbol{\alpha}) \tag{6.4.10}$$

with

$$P(\alpha) = \left(-\frac{1}{4\pi^2}\right) e^{\alpha\beta} \sum_{n,m} C_{nm} n! m! \alpha^{-m-1} \beta^{-n-1} \tag{6.4.11}$$

where C, C' are integration paths enclosing the origin.

Theorem 2: A complex P-representation exists for any operator with an expansion on a bounded range of coherent states, i.e., for

$$\rho = \iint_{DD'} \Lambda(\alpha, \beta) C(\alpha, \beta) d^2\alpha \, d^2\beta, \tag{6.4.12}$$

where D, D' are bounded in each complex plane.
***Proof*:** Application of Cauchy's theorem shows that if

$$P(\alpha) = -\frac{1}{4\pi^2} \iint_{DD'} C(\alpha', \beta') / \left[(\alpha - \alpha')(\beta - \beta')\right] d^2\alpha' \, d^2\beta', \tag{6.4.13}$$

then

$$\rho = \iint_{CC'} \Lambda(\alpha) P(\alpha) d\alpha \, d\beta, \tag{6.4.14}$$

where C, C' enclose D, D' respectively. Hence the complex P-representation exists in this case relative to any bounded expansion in coherent state projection operators.

Theorem 3: A positive P-representation exists for any quantum density operator ρ, with

$$P(\alpha) = (1/4\pi^2) \exp(-|\alpha - \beta^*|^2/4) \langle \tfrac{1}{2}(\alpha + \beta^*)| \, \rho \, |\tfrac{1}{2}(\alpha + \beta^*)\rangle. \tag{6.4.15}$$

***Proof*:** $P(\alpha)$ is positive, since ρ is a density operator, and it is composed of a diagonal matrix element multiplied by a positive function. In order to show that this represents a quantum density operator in the general case, the characteristic function

$$\chi(\lambda, \lambda^*) \equiv \mathrm{Tr}\left\{\rho \, e^{\lambda a^\dagger} e^{-\lambda^* a}\right\} \tag{6.4.16}$$

is used. This has been shown in Sect. 4.4.3 to define the density operator uniquely. In terms of the R-representation for ρ, the characteristic function is

$$\chi(\lambda, \lambda^*) = \frac{1}{\pi} \int R(\alpha^*, \lambda + \alpha) \exp(-\lambda^* \alpha - |\alpha|^2) d^2\alpha. \tag{6.4.17}$$

We now substitute the R-representation for ρ into (6.4.15) which defines $P(\alpha)$ in terms of the diagonal matrix elements of ρ. We then define ρ_P to be given by the positive P-representation form (6.4.4) with $P(\alpha)$ as given by the previous process, calculate the corresponding characteristic function $\chi_P(\lambda, \lambda^*)$ using (6.4.16) and show that this is the same as the original characteristic function for ρ. Thus:

$$\begin{aligned}\chi_P(\lambda,\lambda^*) &= \iint P(\alpha)\exp(\lambda\beta - \lambda^*\alpha)d^2\alpha d^2\beta \\ &= \frac{1}{4\pi^4}\iiiint R(\alpha'^*,\beta')\exp\left[\lambda\beta - \lambda^*\alpha - \tfrac{1}{2}|\alpha|^2 - \tfrac{1}{2}|\beta|^2 - |\alpha'|^2\right. \\ &\qquad \left. -|\beta'|^2 + \tfrac{1}{2}\beta'^*(\alpha+\beta^*) + \tfrac{1}{2}\alpha'^*(\alpha^*+\beta)\right] d^2\alpha\, d^2\beta\, d^2\alpha'\, d^2\beta'.\end{aligned} \tag{6.4.18}$$

We now make a variable change by defining

$$\begin{aligned} &\gamma = (\alpha+\beta^*)/2 \qquad && \delta = (\alpha-\beta^*)/2 \\ &\alpha = (\gamma+\delta) && \beta^* = (\gamma-\delta) \\ &d^2\alpha\, d^2\beta = 4d^2\gamma\, d^2\delta. \end{aligned} \tag{6.4.19}$$

Noting that R is an analytic function, the following identity is useful:

$$R(\alpha^*,\gamma) = \frac{1}{\pi}\int R(\alpha^*,\beta)\exp(\gamma\beta^* - |\beta|^2)d^2\beta. \tag{6.4.20}$$

Hence, the above expression for the characteristic function can be simplified to give

$$\begin{aligned}\chi_P(\lambda,\lambda^*) = \frac{1}{\pi^3}\iiint R(\alpha'^*,\gamma)\exp\left[\lambda(\gamma-\delta)^* - \lambda^*(\gamma+\delta)\right. \\ \left. -|\gamma|^2 - |\delta|^2 - |\alpha'|^2 + \alpha'\gamma^*\right] d^2\gamma\, d^2\delta\, d^2\alpha'\end{aligned} \tag{6.4.21}$$

$$\begin{aligned}= \frac{1}{\pi^2}\iint R(\alpha'^*,\gamma)\exp(|\lambda|^2 + \lambda\gamma^* - \lambda^*\gamma - |\gamma|^2 \\ - |\alpha'|^2 + \alpha'\gamma^*)d^2\gamma\, d^2\alpha'\end{aligned} \tag{6.4.22}$$

$$= \frac{1}{\pi}\int R(\alpha^*,\lambda+\alpha)\exp(-\lambda^*\alpha - |\alpha|^2)d^2\alpha. \tag{6.4.23}$$

Hence,

$$\chi_P(\lambda,\lambda^*) = \mathrm{Tr}\left\{\rho e^{\lambda a^\dagger} e^{-\lambda^* a}\right\} = \chi(\lambda,\lambda^*). \tag{6.4.24}$$

The last step follows from the identity with the characteristic function defined relative to the R-representation in (6.4.16). Thus, we deduce that $\rho_P = \rho$.

6.4.3 Definition of the Positive P-Representation by Means of the Quantum Characteristic Function

The proof in Theorem 3 depends heavily on the use of characteristic functions, and can be used as an implicit way of characterizing the positive P-representation. We can say that $P(\alpha)$ is a positive P-function corresponding to a density operator ρ provided

$$\chi_P(\lambda, \lambda^*) \equiv \int\int P(\alpha)\exp(\lambda\beta - \lambda^*\alpha) d^2\alpha d^2\beta \tag{6.4.25}$$

is identical with the quantum characteristic function of the density operator,

$$\chi(\lambda, \lambda^*) \equiv \mathrm{Tr}\left\{\rho e^{\lambda a^\dagger} e^{-\lambda^* a}\right\}. \tag{6.4.26}$$

The result of Theorem 3 is to establish that for any ρ there exists at least one $P(\alpha)$, that given by (6.4.15). However there can be others—see Sect. 6.4.7b.

a) Convergence Conditions: It would be nice to establish what kinds of $P(\alpha)$ can conceivably exist, and in particular, what conditions on $P(\alpha)$ are necessary to ensure the convergence of the integral in (6.4.25). This depends on how we evaluate the integrals.

Let us require that the integrals be carried out in polar coordinates in α and β separately; i.e. we define χ_P by setting

$$\alpha = re^{i\theta}, \qquad \beta = r'e^{i\theta'}, \qquad \lambda = \mu e^{i\phi} \tag{6.4.27}$$

and

$$\chi_P(\mu, \phi) = \int_0^\infty r\,dr \int_0^\infty r'\,dr' \left\{ \int_0^{2\pi} d\theta \int_0^{2\pi} d\theta' \exp\left\{\mu r' e^{i(\theta'+\phi')}\right\} \right.$$
$$\left. \times \exp\left\{-\mu r e^{i(\theta-\phi)}\right\} P(r, r', \theta, \theta')\right\}. \tag{6.4.28}$$

Since P must be normalized, we can assume the existence of a Fourier series representation

$$P(r, r', \theta, \theta') = \sum_{n=-\infty}^{\infty} \sum_{m=-\infty}^{\infty} c_{nm}(r, r') e^{in\theta} e^{im\theta'}. \tag{6.4.29}$$

Notice that if $P(r, r', \theta, \theta')$ is a bounded positive function this expansion will exist, and

$$c_{nm}(r, r') = \frac{1}{4\pi^2} \int\int d\theta d\theta' e^{-in\theta - im\theta'} P(r, r', \theta, \theta'), \tag{6.4.30}$$

and it follows that $c_{nm}(r, r')$ is also uniformly bounded by $c_{00}(r, r')$ for all n, m. This also means that we can substitute the expansion into (6.4.28) and integrate term by term over θ, θ'. Now put

$$z = \mu r' e^{i(\theta' - \phi)}, \qquad y = -\mu r e^{i(\theta + \phi)} \tag{6.4.31}$$

so that the n, m term is

$$-c_{n,m}(r,r')\left(\mu e^{i\phi}\right)^n\left(-\mu e^{i\phi}\right)^m \oint_{C'}\frac{dz}{z}z^{-m}e^z\oint_C\frac{dy}{y}y^{-n}e^y \tag{6.4.32}$$

where C', C, are circles with centre at the origin, and radii respectively $\mu r'$ and μr. The contour integrals are of course evaluated by Cauchy's theorem, to give

$$\chi_P(\mu,\phi)=4\pi^2\int_0^\infty rdr\int_0^\infty r'dr'\sum_{m,n=0}^{\infty}c_{-n,-m}(r,r')\frac{[\mu re^{-i\phi}]^n[-\mu r'e^{i\phi}]^m}{n!m!}. \tag{6.4.33}$$

A rather strong condition for the existence of $\chi_P(\mu,\phi)$ is that

$$|\,c_{-n,-m}(r,r')| \le c_{0,0}(r,r')(r)^{-n-2-\epsilon}(r')^{-m-2-\epsilon} \tag{6.4.34}$$

for some $\epsilon > 0$. Under these conditions the analytic moments

$$\langle a^{\dagger N}a^M\rangle \equiv \int d^2\alpha d^2\beta\,\alpha^M\beta^N P(\boldsymbol{\alpha}) \tag{6.4.35}$$

exist for all M and N, which is a necessary condition for the analyticity of the quantum characteristic function.

b) Comments: This condition, (6.4.34), is not nearly as strong as the condition for all moments of $P(\boldsymbol{\alpha})$ to exist, which would require all $c_{n,m}$ to vanish faster than any powers of r and r'. It is also much milder than one might expect from (6.4.25), which would lead one to expect that $P(\boldsymbol{\alpha})$ would need to drop off exponentially as $|\alpha|, |\beta| \to \infty$ in the direction of positive $\lambda\beta, -\lambda^*\alpha$.

6.4.4 Operator Identities

From the definitions (6.4.2) of the nondiagonal coherent state projection operators, the following identities can be obtained. Again, $\boldsymbol{\alpha}$ is used to denote (α, β):

$$\begin{aligned}
a\Lambda(\boldsymbol{\alpha}) &= \alpha\Lambda(\boldsymbol{\alpha})\\
a^\dagger\Lambda(\boldsymbol{\alpha}) &= (\beta+\partial/\partial\alpha)\Lambda(\boldsymbol{\alpha})\\
\Lambda(\boldsymbol{\alpha})a^\dagger &= \beta\Lambda(\boldsymbol{\alpha})\\
\Lambda(\boldsymbol{\alpha})a &= (\partial/\partial\beta+\alpha)\Lambda(\boldsymbol{\alpha}).
\end{aligned} \tag{6.4.36}$$

By substituting the above identities into (6.4.1), which defines the generalized P-representation and using partial integration (provided the boundary terms vanish), these identities can be used to generate operations on the P-function depending on the representation.

6.4.7 Positive P-Representation

We assume that the same equation (6.4.50) is being considered but with a positive P-representation. The symmetric matrix can always be factorized into the form

$$\mathbf{D}(\alpha) = \mathbf{B}(\alpha)\mathbf{B}^T(\alpha). \tag{6.4.52}$$

We now write

$$\mathbf{A}(\alpha) = \mathbf{A}_x(\alpha) + \mathrm{i}\mathbf{A}_y(\alpha) \tag{6.4.53}$$

$$\mathbf{B}(\alpha) = \mathbf{B}_x(\alpha) + \mathrm{i}\mathbf{B}_y(\alpha) \tag{6.4.54}$$

where $\mathbf{A}_x, \mathbf{A}_y, \mathbf{B}_x, \mathbf{B}_y$ are real. We then find that the master equation yields

$$\begin{aligned}\frac{\partial\rho}{\partial t} &= \iint d^2\alpha\, d^2\beta \Lambda(\alpha)(\partial P(\alpha)/\partial t) \\ &= \iint P(\alpha)\left[A_x^\mu(\alpha)\partial_\mu^x + A_y^\mu(\alpha)\partial_\mu^y + \tfrac{1}{2}(B_x^{\mu\sigma} B_x^{\nu\sigma}\partial_\mu^x\partial_\nu^x \right.\\ &\quad \left. +B_y^{\mu\sigma} B_y^{\nu\sigma}\partial_\mu^y\partial_\nu^y + 2B_x^{\mu\sigma} B_y^{\nu\sigma}\partial_\mu^x\partial_\nu^y)\right] \Lambda(\alpha) d^2\alpha\, d^2\beta. \end{aligned} \tag{6.4.55}$$

Here we have written, for notational simplicity, $\partial/\partial\alpha_x^\mu = \partial_\mu^x$, etc., and have used the analyticity of $\Lambda(\alpha)$ to make either of the replacements

$$\partial/\partial\alpha^\mu \leftrightarrow \partial_\mu^x \leftrightarrow -\mathrm{i}\partial_\mu^y \tag{6.4.56}$$

in such a way as to yield (6.4.55). Now provided partial integration is permissible, we deduce the Fokker-Planck equation

$$\begin{aligned}\partial P(\alpha)/\partial t = \Big\{&-\partial_\mu^x A_x^\mu(\alpha) - \partial_\mu^y A_y^\mu(\alpha) + \tfrac{1}{2}\left[\partial_\mu^x\partial_\nu^x B_x^{\mu\sigma}(\alpha)B_x^{\nu\sigma}(\alpha)\right.\\ &\left.+2\partial_\mu^x\partial_\nu^y B_x^{\mu\sigma}(\alpha)B_y^{\nu\sigma}(\alpha) + \partial_\mu^y\partial_\nu^y B_y^{\mu\sigma}(\alpha)B_y^{\nu\sigma}(\alpha)\right]\Big\} P(\alpha). \end{aligned} \tag{6.4.57}$$

Again, this is not a unique time-development equation but (6.4.50) is a consequence of (6.4.57).

However, the Fokker-Planck equation (6.4.57) now possesses a positive semidefinite diffusion matrix in a four-dimensional space whose vectors are

$$(\alpha_x^{(1)}, \alpha_x^{(2)}, \alpha_y^{(1)}, \alpha_y^{(2)}) \equiv (\alpha_x, \beta_x, \alpha_y, \beta_y). \tag{6.4.58}$$

We find the drift vector is

$$\mathcal{A}(\alpha) \equiv \left(A_x^{(1)}(\alpha), A_x^{(2)}(\alpha), A_y^{(1)}(\alpha), A_y^{(2)}(\alpha)\right) \tag{6.4.59}$$

and the diffusion matrix is

$$\mathcal{D}(\alpha) = \begin{pmatrix} \mathbf{B}_x\mathbf{B}_x^T, & \mathbf{B}_x\mathbf{B}_y^T \\ \mathbf{B}_y\mathbf{B}_x^T, & \mathbf{B}_y\mathbf{B}_y^T \end{pmatrix}(\alpha) \equiv \mathcal{B}(\alpha)\mathcal{B}^T(\alpha), \tag{6.4.60}$$

where

$$\mathcal{B}(\alpha) = \begin{pmatrix} \mathbf{B}_x, & 0 \\ \mathbf{B}_y, & 0 \end{pmatrix}(\alpha) \tag{6.4.61}$$

and $\mathcal{D}$ is thus explicitly positive semidefinite (and not positive definite). The corresponding Ito stochastic differential equations can be written as

$$\frac{d}{dt}\begin{pmatrix}\alpha_x\\ \alpha_y\end{pmatrix} = \begin{pmatrix}A_x(\alpha)\\ A_y(\alpha)\end{pmatrix} + \begin{pmatrix}\mathbf{B}_x(\alpha)\xi(t)\\ \mathbf{B}_y(\alpha)\xi(t)\end{pmatrix}, \tag{6.4.62}$$

or recombining real and imaginary parts

$$d\alpha/dt = A(\alpha) + \mathbf{B}(\alpha)\xi(t). \tag{6.4.63}$$

Apart from the substitution $\alpha^* \to \beta$, (6.4.63) is just the stochastic differential equation which would be obtained by using the Glauber-Sudarshan representation and naively converting the Fokker-Planck equation with a non-positive-definite diffusion matrix into an Ito stochastic differential equation.

In our derivation, the two formal variables (α, α^*) have been replaced by variables in the complex plane (α, β) that are allowed to fluctuate independently. The positive P-representation as defined here thus appears as a mathematical justification of this procedure.

a) Neglect of Boundary Terms: In the derivation of the positive P-representation Fokker-Planck equation in the previous section, partial integration is used, and it was assumed that boundary terms at infinity could be neglected. The direct derivation in terms of projection operators $\Lambda(\alpha)$ makes it very difficult to assess the magnitude of these. However, defining the positive P-representation through the characteristic function as in Sect. 6.4.3 makes the problem merely one of calculus. The equation of motion for the characteristic function is obtained by the rules

$$\begin{aligned}
\rho a^\dagger &\leftrightarrow \frac{\partial\chi}{\partial\lambda}\\
a\rho &\leftrightarrow -\frac{\partial\chi}{\partial\lambda^*}\\
\rho a &\leftrightarrow \left(\lambda - \frac{\partial}{\partial\lambda^*}\right)\chi\\
a^\dagger\rho &\leftrightarrow \left(-\lambda^* + \frac{\partial}{\partial\lambda}\right)\chi
\end{aligned} \tag{6.4.64}$$

which easily gives the correspondence (6.4.37,41) provided the boundary terms can be neglected. We can estimate these boundary terms for any particular case, provided we know the behaviour of $\mathbf{A}(\alpha), \mathbf{B}(\alpha)$ as $\alpha, \beta \to \infty$. In most cases of interest, these coefficients are low order polynomials in α, β, and we will find that corresponding to the condition (6.4.34) we will require a slightly faster drop off as $r, r' \to \infty$.

b) The Nature of the Equivalence: Of course the equivalence between the SDEs and the Fokker-Planck equation is only an equivalence in distribution, in the sense that only $P(\alpha)$ has any meaning, and it is related to reality only by the characteristic function as shown in Theorem 3, Sect. 6.4.2. Thus the fact that individual trajectories travel in an unphysical region is unimportant—the only things measurable are the moments of physical quantities, like $\langle a^\dagger a\rangle$.

c) Non-Uniqueness: It should be noted that all physical moments involve only α, β, *not* α^*, or β^*. It is possible for two different P-functions to have identical values for all moments of α and β, and yet to be different from each other.

Exercise. Show that if $P(\alpha,\beta) = f(|\alpha|, |\beta|)$ which is such that

$$4\pi^2 \int r\, dr\, r' dr' f(r, r') = 1 \tag{6.4.65}$$

then for all such P-functions

$$\langle (a^\dagger)^m a^n \rangle = \delta_{n0}\delta_{m0}. \tag{6.4.66}$$

This result illustrates the non-uniqueness of the positive P-function rather dramatically.

From the example in the exercise, it is clear that $P(\alpha,\beta)$ could be significant in a very large domain, and yet correspond to exactly the same situation as $\delta^2(\alpha)\delta^2(\beta)$ i.e. the variable being located at the origin.

6.5 Applications of the Generalized P-Representations

In this section, we will consider a number of examples of how to apply the two major generalized P-representations. It will be seen that there are problems where these methods can provide solutions which cannot otherwise be solved.

6.5.1 Complex P-Representation

We consider the example of Sect. 6.4.5, choosing ϵ to be real for simplicity. Using the appropriate operator correspondence, the complex P-representation Fokker-Planck equation is

$$\partial_t P(\alpha,\beta) = \left[-\frac{\partial}{\partial\alpha}(\epsilon - K\alpha^2\beta) - \tfrac{1}{2}\frac{\partial^2}{\partial\alpha^2}(K\alpha^2) \right.$$
$$\left. -\frac{\partial}{\partial\beta}(\epsilon - K\alpha\beta^2) - \tfrac{1}{2}\frac{\partial^2}{\partial\beta^2}(K\beta^2) \right] P(\alpha,\beta). \tag{6.5.1}$$

Rather miraculously, we see that this Fokker-Planck equation satisfies potential conditions of S.M. Sect. 5.3.3. For, in that notation,

$$\mathbf{A} = \begin{pmatrix} \epsilon - K\alpha^2\beta \\ \epsilon - K\beta^2\alpha \end{pmatrix}, \qquad \mathbf{B} = \begin{pmatrix} -K\alpha^2 & 0 \\ 0 & -K\beta^2 \end{pmatrix} \tag{6.5.2}$$

so, using S.M. (5.3.22, 23),

$$\mathbf{Z} = -\frac{2}{K}\begin{pmatrix} \epsilon/\alpha^2 - K\beta + K/\alpha \\ \epsilon/\beta^2 - K\alpha + K/\beta \end{pmatrix}, \tag{6.5.3}$$

$$\frac{\partial Z_\alpha}{\partial\beta} = \frac{\partial Z_\beta}{\partial\alpha} = 2 \tag{6.5.4}$$

and

$$\phi(\alpha,\beta) = -\int (Z_\alpha d\alpha + Z_\beta d\beta) \tag{6.5.5}$$

$$= -\frac{2\epsilon}{K}\left(\frac{1}{\alpha}+\frac{1}{\beta}\right) + 2\log(\alpha\beta) - 2\alpha\beta, \tag{6.5.6}$$

so that

$$P_s(\alpha,\beta) = \mathcal{N}(\alpha\beta)^{-2}\exp\left[2\alpha\beta + \frac{2\epsilon}{K}\left(\frac{1}{\alpha}+\frac{1}{\beta}\right)\right]. \tag{6.5.7}$$

where $\mathcal{N}$ is a normalization factor. The only acceptable contours for this stationary distribution are C, C' which are independent contours in the α and β planes which encircle the essential singularities at $\alpha = 0$ and $\beta = 0$.

A potential solution of this kind is extremely useful and could not be obtained with the Glauber-Sudarshan P-representation. The moments can be obtained from

$$\mathcal{N}\iint d\alpha\, d\beta\, \beta^m\alpha^n(\alpha\beta)^{-2}\exp\left[2\alpha\beta + \frac{2\epsilon}{K}\left(\frac{1}{\alpha}+\frac{1}{\beta}\right)\right] \tag{6.5.8}$$

and we can expand $\exp(2\alpha\beta)$ in power series and contour integrate term by term, to obtain

$$\langle (a^\dagger)^m a^n\rangle = \mathcal{N}\sum_{r=0}^{\infty}\frac{(2\epsilon/K)^{n+m+2r-2}\,2^r}{r!(n+r-1)!(m+r-1)!} \tag{6.5.9}$$

which is an easily computed series. The normalization is now obtained by requiring that $\langle (a^\dagger)^0 a^0\rangle = 1$.

6.6 Applications of the Positive P-Representation

The positive P-representation yields c-number stochastic differential equations, and thus enables genuinely quantum-mechanical problems can be reduced to classical noise problems. In contrast the Glauber-Sudarshan P-representation does this for a more restricted class of systems, in which there is usually significant classical thermal noise—it cannot represent nonclassical effects such as the production of squeezing or antibunching, which require density operators for which there is no Glauber-Sudarshan P-function.

However, because there are twice as many variables as in the classical case, even the deterministic part of the motion in a positive-P stochastic differential equation takes place in a phase space of doubled dimensions. The introduction of the quantum noise usually introduces fluctuations which are not complex conjugate for the equations in α, β, and thus, in simulations, the values of α and β on a trajectory can be far from complex conjugate to each other. Thus, the underlying dynamics is changed fundamentally.

Two procedures for implementing the positive-P representation have been developed and used quite extensively; linearization, and stochastic simulation.

6.6.1 Linear Systems and Linearization

Systems in which the resulting positive-P stochastic differential equations (6.4.49) have *constant* diffusion matrix $\mathbf{B}(\boldsymbol{\alpha})$, and a drift matrix $A(\boldsymbol{\alpha})$ *linear* in the variables $\boldsymbol{\alpha}$ always yield Gaussian processes in the variables $\mathrm{Re}(\boldsymbol{\alpha})$, $\mathrm{Im}(\boldsymbol{\alpha})$, and thus generate positive P-functions which decrease sufficiently rapidly in all directions in the complex plane for the neglect of boundary terms to be valid. Such systems, which we call *linear systems*, therefore can always be represented exactly by a positive P-representation

Exercise—Degenerate Parametric Amplifier. The model of Sect. 6.3.3 gives a positive-P Fokker-Planck equation exactly the same as (6.3.14), but with the substitution $\alpha^* \to \beta$. What are the corresponding SDEs? Show that they are valid for any $\bar{N}$, and not only values which satisfy (6.3.15).

In a non-linear system with sufficiently small noise, it is reasonable to suppose that we can use a small noise expansion, in which the fluctuations are linearized around the deterministic solution. One must be aware of course that any results so derived will be valid only in the small fluctuation regime, but, apart from the problem of transitions from one steady state to another in bistable systems, this is largely what is observable anyway. This approximate linearized system can itself be exactly represented by the positive P-representation.

As an example, consider the problem of Sect. 6.5.1, but now using the positive P-representation. We obtain the stochastic differential equation

$$\begin{pmatrix} d\alpha \\ d\beta \end{pmatrix} = \begin{pmatrix} \epsilon - K\alpha^2\beta \\ \epsilon - K\alpha\beta^2 \end{pmatrix} dt + \mathrm{i}\sqrt{K} \begin{pmatrix} \alpha\, dW_1(t) \\ \beta\, dW_2(t) \end{pmatrix}. \tag{6.6.1}$$

It should be noted that this equation does not contain any very obvious small noise parameter. However, a large driving field limit can be obtained by setting

$$K = \tilde{K}/\epsilon^2, \qquad \alpha = \tilde{\alpha}\epsilon, \qquad \beta = \tilde{\beta}\epsilon \tag{6.6.2}$$

so that

$$\begin{pmatrix} d\tilde{\alpha} \\ d\tilde{\beta} \end{pmatrix} = \begin{pmatrix} 1 - \tilde{K}\tilde{\alpha}^2\tilde{\beta} \\ 1 - \tilde{K}\tilde{\alpha}\tilde{\beta}^2 \end{pmatrix} dt + \mathrm{i}\frac{\sqrt{K}}{\epsilon} \begin{pmatrix} \tilde{\alpha}\, dW_1(t) \\ \tilde{\beta}\, dW_2(t) \end{pmatrix}. \tag{6.6.3}$$

A small noise linearization process can be carried out in this limit of large driving field and small nonlinearity which is, in fact, a situation of practical utility.

Exercise. Use the SDEs (6.6.3) to find the stationary means and variances of the number operator $a^\dagger a$, and compare with the exact results (6.5.9).

6.6.2 Stochastic Simulation

It is natural to try to simulate the positive-P SDEs numerically, and thus use classical stochastics to compute quantum results, but it turns out that the doubling of dimensions induced by the use on non-conjugate α, β variables can cause some technical difficulties. In simulations with a relatively large nonlinearity, a typical behaviour found is that positive-P simulations are well behaved up to a certain time, at which stage they make large excursions in the complex plane, in a direction in which α and β are not complex conjugate. A number of authors found this problem [6.4], and usually simply discarded such trajectories as being "unphysical", or as being the result of numerical instabilities. However, it is now apparent that this procedure is not correct. That the problem was not related to numerical instability was demonstrated by *Smith* and *Gardiner* [6.5], who exhibited a non-linear damping model which could be reliably simulated without discarding trajectories. Nevertheless, incorrect results were obtained when there was very large non-linear damping.

The issue of the validity of positive P-representation simulations is the subject of the remainder of this chapter. It is now known that the equations of motion arising from the positive P-representation are normally valid in an *asymptotic* sense only. The nonlinear terms which typically appear in the *deterministic* part of the SDEs are the principal cause of problems, and the asymptotic validity is with respect to the smallness of this nonlinearity. We will explain the source of the difficulties which appear to arise from escaping trajectories, and show that *asymptotic* sense in which the simulations are valid enables us to give rules for the practical use of the method. We shall start with two examples which exhibit the problems which can occur, and then explain how these problems arise, and how they can be avoided. The exposition here is based on the work of *Gilchrist*, *Gardiner* and *Drummond* [6.6].

6.6.3 The Single Mode Laser

Schack and *Schenzle* [6.7], suggested that it should be quite legitimate to use a positive-P interpretation, and study the growth of the single-mode laser field from the vacuum state, using the master equation for the single mode laser, as given in Sect. 9.3.2. Since the Glauber-Sudarshan P-representation is in this case well defined, the laser equations of motion do not *require* the use of the positive P-representation, but the fact that we can test the positive P-representation on a non-trivial example whose correct behaviour is known gives a very useful insight into the validity of the method.

a) Laser Stochastic Differential Equation: We use the laser model of Sect. 9.3 where, after various approximations, the equations of motion for the laser field could be written as the Glauber-Sudarshan P-function SDE (9.3.78), which we rewrite in terms of the positive P-representation as

$$d\tilde{\alpha} = -\kappa\tilde{\alpha}\left(1 - \frac{C}{1 + \tilde{\alpha}\tilde{\beta}/\tilde{n}_0}\right) dt + \frac{2Q}{\sqrt{N}}\, d\eta(t) \tag{6.6.4}$$

$$d\tilde{\beta} = -\kappa\tilde{\beta}\left(1 - \frac{C}{1 + \tilde{\alpha}\tilde{\beta}/\tilde{n}_0}\right) dt + \frac{2Q}{\sqrt{N}}\, d\eta^*(t). \tag{6.6.5}$$

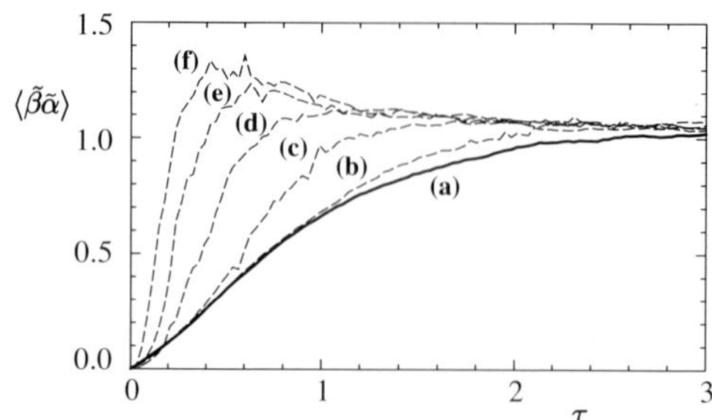

Fig. 6.2 Time dependence of the mean photon number with equivalent initial Gaussian distributions: **(a)** Solid line, variance $\sigma^2 = 0$—the physically correct result. The remaining dashed curves are spurious results, and come from the choices of initial variance: **(b)** $\sigma^2 = 0.2$, **(c)** $\sigma^2 = 0.4$, **(d)** $\sigma^2 = 0.6$, **(e)** $\sigma^2 = 0.8$, **(f)** $\sigma^2 = 1.0$.

Here $d\eta(t)$ is a complex Wiener increment as used in Sect. 6.2.1b, for which $d\eta(t)\,d\eta^*(t) = dt$—we choose the simplified case in which Q is independent of α and β, so that the Ito and Stratonovich forms of the equation are the same. The number of atoms is $\mathcal{N}$, and the other parameters are as in Sect. 9.3.2f. The quantities $\tilde{\alpha}, \tilde{\beta}, \tilde{g}, \tilde{n}_0, Q$ are scaled quantities, which are independent of $\mathcal{N}$ as $n\mathcal{N} \to \infty$. In particular

$$\alpha = \tilde{\alpha}\sqrt{\mathcal{N}}\,, \qquad \beta = \tilde{\beta}\sqrt{\mathcal{N}}\,. \tag{6.6.6}$$

We make an additional scaling by defining

$$\bar{\alpha} = \tilde{\alpha}\,\sqrt{C/\tilde{n}_0(C-1)}\,, \qquad \bar{\beta} = \tilde{\beta}\,\sqrt{C/\tilde{n}_0(C-1)}\,, \tag{6.6.7}$$

$$\tau = \kappa(C-1)t. \tag{6.6.8}$$

Finally, since for the validity of the choice of the form of Q it was assumed that $\tilde{\beta}\tilde{\alpha} \ll \tilde{n}_0$, we can again make use of this assumption to invoke a binomial expansion to first order in the drift terms. The equivalent stochastic differential equations which result from this procedure are,

$$\begin{aligned} d\bar{\alpha} &= (\bar{\alpha} - \bar{\beta}\bar{\alpha}^2)d\tau + \sqrt{\bar{Q}}\; dW(\tau) \\ d\bar{\beta} &= (\bar{\beta} - \bar{\beta}^2\bar{\alpha})d\tau + \sqrt{\bar{Q}}\; dW^*(\tau) \end{aligned} \tag{6.6.9}$$

The complex noise $dW(t)$ here satisfies $dW^*(\tau)dW(\tau) = 2d\tau$, and $\bar{Q} = 2Q/\mathcal{N}$, which is a very small quantity for large $\mathcal{N}$.

b) Interpretation of the Equations: Because the noises here are complex conjugate, the solutions of the stochastic differential equations (6.6.9) with initial conditions such that $\bar{\beta} = \bar{\alpha}^*$ will preserve this condition. This means that the P-function for this case can be treated as a Glauber-Sudarshan P-function, whose simulation causes no problems. Thus, one can choose a vacuum initial condition represented by

$$P(\alpha, \beta, t=0) = \delta^2(\alpha)\delta^2(\beta). \tag{6.6.10}$$

The initial conditions of the SDEs are then complex conjugate, and their solution amounts to solving for a Glauber-Sudarshan P-function. The results depicted in Fig. 6.2 as the solid curve (a) are found, and these known to be correct.

Alternatively, as noted in [6.7], one may make a choice of initial distribution corresponding to that of Theorem 3, (6.4.15) with a density operator $\rho = |0\rangle\langle 0|$. In terms of the *scaled* variables, this initial condition takes the form

$$P(\bar{\alpha}, \bar{\beta}, t = 0) = \frac{1}{4\pi^2\sigma^4} \exp\left\{-\frac{|\bar{\alpha}|^2 + |\bar{\beta}|^2}{2\sigma^2}\right\} \tag{6.6.11}$$

whose variance in each of the four quantities given by the real and imaginary parts of $\bar{\alpha}, \bar{\beta}$ is

$$\sigma^2 \equiv \frac{C}{\mathcal{N} n_0 (C-1)} \, . \tag{6.6.12}$$

For a laser far above threshold this is very small, and, as well, the approximations used to derive the equations are invalid when this is not small. The curves (b–f) in Fig. 6.2 are the results of solving the SDEs with different variances—all of order of magnitude 1, and therefore of course physically unrealistic—and it can clearly be seen that the behaviour found is wrong, and depends on the initial variance. For any physically reasonable variance, however, the result would be quite indistinguishable from the correct result, curve (a).

However, the initial vacuum state can indeed be represented by Gaussian distributions of *arbitrary* variance as shown in the exercise in Sect. 6.4.7b. The anomaly therefore does in principle exist, and is worthy of study, especially as we shall see that it illuminates the general mechanism of the breakdown of the positive P-simulations.

6.6.4 Analytic Treatment via the Deterministic Equation

In this section we will demonstrate that the mechanism for the breakdown of the positive P-simulations exists even in the deterministic equations alone. To show this we concentrate on the closed equation for the photon number, $\bar{N} = \alpha\bar{\beta}$. In the Stratonovich form, which provides the best description for simulations, it is

$$(\mathbf{S})\, d\bar{N} = -2(\bar{N} - a)(\bar{N} - b)d\tau + 2\sqrt{\bar{Q}\bar{N}}\, dW(\tau) \tag{6.6.13}$$

where $dW(\tau)$ is now a real Wiener increment, with $dW(\tau)^2 = d\tau$, and the Stratonovich form of the equation sets a and b to be

$$a = \tfrac{1}{2} - \sqrt{\tfrac{1}{4} + \tfrac{1}{2}\bar{Q}}\,, \qquad b = \tfrac{1}{2} + \sqrt{\tfrac{1}{4} + \tfrac{1}{2}\bar{Q}}\,. \tag{6.6.14}$$

However, our conclusions could be equally well be developed using the Ito equations, in which case a and b are given by a similar formula, but with $\frac{1}{2}\bar{Q} \to \bar{Q}$.

The deterministic part of (6.6.13),

$$\frac{d\bar{N}}{dt} = -2(\bar{N} - a)(\bar{N} - b), \tag{6.6.15}$$

contains the essence of the problem. There are two stationary points, one at b, which is stable (an attractor) and the other at a, which is unstable (a repeller). If $\bar{N}$

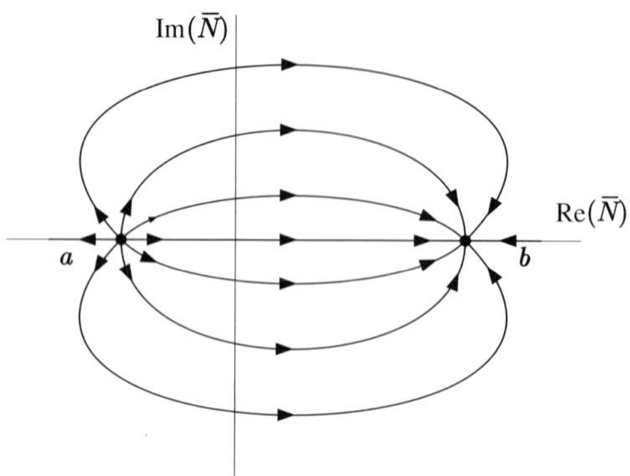

Fig. 6.3 A schematic representation of the phase space motion given by equation (6.6.15). It is of particular importance that there is a single trajectory from a which escapes to $-\infty$, before returning from $+\infty$ to b.

is real and non-negative, corresponding to $\bar{\alpha}$ and $\bar{\beta}$ being complex conjugate, the solutions of this differential equation remain real and non-negative. (This property is also true of the full stochastic equation (6.6.13).) The effect of the nonlinearity is to wrap the trajectories around from the repeller to the attractor; see Fig. 6.3. There is a single trajectory from a (the repeller) which escapes to $-\infty$, before returning from $+\infty$ to b, and which plays the critical role in the validity of the solutions.

a) Solutions of the Deterministic Differential Equation: The solution of (6.6.15) with initial condition $\bar{N}(t=0) = n$ is

$$\bar{N}(t,n) = a + \frac{(b-a)(n-a)}{n(1-\mathrm{e}^{\lambda t}) + b\mathrm{e}^{\lambda t} - a} \tag{6.6.16}$$

where $\lambda = 2(b-a)$. The solution clearly has a singularity as a function of n at

$$n = \frac{a - b\mathrm{e}^{-\lambda t}}{1 - \mathrm{e}^{-\lambda t}}, \tag{6.6.17}$$

which starts off at negative infinity when $t = 0$ and moves along the negative real axis reaching the critical point a at $t = \infty$.

b) Solution with an Initial Positive-P Function: The Gaussian initial condition (6.6.11) can be thought of as an appropriately weighted sum of radially uniform distributions on concentric rings of radius $|n| = \bar{R}$. The average photon number at a later time t, averaged over an initial ring of radius $\bar{R}$ is given by

$$\langle \bar{N}(t) \rangle_{\bar{R}} = \frac{1}{2\pi i} \oint_{\mathrm{C}_{\bar{R}}} \frac{dn}{n} \bar{N}(t,n) \tag{6.6.18}$$

where the contour of the integral $\mathrm{C}_{\bar{R}}$ is over the circle $|n| = \bar{R}$. As long as $\bar{N}(t,n)$ is analytic in n inside $\mathrm{C}_{\bar{R}}$ we can use contour integration to get

$$\langle \bar{N}(t) \rangle_{\bar{R}} = \bar{N}(t,0). \tag{6.6.19}$$

Thus the mean value will be the solution for the deterministic equation with the vacuum initial condition. However, $\bar{N}(t,n)$ *does* have a singularity (equation (6.6.17)), and this will hit a circle of radius $\bar{R}$ at the point $n = -\bar{R}$, at the time

$$\begin{aligned} t_e &= \frac{1}{\lambda} \ln \left| \frac{\bar{R}+a}{\bar{R}+b} \right| && \text{for } \bar{R} > -a \\ &= \infty && \text{for } \bar{R} \le -a. \end{aligned} \tag{6.6.20}$$

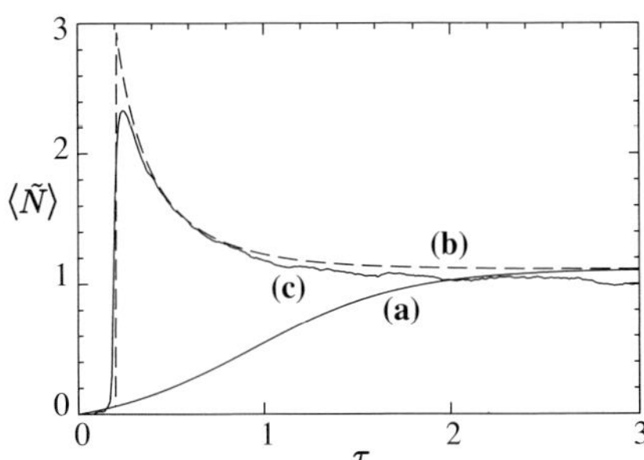

Fig. 6.4 Comparison of analytic treatment for the deterministic equation with the simulation of the full stochastic equation for the laser. **(a)** Mean photon number with an initial delta distribution centred at the origin, **(b)** (dashed line) analytic curve from (6.6.21), **(c)** Simulated mean photon number over trajectories from an initial 1,000 points distributed on a circle.

At the same time t_e we find the solution $\bar{N}(t, -\bar{R})$ escapes to infinity. Taking into account that the singularity is now within the contour, we can evaluate the contour integral to get the discontinuous solution,

$$\langle \bar{N}(t)\rangle_{\bar{R}} = \bar{N}(t,0) + u(t - t_e)u(\bar{R} + a)\frac{e^{\lambda t}(a - b)^2}{(1 - e^{\lambda t})(ae^{\lambda t} - b)}. \tag{6.6.21}$$

Summary:

i) If the circle $|\bar{N}| = \bar{R}$ does not enclose the repeller at $\bar{N} = a$, the average over the ring distribution is always equal to the true average.

ii) Otherwise the ring distribution evolves so that at the time t_e it passes through a point at infinity. At this time it is no longer valid to drop the boundary terms. Prior to this time the boundary terms are necessarily negligible since the distribution is bounded.

iii) In this case the solution of the positive P-representation is correct up to the time t_e and is incorrect thereafter. Only one trajectory—a set of measure zero—actually escapes to infinity and in practice this trajectory never appears in simulations. However trajectories close to this singular trajectory always appear, and undergo large excursions.

iv) If a simulation of the dynamics is made by averaging trajectories corresponding to $\bar{N}(t, n)$ for a number of initial values n uniformly distributed on $C_{\bar{R}}$, we obtain the curve like (c) of Fig. 6.4. This is very close to the curve (b), which corresponds to (6.6.21), itself an analytic average over all curves, including the singular curve.

v) By using the full Gaussian initial condition, (6.6.11), we obtain a weighted average of behaviours for all $\bar{R}$. If the variance σ^2 of the Gaussian is sufficiently small, the part arising from the spurious term in (6.6.21) will be of order of magnitude $\exp\left(-1/2\sigma^2\right)$ smaller than the other (correct term), and thus negligible.

vi) Problems appear at the *earliest time a deterministic trajectory can escape*. Trajectories near to one which actually escapes give an indication of the time the distribution results break down.

6.6.5 Full Stochastic Case

Consider now the full stochastic equations (6.6.13). The influence of the noise term will distort the circle so that we would expect to see a broadening of the jump in the

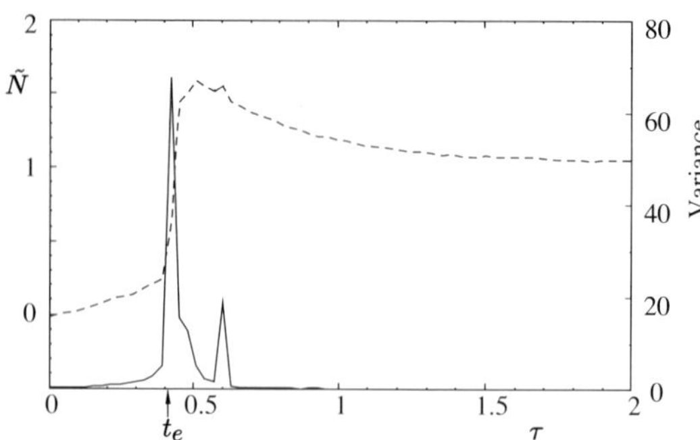

Fig. 6.5 The variance of the distribution (dashed line) increases dramatically at the time t_e when the mean photon number simultaneously undergoes a large change (solid line). There were 3,000 trajectories from an initial ring distribution of radius r = 0.8, and simulation parameters: $\Delta t = 10^{-3}$, $\epsilon = 1$ and $\bar{Q} = 0.25$.

mean. The equations (6.6.13) have been integrated by Gilchrist *et al.* [6.6] with trajectories starting from randomly chosen points along a circle centred at the origin. Large excursions into "unphysical" regions of phase space (spikes) were present in the simulations, looping around and back towards the attractor. Fig. 6.4 compares this full stochastic case with the results of the analytic treatment of the previous section, and demonstrates that the discontinuity in the average photon number is still present and occurs approximately at the time t_e.

The results of Fig. 6.2, for trajectories starting from an initial Gaussian, are reproduced by summing over appropriately weighted results for various rings—the failure of the broad Gaussian initial distribution arises from the rings larger than the critical radius, so that boundary terms are no longer negligible after the time t_e.

6.6.6 Numerical Signatures

The presence of boundary terms is manifested in three different numerical signatures, as follows.

a) Presence of Spikes: The stochastic trajectories occasionally make large excursions into regions of phase space near the unstable manifold, that is "spikes" will occur. The *earliest* time at which one of these spikes occurs is numerically found to be almost exactly the analytically calculated time t_e based purely on the deterministic equation.

b) Increase in the Statistical Error: The large excursions into phase space lead to an increase in the variance of the distribution. As shown in Fig. 6.5, starting from an initial distribution on a ring, the mean photon number and the variance increase dramatically at the time t_e, the time of the discontinuity in the deterministic equation.

c) Development of a Power-Law Tail: The behaviour of the tails of the distribution with time can be explored by binning the trajectories into a set of concentric bins with increasing values of $\bar{R} = |\bar{N}|$, giving an estimate of the probability of a trajectory reaching a certain radius. Fig. 6.6 shows how the tail of the distribution for the laser begins to fall off as a power-law at the same time t_e that the onset of spiking is observed.

The appearance of a power-law tail at the time t_e can be expected to invalidate the partial integration required to derive the Fokker-Planck equation from a master equation. From this we can conclude that the solution given *after* the time t_e is not

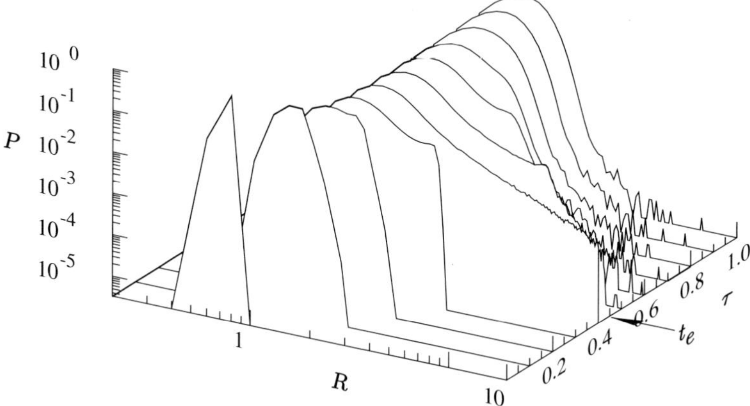

Fig. 6.6 The locations of trajectories were binned according to their radii ($R = |\bar{N}|$) at specific instances in time, yielding an estimate for the probability. This figure shows the behaviour of the tails of the probability distribution. A power law distribution will be a straight line (indicated by the data at the time t_e with a power of about $n = 3$). The time arrowed is the time t_e described in the text.

a valid solution to the master equation with the given initial condition, even though the power-law tail quickly disappears after this time.

6.7 The Anharmonic Oscillator

We consider now the damped undriven anharmonic oscillator. This exactly soluble problem is the simplest possible physical system with nonlinearity and damping present. Although the positive P-representation gives reliable results for this model, there are practical numerical difficulties to be surmounted and care must be taken in interpreting the numerical signatures. We start from the following master equation for the reduced density operator, with an appropriately scaled time variable τ,

$$\frac{\partial \rho}{\partial \tau} = \tfrac{1}{2}\gamma(2a\rho a^\dagger - a^\dagger a\rho - \rho a^\dagger a) - \tfrac{1}{2}\mathrm{i}[a^{\dagger 2}a^2, \rho]. \tag{6.7.1}$$

Exercise. Show that this master equation is exactly soluble by solving the equations of motion for the number state matrix elements of the density operator.

The mean photon number exactly obeys the equation

$$\frac{d\langle a^\dagger a\rangle}{d\tau} = \mathrm{Tr}\{a^\dagger a\rho\} = -\gamma\langle a^\dagger a\rangle. \tag{6.7.2}$$

The positive P-representation gives the stochastic differential equations

$$\begin{aligned}(\mathbf{I})\ d\alpha &= (-\tfrac{1}{2}\gamma\alpha - \mathrm{i}\alpha^2\beta)d\tau + \sqrt{-\mathrm{i}}\ \alpha\, dW_1(\tau)\\ (\mathbf{I})\ d\beta &= (-\tfrac{1}{2}\gamma\beta + \mathrm{i}\beta^2\alpha)d\tau + \sqrt{\mathrm{i}}\ \beta\, dW_2(\tau)\end{aligned} \tag{6.7.3}$$

where dW_i are real Gaussian stochastic processes satisfying $dW_i(\tau)dW_j(\tau) = \delta_{ij}d\tau$. From (6.7.3) we can write a single stochastic differential equation in the variable $N = \beta\alpha$,

$$(\mathbf{I})\, dN = -\gamma N d\tau + N\left(\sqrt{-\mathrm{i}}\; dW_1(\tau) + \sqrt{\mathrm{i}}\; dW_2(\tau)\right), \tag{6.7.4}$$

from which it follows that

$$d\langle N\rangle = -\gamma\langle N\rangle d\tau \tag{6.7.5}$$

which is the same as (6.7.2)—the exact master equation solution is recovered for the photon number.

Exercise. Show that the SDE also gives the correct equations of motion for all of the moments $\langle N^m\rangle$

However, (6.7.4) has noise terms and hence diffusion in the phase space, unlike (6.7.5). This can lead to problems, since at some stage the diffusive spreading of the distribution may be great enough to cause the statistical errors to become significant in a simulation. In order to treat this, consider the new stochastic variable $M = R^2 = |N|^2 = \alpha\alpha^*\beta\beta^*$. This has no corresponding physical observable, because it is not an analytic function of N, but $\sqrt{\langle M\rangle}$ gives a measure of the mean distribution radius in complex N-space. Following the usual Ito rules for products of stochastic variables, $\langle M\rangle$ satisfies the equation:

$$d\langle M\rangle = 2(1-\gamma)\langle M\rangle d\tau. \tag{6.7.6}$$

We see here an indication of difficulties that can occur in stochastic simulations, even when there is a complete theoretical equivalence of the Fokker-Planck and master equations. If $\gamma < 1$, then the mean distribution radius grows in time. This has no effect on ensemble averages for infinite ensembles, but clearly can give rise to increased sampling errors at long times, for the finite samples used in numerical simulations. In fact, it is easy to show that all moments of the distribution in N remain finite if initially finite. Thus, the distribution is at least exponentially bounded, and hence has no boundary terms. This is compatible with the fact that the stochastic equations give exactly identical results to the master equation, as expected. However, each radial moment has a critical damping, below which it exhibits steady growth in time.

6.7.1 Numerical Signatures

The anharmonic oscillator is an example where the growth of sampling error with time (for small or zero linear damping) can mimic the effect of boundary terms. Examination of the same numerical signatures as for the laser yields the following.

a) Presence of Spikes: Trajectories which looked like spikes were observed early in the simulation—with no apparent lower bound on the first "spike" time, for large samples. However, in this case the large radius trajectories belong to a distribution sufficiently well bounded to give rise to correct results—hence there is no contribution from boundary terms.

b) Increase in the Statistical Error: The quantity $\langle |N|^2 \rangle$ increases in keeping with the predictions from the analytic equations. There is no threshold time for this to take place, since it is due to a diffusive process.
c) Development of a Power-Law Tail: Because of the diffusive spreading, for *any* given distance scale there is a characteristic time after which the tails of the radial distribution can take on the character of a power-law. On larger distance scales, exponentially bounded tails are recovered and there is no *sudden* appearance of a power-law tail.
d) Increasing the Linear Damping: Finally as the amount of linear damping is increased, the numerically difficult quantum features in the simulation gradually disappear. That is, "spiking" is no longer observed, the statistical error stays small, and the distribution tails are reduced.

6.8 Theoretical Framework of the Problem

In the previous sections two features were found to be correlated with the failure of the positive P-representation:
i) The appearance of power law tails
ii) The correlation of the earliest spike time with the failure of the method.
In this section we will examine the derivation of the positive-P stochastic differential equations, and show how this kind of behaviour can arise.

6.8.1 The Power-Law Tails

The appearance of a power-law tail in the averaged radial distribution of the P-function almost certainly means that the boundary terms are significant and cannot be simply neglected.

The choice of the order of integration can therefore be significant, since the mere fact that $P(\alpha, \beta)$ can be normalized is *not* sufficient to guarantee the existence of $\chi_P(\lambda, \lambda^*)$. The condition (6.4.34) of Sect. 6.4.3 for the existence of $\chi_P(\lambda, \lambda^*)$, which is rather strong, is already weaker than the condition for all moments of $P(\alpha, \beta)$ to exist, which would require all $c_{n,m}$ to vanish faster than any power of r and r'.

A $P(\alpha, \beta)$ which falls off as a Gaussian will not cause any problems and hence the simulations will be reliable. On the other hand a $P(\alpha, \beta)$ which falls off as a power law comes dangerously close to condition (6.4.34). The degree of the power necessary so that $\chi_P(\lambda, \lambda^*)$ fails to exist is a problem that depends heavily on the particular situation under study, but the simulations of [6.6] for the averaged radial distribution in the variable $R = \sqrt{r^2 + r'^2}$, indicate that a small power will inevitably lead to problems.

6.8.2 The Earliest "Spike" Time

It is possible to derive the stochastic differential equations using a method in which no integration by parts is necessary, as follows. From the definition (6.4.26) of the quantum characteristic function, an equation of motion for $\chi(\lambda, \lambda^*, t)$ can be

satisfies the generalized Lipshitz condition but not the growth condition, and has the explicit solution (found by substituting $y = 1/x$)

$$x(t) = 1/\left(x(0)^{-1} + W(t) - W(0)\right). \tag{6.8.14}$$

Since $W(t) - W(0)$ is Gaussian with variance t, for any choice of $x(0)$ the denominator in (6.8.14) can vanish at any $t > 0$, although the probability of this will be small for $t \ll 1/\sqrt{x(0)}$

a) The Deterministic Equation: The large α, β behaviour of the solutions of the stochastic differential equations (6.8.9) is dominated by the deterministic term, whose solutions are easy to find explicitly—it is sufficient to note that a solution for $\beta\alpha$, when the noise terms are omitted is

$$\beta\alpha = 1/(A + 2t) \tag{6.8.15}$$

where $1/A$ is the initial value of $\beta\alpha$. In the positive P-representation this may be an arbitrary complex number, and in particular may be negative. But if A is negative the solution will cease to exist when $t = -A/2$. This kind of behaviour is in general still present in the full equations including the noise terms—that is, if we start with an initial ensemble of points (α_i, β_i) contained within a bounded region of the phase space, then this kind of equation does not satisfy the conditions necessary for the existence of a solution with a given initial condition in every finite interval $[0, T]$ after $t = 0$. Thus with any initial ensemble we may expect that there will be realizations of $W_1(t)$ and $W_2(t)$ which are such that the solution of the stochastic differential equations (6.8.9) with a particular initial condition chosen from the ensemble reach infinity in a *finite* time. Suppose that there is a lower bound t_e to all such times, and suppose the initial ensemble is such that $\int d^2\alpha d^2\beta |\alpha|^q |\beta|^{q'} P(\alpha, \beta, 0)$ exists for all q, q'. Then we conjecture that for $t < t_e$:

i) The characteristic function χ_P exists;

ii) All λ, λ^* differentiations can be carried out under the integral signs;

iii) Hence χ and χ_P obey the same partial differential equation (6.8.6) with the same initial condition, and hence for $t < t_e$ we can conclude that $\chi_P = \chi$.

This will mean that for $t < t_e$ a positive P-function simulation would give correct results. We cannot conclude that for $t \geq t_e$ that the results will be wrong, but we have given ample evidence in the previous sections that this is very often the case. At $t = t_e$ the P-function becomes unnormalizable, so that at that time the arguments that we can integrate by parts to get the P-function Fokker-Planck equation, and again to get the stochastic differential equations, also become invalid.

b) Exponential Suppression: In principle, wherever a trajectory can escape to infinity in a finite time, it would appear the the positive P-representation must fail. However, experience with the single mode laser shows that very accurate results can be obtained as long as there is a very small chance of any trajectory in the simulation approaching such a trajectory. To explain this, notice that in Fig. 6.3 a point initially near the positive real axis can only approach the repeller at a by proceeding, driven by the noise, against the deterministic flow. The "dangerous region" near a is thus almost inaccessible if the noise is small—in fact, when $\bar{Q} \ll 1$, the point a is finitely separated from the origin. Moreover, the noise in the equation for N (6.6.13) is proportional to $\sqrt{N}$, and thus vanishes at the origin. What we have therefore is a barrier surmounting problem such as is treated in S.M. Chap.9. The typical time to surmount such a barrier and reach a "dangerous" trajectory is $T_{\text{barrier}} \sim \exp(\text{const}/\bar{Q})$, and goes to infinity as $\bar{Q} \to 0$. The actual time for an escape to infinity will be thus of the order of magnitude $T_{\text{barrier}} + t_e$. In our simulations with $\sigma^2 \approx 1$, we find that $T_{\text{barrier}} \approx 0$, because the point a lies well within the initial probability distribution, but in practical situations Q is so small that the point a lies on the far fringes of the initial distribution, and contribution of escapes to infinity is utterly negligible, though non-zero.

The results of such simulations are thus only asymptotically accurate, in a limit in which the effective noise becomes small. In many cases we are only interested in the situation of very low temperature, which in a P-representation corresponds to zero thermal noise. In this case the relevant noise can arise only from the nonlinear terms such as the nonlinear term in the anharmonic oscillator as in (6.7.1), or from nonlinear damping, such as two-photon absorption, so that the limit of validity is that of small nonlinearity. In practice this is almost invariably satisfied.

6.9 Conclusions

The conclusion which we reach is that for sufficiently nonlinear problems that involve small photon numbers the positive P simulation is only valid up to the earliest time at which a solution of the stochastic differential equations with initial values chosen from the given initial ensemble can reach infinity. At that time the P-function becomes unnormalizable, and all arguments based on integration by parts and discarding surface terms at infinity also become invalid. However, the range of dynamical systems where the positive P-representation has been demonstrated to fail is quite limited, and in any case these problems are better dealt with by other methods. Directly simulating the master equation often yields results far more easily than the positive P-representation in these extreme parameter regimes. Furthermore, and more importantly, when the positive P simulations break down they do so in a predictable, and in an easily verifiable way.

6.9.1 Guidelines for Simulations

When the positive P-representation equations become inaccurate, they do so in a predictable way. Specifically, the presence of the following signatures indicate non-negligible boundary conditions:

i) The presence of "spikes", typically large excursions into regions of phase space with $\text{Re}(N) < 0$, that are associated with singular deterministic trajectories. If spikes are seen, the solutions are not reliable for for all times after the earliest appearance of such spikes. It is not correct to simply discard such trajectories.
ii) An increase in the statistical error of the distribution. This in essence corroborates the first signature.
iii) The probability distribution develops power-law tails. This is the most quantitative indicator, but requires more numerical work to test.

Provided sufficient care is taken to search for the numerical indicators above the positive P-representation can be a useful and reliable tool.

6.10 Example—Quantum Noise in the Parametric Oscillator

Wolinsky and *Carmichael* [6.9] have given an example in which certain solutions of positive P-representation equations are non-zero only in a bounded domain—thus there is no question of neglected boundary terms at infinity. One considers a simple model of the degenerate parametric amplifier, in which two quantized electromagnetic field modes, with frequencies $\omega, 2\omega$, interact with a classical pump field in a cavity. This is a slightly more realistic model than that described in Sect. 6.3.3 in that the pump field is assumed to drive a cavity, which in turn drives the non-linear crystal, as is usual in practice, since higher driving field strengths can then be attained.

Both fields have losses out of the cavity; the Hamiltonian takes the form (in the interaction picture)

$$H = \tfrac{1}{2}\mathrm{i}\hbar\bar{g}\left(a^{\dagger 2}b - a^2 b^\dagger\right) + \mathrm{i}\hbar\mathcal{E}(b^\dagger - b) + H_{\mathrm{loss}}, \tag{6.10.1}$$

The term H_{loss} leads to the usual loss terms in the master equation (assumed to correspond to zero temperature), which are written

$$\mathcal{L}_{\mathrm{loss}}\rho = \frac{\kappa_1}{2}\left(2a\rho a^\dagger - a^\dagger a\rho - \rho a^\dagger a\right) + \frac{\kappa_2}{2}\left(2b\rho b^\dagger - b^\dagger b\rho - \rho b^\dagger b\right) \tag{6.10.2}$$

Exercise. Derive Fokker-Planck equations in the positive P-representation corresponding to (6.10.1), and then the corresponding Langevin equations. Assume that the time constant for the b mode, κ_2, is large, so that we can adiabatically eliminate the b variable, (by naive methods, i.e., set the derivatives in the equations for the b positive-P variable equal to zero, and substitute the resulting solution in the other equations) and rescale variables according to

$$\alpha \to \alpha/g, \qquad \text{where } g = \bar{g}/\sqrt{2\kappa_1\kappa_2} \tag{6.10.3}$$

$$t \to \kappa_1 t \tag{6.10.4}$$

$$\lambda = -g\mathcal{E}/\kappa_1\kappa_2 \tag{6.10.5}$$

to obtain SDEs.

$$\begin{aligned} d\alpha &= \left[-\alpha + \beta(\lambda - \alpha^2)\right] dt + g\sqrt{\lambda - \alpha^2}\, dW_1(t) \\ d\beta &= \left[-\beta + \alpha(\lambda - \beta^2)\right] dt + g\sqrt{\lambda - \beta^2}\, dW_2(t). \end{aligned} \tag{6.10.6}$$

where $dW_1(t)$ and $dW_2(t)$ are Wiener increments.

If one chooses an initial condition with α and β real and $|\alpha|, |\beta| \le \sqrt{\lambda}$, the solution of (6.10.6) cannot leave this region. Hence no escapes to infinity can happen, and the positive P-representation gives an exact representation of the physics.

a) Bistable Behaviour: *Kinsler* and *Drummond* [6.11] have used this method to analyse the bistable behaviour which is apparent in the stationary solutions of the deterministic equations, $\alpha = \beta = \pm\sqrt{\gamma - 1}$, which are stable for $\gamma > 1$. This bistable behaviour was noted by *Graham* [6.12] before the invention of the positive P-representation. Graham used the Wigner function to analyse this model, but no exact treatment could be found, since the Wigner function equations of motion gave third order derivatives.

Exercise. Derive the Wigner function equations of motion corresponding to those derived for the positive P-function in the first exercise of this section. Show that by neglecting third order derivatives in the equations obtained *before any adiabatic elimination*, that a stochastic equation with positive definite diffusion matrix is obtained.

However *Drummond* [6.13] has shown that the neglect of third order derivatives can be very dangerous if one wishes to compute the switching time between one maximum and another in bistable problems. He noted that the switching time must be the inverse of the smallest non-zero eigenvalue of the Fokker-Planck operator, and gave an asymptotic method of computing this in bistable systems. His method involves an expansion in a parameter which measures the distance of the physical situation from a critical point, i.e., that configuration of parameters at which bistability emerges. Drummond and Kinsler investigated the switching time in this parametric amplifier problem by several methods.

i) Numerical simulation of the stochastic equations (6.10.6).

ii) Standard stochastic analytic approximations (as in S.M. Ch.9) applied to the positive-P Fokker-Planck equations.

iii) Numerical solution of the eigenvalue problem for the master equation. This means that one computes what the master equation is equivalent to in a number state basis, as in Sect. 6.1. However in this case we get infinite matrix equations, which must be truncated at some maximum photon number. With sufficient computing power, this maximum can be made large enough to get good results.

iv) The methods i) and ii) can be applied to the Wigner function equations.

7. Amplifiers and Measurement

This chapter consists of two parts, not closely related, but conceptually close. It is largely a chapter on the applications of input-output theory. The first part, on the theory of linear amplifiers, is a direct application of input-output theory, since amplifiers do of necessity have inputs and outputs, and the relation of one to the other is the central problem under study.

The second part is a partial answer to the problems raised in Chap.2 , on the foundations of the von Neumann theory of measurement. By means of a few specific examples, it is possible to show how in practice the existence of macroscopic quantum superposition states is ruled out in almost any reasonable situation. Finally, it is shown how the mechanism which destroys the coherences which would characterize any macroscopic quantum superposition state can also give a very plausible explanation of the von Neumann measurement postulate. If a small system is measured by a macroscopic measuring, apparatus, which is of necessity in contact with an environment (i.e., a heat bath), then the indirect effect of the heat bath is to feed noise back into the system in such a way as to destroy any coherences—thus the system density operator is diagonalized. One can then say that, after the measurement, the system is left in a particular quantum state.

7.1 Input-Output Theory of Amplifiers and Attenuators

An attenuator for a signal is a rather simple concept, implemented optically by a piece of grey glass. It is a passive device, and can easily be modelled by the following system. We have a harmonic oscillator, represented by the destruction operator a, coupled to two heat baths. One of the heat baths, with operators $b(\omega)$, is viewed as a device which introduces the input signal, and carries away the output signal. The other, with operators $h(\omega)$, represents the bath which absorbs the power from the signal. Thus the loss mechanism and the mechanism for carrying the signal in and out are of the same kind. The loss part of the Hamiltonian is

$$H_{\text{Loss}} = \mathrm{i}\hbar \int d\omega\, \gamma(\omega)\{h(\omega)a^\dagger - ah^\dagger(\omega)\} \tag{7.1.1}$$

while the input-output part of the Hamiltonian is

$$H_{\text{Sig}} = \mathrm{i}\hbar \int d\omega\, \kappa(\omega)\{b(\omega)a^\dagger - ab^\dagger(\omega)\}. \tag{7.1.2}$$

The quantum Langevin equation for the oscillator is (in the interaction picture)

$$\dot{a} = -\tfrac{1}{2}\gamma a - \sqrt{\gamma}\, h_{\text{in}}(t) - \tfrac{1}{2}\kappa a - \sqrt{\kappa}\, b_{\text{in}}(t) \tag{7.1.3}$$

which has the solution

$$a(t) = a(0)\mathrm{e}^{-(\kappa+\gamma)t/2} - \int_0^t dt'\, \mathrm{e}^{-(\kappa+\gamma)(t-t')/2}\{\sqrt{\gamma}\, h_{\mathrm{in}}(t') + \sqrt{\kappa}\, b_{\mathrm{in}}(t')\}. \tag{7.1.4}$$

Let us assume that $(\kappa + \gamma)^{-1}$ is much smaller than the time scales of interest in $b_{\mathrm{in}}(t)$; then we can approximate the stationary solution by dropping the initial value term and neglecting the time dependence of the signal and loss bath operators in the integral, to get

$$a(t) = -\frac{2}{\gamma+\kappa}\{\sqrt{\gamma}\, h_{\mathrm{in}}(t) + \sqrt{\kappa}\, b_{\mathrm{in}}(t)\}. \tag{7.1.5}$$

The boundary condition of the signal is

$$b_{\mathrm{out}}(t) = b_{\mathrm{in}}(t) + \sqrt{\kappa}\, a(t) \tag{7.1.6}$$

so that

$$b_{\mathrm{out}}(t) = \frac{\gamma-\kappa}{\gamma+\kappa} b_{\mathrm{in}}(t) - \frac{2\sqrt{\gamma\kappa}}{\gamma+\kappa} h_{\mathrm{in}}(t). \tag{7.1.7}$$

The output term then consists of an attenuated version of the input signal, as well as an additional term which represents the noise which arises from the loss medium.

7.2 Amplifiers

To obtain an amplifier we can follow the same procedure, but will need a medium with gain, rather than loss. We have already seen that a master equation can be easily set up which gives gain, namely in Sect. 5.1.4, the master equation for an assembly of two level atoms in which $N_a^+ > N_a^-$. We would like to find a QSDE which corresponds to this master equation, but comparison with the master equation (5.4.12) shows that this would require $\bar{N} > \bar{N} + 1$ so this cannot be done straightforwardly.

7.2.1 The Inverted Oscillator Heatbath

To obtain gain rather than damping from a medium requires an inversion, and this cannot be achieved with a harmonic oscillator heat bath. To see this, one need only note the damped harmonic oscillator equations of motion (5.3.26), which give damping for the mean independent of the state of the heat bath.

A device introduced by Glauber approximates an inverted two level system by a harmonic oscillator with energy levels

$$E_n = -(n + \tfrac{1}{2})\hbar\omega \tag{7.2.1}$$

maintained at a *negative* temperature $-T$, as illustrated in Fig.7.1. The populations of the energy levels will be given by

$$\frac{\mathrm{e}^{-n\hbar\omega/kT}}{1-\mathrm{e}^{-\hbar\omega/kT}} \tag{7.2.2}$$

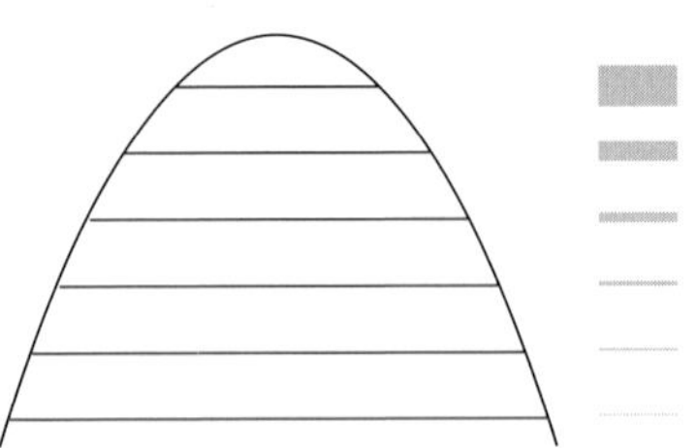

Fig. 7.1 The inverted oscillator with the populations corresponding to a negative temperature.

and if T is not too large, nearly all the population is in levels 1 and 2, with most in level 1, thus giving an approximation to an inverted two level atom. Since the raising operator for the energy decreases the number of quanta in the inverted oscillator, the role of creation and destruction operators is reversed.

Thus, the Hamiltonian for an inverted bath may be written in terms of operators $h(\omega)$, $h^\dagger(\omega)$, which have commutation relations

$$[h(\omega), h^\dagger(\omega')] = -\delta(\omega - \omega'), \tag{7.2.3}$$

that is, the roles of h and $h^\dagger$ are reversed. However the bath Hamiltonian is

$$H_H = -\int d\omega\, \hbar\omega \big(h(\omega)h^\dagger(\omega) + \tfrac{1}{2}\big) \tag{7.2.4}$$

and we find, in the negative temperature state

$$\begin{aligned} \langle h^\dagger(\omega)h(\omega')\rangle &= \big(\bar{N}(\omega) + 1\big)\delta(\omega - \omega') \\ \langle h(\omega)h^\dagger(\omega')\rangle &= \bar{N}(\omega)\delta(\omega - \omega') \end{aligned} \tag{7.2.5}$$

with $\bar{N}(\omega)$ as in (5.3.40), in which $-T$ is the value of the negative temperature.

In terms of the inputs and outputs, this means that we can write

$$\begin{aligned} \langle h_{\text{in}}^\dagger(t)h_{\text{in}}(t')\rangle &= (\bar{N} + 1)\delta(t - t') \\ \langle h_{\text{in}}(t)h_{\text{in}}^\dagger(t')\rangle &= \bar{N}\delta(t - t') \end{aligned} \tag{7.2.6}$$

to achieve the quantum white noise limit. The effect on the QSDEs and master equations is thus simply to interchange the coefficients of $\bar{N}$ and $\bar{N} + 1$. We then have a simple model of a medium which gives linear gain.

7.2.2 The Amplifier Model

The QSDE for an amplifier can be made from the following model. A gain medium (perhaps composed of an assembly of inverted atoms as in Sect. 5.1.4) is enclosed in an optical cavity. The mode inside the cavity thus experiences the effect of the gain medium, the input optical signal, and the losses via the output optical signal. However, we model the inverted atomic medium by an inverted oscillator heat bath.

The gain part of the Hamiltonian is thus obtained by writing an interaction Hamiltonian

$$H_{\text{gain}} = \mathrm{i}\hbar \int d\omega \gamma(\omega)\big[h(\omega)a^\dagger - a h^\dagger(\omega)\big]. \tag{7.2.7}$$

The input-output electromagnetic field is represented by

$$H_{\text{loss}} = \mathrm{i}\hbar \int d\omega\, \kappa(\omega)\big[b(\omega)a^\dagger - ab^\dagger(\omega)\big] \tag{7.2.8}$$

as in (5.3.3).

Putting these together the quantum Langevin equation for a is (setting $H_{\text{sys}} = 0$ for simplicity)

$$\dot{a} = \frac{\gamma}{2}a + \sqrt{\gamma}\, h_{\text{in}}(t) - \frac{\kappa}{2}a - \sqrt{\kappa}\, b_{\text{in}}(t) \tag{7.2.9}$$

which has the solution

$$a(t) = a(0)\mathrm{e}^{-(\kappa-\gamma)t/2} + \int_0^t dt'\mathrm{e}^{-(\kappa-\gamma)(t-t')/2}\big[\sqrt{\gamma}\, h_{\text{in}}(t') - \sqrt{\kappa}\, b_{\text{in}}(t')\big]. \tag{7.2.10}$$

If the loss, κ, exceeds the gain, then this eventually comes to a steady state in which the initial value term vanishes.

These equations are valid for any statistics of $b_{\text{in}}(t)$ and $h_{\text{in}}(t)$ so we can consider situations in which the damping and gain are sufficiently strong that $(\kappa - \gamma)^{-1}$ is a much smaller time than the time scale of the fluctuations in $b_{\text{in}}(t)$ or $h_{\text{in}}(t)$. In that case we can set $b^\dagger_{\text{in}}(t') \to b^\dagger_{\text{in}}(t)$, $h_{\text{in}}(t') \to h_{\text{in}}(t)$ in (7.2.10) and obtain approximately

$$a(t) = \frac{2}{\kappa - \gamma}\{\sqrt{\gamma}\, h_{\text{in}}(t) - \sqrt{\kappa}\, b_{\text{in}}(t)\}. \tag{7.2.11}$$

The input-output boundary condition on the b operators, is from (5.3.24),

$$b_{\text{out}}(t) = b_{\text{in}}(t) + \sqrt{\kappa}\, a(t) \tag{7.2.12}$$

so that

$$b_{\text{out}}(t) = \frac{\gamma + \kappa}{\gamma - \kappa} b_{\text{in}}(t) - \frac{2\sqrt{\gamma\kappa}}{\gamma - \kappa} h_{\text{in}}(t). \tag{7.2.13}$$

The output then consists of an amplified version of the input, as well as an additional term. If the statistics of the gain medium are effectively random, this is an added noise term.

7.2.3 Added Noise

It is easy to see that (7.2.13) guarantees that the "out" fields have the right commutation relations. But the presence of the term $h_{\text{in}}(t)$ says that there will be added noise to the amplified signal. To investigate the noise, however, the equations are best solved exactly by Fourier transform. If $\tilde{a}(\omega), \tilde{h}_{\text{in}}(\omega), \tilde{b}_{\text{in}}(\omega)$, are the Fourier components of the various quantities involved , then it is not difficult to show that

$$\tilde{a}(\omega) = \frac{\sqrt{\gamma}\, \tilde{h}_{\text{in}}(\omega) - \sqrt{\kappa}\, \tilde{b}_{\text{in}}(\omega)}{\mathrm{i}\omega + (\kappa - \gamma)/2} \tag{7.2.14}$$

$$\tilde{b}_{\text{out}}(\omega) = \tilde{b}_{\text{in}}(\omega)\frac{\gamma + \kappa - 2i\omega}{\gamma - \kappa - 2i\omega} - \tilde{h}_{\text{in}}(\omega)\frac{2\sqrt{\gamma\kappa}}{\gamma - \kappa - 2i\omega}. \tag{7.2.15}$$

We now suppose that there is an input *signal* given by $\langle \tilde{b}_{\text{in}}(\omega)\rangle$, but that the amplifier bath has simply thermal statistics, so that $\langle \tilde{h}_{\text{in}}(\omega)\rangle = 0$. Then

$$\langle \tilde{b}_{\text{out}}(\omega)\rangle = \langle \tilde{b}_{\text{in}}(\omega)\rangle \frac{\gamma + \kappa - 2i\omega}{\gamma - \kappa - 2i\omega}. \tag{7.2.16}$$

The amplifier has a frequency dependent gain $G(\omega)$ given by

$$G(\omega) = \frac{\gamma + \kappa - 2i\omega}{\gamma - \kappa - 2i\omega} \tag{7.2.17}$$

which is a maximum at zero frequency. If $\gamma < \kappa$, there is no stationary situation, and the amplifier gives a signal which increases without limit. Essentially, the power being fed into the cavity cannot escape fast enough. (Of course the idea of an inverted medium which maintains its inversion independent of power output is not exactly valid, and depletion effects will then need to be considered. The system is then essentially a laser).

The noise is given by the mean square fluctuation power per unit bandwidth. We use the notation for the noise power, $\mathbf{N}\{\tilde{b}_{\text{in}}(\omega)\}$, defined by

$$\begin{aligned}&\mathbf{N}\{\tilde{b}_{\text{in}}(\omega)\}\,\delta(\omega - \omega') \\ &\qquad = \hbar\omega \left\langle \left(\tilde{b}^\dagger_{\text{in}}(\omega) - \langle b^\dagger_{\text{in}}(\omega)\rangle\right)\left(\tilde{b}_{\text{in}}(\omega') - \langle \tilde{b}_{\text{in}}(\omega')\rangle\right)\right\rangle\end{aligned} \tag{7.2.18}$$

with a similar notation for $\tilde{b}_{\text{out}}(\omega)$ and $\tilde{h}_{\text{in}}(\omega)$. From (7.2.15,17) we can see that

$$\mathbf{N}\{\tilde{b}_{\text{out}}(\omega)\} = \mathbf{N}\{\tilde{b}_{\text{in}}(\omega)\}|G(\omega)|^2 + \mathbf{N}\{\tilde{h}_{\text{in}}(\omega)\}\{|G(\omega)|^2 - 1\}. \tag{7.2.19}$$

The *added noise*, $A(\omega)$, is that amount of noise in the input which would produce the last term on the right hand side, thus

$$A(\omega) = \mathbf{N}\{\tilde{h}_{\text{in}}(\omega)\}\{1 - |G(\omega)|^{-2}\}. \tag{7.2.20}$$

If the amplifier bath is thermal at temperature $-T$, then

$$\boxed{A(\omega) = \left\{\frac{\hbar\omega}{e^{\hbar\omega/kT} - 1} + \hbar\omega\right\}\{1 - |G(\omega)|^{-2}\}} \tag{7.2.21}$$

and the minimum added noise, which occurs at $T = 0$, is

$$A(\omega)_{T=0} = \hbar\omega\{1 - |G(\omega)|^{-2}\}. \tag{7.2.22}$$

Thus

$$0 \le A(\omega)_{T=0} \le \hbar\omega \tag{7.2.23}$$

with zero being attained only at unit gain.

Thus, even at absolute zero, quantum noise will be added to the signal unless there is no gain at all.

7.2.4 Signal to Noise Ratio

Using a similar notation for the signal power,

$$\mathbf{S}\{\tilde{b}_{\text{in}}(\omega)\} = \hbar\omega|\langle\tilde{b}_{\text{in}}(\omega)\rangle|^2 \tag{7.2.24}$$

and the corresponding definition of the "out" signal, we are now in a position to consider the signal to noise ratios $(\mathbf{S}/\mathbf{N})$. Since from (7.2.16,24) we know

$$\mathbf{S}\{\tilde{b}_{\text{out}}(\omega)\} = |G(\omega)|^2\mathbf{S}\{\tilde{b}_{\text{in}}(\omega)\}, \tag{7.2.25}$$

the signal to noise ratios are related by

$$(\mathbf{S}/\mathbf{N})_{\text{out}}^{-1} = (\mathbf{S}/\mathbf{N})_{\text{in}}^{-1} + \frac{\mathbf{N}\{\tilde{h}_{\text{in}}(\omega)\}}{\mathbf{S}_{\text{in}}}\left\{1 - |G(\omega)|^{-2}\right\}. \tag{7.2.26}$$

If we assume that the gain medium is thermal and has temperature $-T$ so that $\mathbf{N}\{\tilde{h}_{\text{in}}(\omega)\} = 1 + \bar{N}(\omega)$, and that $|G(\omega)| \to \infty$, we find that

$$(\mathbf{S}/\mathbf{N})_{\text{out}}^{-1} \approx (\mathbf{S}/\mathbf{N})_{\text{in}}^{-1} + \frac{\hbar\omega\{1 + \bar{N}(\omega)\}}{\mathbf{S}_{\text{in}}}. \tag{7.2.27}$$

As expected, the signal to noise ratio is degraded even at absolute zero by the added vacuum fluctuations from the oscillator heat bath.

7.2.5 "Noise Temperature" of an Amplifier

It is traditional to express the noise produced in an amplifier in terms of the concept of "noise temperature". The idea behind the concept is to express the amplifier noise as the amount of extra noise which would be produced by increasing the temperature of the input by an amount called the noise temperature.

In fact the noise temperature is not really a very convenient measure, and is only included here for completeness. In (7.2.20) the added noise term in the input is $\mathbf{N}\{\tilde{h}_{\text{in}}(\omega)\}\{1 - |G(\omega)|^{-2}\}$. The *noise temperature* will be defined as T_n, where

$$\frac{\hbar\omega}{e^{\hbar\omega/kT_n} - 1} = \mathbf{N}\{\tilde{h}_{\text{in}}(\omega)\}\{1 - |G(\omega)|^{-2}\} \tag{7.2.28}$$

so that

$$T_n = \frac{\hbar\omega/k}{\log\left[1 + \hbar\omega/\left[\left(1 - |G(\omega)|^2\right)\mathbf{N}\{\tilde{h}_{\text{in}}(\omega)\}\right]\right]}. \tag{7.2.29}$$

Notice that

i) As $|G| \to 1$, $T_n \to 0$; there is no extra noise in the case of unit gain.

ii) If the gain medium is thermal at temperature $-T$, then

$$T_n = \frac{\hbar\omega}{k}\left[\log\left(1 + \frac{1 - e^{-\hbar\omega/kT}}{1 - |G(\omega)|^{-2}}\right)\right]^{-1}. \tag{7.2.30}$$

iii) If $T \to 0$, giving a perfect inversion, then

$$T_n = \frac{\hbar\omega}{k}\left[\log\left(\frac{2|G(\omega)|^2 - 1}{|G(\omega)|^2 - 1}\right)\right]^{-1} \tag{7.2.31}$$

so that in this case we deduce that

$$0 \le T_n \le \frac{\hbar\omega}{k\log 2}. \tag{7.2.32}$$

iv) An alternative definition can be obtained if, instead of the definition (7.2.18) of noise power in terms of the normally ordered average, we use the symmetrically ordered average, and define the noise temperature as that temperature at which the added noise equals the increase in noise compared to absolute zero. In this case, assuming $T = 0$, we find (using T_n' for this alternative definition)

$$T_n' = \frac{\hbar\omega}{k}\left[\log\left(\frac{3|G(\omega)|^2 - 1}{|G(\omega)|^2 - 1}\right)\right]^{-1} \tag{7.2.33}$$

and

$$0 \le T_n' \le \frac{\hbar\omega}{k\log 3}. \tag{7.2.34}$$

v) The expressions for noise temperature give the impression of having deep significance, but their form is really only a reflection of an inappropriate way of characterizing noise, which I feel should now be replaced by the concept of added noise.

7.2.6 QSDEs in the Case of a Negative Temperature Bath

Exercise. Work through the derivation of the quantum Langevin equation in Sect. 5.3.2 and show that, in the case of a negative temperature bath, as in Sect. 7.2.1, the quantum Langevin equation (5.3.15) becomes

$$\dot{a} = -\frac{\mathrm{i}}{\hbar}[a, H_{\mathrm{sys}}] + [a, c^\dagger]\left[\frac{\gamma}{2}c - \sqrt{\gamma}\, h_{\mathrm{in}}(t)\right] - \left[\frac{\gamma}{2}c^\dagger - \sqrt{\gamma}\, h_{\mathrm{in}}^\dagger(t)\right][a, c]. \tag{7.2.35}$$

Show that the corresponding Ito QSDE takes the form

$$\begin{aligned}(\mathbf{I})da = &-\frac{\mathrm{i}}{\hbar}[a, H_{\mathrm{sys}}]dt + \frac{\gamma}{2}\bar{N}(2c^\dagger a c - a c^\dagger c - c^\dagger c a)dt \\ &+\frac{\gamma}{2}(\bar{N}+1)(2 c a c^\dagger - a c c^\dagger - c c^\dagger a)dt \\ &-\sqrt{\gamma}\,[a, c^\dagger]dH(t) + \sqrt{\gamma}\, dH^\dagger(t)[a, c]\end{aligned} \tag{7.2.36}$$

in which

$$\begin{aligned}&dH(t)^2 = dH^\dagger(t)^2 = 0\\ &dH^\dagger(t)dH(t) = (\bar{N}+1)\,dt\\ &dH(t)dH^\dagger(t) = \bar{N}\,dt.\end{aligned} \tag{7.2.37}$$

7.2.7 Ito QSDEs for Positive and Negative Temperature

Notice that the Ito equation can be written in a form which is valid for both positive and negative temperature baths as follows: Define for each case

$$
\begin{aligned}
&T > 0 && ; \quad T < 0 \\
&dW(t) = \sqrt{\gamma}\; dB^\dagger(t) && ; \quad = \sqrt{\gamma}\; dH(t) \\
&dW^\dagger(t) = \sqrt{\gamma}\; dB(t) && ; \quad = \sqrt{\gamma}\; dH^\dagger(t) \\
&\lambda = \gamma\bar{N} && ; \quad = \gamma(\bar{N}+1) \\
&\mu = \gamma(\bar{N}+1) && ; \quad = \gamma\bar{N}.
\end{aligned}
\tag{7.2.38}
$$

Then

$$
\begin{aligned}
&dW(t)^2 = dW^\dagger(t)^2 = 0 \\
&dW^\dagger(t)dW(t) = \lambda dt \\
&dW(t)dW^\dagger(t) = \mu dt
\end{aligned}
\tag{7.2.39}
$$

and

$$
\begin{aligned}
(\mathbf{I})da = & -\frac{\mathrm{i}}{\hbar}[a, H_{\mathrm{sys}}] + \frac{\mu}{2}(2c^\dagger ac - ac^\dagger c - c^\dagger ca)dt \\
& + \frac{\lambda}{2}(2cac^\dagger - acc^\dagger - cc^\dagger a)dt - [a, c^\dagger]dW(t) + dW^\dagger(t)[a, c].
\end{aligned}
\tag{7.2.40}
$$

The different cases of negative temperature and positive temperature depend on the relative sizes of μ and λ; for positive temperature $\lambda > \mu$, for negative temperature $\lambda < \mu$.

7.2.8 Phase Conjugating Amplifier

Exercise. Consider a modification of the amplifier model of Sect. 7.2.9, made by using the couplings

$$
H = \mathrm{i}\hbar \int d\omega\, \gamma(\omega)[h(\omega)a - a^\dagger h^\dagger(\omega)] + \mathrm{i}\hbar \int d\omega\, \kappa(\omega)[b(\omega)a^\dagger - ab^\dagger(\omega)].
\tag{7.2.41}
$$

The second term is quite normal, but the first term can be seen to create a bath quantum and a system quantum at the same time. Both $h(\omega)$ and $b(\omega)$ represent ordinary non-inverted heat baths. The first term could in practice be approximated by a pumped Hamiltonian

$$
H = \mathrm{i}\hbar \int d\omega\, \mu(\omega)[E^*(2\omega)h(\omega)a - a^\dagger h^\dagger(\omega)E(2\omega)],
\tag{7.2.42}
$$

in which $E(2\omega)$ is in a coherent state created by a strong pump field, and can thus be approximated by a classical field $\mathcal{E}(2\omega)$, and $\gamma(\omega) = \mu(\omega)\mathcal{E}(2\omega)$.

Using the same general notation as for the ordinary amplifier show that the equations of motion are

$$
\dot{a} = \frac{\gamma - \kappa}{2}a + \sqrt{\gamma}\; h_{\mathrm{in}}^\dagger(t) - \sqrt{\kappa}\; b_{\mathrm{in}}(t).
\tag{7.2.43}
$$

Taking $b_{\text{in}}(t)$ to be the input field, and $h_{\text{out}}(t)$ to be the output, show that this represents an amplifier such that

$$h_{\text{out}}(t) = Gb_{\text{in}}^\dagger(t) + \sqrt{G^2+1}\; h_{\text{in}}(t) \tag{7.2.44}$$

with

$$G = \frac{2\sqrt{\kappa\gamma}}{\kappa-\gamma}. \tag{7.2.45}$$

Analyse the noise performance of this amplifier.

The amplifier is called "phase conjugating" since a change in the phase $e^{i\phi}$ of the input $b_{\text{in}}(t)$ results in a phase change $e^{-i\phi}$ in the output $h_{\text{out}}(t)$.

Notice that (7.2.44) is very similar to the inverted bath amplifier (7.2.13)—each represents essentially a mixture of creation and destruction operators. This occurs in (7.2.13) because h and $h^\dagger$ have roles which are reversed, as a result of the inverted bath; in (7.2.44) the creation operator term arises because of the abnormal coupling in (7.2.41). However the output signal in the phase conjugating amplifier is in a different channel from the input.

7.2.9 The Degenerate Parametric Amplifier

In the previous example the signal to noise ratio of the amplified input was degraded by the added term arising from the amplifier heat bath. It is possible to have amplification which does not depend on a bath, as was noted by *von Neumann* [7.1], and in this case no added noise occurs.

In this amplifier the gain comes from a *Hamiltonian* term

$$H_{\text{gain}} = \frac{i\hbar\epsilon}{4}(a^{\dagger 2} - a^2) \tag{7.2.46}$$

but the arrangement is otherwise the same as in the previous example. At this stage we will not be concerned with how to create such a Hamiltonian in practice—see Sect. 10.2for an explanation. The quantum Langevin equation now becomes

$$\dot{a} = \frac{\epsilon}{2}a^\dagger - \frac{\kappa}{2}a - \sqrt{\kappa}\; b_{\text{in}}(t). \tag{7.2.47}$$

We can solve this equation by introducing the *quadrature phases*

$$X = \frac{1}{2}(a + a^\dagger), \qquad Y = \frac{1}{2i}(a - a^\dagger) \tag{7.2.48}$$

$$B_{\text{in}}^X(t) = \frac{1}{2}[b_{\text{in}}(t) + b_{\text{in}}^\dagger(t)], \qquad B_{\text{in}}^Y(t) = \frac{1}{2i}[b_{\text{in}}(t) - b_{\text{in}}^\dagger(t)], \tag{7.2.49}$$

and these have the commutation relations

$$[B_{\text{in}}^X(t), B_{\text{in}}^Y(t')] = \frac{i}{2}\delta(t-t'). \tag{7.2.50}$$

The equations of motion are then

$$\dot{X} = \frac{\epsilon-\kappa}{2}X - \sqrt{\kappa}\; B_{\text{in}}^X(t), \qquad \dot{Y} = -\frac{\epsilon+\kappa}{2}Y - \sqrt{\kappa}\; B_{\text{in}}^Y(t). \tag{7.2.51}$$

a) Approximate Solution: Using the same techniques as in the previous example, one finds the stationary solutions are

$$X = \frac{2\sqrt{\kappa}}{\epsilon - \kappa} B_{\rm in}^X(t), \qquad Y = -\frac{2\sqrt{\kappa}}{\epsilon + \kappa} B_{\rm in}^Y(t) \tag{7.2.52}$$

so that

$$\begin{aligned} B_{\rm out}^X(t) &= \frac{\epsilon + \kappa}{\epsilon - \kappa} B_{\rm in}^X(t) \\ B_{\rm out}^Y(t) &= \frac{\epsilon - \kappa}{\epsilon + \kappa} B_{\rm in}^Y(t). \end{aligned} \tag{7.2.53}$$

Thus $B_{\rm in}^X(t)$ is amplified with a gain $G = (\epsilon + \kappa)/(\epsilon - \kappa) > 1$, whereas $B_{\rm in}^Y(t)$ is attenuated by a factor G^{-1}. In terms of the $b, b^\dagger$ variables, we can write

$$b_{\rm out}(t) = b_{\rm in}(t)\frac{G + G^{-1}}{2} + b_{\rm in}^\dagger(t)\frac{G - G^{-1}}{2}. \tag{7.2.54}$$

b) Commutation Relations: It is easy to check that the commutation relations are preserved—no added noise term is necessary since the amplification mixes $b_{\rm in}(t)$ and $b_{\rm in}^\dagger(t)$. This means that the gain is only in half of the signal, the X quadrature $B_{\rm in}^X(t)$, and the Y quadrature $B_{\rm in}^Y(t)$ is attenuated.

c) No Added Noise: We define a Fourier transform variable

$$X(t) = \frac{1}{\sqrt{2\pi}} \int d\omega\, {\rm e}^{{\rm i}\omega t} \tilde{X}(\omega) \tag{7.2.55}$$

with a similar definition for the other variables. Then exact solutions are

$$\begin{aligned} \tilde{B}_{\rm out}^X(\omega) &= \tilde{B}_{\rm in}^X(\omega)\frac{{\rm i}\omega - (\epsilon + \kappa)/2}{{\rm i}\omega - (\epsilon - \kappa)/2} \\ \tilde{B}_{\rm out}^Y(\omega) &= \tilde{B}_{\rm in}^Y(\omega)\frac{{\rm i}\omega + (\epsilon - \kappa)/2}{{\rm i}\omega + (\epsilon + \kappa)/2}. \end{aligned} \tag{7.2.56}$$

Since there is no added term in either of these, the signal to noise ratio for the output will be exactly the same as that of the input. Nevertheless, even though one quadrature is attenuated and the other amplified, the power put into the amplified quadrature more than compensates for that lost in the attenuated signal.

Exercise. Show that the power gain is $\frac{1}{2}\{|G(\omega)|^2 + |G(\omega)|^{-2}\} > 1$, where $G(\omega) = (2{\rm i}\omega - \epsilon - \kappa)/(2{\rm i}\omega - \epsilon + \kappa)$.

d) Squeezing: The time integrals of the field operators

$$I_{\rm in}^X(t) = \int_t^{t+\Delta t} B_{\rm in}^X(t')dt', \qquad I_{\rm in}^Y(t) = \int_t^{t+\Delta t} B_{\rm in}^Y(t')dt' \tag{7.2.57}$$

have canonical commutation relations

$$\left[I_{\rm in}^X(t), I_{\rm in}^Y(t)\right] = \frac{\rm i}{2}\Delta t. \tag{7.2.58}$$

Hence, if there is no mean field,

$$\langle I_{\rm in}^X(t)^2\rangle\langle I_{\rm in}^Y(t)^2\rangle \geq \Delta t^2/16. \tag{7.2.59}$$

If the input field is the vacuum, then

$$\langle I_{\rm in}^X(t)^2\rangle = \langle I_{\rm in}^Y(t)^2\rangle = \Delta t/4 \tag{7.2.60}$$

and we can see that the vacuum is a minimum uncertainty state.

From the approximate form of solution (7.2.53), we see that (7.2.59) is still true for the outputs, but

$$\begin{aligned}\langle I_{\rm out}^X(t)^2\rangle &= G^2\Delta t/4\\ \langle I_{\rm out}^Y(t)^2\rangle &= G^{-2}\Delta t/4.\end{aligned} \tag{7.2.61}$$

The uncertainty in the Y quadrature can be made as small as one pleases by letting $\epsilon \to \kappa$, with of course a corresponding increase in the uncertainty of the X quadrature. The output field is said to be *squeezed*.

Exercise. The canonical variables X and Y describe the field inside the cavity. The exact stationary solutions of (7.2.51) are

$$\begin{aligned}X(t) &= -\sqrt{\kappa}\int_{-\infty}^{t} dt' \mathrm{e}^{-(\kappa-\epsilon)(t-t')/2} B_{\rm in}^X(t')\\ Y(t) &= -\sqrt{\kappa}\int_{-\infty}^{t} dt' \mathrm{e}^{-(\kappa+\epsilon)(t-t')/2} B_{\rm in}^Y(t').\end{aligned} \tag{7.2.62}$$

Show explicitly that $[X, Y] = \mathrm{i}/2$, and that

$$\langle X^2\rangle\langle Y^2\rangle = \kappa^2/16(\kappa^2 - \epsilon^2) \tag{7.2.63}$$

so that the internal field is *not* a minimum uncertainty state. In fact the minimum value of $\langle Y^2\rangle$ is $\frac{1}{8}$, not zero as is the case for the "out" field; see [7.2].

7.3 The Macroscopic Limit in Open Quantum Systems

In various examples throughout this book we have met situations in which the connection of a system to a heat bath will tend to destroy superposition states. One example is found in Sect. 3.6.2c where it was shown that the interference between plane waves was destroyed by damping and another occurs in the phase damped oscillator, in Sect. 6.2.1, where we noted that off-diagonal elements are rapidly damped, and only probabilities play a significant role in the further time evolution. Although this damping of coherence seems to happen very frequently, it is not an automatic consequence of damping, but depends on the basis we choose to measure with. An example can perhaps clarify this.

7.3.1 Example—Quantum Brownian Motion

Consider the quantum Brownian motion master equation (3.6.28)

$$\frac{d\rho}{dt} = -\frac{\mathrm{i}}{\hbar}\left[H_{\mathrm{sys}}, \rho\right] - \frac{\mathrm{i}\gamma}{2\hbar}\left[X, [\dot{X}, \rho]_{+}\right] - \frac{\gamma kT}{\hbar^2}\left[X, [X, \rho]\right]. \tag{7.3.1}$$

We shall not, however, specify the coupling operator X, or H_{sys} at this stage. Suppose, however, that X commutes with H_{sys}, so that it is a constant of the motion of the undamped system. Then the equation of motion for the matrix elements of ρ in a basis of the eigenstate $|x\rangle$ of X, is

$$\frac{d}{dt}\langle x|\rho|y\rangle = \left\{-\mathrm{i}\omega_{xy} - \frac{\gamma kT}{\hbar^2}(x-y)^2\right\}\langle x|\rho|y\rangle \tag{7.3.2}$$

where

$$\hbar\omega_{xy} = \langle x|H_{\mathrm{sys}}|x\rangle - \langle y|H_{\mathrm{sys}}|y\rangle. \tag{7.3.3}$$

It is very clear that the solutions of (7.3.2) are damped exponentially, with a time constant proportional to the square of the distance from the diagonal, $x = y$. Further, this constant has $\hbar^2$ in the denominator, so if $x - y$ has a macroscopic scale, the damping will be exceedingly fast. Only the matrix elements of ρ close to the diagonal will remain, and these do not change with time.

Exercise. It has been assumed that $|x\rangle$ characterize the system—consider the situation in which the states are degenerate, $|x, i\rangle$, and show that coherence remains within the same eigenbasis $|x, i\rangle$, for different i, but the same x.

a) Destruction of Macroscopic Coherence: We see from this example that the density operator rapidly becomes a diagonal density operator, represented therefore only by the probabilities $\langle x|\rho|x\rangle$, and of course the basis in which the diagonalization occurs is determined by the coupling operator X. That operator which is coupled to the bath, or environment, is diagonalized rapidly—it therefore behaves as a classical variable. A superposition state, represented by a density operator $\frac{1}{2}\{|x_1\rangle + |x_2\rangle\}\{\langle x_1| + \langle x_2|\}$ would rapidly become a mixture, $\frac{1}{2}|x_1\rangle\langle x_1| + \frac{1}{2}|x_2\rangle\langle x_2|$. Notice that this diagonalization is indeed macroscopic, since it happens rapidly when $\gamma kT(x-y)^2/\hbar^2$ is large. If the difference between x and y is sufficiently small, the decay of these off-diagonal elements will not happen very rapidly.

b) Competing Diagonalizations: Suppose there are two couplings, for example, to two operators with canonical commutation relations, $[X, P] = \mathrm{i}\hbar$. Then there is no simultaneous diagonal basis, and exact diagonalization in both bases simultaneously is impossible. However, the damping term

$$-\frac{\gamma_1 kT}{\hbar^2}[X, [X, \rho]] - \frac{\gamma_2 kT}{\hbar^2}[P, [P, \rho]] \tag{7.3.4}$$

becomes for the x matrix elements

$$\left(-\frac{\gamma_1 kT}{\hbar^2}z^2 + \gamma_2 kT\frac{\partial^2}{\partial z^2}\right)\langle u + \tfrac{1}{2}z|\rho(t)|u - \tfrac{1}{2}z\rangle \tag{7.3.5}$$

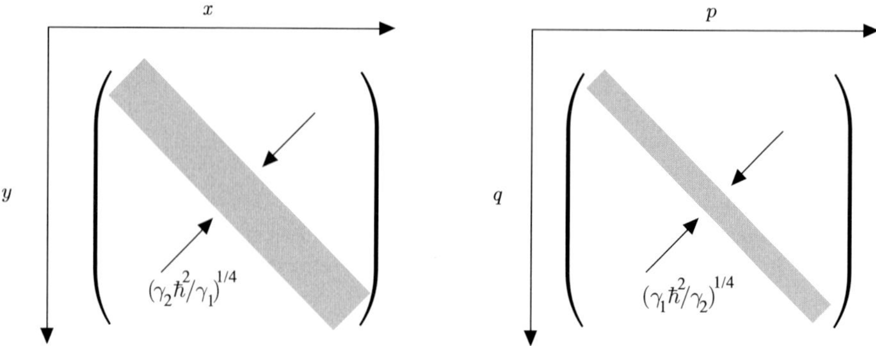

Fig. 7.2 Competing diagonalizations of the density operator by X and P damping

with $z = x - y$ and $u = (x + y)/2$. The operator in (7.3.5) is the same as that for the harmonic oscillator, so that the time development will be given by exponentials of the form $\exp(-\lambda_n t)$, with

$$\lambda_n = (2n + 1)\sqrt{\gamma_1\gamma_2}\,/\hbar \tag{7.3.6}$$

with the eigenfunctions the same as those of the harmonic oscillator, with a characteristic distance scale $(\gamma_2\hbar^2/\gamma_1)^{1/4} \equiv l_c$. Thus, the density operator will diagonalize to a width l_c, and will be a Gaussian in the variable z. A complementary result will hold for the variable P; in essence, we reach a minimum uncertainty wavefunction in the conjugate variables. The degree to which this wavefunction is diagonal in P or X depends on the relative strengths of the damping constants γ_1 and γ_2. The density operator becomes as diagonal as it can in both variables simultaneously, and if the system is macroscopic, this amounts essentially to well defined values of P and X. The situation is illustrated in Fig.7.2.

c) Meaning of the Diagonalization: Once the damping has "diagonalized" the density operator, all we are left with is a probability distribution. If all coupling operators, like X and P, are conserved, the resulting probability distribution is the same as that implied by the original density operator—but no superposition effects will be observable on a macroscopic scale in this basis.

d) Suppose X is not Conserved: The example of Sect. 3.6.2 in fact is such a case; even for a free particle, the position is not conserved, and the degree to which it is not conserved depends on the mass of the particle, and on the value of the momentum, by Newton's law $\dot{x} = p/m$. If m is very large, then we can regard the position as almost constant on the time scale of damping, and all the above conclusions will still be true.

Exercise— Scaling with Mass. Take the formula (3.6.60), and consider the limit $m \to$ large, with fixed damping time, γ/m, and for a superposition state consisting of fixed *velocities* $v_i = \hbar k_i/m$. Show that $\eta(t) \sim m$ for large m. Thus the classical limit corresponds to the situation of very weak H_{sys}. However, notice that if we fix the *momenta*, $p_i = \hbar k_i$, the damping is weaker at large m.

7.3.2 Example—The Quantum Optical Situation

Consider a damped harmonic oscillator at absolute zero, for which the master equation is

$$\dot{\rho} = \gamma \left(2a\rho a^\dagger - a^\dagger a\rho - \rho a^\dagger a\right). \tag{7.3.7}$$

Using the positive P-representation, we can deduce the equivalent SDEs

$$\begin{aligned} d\alpha &= -\gamma\alpha \, dt \\ d\beta &= -\gamma\beta \, dt \end{aligned} \tag{7.3.8}$$

there being no fluctuating term because this is a zero temperature situation. For an initial superposition of coherent states $\left(|\alpha_1\rangle + |\alpha_2\rangle\right)/\sqrt{2}$, we find that

$$\begin{aligned} P(\alpha,\beta,0) = {} & \tfrac{1}{2}\delta^2(\alpha-\alpha_1)\delta^2(\beta-\alpha_1^*) + \tfrac{1}{2}\delta^2(\alpha-\alpha_2)\delta^2(\beta-\alpha_2^*) \\ & + \tfrac{1}{2}\langle\alpha_2|\alpha_1\rangle\delta^2(\alpha-\alpha_1)\delta^2(\beta-\alpha_2^*) \\ & + \tfrac{1}{2}\langle\alpha_1|\alpha_2\rangle\delta^2(\alpha-\alpha_2)\delta^2(\beta-\alpha_1^*). \end{aligned} \tag{7.3.9}$$

Notice the normalization factors in the second two terms which occur because the positive P-function is an expansion in $\Lambda(\alpha,\beta) = |\alpha\rangle\langle\beta^*|/\langle\beta^*|\alpha\rangle$. If α and β are very different, this makes $\Lambda(\alpha,\beta)$ very large, since $\langle\alpha|\beta\rangle$ is very small. Correspondingly, the coefficient of the delta function in (7.3.9) is very small. With time evolution, (7.3.9) will, according to (7.3.8) become

$$\begin{aligned} P(\alpha,\beta,t) = {} & \tfrac{1}{2}\delta^2(\alpha-\alpha_1 e^{-\gamma t})\delta^2(\beta-\alpha_1^* e^{-\gamma t}) \\ & + \tfrac{1}{2}\delta^2(\alpha-\alpha_2 e^{-\gamma t})\delta^2(\beta-\alpha_2^* e^{-\gamma t}) \\ & + \tfrac{1}{2}\langle\alpha_2|\alpha_1\rangle\delta^2(\alpha-\alpha_1 e^{-\gamma t})\delta^2(\beta-\alpha_2^* e^{-\gamma t}) \\ & + \tfrac{1}{2}\langle\alpha_1|\alpha_2\rangle\delta^2(\alpha-\alpha_2 e^{-\gamma t})\delta^2(\beta-\alpha_1^* e^{-\gamma t}) \end{aligned} \tag{7.3.10}$$

and hence the density operator is

$$\begin{aligned} \rho = {} & \tfrac{1}{2}\left\{|\alpha_1 e^{-\lambda t}\rangle\langle\alpha_1 e^{-\lambda t}| + |\alpha_2 e^{-\lambda t}\rangle\langle\alpha_2 e^{-\lambda t}|\right\} \\ & + \tfrac{1}{2}\frac{\langle\alpha_1|\alpha_2\rangle}{\langle\alpha_1 e^{-\gamma t}|\alpha_2 e^{-\gamma t}\rangle}|\alpha_2 e^{-\gamma t}\rangle\langle\alpha_1 e^{\gamma t}| \\ & + \tfrac{1}{2}\frac{\langle\alpha_2|\alpha_1\rangle}{\langle\alpha_2 e^{-\gamma t}|\alpha_1 e^{-\gamma t}\rangle}|\alpha_1 e^{-\gamma t}\rangle\langle\alpha_2 e^{-\gamma t}|. \end{aligned} \tag{7.3.11}$$

Notice the extra coefficient in the off diagonal terms. The numerator is in both cases smaller than the denominator, because with the exponential decay, the two states approach each other. Using (4.3.6), we find the modulus of the coefficient of the diagonal terms is

$$\exp\left\{-\tfrac{1}{2}|\alpha_1-\alpha_2|^2\left(1-e^{-2\gamma t}\right)\right\}. \tag{7.3.12}$$

If α_1 and α_2 are macroscopic, then $|\alpha_1|^2$ and $|\alpha_1|^2$ are approximately the number of quanta, so that $|\alpha_1 - \alpha_2|^2$ is enormous; thus, for γt the slightest bit different from zero, (7.3.12) is essentially zero—the macroscopic superposition has been destroyed.

However, since they are coherent states, these states are only almost diagonal. This is because of the rotating wave interaction which gives rise to the master equation is essentially a damping which is coupled to *two* operators, $\frac{1}{2}(a + a^\dagger)$, and $\frac{1}{2}\mathrm{i}(a - a^\dagger)$, and the situation is thus similar to that in Sect. 7.3.1b.

7.3.3 Application to a Model of Quantum Measurement

A simple model of a quantum measurement was introduced by *Walls et al.* [7.3]. The basic idea is based on an exposition by *Zurek* [7.4].

A naive view of measurement is given by considering a system which is to be measured, coupled by some Hamiltonian to a meter, which is some other quantum system. With the evolution of time, the states of the meter and the system become correlated, thus knowledge of the state of the meter will give information about the system.

An ideal measurement should not disturb the system being measured—this is of course impossible. The von Neumann hypothesis suggests that the degree of disturbance should be such as not to disturb the variable being measured; in other words, we would expect any reasonable measurement to be such that the variable being measured would commute with the total Hamiltonian.

a) An Optical Example: Let us consider a system and meter, described by harmonic oscillator variables, a, b, with

$$\begin{aligned} H &\equiv H_{\text{sys}} + H_{\text{M}} + H_{\text{Int}} \\ &\equiv \hbar\omega a^\dagger a + \hbar\omega b^\dagger b - \mathrm{i}\hbar a^\dagger a(b\epsilon^* - b^\dagger\epsilon). \end{aligned} \tag{7.3.13}$$

This could describe two modes of the electromagnetic field, coupled by a classical pump field ϵ, which induces a four wave mixing in some intracavity non-linear medium.

Notice that H commutes with $a^\dagger a$, the system photon number, so that $a^\dagger a$ is a candidate for the operator being measured. We also assume that the b field, which carries out the measurement, is itself damped.

This is done by coupling meter variable b to an environment, represented by a heat bath. We can then write, in an interaction picture, QSDE's for a and b, including the damping of the b field,

$$\begin{aligned} da &= a(b\epsilon^* - b^\dagger\epsilon)dt \\ db &= \left(\epsilon a^\dagger a - \gamma b\right) dt + \sqrt{2\gamma}\, dB(t). \end{aligned} \tag{7.3.14}$$

These equations can be considered in either Ito or Stratonovich form, since the noise term is independent of a and b. Notice that the equation implies that $a^\dagger a$ is a constant of the motion, and to emphasize this we define $N \equiv a^\dagger a$.

b) Solution of the Equations: The equations can be solved exactly, firstly the b equation

$$b(t) = \frac{\epsilon N}{\gamma}\left(1 - \mathrm{e}^{-\gamma t}\right) + b(0)\mathrm{e}^{-\gamma t} + \sqrt{2\gamma}\int_0^t \mathrm{e}^{-\gamma(t-t')} dB(t'). \tag{7.3.15}$$

This solution represents the measurement process. One can see that if ϵ is sufficiently large, the first term can have a macroscopic scale, and is proportional to N, the operator to be measured. Thus the measurement process has gain—this will be seen to be an essential feature of the theory.

We now substitute this solution into the first equation;

$$\frac{da}{dt} = aX(t) \tag{7.3.16}$$

with

$$X(t) = \left[\epsilon^* b(0) - \epsilon b^\dagger(0)\right] e^{-\gamma t} + \sqrt{2\gamma} \int_0^t e^{-\gamma(t-t')} \left[\epsilon^* dB(t') - \epsilon dB^\dagger(t')\right]. \tag{7.3.17}$$

Notice that $\left[X(t), X(t')\right] = 0$, so that

$$a(t) = a(0) \exp\left[\int_0^t dt' X(t')\right] = \exp\left[\int_0^t dt' X(t')\right] a(0). \tag{7.3.18}$$

Notice that $X(t)$ is in fact completely independent of N; the dependence given by the solution (7.3.15) for $b(t)$ cancels. By construction $X(t)$ is quantum Gaussian, so that

$$\left\langle \exp\left[\int_0^t dt' X(t')\right]\right\rangle = \exp\left\{\tfrac{1}{2}\left\langle\left[\int_0^t dt' X(t')\right]^2\right\rangle\right\} \tag{7.3.19}$$

$$= \exp\left\{\tfrac{1}{2}\left\langle \int_0^t dt' \int_0^t ds\, X(t')X(s)\right\rangle\right\}. \tag{7.3.20}$$

Exercise. Assume that the bath is at zero temperature, and compute that, in the steady state,

$$\langle X(t')X(t)\rangle = -|\epsilon|^2 e^{-\gamma|t-t'|}. \tag{7.3.21}$$

c) Phase Destruction: We can now see that the phase of $a(t)$ is randomized through the solution (7.3.18), where $X(t)$ behaves like a phase noise, which arises as a result of the meter field $b(t)$. This can be called the “back action” of the meter.

d) Quantum Characteristic Function: Let us compute the joint quantum characteristic function of the system, assuming that it was initially in a state

$$\rho(0) = \sum p_{nm} \left(|n\rangle\langle m|\right)_{\rm sys} \otimes \left(|0\rangle\langle 0|\right)_{\rm meter}. \tag{7.3.22}$$

This will involve terms like

$$\langle 0|\langle n| \exp\left[\lambda b^\dagger(t)\right] \exp\left[\mu a^\dagger(t)\right] \exp\left[-\mu^* a(t)\right] \exp\left[-\lambda^* b(t)\right] |m\rangle|0\rangle. \tag{7.3.23}$$

We now use the solution (7.3.15) to replace N in $b, b^\dagger$, with m or n as appropriate, and bring the remainder outside the $\langle n| \quad |m\rangle$ mean; thus (7.3.23) becomes

$$\langle 0|\exp\left[\lambda b_n^\dagger(t)\right]\langle n|\exp\left[\mu a^\dagger(t)\right]\exp\left[-\mu^* a(t)\right]|m\rangle\exp\left[-\lambda^* b_m(t)\right]|0\rangle \tag{7.3.24}$$

Now :

i) Write

$$|n,t\rangle = \frac{a^\dagger(t)^n}{\sqrt{n!}}|0\rangle = \exp\left[-n\int_0^t dt' X(t')\right]|n\rangle \tag{7.3.25}$$

and note that $X(t)$ is an operator independent of a.

ii) Note

$$\exp\left[-\lambda b_m(t)\right]|0\rangle = \exp\left[-\frac{\lambda\epsilon m}{\gamma}\left(1-\mathrm{e}^{-\gamma t}\right)\right]|0\rangle \tag{7.3.26}$$

and similarly for n, and write

$$\begin{aligned}(7.3.24) = {} & \chi_{n,m}(\mu,\mu^*)\langle 0|\exp\left[(n-m)\int_0^t dt' X(t')\right]|0\rangle \\ & \times\exp\left\{-\frac{\lambda^*\epsilon m}{\gamma}\left(1-\mathrm{e}^{-\gamma t}\right)\right\}\times\exp\left\{-\frac{\lambda\epsilon^* n}{\gamma}\left(1-\mathrm{e}^{-\gamma t}\right)\right\}\end{aligned} \tag{7.3.27}$$

Here, $\chi_{nm}(\mu,\mu^*)$ is the quantum characteristic function for $|m,t\rangle\langle n,t|$, in terms of the operators $a(t)$, $a^\dagger(t)$, which is thus exactly the same as that for $|m\rangle\langle n|$, since $a(t)$ and $a(0)$ are unitarily related.

Finally, the average can be evaluated using (7.3.21). We find

$$\begin{aligned}\chi(\lambda,\lambda^*,\mu,\mu^*) = {} & \chi_{n,m}(\mu,\mu^*)\exp\left\{\frac{|\epsilon|^2(n-m)^2}{\gamma^2}\left(1-\gamma t-\mathrm{e}^{-\gamma t}\right)\right\} \\ & \times\exp\left\{-\frac{\lambda^*\epsilon m}{\gamma}(1-\mathrm{e}^{-\gamma t})-\frac{\lambda\epsilon^* n}{\gamma}(1-\mathrm{e}^{-\gamma t})\right\}.\end{aligned} \tag{7.3.28}$$

e) Density Operator Solution: This characteristic function corresponds to a state of the form

$$|\alpha_m(t)\rangle\langle\alpha_n(t)|\otimes|n\rangle\langle m|\times R(n,m,t) \tag{7.3.29}$$

where we can compute that

$$|R(n,m,t)| = \exp\left\{-\frac{|\epsilon|^2(n-m)^2}{2\gamma^2}\left[\gamma t-(1-\mathrm{e}^{-\gamma t})-(1-\mathrm{e}^{-\gamma t})^2\right]\right\}. \tag{7.3.30}$$

Here the $(1 - e^{-\gamma t})^2$ term comes from the fact that the normalization of $|\alpha\rangle\langle\beta|$ is not 1, and

$$\alpha_m(t) = \frac{\epsilon m}{\gamma}(1 - e^{-\gamma t}). \tag{7.3.31}$$

f) Interpretation—The Projection Postulate: We see that the measurement process drives the meter into a coherent state $|\alpha_m(t)\rangle$, whose mean is proportional to m. However, if this value is macroscopic, this will also mean that $|\epsilon|^2(n-m)^2/2\gamma^2$ is also macroscopic, and thus for any γt significantly different from zero,

$$|R(n, m, t)| \to \delta_{n,m}. \tag{7.3.32}$$

Thus, the state after the measurement is essentially fully diagonal,

$$\rho \to \sum_{n,n} |n\rangle\langle n| \otimes |\alpha_n(t)\rangle\langle\alpha_n(t)|. \tag{7.3.33}$$

This is the von Neumann projection postulate; the collapse of the wavefunction. Let us suppose the values of n are such that we can easily distinguish the macroscopic states $|\alpha_n(t)\rangle$ and $|\alpha_m(t)\rangle$ for $m \neq n$. Then if we see that the meter is in a state corresponding to $\alpha_n(t)$, we *know* that the meter-system state is

$$|n\rangle\langle n| \otimes |\alpha_n(t)\rangle\langle\alpha_n(t)|. \tag{7.3.34}$$

The calculation we have done shows that the coherences will be destroyed by the action of the macroscopic meter; the problem is then purely probabilistic. No prediction is given for what value is measured, but the diagonalization of the density operator ensures that if we measure, then we know that the system must be in the state $|n\rangle\langle n|$.

g) Concluding Remarks: Notice that the reason that the density operator takes on this "classical" form is ultimately the commutation relations of $dB(t)$ and $dB^\dagger(t)$, and the quantum noise driving them. The phase destruction factor comes from (7.3.21), and the fact that this is non-zero even though the state of $dB(t)$ and $dB^\dagger(t)$ is the vacuum. (If additional thermal noise is included, the diagonalization will be faster; but the result a thermal state).

The phase destruction arises from the fact that the coupling to the meter is a Hamiltonian interaction—the same term that allows the system to influence the meter makes the meter influence the system. Because of the coupling chosen, the variable being measured is not itself affected, but variables corresponding to phases are, and the result is the collapse of the wavefunction.

Notice also that it is important for a fast collapse that $\epsilon n/\gamma$ be extremely large, i.e., macroscopic. We thus conclude that quantum mechanics has within it a mechanism which ensures indeterminacy at a microscopic level, but this very mechanism also acts to conceal the effects of quantum mechanical superposition in macroscopic systems.

8.1.2 Expansion in Mode Functions

We separate the vector potential into two complex terms

$$\mathbf{A}(\mathbf{r},t) = \mathbf{A}^{(+)}(\mathbf{r},t) + \mathbf{A}^{(-)}(\mathbf{r},t). \tag{8.1.7}$$

Here $\mathbf{A}^{(+)}(\mathbf{r},t)$ contains only Fourier components with positive frequency, i.e., only terms which vary as $e^{-i\omega t}$ for $\omega > 0$, and $\mathbf{A}^{(-)}(\mathbf{r},t)$ contains amplitudes which vary as $e^{i\omega t}$. We take $\mathbf{A}$ to be real (or quantum mechanically, Hermitian) so $\mathbf{A}^{(-)}(\mathbf{r},t) = \{\mathbf{A}^{(+)}(\mathbf{r},t)\}^*$. It is more convenient to deal with a discrete set of variables rather than the whole continuum. We shall therefore describe the field restricted to a certain volume of space and expand the vector potential in terms of a discrete set of orthogonal mode functions

$$\mathbf{A}^{(+)}(\mathbf{r},t) = \sum_k c_k \mathbf{u}_k(\mathbf{r}) e^{-i\omega_k t} \tag{8.1.8}$$

where the Fourier coefficients c_k are constant for a free field. The set of vector mode functions $\mathbf{u}_k(\mathbf{r})$ which correspond to frequency ω_k will satisfy the wave equation

$$\left(\nabla^2 + \frac{\omega_k^2}{c^2}\right) \mathbf{u}_k(\mathbf{r}) = 0. \tag{8.1.9}$$

The mode functions are also required to satisfy the transversality condition which arises from (8.1.5)

$$\nabla \cdot \mathbf{u}_k(\mathbf{r}) = 0. \tag{8.1.10}$$

They also form an orthonormal set

$$\int_V \mathbf{u}_k^*(\mathbf{r}) \cdot \mathbf{u}_{k'}(\mathbf{r}) d^3\mathbf{r} = \delta_{kk'} \tag{8.1.11}$$

which is complete within the chosen volume. The mode functions depend on the boundary conditions of the physical volume under consideration, e.g., periodic boundary conditions corresponding to travelling wave modes or conditions appropriate to reflecting walls which lead to standing waves. Plane wave mode functions appropriate to a cubical volume of side L with periodic boundary conditions may be written as

$$\mathbf{u}_k(\mathbf{r}) = L^{-\frac{3}{2}} \hat{\mathbf{e}}^{(\lambda)} \exp(i\mathbf{k}\cdot\mathbf{r}) \tag{8.1.12}$$

where $\hat{\mathbf{e}}^{(\lambda)}$ are the unit polarization vectors satisfying

$$\mathbf{k} \cdot \hat{\mathbf{e}}^{(\lambda)} = 0, \qquad \hat{\mathbf{e}}^{(\lambda)} \cdot \hat{\mathbf{e}}^{(\lambda')} = \delta_{\lambda\lambda'}. \tag{8.1.13}$$

The mode index k describes several discrete variables, the polarization index ($\lambda = 1, 2$) and the three Cartesian components of the propagation vector $\mathbf{k}$. Each component of the wave vector **k** takes the values

$$\begin{aligned} k_x &= \frac{2\pi n_x}{L} \\ k_y &= \frac{2\pi n_y}{L} \\ k_z &= \frac{2\pi n_z}{L} \\ n_x, n_y, n_z &= 0, \pm 1, \pm 2 \ldots \end{aligned} \tag{8.1.14}$$

and as a consequence of the equation (8.1.9)

$$\omega_k = c|\mathbf{k}|. \tag{8.1.15}$$

The polarization vector $\hat{\mathbf{e}}^{(\lambda)}$ is required to be perpendicular to **k** by the transversality condition (8.1.10). The vector potential may now be written in the form

$$\mathbf{A}(\mathbf{r}, t) = \sum_k \left(\frac{\hbar}{2\omega_k \epsilon_0}\right)^{\frac{1}{2}} \left(a_k \mathbf{u}_k(\mathbf{r}) \mathrm{e}^{-\mathrm{i}\omega_k t} + a_k^\dagger \mathbf{u}_k^*(\mathbf{r}) \mathrm{e}^{\mathrm{i}\omega_k t}\right). \tag{8.1.16}$$

The corresponding form for the electric field is

$$\mathbf{E}(\mathbf{r}, t) = \mathrm{i} \sum_k \left(\frac{\hbar\omega_k}{2\epsilon_0}\right)^{\frac{1}{2}} \left(a_k \mathbf{u}_k(\mathbf{r}) \mathrm{e}^{-\mathrm{i}\omega_k t} - a_k^\dagger \mathbf{u}_k^*(\mathbf{r}) \mathrm{e}^{\mathrm{i}\omega_k t}\right). \tag{8.1.17}$$

The normalization factors have been chosen such that the amplitudes a_k and $a_k^\dagger$ are dimensionless.

8.1.3 Quantization by Commutation Relations

The Hamiltonian for the electromagnetic field is given by

$$H = \int \left(\frac{\epsilon_0 \mathbf{E}^2}{2} + \frac{\mathbf{B}^2}{2\mu_0}\right) d^3\mathbf{r}. \tag{8.1.18}$$

By substituting for **E** and **B** using the expression (8.1.17) for **E** and a similar one for **B**, and by making use of the conditions (8.1.9,10), this Hamiltonian can be reduced to the form

$$H = \tfrac{1}{2} \sum_k \hbar\omega_k \left(a_k^\dagger a_k + a_k a_k^\dagger\right). \tag{8.1.19}$$

This is of course exactly the same as the Hamiltonian for an assembly of independent harmonic oscillators. Quantization is accomplished by assuming the same commutation relations for $a_k, a_k^\dagger$ as would be found from quantizing such an assembly, namely

$$\left[a_k, a_{k'}\right] = \left[a_k^\dagger, a_{k'}^\dagger\right] = 0, \qquad \left[a_k, a_{k'}^\dagger\right] = \delta_{kk'}. \tag{8.1.20}$$

The dynamical states of the electromagnetic field may then be described by assigning an appropriate quantum state to each of the modes—these modes may be assigned and described independently, and all the apparatus developed for harmonic oscillators applied to them.

8.1.4 Quantization in an Infinite Volume

While quantization in a finite box is perfectly legitimate, provided the box is very much larger than the dimensions under consideration, its use is nevertheless slightly deceptive, for two reasons.

i) In quantum optics the quantization volume is sometimes chosen finite, for physical reasons, since many optical experiments take place either in a cavity, or between highly reflecting mirrors. In such a situation, only a discrete number of field modes can occur, and the description by quantization of the discrete cavity modes is appropriate. Of course if we want to describe the rest of the universe, we must choose a box which is very large compared to the cavity. It is conceptually simpler to think of the outside world as being actually infinite, than to have two quantization volumes.

ii) In a box irreversibility does not occur, since quanta cannot be lost forever—they bounce back from the walls of the box, or return from the other side if periodic boundary conditions are imposed. Of course for a large enough box this may take a long time, but the pure vastness of infinite space into which light can disappear without trace yields the concept of irreversibility directly.

The transition to an infinite volume decomposition is achieved by letting the quantization volume become infinite. We note that as n_x, n_y, n_z increase by 1,

$$k_x \to k_x + \frac{2\pi}{L}, \text{etc.,} \tag{8.1.21}$$

so that

$$\frac{1}{L^3}\sum_k \to \frac{1}{(2\pi)^3}\sum_\lambda \int d^3\mathbf{k}. \tag{8.1.22}$$

If we also let

$$a_k \to \left(\frac{2\pi}{L}\right)^{\frac{3}{2}} a^\lambda(\mathbf{k}) \tag{8.1.23}$$

we find

$$\left[a^\lambda(\mathbf{k}), a^{\lambda'}(\mathbf{k}')^\dagger\right] = \delta^3(\mathbf{k}-\mathbf{k}')\delta_{\lambda\lambda'} \tag{8.1.24}$$

and

$$\mathbf{A}^{(+)}(\mathbf{r},t) = \frac{1}{(2\pi)^{\frac{3}{2}}}\int d^3\mathbf{k}\left(\frac{\hbar}{2\omega_k\epsilon_0}\right)^{\frac{1}{2}}\sum_\lambda \hat{\mathbf{e}}^\lambda \exp(i\mathbf{k}\cdot\mathbf{r} - i\omega_k t)a^\lambda(\mathbf{k}) \tag{8.1.25}$$

with corresponding expressions for $\mathbf{A}(\mathbf{r},t)$, $\mathbf{E}(\mathbf{r},t)$, etc.

8.1.5 Optical Electromagnetic Fields

It is very rare in optical situations to have an electromagnetic field whose bandwidth is anything more than an infinitesimal fraction of the frequency being used. For example, a typical optical frequency is 6×10^{14} Hz , whereas the light from an atomic transition never has a bandwidth of more than one thousandth of this. This means that there will not be much difference between quantities calculated using $\mathbf{A}(\mathbf{r},t)$ or $\mathbf{E}(\mathbf{r},t)$, and quantities calculated using *approximate* expressions in which the $\sqrt{\omega_{\mathbf{k}}}$ factors which occur in the mode representations (8.1.16 ,17) are replaced by $\sqrt{\Omega}$, where Ω is the frequency being used. This means that we could write, approximately,

$$\begin{aligned} \mathbf{E}^{(+)}(\mathbf{r},t) &\approx \mathrm{i}\Omega\mathbf{A}^{(+)}(\mathbf{r},t) \\ &= \frac{\mathrm{i}}{(2\pi)^{\frac{3}{2}}}\left(\frac{\hbar\Omega}{2\epsilon_0}\right)^{\frac{1}{2}}\int d^3\mathbf{k}\sum_\lambda \hat{\mathbf{e}}^\lambda \mathrm{e}^{\mathrm{i}(\mathbf{k}\cdot\mathbf{r}-\omega_{\mathbf{k}}t)}a^\lambda(\mathbf{k}). \end{aligned} \tag{8.1.26}$$

The independent variables in this approximation are now the positive and negative frequency parts of $\mathbf{E}(\mathbf{r},t)$, and these satisfy the equal times commutation relations,

$$\left[E_i^{(+)}(\mathbf{r},t), E_j^{(-)}(\mathbf{r}',t)\right] = \frac{\hbar\Omega}{2\epsilon_0}\delta_{ij}^T(\mathbf{r}-\mathbf{r}'). \tag{8.1.27}$$

The quantity $\delta_{ij}^T(\mathbf{r}-\mathbf{r}')$ is called the *transverse delta function* and is defined by

$$\delta_{ij}^T(\mathbf{r}-\mathbf{r}') = \left(\delta_{ij} - \frac{\partial_i\partial_j}{\nabla^2}\right)\delta(\mathbf{r}-\mathbf{r}'). \tag{8.1.28}$$

It arises in the commutator because of the identity

$$\sum_\lambda \hat{\mathrm{e}}_i^{(\lambda)}(\mathbf{k})\hat{\mathrm{e}}_j^{(\lambda)}(\mathbf{k}) = \delta_{ij} - k_i k_j/\mathbf{k}^2, \tag{8.1.29}$$

which follows from the transversality of the electromagnetic field. By construction, it follows that

$$\partial_i\delta_{ij}^T(\mathbf{r}-\mathbf{r}') = 0 \tag{8.1.30}$$

which means that the commutation relation is then consistent with $\nabla\cdot\mathbf{E} = 0$.

The Hamiltonian is now written as in (8.1.19), and yields the equation of motion

$$\dot{\mathbf{E}}^{(+)}(\mathbf{r},t) = \frac{\mathrm{i}}{\hbar}\left[H, \mathbf{E}^{(+)}(\mathbf{r},t)\right] = -\mathrm{i}c\sqrt{-\nabla^2}\,\mathbf{E}^{(+)}(\mathbf{r},t). \tag{8.1.31}$$

The $\sqrt{-\nabla^2}$ is to be interpreted as a factor $|\mathbf{k}|$ in the Fourier transform. This means that the equation (8.1.31) has only positive frequency solutions, as should be the case.

8.1.6 The Photon

From almost the very beginning of quantum theory, the idea of quantization has been completely bound up with the idea that light consists of discrete "bundles" of energy $h\nu$, called photons. Nevertheless, in the days of the "old quantum theory" almost all physicists had great difficulty in reconciling the idea of the photon with Maxwell's electromagnetic field equations. As an exact concept, the idea of a photon with energy $h\nu$ does present difficulties, since an exact frequency ν can only be specified in a wave field which lasts for an infinite time—thus practical photons, which appear to be moderately well localized in space and time, cannot have a well defined frequency ν.

Furthermore, looking at the Hamiltonian formula (8.1.19), it is obvious that the energy eigenvalues of the system have the form

$$\sum_k \hbar\omega_k(n_k + \tfrac{1}{2}) \tag{8.1.32}$$

which says that the only observable energies are equivalent to those of a number of photons of energies $\hbar\omega_k = h\nu_k$. However, when we consider a genuinely infinite system this simple interpretation is no longer possible, and one must talk about ranges of energies etc., which, while quite straightforward mathematically, are not very enlightening. And even with a finite system, the quantization volume must be so large that the frequency of any practical photon is only defined up to an accuracy which would encompass an enormous number of very finely spaced possible values of $\hbar\omega_k$.

The basic conceptual stumbling block lies in the formula $E = h\nu$, the foundation stone of quantum theory. It is perfectly straightforward to define a photon which does not have a definite energy, and we can show that all states of the electromagnetic field are linear combinations of states with a finite number of photons.

Let us take a complete set of complex solutions of Maxwell's equations; the vector potential $\boldsymbol{\mathcal{A}}^\alpha(\mathbf{r},t)$, and the electric field $\boldsymbol{\mathcal{E}}^\alpha(\mathbf{r},t)$, which are assumed to have only positive frequency. Noting that these functions must be transverse, we assume the normalization and completeness properties

$$\sum_\alpha \mathcal{A}_i^\alpha(\mathbf{r},t)\mathcal{E}_j^{\alpha *}(\mathbf{r},t) = \frac{i\hbar}{2\epsilon_0}\delta_{ij}^T(\mathbf{r}-\mathbf{r}') \tag{8.1.33}$$

$$\int d^3\mathbf{r}\,\boldsymbol{\mathcal{A}}^\alpha(\mathbf{r},t)\cdot\boldsymbol{\mathcal{E}}^{\beta *}(\mathbf{r},t) = \frac{2\epsilon_0}{i\hbar}\delta_{\alpha\beta}. \tag{8.1.34}$$

Because $\boldsymbol{\mathcal{E}}^\alpha(\mathbf{r},t)$, $\boldsymbol{\mathcal{A}}^\alpha(\mathbf{r},t)$ are solutions of Maxwell's equations, we can show that these conditions are preserved as time evolves; that is, the left hand sides are guaranteed to be time independent.

The electromagnetic field operators are then able to be expanded as

$$\mathbf{A}^{(+)}(\mathbf{r},t) = \sum_\alpha a_\alpha \boldsymbol{\mathcal{A}}^{\alpha *}(\mathbf{r},t) \tag{8.1.35}$$

$$\mathbf{E}^{(+)}(\mathbf{r},t) = \sum_\alpha a_\alpha \boldsymbol{\mathcal{E}}^{\alpha *}(\mathbf{r},t) \tag{8.1.36}$$

and the commutation relations between **E** and **A**, as required by the expansions (8.1.16,17) are

$$[A_i(\mathbf{r},t), E_j(\mathbf{r}',t)] = -\frac{\mathrm{i}\hbar}{\epsilon_0}\delta^T_{ij}(\mathbf{r}-\mathbf{r}'). \tag{8.1.37}$$

From these commutation relations and the orthogonality and completeness relations, (8.1.33,34), we can deduce the canonical commutation relation

$$[a_\alpha, a^\dagger_\beta] = \delta_{\alpha\beta}. \tag{8.1.38}$$

These operators a_α, $a^\dagger_\alpha$ are thus destruction and creation operators for the *mode* whose mode functions are $\mathcal{A}^\alpha$, $\mathcal{E}^\alpha$. The states of the electromagnetic field can thus be written in terms of n photon states such as $a^\dagger_{\alpha_1}a^\dagger_{\alpha_2}\dots a^\dagger_{\alpha_n}|0\rangle$. This will be a Heisenberg picture state, with one photon in each of the modes $\alpha_1, \alpha_2, \dots, \alpha_n$, with electromagnetic fields $\mathcal{E}^{\alpha_i *}(\mathbf{r},t)$, $\mathcal{E}^{\alpha_i *}(\mathbf{r},t)$ etc., which can describe a wide variety of possible situations. In particular, it is possible to speak of a one photon wavepacket when the mode functions correspond to those of a wavepacket.

Thus the concept of a photon is not limited to that of a quantum with energy $h\nu$. Any solution of Maxwell's equations can be chosen as a mode function, after being appropriately normalized as in (8.1.33). The modes orthogonal to this mode can then be constructed if required. The modes corresponding to definite frequency are of course those which diagonalize the Hamiltonian, in which the energy eigenvalues take on the form $\sum \hbar\omega_k[n(k)+\frac{1}{2}]$, but this is not possible in a general kind of mode.

8.1.7 Beams of Light

It is also of interest to consider a beam propagating in one dimension, say along the z axis. If the beam is quite wide compared to the wavelengths of interest, we can consider

$$\bar{\mathbf{E}}^{(+)}(z,t) \equiv \frac{2\pi}{A}\int_A dx\,dy\,\mathbf{E}^{(+)}(\mathbf{r},t). \tag{8.1.39}$$

We can show that, if A is very large

$$\bar{\mathbf{E}}^{(+)}(z,t) = \frac{\mathrm{i}}{\sqrt{2\pi}}\left(\frac{\hbar\Omega}{2\epsilon_0}\right)^{\frac{1}{2}}\int_{-\infty}^{\infty} dq \sum_\lambda \hat{\mathbf{e}}^\lambda\,\mathrm{e}^{\mathrm{i}(qz-\omega_q t)}\,\bar{a}^\lambda(q) \tag{8.1.40}$$

where

$$\bar{a}^\lambda(q) \approx \frac{(2\pi)^2}{A}a^\lambda(\{0,0,q\})$$

and

$$\left[\bar{a}^\lambda(q), \bar{a}^{\lambda'}(q')^\dagger\right] = \delta_{\lambda\lambda'}\delta(q-q'). \tag{8.1.41}$$

the origin of the occurrence of $E^{(+)}(\mathbf{r}, t)$ only. There are considerable technical problems with any treatment which does not make this approximation. Such a treatment is necessary in situations in which the detection takes place for only very short times. The fundamental problem is the same as that noted in the treatment of the harmonic oscillator in Sect. 3.4.3. There we assumed that the system was initially in a state which was factorized into a bath and an oscillator term. As soon as the time evolves, a very large transient occurs. In this case, we would find an infinite counting rate would initially occur, unless the rotating wave approximation is made in the initial coupling Hamiltonian. But this initial infinite counting rate is caused by our assumption that the field and the atom are initially independent, which is not the case, because of vacuum fluctuations. If we assume the atom-field system is initially in the true ground state (which may not be easy to determine) no infinite counting rates occur, and in the long time limit the rotating wave approximation result is obtained. A discussion of this problem, and a solution by means of a canonical transformation, is given by *Drummond* [8.5].

8.2.2 Coherence and Correlation Functions

It is convenient to define the nth order correlation functions by

$$\begin{aligned} &G^n(x_1, \ldots, x_n, x_{n+1}, \ldots, x_{2n}) \\ &\quad = \mathrm{Tr}\left\{\rho E^{(-)}(x_1) \ldots E^{(-)}(x_n) E^{(+)}(x_{n+1}) \ldots E^{(+)}(x_{2n})\right\} \end{aligned} \tag{8.2.6}$$

where

$$x = (\mathbf{r}, t). \tag{8.2.7}$$

This is more general than that measurable by a strictly local detection process, for which $x_i = x_{2n+1-i}$. The more general kind of correlation function will arise in case the measurement process is non-local, i.e., when we have $\int d^3\mathbf{r}\, \epsilon(\mathbf{r} - \mathbf{r}')E^{(+)}(\mathbf{r}', t)$ instead of $E^{(+)}(\mathbf{r}, t)$.

A link to the classical correlation functions can be provided by using a P-representation for the various modes of the electromagnetic field. We consider a situation in which the radiation field is described by a coherent state in every mode, $|\{\alpha\}\rangle$ so that, for all a_k,

$$a_k|\{\alpha\}\rangle = \alpha_k|\{\alpha\}\rangle. \tag{8.2.8}$$

Since $E^{(+)}(x)$ is a function only of destruction operators, it is also true that

$$E^{(+)}(x)|\{\alpha\}\rangle = \mathcal{E}^{(+)}(x)|\{\alpha\}\rangle \tag{8.2.9}$$

where

$$\mathcal{E}^{(+)}(x) = \mathrm{i}\sum_k \left(\frac{\hbar\omega_k}{2\epsilon_0}\right)^{\frac{1}{2}} \alpha_k \mathbf{u}_k(\mathbf{r}) \mathrm{e}^{-\mathrm{i}\omega_k t} \tag{8.2.10}$$

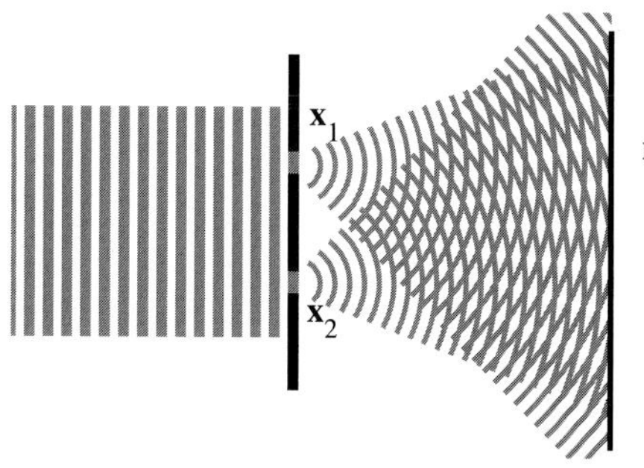

Fig. 8.1 Interference in a double slit experiment

is the positive frequency part of a classical electromagnetic field $\mathcal{E}(x)$. The density operator corresponding to $|\{\alpha\}\rangle$ is of course $|\{\alpha\}\rangle\langle\{\alpha\}|$, and substituting into (8.2.6), we see that

$$\begin{aligned} G^{(n)}\left(x_1, \ldots, x_n, x_{n+1}, \ldots, x_{2n+1}\right) \\ = \mathcal{E}^{(-)}(x_1)\ldots\mathcal{E}^{(-)}(x_n)\mathcal{E}^{(+)}(x_{n+1})\ldots\mathcal{E}^{(+)}(x_{2n+1}), \end{aligned} \tag{8.2.11}$$

that is, the correlation functions all factorized. This property of factorization is intimately connected with the visibility of the interference fringes.

We can imagine combining together the fields at points $\mathbf{x}_1$ and $\mathbf{x}_2$ by the classic two slit experiment, as in Fig.8.1. The field at point $\mathbf{r}$ is given (up to geometrical factors, in the case that $|\mathbf{x}_1 - \mathbf{r}| \simeq |\mathbf{x}_2 - \mathbf{r}|$) by

$$E^{(+)}(\mathbf{r}, t) = E^{(+)}(\mathbf{x}_1, t)\mathrm{e}^{\mathrm{i}k|\mathbf{r}-\mathbf{x}_1|} + E^{(+)}(\mathbf{x}_2, t)\mathrm{e}^{\mathrm{i}k|\mathbf{r}-\mathbf{x}_2|} \tag{8.2.12}$$

so that the average rate of photodetection at $\mathbf{r}$, where the beams are combined, is

$$P(\mathbf{r}, t) = \epsilon\langle E^{(-)}(\mathbf{r}, t)E^{(+)}(\mathbf{r}, t)\rangle. \tag{8.2.13}$$

If we assume the distance between the slits is very small compared to the distance to the point of detection $\mathbf{r}$, we can then write approximately

$$\begin{aligned} |\mathbf{r} - \mathbf{x}_1| &\approx r - \hat{\mathbf{r}}\cdot\mathbf{x}_1 \\ |\mathbf{r} - \mathbf{x}_2| &\approx r - \hat{\mathbf{r}}\cdot\mathbf{x}_2 \end{aligned} \tag{8.2.14}$$

where $\hat{\mathbf{r}} = \mathbf{r}/r$, i.e,. is a unit vector in the direction $\mathbf{r}$. Thus

$$\begin{aligned} P(\mathbf{r}, t) \approx \epsilon\Bigg[&\langle E^{(-)}(\mathbf{x}_1, t)E^{(+)}(\mathbf{x}_1, t)\rangle + \langle E^{(-)}(\mathbf{x}_2, t)E^{(+)}(\mathbf{x}_2, t)\rangle \\ &+2\mathrm{Re}\left\{\langle E^{(-)}(\mathbf{x}_2, t)E^{(+)}(\mathbf{x}_1, t)\rangle\mathrm{e}^{-\mathrm{i}k\hat{\mathbf{r}}\cdot(\mathbf{x}_1-\mathbf{x}_2)}\right\}\Bigg]. \end{aligned} \tag{8.2.15}$$

We can now consider various cases.

a) Electromagnetic Field in a Coherent State: In this case, as noted above, the correlation functions factorized, and

$$P(\mathbf{r}, t) = \epsilon|\mathcal{E}(\mathbf{x}_1, t) + \mathcal{E}(\mathbf{x}_2, t)\mathrm{e}^{-\mathrm{i}k\hat{\mathbf{r}}\cdot(\mathbf{x}_1-\mathbf{x}_2)}|^2. \tag{8.2.16}$$

If we now assume that $|\mathcal{E}(\mathbf{x}_1,t)|^2 = |\mathcal{E}_2(\mathbf{x}_2,t)|^2 = \mathcal{E}^2$, a constant, so that

$$\mathcal{E}(\mathbf{x}_1,t) = \mathcal{E}\,\mathrm{e}^{\mathrm{i}\phi(\mathbf{x}_1,t)}, \qquad \mathcal{E}(\mathbf{x}_2,t) = \mathcal{E}\,\mathrm{e}^{\mathrm{i}\phi(\mathbf{x}_2,t)}, \tag{8.2.17}$$

then

$$P(\mathbf{r},t) = 4\epsilon|\mathcal{E}|^2\cos^2\left(\tfrac{1}{2}\left[k\hat{\mathbf{r}}\cdot(\mathbf{x}_1-\mathbf{x}_2)+\phi(\mathbf{x}_1,t)-\phi(\mathbf{x}_2,t)\right]\right) \tag{8.2.18}$$

and it can be seen that as $\hat{\mathbf{r}}$ varies, the $\cos^2$ function will vary between 1 and 0—this gives rise to interference fringes, which can thus be said to be of maximum visibility. This corresponds to the classical concept of *coherence*—the fields at $\mathbf{x}_1$ and $\mathbf{x}_2$ are said to be maximally coherent.

b) Cross Correlation Function Zero: By this we mean that the correlation function $\langle E^{(-)}(\mathbf{x}_2,t)E^{(+)}(\mathbf{x}_1,t)\rangle = 0$, and thus

$$P(\mathbf{r},t) = \langle E^{(-)}(\mathbf{x}_1,t)E^{(+)}(\mathbf{x}_1,t)\rangle + \langle E^{(-)}(\mathbf{x}_2,t)E^{(+)}(\mathbf{x}_2,t)\rangle, \tag{8.2.19}$$

that is, the intensity at $\mathbf{r}$ is simply given by the sum of the two intensities, and no fringes are detectable in the photocount distribution.

8.2.3 Normalized Correlation Functions

It is convenient to define normalized correlation functions by

$$g^{(1)}(x_1,x_2) = \frac{G^{(1)}(x_1,x_2)}{\sqrt{G^{(1)}(x_1,x_1)G^{(1)}(x_2,x_2)}} \tag{8.2.20}$$

$$g^{(2)}(x_1,x_2) = \frac{G^{(2)}(x_1,x_2,x_2,x_1)}{G^{(1)}(x_1,x_1)G^{(1)}(x_2,x_2)}. \tag{8.2.21}$$

(Here the notation $x \equiv (\mathbf{x},t)$, introduced in (8.2.7) is again utilized for compactness.) The first order normalized correlation function determines the visibility of the interference fringes seen in the experiment of the previous section. There are inequalities for correlation functions which arise from the Schwartz inequality

$$|\langle AB^\dagger\rangle|^2 \le \langle A^\dagger A\rangle\langle B^\dagger B\rangle. \tag{8.2.22}$$

In particular, they imply that

$$|g^{(1)}(x_1,x_2)| \le 1. \tag{8.2.23}$$

In a coherent state, this becomes a equality, and we see that if $g^{(1)}(x_1,x_2) = 1$, the interference fringes of (8.2.15) are maximally visible. We define the fringe visibility, $\mathbf{V}$, by

$$\mathbf{V} = \frac{P(\mathbf{r},t)_{\max} - P(\mathbf{r},t)_{\min}}{P(\mathbf{r},t)_{\max} + P(\mathbf{r},t)_{\min}} \tag{8.2.24}$$

$$= \frac{2|g^{(1)}(\mathbf{x}_2,t;\mathbf{x}_1,t)|}{\sqrt{I_1/I_2} + \sqrt{I_2/I_1}} \tag{8.2.25}$$

where $I_1 = \langle E^{(-)}(\mathbf{x}_1, t)E^{(+)}(\mathbf{x}_1, t)\rangle$ is the intensity at $\mathbf{x}_1$, and I_2 is similarly defined. When $|g^{(1)}| = 1$, the visibility is maximal for given I_1 and I_2, and when $g^{(1)} = 0$, there are no fringes.

8.3 Photon Counting Formulae

Although the formula (8.2.4) for the probability of detecting photons at various times and places is completely general, it does not give directly a practical formula for analysing photodetection. The major problem lies in the effect that the detection process has on the light field being measured. The detection formula requires the knowledge of the field at the places where the photodetection takes place, and in a thick detector, or a highly efficient detector, this field is attenuated by an amount which must be determined.

It is possible to formulate a theory of weak detection, in which the field is not considered to be appreciably attenuated by the detection process, but this is hardly appropriate in practical situations, where the efficiency can be quite high. *Srinivas* and *Davies* [8.6] have developed a method of formulating photodetection which uses measurement theory as its basis, and uses a Markovian view of the processes involved. Using this formulation, it is possible to compute the absorption induced by photodetection, and to develop quite simple formulae, in a way which makes their generality clear.

8.3.1 Development of the Formulae

We want to consider a counting process with a detector taking up a finite volume V, so that the probability of detecting a single photon in the infinitesimal time interval dt is, (in the Schrödinger picture in which $\mathbf{E}$ does not depend on time)

$$P(1,t)dt = \epsilon\, dt \int_V d^3\mathbf{r}\, \mathrm{Tr}\left\{E^{(-)}(\mathbf{r})\, E^{(+)}(\mathbf{r})\, \rho(t)\right\}. \tag{8.3.1}$$

We want to derive a formula for the probability of counting a finite number of photons in a *finite* time interval $(t, t+\tau)$. We again use the Schrödinger picture. The processes under consideration are as follows. We can consider detecting m photons in $(t, t+\tau)$, and introduce the operation $N_{t+\tau,t}(m)$, which represents this process by:

i) After measuring m photons in this interval, the density operator is

$$\frac{N_{t+\tau,t}(m)\rho}{\mathrm{Tr}\left\{N_{t+\tau,t}(m)\rho\right\}}. \tag{8.3.2}$$

ii) The probability that m photons are measured is

$$P(t+\tau, t, m) = \mathrm{Tr}\left\{N_{t+\tau,t}(m)\rho\right\}. \tag{8.3.3}$$

This formula and (8.2.3) tell us that, for example, we may choose

$$N_{t+dt,t}(1)\rho = \epsilon\, dt \int d^3\mathbf{r}\, E^{(+)}(\mathbf{r})\rho(t)E^{(-)}(\mathbf{r}) \tag{8.3.4}$$

and this formula will be fundamental to the final result. The exact nature of $N_{t+dt,t}(1)\rho$ however, is not a fundamental assumption, but a description of the kind of detector being used.

a) Markov Assumption: From the very definition of counting, we can write

$$N_{t_1,t_0}(m) = \sum_{m_1+m_2=m} N_{t_1,s}(m_1)N_{s,t_0}(m_2) \qquad (t_1 > s > t_0) \tag{8.3.5}$$

which simply says that we can partition the interval (t_1, t_0) anywhere, and count some of the photons in the first interval, and some in the second. The processes of counting in the two intervals are *independent*, as is shown by the fact that the operations in each of the two intervals do not involve the number counted in the other interval. Because of this assumption, the resulting equations will have a Markov property. If the medium which causes the absorption has any kind of "memory" of how many photons it has absorbed, then this need not be true.

b) Semigroup Equations: From this equation we can derive two evolution equations. We define

$$S_{t+\tau,t} = N_{t+\tau,t}(0) \tag{8.3.6}$$

$$T_{t+\tau,t} = \sum_{m=0}^{\infty} N_{t+\tau,t}(m). \tag{8.3.7}$$

Then from (8.3.5) we find that both form semigroups, i.e.,

$$S_{t_1,t_0} = S_{t_1,s}S_{s,t_0} \tag{8.3.8}$$

$$T_{t_1,t_0} = T_{t_1,s}T_{s,t_0}. \tag{8.3.9}$$

The significance of these two operations lies in the fact that S_{t_1,t_0} describes the process of counting no photons, while T_{t_1,t_0} describes the process of counting an unknown number of photons. More precisely, we see from (8.3.2) that $S_{t_1,t_0}\rho(t_0)$ is the density operator at time t, under the condition that the detector was active during the time interval (t_0, t_1) and *no* photons were counted.

From (8.3.2,3) we can also see that

$$T_{t+\tau,t}\rho(t) = \sum_m \frac{N_{t+\tau,t}(m)\rho(t)}{\mathrm{Tr}\left\{N_{t+\tau,t}(m)\rho(t)\right\}} P(t+\tau, t, m) \tag{8.3.10}$$

which is the average of the density operators for all possible numbers of photons counted. Thus, this gives the field density operator as time evolves and photons are counted, in the situation when we do not know how many photons were counted.

c) Infinitesimal Forms: For sufficiently small τ we know that the probability of counting m photons is proportional to τ^m; this follows from (8.2.4). This means that we can write (since the operation of counting no photons in an infinitesimal time interval must approach the identity)

$$S_{t+\tau,t} = N_{t+\tau,t}(0) = 1 - \tau A(t) \tag{8.3.11}$$

when τ is infinitesimal, and where $A(t)$ is an operation to be determined. We can also write

$$N_{t+\tau,t}(1) = \tau B(t) \tag{8.3.12}$$

where in this case we shall assume that

$$B(t)\rho(t) = \epsilon \int d^3\mathbf{r} E^{(+)}(\mathbf{r})\rho(t)E^{(-)}(\mathbf{r}), \tag{8.3.13}$$

and we see that $B(t)$ is actually independent of t—thus we shall omit the t dependence. All other $N_{t+\tau,t}(m)$ vanish faster than τ, and can be determined as follows.

Using the limits (8.3.11,12) we find, from (8.3.5) that

$$\dot{S}_{t_1,t_0} = A(t_1)S_{t_1,t_0} \tag{8.3.14}$$

$$\dot{T}_{t_1,t_0} = \{A(t_1) + B\}T_{t_1,t_0}. \tag{8.3.15}$$

We can now determine $A(t_1)$ by consistency arguments.

i) Equation (8.3.10) requires that $\mathrm{Tr}\left\{\dot{T}_{t_1,t_0}\rho\right\} = 0$, so that the trace of ρ will be preserved equal to 1. Thus from (8.3.15) we can see that

$$\mathrm{Tr}\left\{A(t)\rho\right\} = -\mathrm{Tr}\left\{B\rho\right\} = -\epsilon \int d^3\mathbf{r}\mathrm{Tr}\left\{E^{(+)}(\mathbf{r})\rho E^{(-)}(\mathbf{r})\right\}. \tag{8.3.16}$$

This means that we can write

$$A(t)\rho = -\tfrac{1}{2}\epsilon \int d^3\mathbf{r}\left\{E^{(-)}(\mathbf{r})E^{(+)}(\mathbf{r})\rho + \rho E^{(-)}(\mathbf{r})E^{(+)}(\mathbf{r})\right\}$$
$$-\frac{i}{\hbar}\left[H(t),\rho\right] + \mathcal{L}(t)\rho. \tag{8.3.17}$$

ii) Here $H(t)$ is any Hermitian operator, and $\mathcal{L}(t)\rho$ is any operation which preserves ρ positive definite and with trace 1, and does not contain any Hamiltonian term. Because of the arbitrariness of $H(t)$ and $\mathcal{L}(t)$, this is in fact the most general operation whose trace satisfies to (8.3.16). From the results of Sect. 5.1.1, this means that $\mathcal{L}(t)$ is some kind of master equation operator. Thus $\rho(t) \equiv T(t,t_0)\rho(t_0)$ obeys a master equation which can be written

$$\dot{\rho}(t) = \tfrac{1}{2}\epsilon \int d^3\mathbf{r}\left\{2E^{(+)}(\mathbf{r})\rho E^{(-)}(\mathbf{r}) - E^{(-)}(\mathbf{r})E^{(+)}(\mathbf{r})\rho - \rho E^{(-)}(\mathbf{r})E^{(+)}(\mathbf{r})\right\}$$
$$-\frac{i}{\hbar}\left[H(t),\rho\right] + \mathcal{L}(t)\rho. \tag{8.3.18}$$

iii) The quantities $H(t)$ and $\mathcal{L}(t)$ are now determined by the physics of the situation. Obviously $H(t)$ determines the free evolution of the field, and must be the electromagnetic Hamiltonian,

$$H_{\mathrm{EM}}(t) = \tfrac{1}{2}\int d^3\mathbf{r}\{\epsilon_0\mathbf{E}^2 + \mathbf{B}^2/\mu_0\} \tag{8.3.19}$$

unless we want to modify it to take account of a refractive index.

The master equation operator in the first line of (8.3.18) can fairly be described as the evolution operator associated with the detection process, and $\mathcal{L}(t)$ describes any other Markovian evolution that may be necessary. For example, it could contain terms describing such absorption of electromagnetic radiation as does not lead to photoelectrons, or it could contain the terms necessary to take account of the fact that the detector may not be at zero temperature.

If we want to consider an ideal Markovian photodetector, we set $\mathcal{L}(t) = 0$, as we shall mostly from now on.

8.3.2 Master Equation and Quantum Stochastic Differential Equations

Since the field density operator obeys the master equation (8.3.18) we can write a corresponding QSDE for the field operators. Since there is a continuum of damping terms, one at each point in space, there will be a continuum of quantum noise input terms. The required generalization is quite straightforward, and the QSDE (in the Stratonovich form) takes the form, for any field operator $F(\mathbf{r}, t)$,

$$\begin{aligned} dF(\mathbf{r}, t) = &-\frac{\mathrm{i}}{\hbar}[F(\mathbf{r}, t), H_{EM}]\,dt \\ &-\int d^3\mathbf{r}' \Big\{ \sum_i \Big[F(\mathbf{r}, t), E_i^{(-)}(\mathbf{r}', t)\Big] \Big(\frac{\epsilon}{2}E_i^{(+)}(\mathbf{r}', t)dt + \sqrt{\epsilon}\, dB_i(\mathbf{r}', t)\Big) \\ &-\sum_i \Big(\frac{\epsilon}{2}E_i^{(-)}(\mathbf{r}', t)dt + \sqrt{\epsilon}\, dB_i^\dagger(\mathbf{r}', t)\Big) \Big[F(\mathbf{r}, t), E_i^{(+)}(\mathbf{r}', t)\Big] \Big\} \end{aligned} \tag{8.3.20}$$

where the spatially distributed quantum white noises are at zero temperature, commute at *different* times, and satisfy

$$\Big[dB_i(\mathbf{r}, t), dB_j^\dagger(\mathbf{r}', t)\Big] = \delta_{ij}\delta(\mathbf{r} - \mathbf{r}')dt \tag{8.3.21}$$

$$\langle dB_i^\dagger(\mathbf{r}, t) dB_j(\mathbf{r}', t')\rangle = 0, \tag{8.3.22}$$

which corresponds to zero temperature of the bath. Notice that the full form of the photodetector interaction as in (8.2.2) has been used explicitly.

The QSDE (8.3.20) is most easily interpreted when the approximations of Sect. 8.1.7 are made—using the commutation relations in the form (8.1.27), the equation of motion analogous to (8.1.31) is

$$\dot{\mathbf{E}}^{(+)}(\mathbf{r}, t) = -\mathrm{i}c\sqrt{-\nabla^2}\,\mathbf{E}^{(+)}(\mathbf{r}, t) - \frac{\hbar\Omega\epsilon}{4\epsilon_0}\mathbf{E}^{(+)}(\mathbf{r}, t) - \frac{\hbar\Omega\sqrt{\epsilon}}{2\epsilon_0}\mathbf{b}(\mathbf{r}, t), \tag{8.3.23}$$

where $\mathbf{b}(\mathbf{r}, t)$ is a zero temperature *transverse* vector quantum white noise which satisfies

$$\Big[b_i(\mathbf{r}, t), b_j^\dagger(\mathbf{r}', t')\Big] = \delta_{ij}^T(\mathbf{r} - \mathbf{r}')\delta(t - t') \tag{8.3.24}$$

$$\langle b_i^\dagger(\mathbf{r}, t) b_j(\mathbf{r}', t')\rangle = 0 \tag{8.3.25}$$

and is defined by

$$b_i(\mathbf{r},t)dt = \sum_j \int d^3\mathbf{r}'\delta^T_{ij}(\mathbf{r}-\mathbf{r}')dB_j(\mathbf{r}',t). \tag{8.3.26}$$

The equation (8.3.23) for $\mathbf{E}^{(+)}$ is a wave equation, written in a form that allows positive frequency solutions only, supplemented by a damping term and a quantum noise term, both of which are the direct result of the detection process.

a) Solutions of the Propagation Equation: The propagation equation (8.3.23) is not straightforwardly amenable to intuitive comprehension because of the $\sqrt{-\nabla^2}$. However, it can be understood in two ways. First, notice that because the photodetection medium is contained within the volume V, the equation is only valid within V; outside V we should set ϵ equal to zero. To take this into account, ϵ should be written explicitly as a function of $\mathbf{r}$, $\epsilon(\mathbf{r})$. For compactness, let us also define

$$\mu = \frac{\hbar\Omega}{4\epsilon_0} \tag{8.3.27}$$

and an operator D by

$$D = -\mathrm{i}c\sqrt{-\nabla^2}\ , \tag{8.3.28}$$

so that the propagation equation can be written formally as

$$\dot{\mathbf{E}}^{(+)}(\mathbf{r},t) = \left[D - \mu\epsilon(\mathbf{r})\right]\mathbf{E}^{(+)}(\mathbf{r},t) - 2\mu\sqrt{\epsilon(\mathbf{r})}\ \mathbf{b}(\mathbf{r},t). \tag{8.3.29}$$

This has a formal solution in the form

$$\mathbf{E}^{(+)}(\mathbf{r},t) = \mathbf{E}^{(+)}_{\mathrm{SIG}}(\mathbf{r},t) + \mathbf{E}^{(+)}_{\mathrm{NOISE}}(\mathbf{r},t) \tag{8.3.30}$$

since it is a first order (operator) differential equation in the time variable. The "SIG" term depends on $\mathbf{E}^{(+)}(\mathbf{r},t)$ for various $\mathbf{r}$ and some initial value t_0, and the "NOISE" term depends on $\mathbf{b}(\mathbf{r},t')$ for $t_0 < t' < t$. This division into dependency on initial condition at t_0 and noise at later times is important in using the photodetection formulae.

As a method of actually constructing a solution (8.3.29) and (8.3.30) are almost useless. However, if we differentiate (8.3.23) again we can derive a simpler equation

$$\begin{aligned}\ddot{\mathbf{E}}^{(+)}(\mathbf{r},t) &- c^2\nabla^2\mathbf{E}^{(+)}(\mathbf{r},t) + 2\mu\epsilon\dot{\mathbf{E}}^{(+)}(\mathbf{r},t) + (\mu\epsilon)^2\mathbf{E}^{(+)}(\mathbf{r},t)\\ &= -2\mu\epsilon^{\frac{3}{2}}\mathbf{b}(\mathbf{r},t) - 2(\partial_t - \mathrm{i}c\sqrt{-\nabla^2}\ + \mu\epsilon)\mu\sqrt{\epsilon}\ \mathbf{b}(\mathbf{r},t),\end{aligned} \tag{8.3.31}$$

where ϵ has been assumed constant. To match between the vacuum and the detector, one must match $\mathbf{B}$ and $\mathbf{E}$ as usual—see the next exercise.

b) Absorption as a Result of Detection: The homogeneous part of (8.3.31) clearly exhibits the absorption which arises as a consequence of the detection. Provided we neglect the last term (which should be very small if absorption is weak) in the first line of this equation, it has a form would result classically from an electrical conductivity $2\epsilon_0\mu\epsilon = \hbar\Omega\epsilon/2$. The noise term is very complicated, but can be simplified by assuming that only those frequency and wavelength components of the noise of magnitude Ω and Ω/c respectively will contribute, so that we can reduce equation (8.3.31) approximately to

$$\ddot{\mathbf{E}}^{(+)}(\mathbf{r},t) - c^2\nabla^2\mathbf{E}^{(+)}(\mathbf{r},t) + 2\mu\epsilon\dot{\mathbf{E}}^{(+)}(\mathbf{r},t) = -2\mu\sqrt{\epsilon}\,(\mu\epsilon + 2\mathrm{i}\Omega)\mathbf{b}(\mathbf{r},t). \quad (8.3.32)$$

This equation is only approximate, and its solution will not exactly satisfy the dependences on signal and noise just mentioned.

Exercise. Consider a one-dimensional version of (8.3.32) and find the solutions ignoring the quantum noise term, which as noted above does not contribute to the counting formulae. Assume a counting medium in the range $(-\infty, 0)$, and specify the incoming field as

$$A^{(+)}(t,x) = a^{(+)}_{\rm in}(t + x/c) + a^{(+)}_{\rm out}(t - x/c) \quad (8.3.33)$$

with

$$B^{(+)}(t,x) = \partial_x A^{(+)}(t,x) \quad (8.3.34)$$

$$E^{(+)}(t,x) = -\partial_t A^{(+)}(t,x) \quad (8.3.35)$$

being required to be continuous at $x = 0$.

Introduce Fourier transforms by

$$E^{(+)}(t,x) = \frac{1}{\sqrt{2\pi}}\int_{-\infty}^{\infty} d\omega\,\tilde{E}(\omega,x)\mathrm{e}^{-\mathrm{i}\omega t} \quad (8.3.36)$$

with similar definitions for other variables. If

$$\nu = \sqrt{\omega^2 + 2\mathrm{i}\mu\epsilon\omega} \quad (8.3.37)$$

is the root with *negative* imaginary part, show that

$$\tilde{E}_{\rm out}(\omega,0) = \left(\frac{\omega - \nu}{\omega + \nu}\right)\tilde{E}_{\rm in}(\omega,0) \quad (8.3.38)$$

$$\tilde{E}(\omega,x) = \frac{\omega/2}{\omega+\nu}\tilde{E}_{\rm in}(\omega,0)\mathrm{e}^{\mathrm{i}\nu x/c} \quad (8.3.39)$$

$$\approx \tilde{E}_{\rm in}(\omega,0)\mathrm{e}^{\mathrm{i}\omega x/c + \mu\epsilon x/c} \quad (8.3.40)$$

when $\mu\epsilon \ll \omega$.

8.3.3 Photon Counting Probabilities

There are two counting probability densities which are of interest:

a) Coincidence Probability Density: This is the probability density that one count is observed at each of the times $t_1, t_2, \ldots, t_n$ together with other possible counts in the time interval $(0,t)$, such that

$$0 < t_1 < t_2 < \ldots < t_n < t, \quad (8.3.41)$$

and it is given by

$$P(t_1, t_2, \ldots, t_n) = \mathrm{Tr}\left\{T_{t,t_n} B T_{t_n,t_{n-1}} B \ldots B T_{t_1,0}\rho\right\}. \tag{8.3.42}$$

This can be seen as follows; at the times $t_1, t_2 \ldots, t_n$ the operator B acts to carry out the measurement, but between times the evolution operator $T_{t_1,t_{r-1}}$ acts, which corresponds to evolution with an unknown number of possible counts in between.

This is a coincidence measurement—or more accurately, a probability density for delayed coincidences. The quantity $P(t_1, t_2, \ldots, t_n)dt_1 dt_2 \ldots dt_n$ is the probability that one count occurs in each of the pre-assigned intervals $dt_1, dt_2, \ldots, dt_n$, without specifying what happens in any other time intervals. This kind of measurement can be quite practicably carried out as long as not too many times are being considered.

b) Elementary Probability Density: If we set

$$Q(t_1, t_2, \ldots, t_n) = \mathrm{Tr}\left\{S_{t,t_n} B S_{t_n,t_{n-1}} B \ldots B S_{t_1,0}\rho\right\} \tag{8.3.43}$$

This represents the probability density that counts occur as before at $t_1, t_2, \ldots, t_n$; but *no counts* occur between these intervals. It can give the distribution of time intervals between counts. For example, the probability of a count between t_2 and $t_2 + dt_2$ given that the last count was between t_1 and $t_1 + dt_1$ is

$$G(t_2)dt_2 = Q(t_2, t_1)dt_2/Q(t_1). \tag{8.3.44}$$

This is also a measurement that can be practicably carried out.

c) Relation Between Elementary and Coincidence Probability Densities: From the definition, the coincidence probability density can be expressed as the sums of probabilities for n counts and a further $0, 1, 2, 3, \ldots$ more counts at some other points in the interval. It is helpful to extend the definitions of $P(t_1, t_2, \ldots t_n)$ and $Q(t_1, t_2, \ldots t_n)$ as follows. As defined, in both cases it is required that $t_1 < t_2 < \ldots < t_n$. Let us consider extending their definitions to allow all of the t_i to extend over $(0, t)$, but require them to be totally symmetric functions of $t_1, t_2, \ldots, t_n$. Using this meaning, we are led to

$$P(t_1) = \sum_{s=1}^{\infty} \frac{1}{(s-1)!} \int_0^t d\tau_2 \ldots \int_0^t d\tau_s Q(t_1, \tau_2, \ldots, \tau_s) \tag{8.3.45}$$

$$P(t_1, t_2) = \sum_{s=2}^{\infty} \frac{1}{(s-2)!} \int_0^t d\tau_3 \ldots \int_0^t d\tau_s Q(t_1, t_2, \tau_3, \ldots, \tau_s) \tag{8.3.46}$$

$$\vdots \qquad\qquad \vdots$$

$$P(t_1, t_2, \ldots, t_n) = \sum_{s=n}^{\infty} \frac{1}{(s-n)!} \int_0^t d\tau_{n+1} \ldots \int_0^t d\tau_s Q(t_1, t_2, \ldots, t_n, \tau_{n+1}, \ldots, \tau_s) \tag{8.3.47}$$

These formulae simply follow from adding up the probabilities of all events involving exactly the number s of counts, where s is at least the number of counts measured in the coincidence probability.

It is also useful to have an inverse relation. We define generating functionals

$$G(f) = \sum_{n=0}^{\infty} \frac{1}{n!} \int_0^t dt_1 \ldots \int_0^t dt_n f(t_1) f(t_2) \ldots f(t_n) Q(t_1, t_2, \ldots, t_n) \qquad (8.3.48)$$

and

$$H(f) = \sum_{n=0}^{\infty} \frac{1}{n!} \int_0^t dt_1 \ldots \int_0^t dt_n f(t_1) f(t_2) \ldots f(t_n) P(t_1, t_2, \ldots, t_n). \qquad (8.3.49)$$

Then using the symmetry of $P(t_1, \ldots, t_n)$ and $Q(t_1, \ldots, t_n)$ the definition (8.3.47) shows

$$H(f) = G(f+1) \qquad (8.3.50)$$

where by $f+1$, we mean the function $f(t)+1$. The inversion is then simply a matter of writing $G(f) = H(f-1)$, which, when expanded explicitly in (8.3.48), gives the inverse formula

$$Q(t_1, t_2 \ldots t_n) = \sum_{s=n}^{\infty} \frac{(-1)^{n-s}}{(s-n)!} \int_0^t d\tau_{n+1} \ldots \int_0^t d\tau_s P(t_1, t_2, \ldots, t_n, \tau_{n+1}, \ldots \tau_s). \qquad (8.3.51)$$

Exercise. By considering all possible ways in which s counts can occur either before or after t_1, show

$$P(t_1) = \sum_{s=1}^{\infty} \sum_{r=1}^{s-1} \int_{t_1}^{t} d\tau_s \int_{t_1}^{\tau_s} d\tau_{s-1} \ldots \int_{t_1}^{\tau_{r+2}} d\tau_{r+1} \int_0^{t_1} d\tau_{r-1} \int_0^{\tau_{r-1}} d\tau_{r-2} \ldots \int_0^{\tau_2} d\tau_1$$
$$\times Q(\tau_1, \tau_2, \ldots, \tau_{r-1}, t_1, \tau_{r+1}, \ldots, \tau_s) \qquad (8.3.52)$$

and then using the symmetry of Q in all its arguments, show the truth of (8.3.45), and the other formulae.

d) Coincidence Probability Density in Terms of QSDE Solutions: The definition of the evolution operator T_{t_1,t_2} corresponds exactly to that used for the evolution operator in Sect. 5.2, and by using the formula (5.2.20) for multitime correlation functions, we can see that equation (8.3.42) can be interpreted as

$$P(t_1,t_2,\dots,t_n)=\mathrm{Tr}\Big\{\epsilon^n\int d^3\mathbf{r}_n E^{(+)}(\mathbf{r}_n,t_n)\int d^3\mathbf{r}_{n-1}E^{(+)}(\mathbf{r}_{n-1},t_{n-1})$$
$$\dots\int d^3\mathbf{r}_1 E^{(+)}(\mathbf{r}_1,t_1)\rho$$
$$\times\int d^3\mathbf{r}_1 E^{(-)}(\mathbf{r}_1,t_1)\dots\int d^3\mathbf{r}_n E^{(-)}(\mathbf{r}_n,t_n)\Big\} \tag{8.3.53}$$

where the $E^{(+)}(\mathbf{r},t)$ are *solutions of the QSDE* as given approximately by the solutions of (8.3.32). This corresponds exactly to the formula (8.2.4), but we now know the solutions for $E^{(+)}(\mathbf{r}_n,t_n)$. The correlation function

$$\mathrm{Tr}\left\{E^{(+)}(\mathbf{r}_1,t_1)E^{(+)}(\mathbf{r}_2,t_2)\dots E^{(+)}(\mathbf{r}_n,t_n)\rho E^{(-)}(\mathbf{r}_n,t_n)\dots E^{(-)}(\mathbf{r}_1,t_1)\right\} \tag{8.3.54}$$

is therefore the central object of study.

As noted in the previous section, we can write each $E^{(\pm)}(\mathbf{r},t)$ in the form

$$E^{(\pm)}(\mathbf{r},t)=E^{(\pm)}_{\mathrm{SIG}}(\mathbf{r},t)+E^{(\pm)}_{\mathrm{NOISE}}(\mathbf{r},t) \tag{8.3.55}$$

where the "NOISE" term depends on the $E^{(\pm)}(\mathbf{r}',t)$ for various $\mathbf{r}'$ and for $t>0$, and the "SIG" term depends on $E^{(\pm)}(\mathbf{r}',0)$ for various $\mathbf{r}'$. Now all the $\mathbf{b}(\mathbf{r},t)$ will commute with all the $E^{(-)}(\mathbf{r}',0)$, and therefore can be commuted with all the $E^{(-)}$ to stand against ρ, where $\mathbf{b}(\mathbf{r},t)\rho=0$, since the bath is at zero temperature. Thus, the correlation function can be written as

$$\langle E^{(-)}_{\mathrm{SIG}}(\mathbf{r}_n,t_n)E^{(-)}_{\mathrm{SIG}}(\mathbf{r}_{n-1},t_{n-1})\dots E^{(+)}_{\mathrm{SIG}}(\mathbf{r}_1,t_1)E^{(+)}_{\mathrm{SIG}}(\mathbf{r}_1,t_1)\dots E^{(+)}_{\mathrm{SIG}}(\mathbf{r}_n,t_n)\rangle. \tag{8.3.56}$$

Here the $E^{(-)}_{\mathrm{SIG}}$ quantities are computed by the first part of (8.3.55). But this simply means that, to compute the correlation function one should calculate the quantum field at each location by using the boundary conditions and values of $\epsilon(\mathbf{r})$ and the propagation equation (8.3.31), but *omitting the noise* term on the right.

e) Elementary Probability Density: If we define

$$I(t)=\int_V d^3\mathbf{r}\,\mathbf{E}^{(-)}_{\mathrm{SIG}}(\mathbf{r},t)\cdot\mathbf{E}^{(+)}_{\mathrm{SIG}}(\mathbf{r},t) \tag{8.3.57}$$

(where the full form is now used) then the coincidence probability density for counting photons in the volume V is

$$\boxed{P(t_1,t_2\dots t_n)=\epsilon^n\langle{:}I(t_1)I(t_2)\dots I(t_n){:}\rangle} \tag{8.3.58}$$

and using (8.3.51), the elementary probability density can be written

$$\boxed{Q(t_1,t_2,\dots t_n)=\epsilon^n\left\langle{:}I(t_1)I(t_2)\dots I(t_n)\exp\left\{-\epsilon\int_0^t dtI(t)\right\}{:}\right\rangle.} \tag{8.3.59}$$

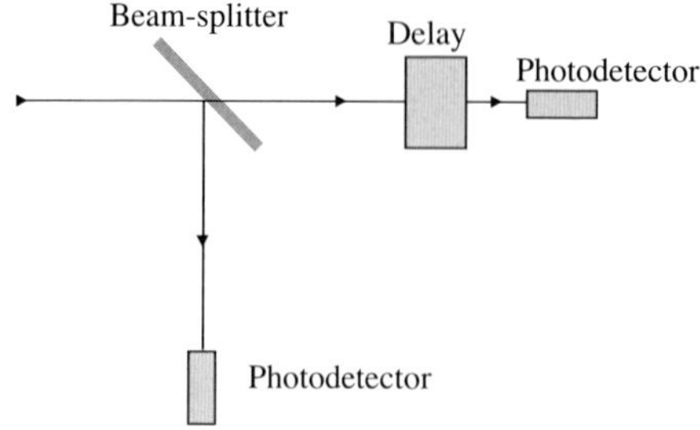

Fig. 8.2 Apparatus used in the Hanbury-Brown Twiss experiment. The beam is divided by the beam splitter, and delayed in one arm

8.3.4 Intensity Correlations and the Hanbury-Brown Twiss Experiment

This experiment provides a measure of the two-time coincidence probability density, $P(t_1, t_2) = \epsilon^2\langle{:}I(t_1)I(t_2){:}\rangle$. The experiment is put into effect by the apparatus of Fig.8.2. The detectors are arranged so that there is a time delay τ between them. Thus the coincidence probability density is $\epsilon^2\langle{:}I(t_1)I(t_2){:}\rangle$. Typical results are shown in Fig.8.3.

a) Thermal Light: In the case of thermal light we can expect the statistics to be quantum Gaussian. Assuming that the light is excited in only one mode $a(t)$, the coincidence probability is proportional to

$$G^{(2)}(t, t+\tau) = \langle a^\dagger(t)a^\dagger(t+\tau)a(t+\tau)a(t)\rangle. \tag{8.3.60}$$

We can use the result of Sect. 4.4.5 to note that if we define

$$g^{(2)}(t, t+\tau) = \frac{\langle a^\dagger(t)a^\dagger(t+\tau)a(t+\tau)a(t)\rangle}{\langle a^\dagger(t)a(t)\rangle\langle a^\dagger(t+\tau)a(t+\tau)\rangle} \tag{8.3.61}$$

$$g^{(1)}(t, t+\tau) = \frac{\langle a^\dagger(t)a(t+\tau)\rangle}{\sqrt{\langle a^\dagger(t)a(t)\rangle\langle a^\dagger(t+\tau)a(t+\tau)\rangle}} \tag{8.3.62}$$

then

$$g^{(2)}(t, t+\tau) = 1 + |g^{(1)}(t, t+\tau)|^2 \tag{8.3.63}$$

and obviously

$$g^{(1)}(t, t) = 1. \tag{8.3.64}$$

In all real situations, for sufficiently large τ, $g^{(1)}(t, t+\tau)$ factorizes. Thus

$$g^{(1)}(t, t+\tau) \to \frac{\langle a^\dagger(t)\rangle\langle a^\dagger(t+\tau)\rangle}{\sqrt{\langle a^\dagger(t)a(t)\rangle\langle a^\dagger(t+\tau)a(t+\tau)\rangle}} \to 0 \tag{8.3.65}$$

in the case of thermal light, for which the mean amplitude must be zero—only the intensity is non-zero.

Thus we find that $g^{(2)}(t, t) = 2$, and $g^{(2)}(t, \infty) = 1$. If, for example, the first order correlation function has an exponential behaviour,

$$g^{(1)}(t, t+\tau) = \mathrm{e}^{-\gamma\tau}, \tag{8.3.66}$$

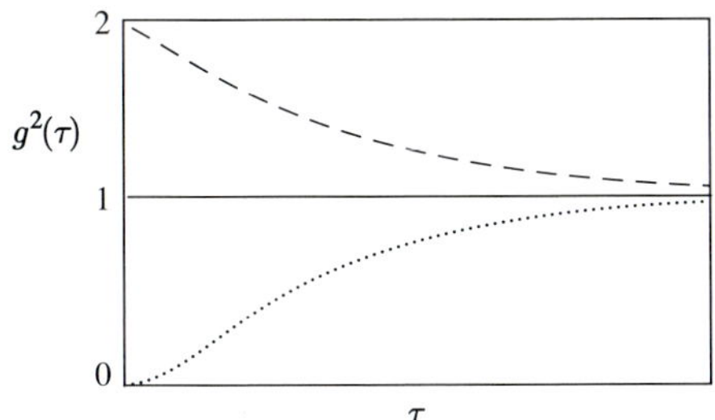

Fig. 8.3 Plots of the intensity correlation functions for thermal light (dashed line), laser light (solid line), and antibunched light (dotted line)

then the second order correlation function is given by

$$g^{(2)}(t, t+\tau) = 1 + \mathrm{e}^{-2\gamma\tau}. \tag{8.3.67}$$

b) Laser Light: On the other hand, if we have an ideal coherent source (which is well approximated by laser light) then the factorization property of the correlation functions means that

$$g^{(2)}(t, t+\tau) = 1. \tag{8.3.68}$$

Thus we find the rather paradoxical result, that maximum coherence yields no correlation at $\tau = 0$, while what one would expect is the most random situation yields a correlation at $\tau = 0$.

c) General Situation: There do exist optical fields with $g^{(2)}(t,t) = 0$, in marked distinction to the Gaussian situation. For example, in Sect. 9.2.3, it is shown that light produced by fluorescence from a two level atom driven by an optical field always has this property. Such light is said to be *antibunched*, while Gaussian light can be said to be *bunched*. In the case of bunched light, the fact that $g^{(2)}(t,t) > 1$ indicates that there is a high probability of detecting another photon immediately after one has been detected; thus the photons have a tendency to arrive together, i.e., in bunches.

d) Antibunching: In contrast, in the case of antibunching, the probability of detecting a photon immediately after one has been detected is zero; the photons arrive with some time interval between them, in a rather regular way—hence the name antibunching. Such a situation cannot occur classically, as can be seen from the following argument. Let us suppose that there is a joint P-function $P(\alpha_1, t_1, \alpha_2, t_2)$, whereby we can write

$$G^{(1)}(t_1, t_2) = \int\int d^2\alpha_1 d^2\alpha_2 \alpha_1^* \alpha_2 P(\alpha_1, t_1, \alpha_2, t_2) \tag{8.3.69}$$

$$G^{(2)}(t_1, t_2) = \int\int d^2\alpha_1 d^2\alpha_2 |\alpha_1|^2 |\alpha_2|^2 P(\alpha_1, t_1, \alpha_2, t_2). \tag{8.3.70}$$

The simplest situation is one in which $P(\alpha_1, t_1, \alpha_2, t_2) = \delta[\alpha_1 - \alpha(t_1)]\delta[\alpha_2 - \alpha(t_2)]$ so that

$$\begin{aligned} G^{(1)}_{\text{coherent}}(t_1, t_2) &= \alpha(t_1)^* \alpha(t_2) \\ G^{(2)}_{\text{coherent}}(t_1, t_2) &= |\alpha(t_1)|^2 |\alpha(t_2)|^2. \end{aligned} \tag{8.3.71}$$

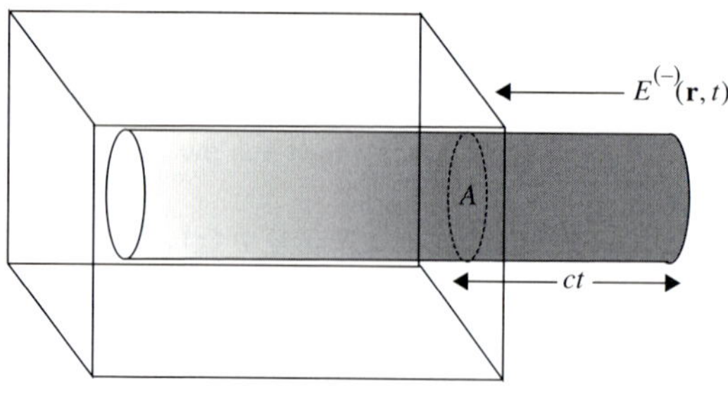

Fig. 8.4 Photodetection in a spatially distributed detector

c) Mandel's Formula as a Photon Counting Formula: Under the condition that the detector is thin and very efficient, the number of photocounts corresponds to the measurement of the photon number operator. We assume that the detector is located at $z < 0$, and the absorption length γ^{-1} of the detector material is so small that we can approximate the time dependence by $\mathrm{e}^{-i\omega t}$ for times less than $(\gamma c)^{-1}$. If $\mathcal{T}$ is the transmission coefficient of the light into the detector medium, then we can say

$$E^{(-)}(\mathbf{r},t) = \mathcal{T} E^{(-)}(x,y,0)\mathrm{e}^{\gamma z - i\omega z/c}. \tag{8.3.85}$$

Here $E^{(-)}$ on the RHS represents the value of the incoming field to the immediate right of the detector surface. The situation is illustrated in Fig.8.4. Using the one dimensional model solution of (8.3.40), we can see that

$$\gamma = \mu\epsilon/c = \hbar\Omega\epsilon/4\epsilon_0 c = \bar{\epsilon}/2c. \tag{8.3.86}$$

Assuming also that the light is travelling along the z axis, we find that

$$\bar{\epsilon}\int_0^t N(V,t')_{\mathrm{SIG}}dt' = |\mathcal{T}|^2 \int_A dx\,dy \int_0^{ct} dz\, E^{(-)}(\mathbf{r},0)E^{(+)}(\mathbf{r},0) \tag{8.3.87}$$

$$= |\mathcal{T}|^2 N_{\mathrm{IN}}(V,0) \tag{8.3.88}$$

where V is the volume of a tube of cross sectional area A and length ct, where t is the time for which one counts. (It has been assumed that the length of the detector along the z direction is much greater than the attenuation length, allowing the integration of a factor $\mathrm{e}^{2\gamma z}$.)

If the counting is not perfectly efficient, due to other absorption as in b) above, then $|\mathcal{T}|^2$ is replaced by e, which is the product of $|\mathcal{T}|^2$ and the intrinsic efficiency of the detector. Thus, Mandel's formula may be written

$$p(n,[0,t]) = \frac{e^n}{n!}\langle{:}N_{\mathrm{IN}}(V,0)^n \exp\{-eN_{\mathrm{IN}}(V,0)\}{:}\rangle. \tag{8.3.89}$$

Since the photon number operator is formed out of creation and destruction operators, we can consider eigenvalues of $N_{\mathrm{IN}}(V,0)$ with a definite number of photons;

these can be written in general as $|\mathbf{n}\rangle$, where n_i are the eigenstates of the operator $a^\dagger(\mathbf{f}_i,t)a(\mathbf{f}_i,t)$. Using the creation and destruction operators of Chap.4, it is easy to show that

$$\frac{e^n}{n!}{:}(a^\dagger a)^n \exp(-ea^\dagger a){:}|r\rangle = \frac{r!}{n!(r-n)!}e^n(1-e)^{r-n}|r\rangle \tag{8.3.90}$$

or

$$\frac{e^n{:}(a^\dagger a)^n \exp(-ea^\dagger a){:}}{n!} = \sum_{r=n}^{\infty} e^n(1-e)^{r-n}\frac{r!}{n!(r-n)!}|r\rangle\langle r|. \tag{8.3.91}$$

The special case of $e = 1$, perfect efficiency, gives

$$\frac{{:}(a^\dagger a)^n \exp(ea^\dagger a){:}}{n!} = |n\rangle\langle n|. \tag{8.3.92}$$

Exercise. We can set $\hat{N} = a^\dagger a$ and write (8.3.91) as

$$\frac{e^n{:}\hat{N}^n \exp(-e\hat{N}){:}}{n!} = \frac{e^n(1-e)^{\hat{N}-n}\hat{N}!}{n!\left(\hat{N}-n\right)!} \tag{8.3.93}$$

The special case of $n = 0$ and $e = 1 - \exp(-\lambda)$ gives

$${:}\exp\left\{\left(1-e^{-\lambda}\right)\hat{N}\right\}{:} = \exp\left(-\lambda\hat{N}\right). \tag{8.3.94}$$

Exercise. Generalize (8.3.92) to the case of N modes, to show that

$$\frac{{:}\left\{\sum_{i=1}^N a_i^\dagger a_i\right\}^n \exp\left\{-\sum_{i=1}^N a_i^\dagger a_i\right\}{:}}{n!} = P_n \tag{8.3.95}$$

where P_n is the projector into states with total photon number equal to n.

Mandel's formula can thus be written, in the case of 100% efficiency

$$p(n,[0,t]) = \langle P_n\rangle = \sum_{\mathbf{n}} \delta(n, \sum n_i)\langle\mathbf{n}|\rho|\mathbf{n}\rangle. \tag{8.3.96}$$

Thus, the probability of n photocounts in time t is the probability that there are n photons in the incoming volume V of light counted. *This means that a photodetector in fact measures the photon number operator for the number of photons in the volume V. We can thus work out the moments of the photocount distribution simply by computing the moments of $N_{\text{in}}(V,0)$.*

8.3.6 Applications to Particular States

a) Number State: Assuming only one mode is excited in the volume V, and the state is a number state $|r\rangle$, then

$$\begin{aligned} p(n,[0,t]) &= \frac{r!}{n!(r-n)!}e^n(1-e)^{r-n} \qquad && n \le r \\ &= 0 && n > r. \end{aligned} \tag{8.3.97}$$

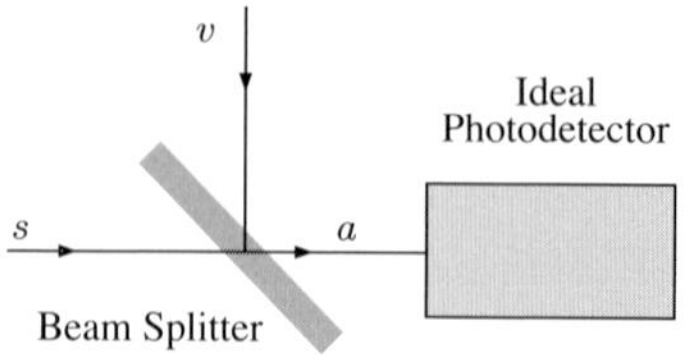

Fig. 8.5 Model of an inefficient photodetector formed by a perfect photodetector preceded by a beam splitter

This is a Bernoulli distribution. When the counting efficiency is perfect, we find

$$p(n,[0,t]) = \delta_{n,r} \tag{8.3.98}$$

which means that we can only get the result that the number of photons counted is equal to the number of photons in the volume.

b) Coherent State $|\alpha\rangle$:

$$p(n,[0,t]) = \frac{(e|\alpha|^2)^n}{n!}\exp\left(-e|\alpha|^2\right). \tag{8.3.99}$$

This is a Poisson distribution for any value of e, the efficiency. In fact, since the system is essentially a harmonic oscillator damped by a zero temperature bath, this corresponds to the optical state remaining a coherent state as it is attenuated or partially transmitted—this is essentially the result of Sect. 9.3.2, that a coherent state remains coherent as it is linearly attenuated.

c) Thermal State: Here ρ_T is given by (4.4.13), and we find that

$$p(n,[0,t]) = \frac{\left[e\bar{N}(T)\right]^n}{\left[1+e\bar{N}(T)\right]^{n+1}} \tag{8.3.100}$$

where, as usual

$$\bar{N}(T) = \left(\mathrm{e}^{\hbar\omega/kT} - 1\right)^{-1}. \tag{8.3.101}$$

We can work out the moments of the photocurrent distribution by simply computing the moments of $N_{\text{in}}(V,0)$.

8.3.7 Model for Efficiency Less Than 100%

An inefficient detector may be modelled by a perfect detector with a beam splitter, of amplitude transmissivity e in front of it as in Fig.8.5 The detected beam can be written

$$a = s\sqrt{e} + v\sqrt{1-e} \tag{8.3.102}$$

Since v is in the vacuum state, all of its normally ordered moments are zero, which means that the normally ordered moments of $a^\dagger a$ are

$$\langle :(a^\dagger a)^n:\rangle = e^n\langle :(s^\dagger s)^n:\rangle \tag{8.3.103}$$

which gives the counting distribution (8.3.89) directly for a; namely that of an inefficient detector. This result is very useful when one needs to model an inefficient detector.

8.4 Homodyne and Heterodyne Detection

In practical detection of light beams considerable advantages can ensue by mixing the signal to be detected with a strong coherent signal, called a *local oscillator*, before detection. The mixing is linear in the fields being mixed, whereas detection is proportional to intensity, so this procedure mixes the signals. Two kinds of detection are in use; *homodyne detection*, in which the local oscillator has the same frequency as that of the detected signal, and *heterodyne detection*, where the local oscillator has a different frequency.

8.4.1 Schematic Setup of Homodyne and Heterodyne Detection

The basic concept is illustrated in Fig.8.6. The local oscillator and signal are mixed in a beam splitter, and then detected. The output photocurrent is then multiplied by a $\cos(\omega t)$ function, and the result integrated for a time T, very much larger than ω^{-1}. The photodetector will not normally be 100% efficient, but we will model it as an ideal photodetector preceded by a beam splitter.

8.4.2 General Formulae

The signal at the photodetector is

$$b(t) = r\alpha(t) + rv(t) + ta(t). \tag{8.4.1}$$

Here $\alpha(t)$ is the *mean value* of the local oscillator field, thus $\langle v(t)\rangle = 0$. The quantity being studied is thus

$$N(\omega, T) = \int_0^T dt b^\dagger(t)b(t) \cos(\omega t). \tag{8.4.2}$$

The assumption is that $\cos(\omega t)$ is rather slowly varying, and (8.4.2) is effectively got by measuring the photon number operator repeatedly. These values are then multiplied by $\cos(\omega t)$, (or $\sin(\omega t)$ as appropriate), and added. We shall only carry out the calculations explicitly for $\cos(\omega t)$.

We now assume that $r\alpha(t)$ is very large, and expanding (8.4.2) to the first two orders only

$$N(\omega, T) - \int_0^T dt \left\{|r|^2|\alpha(t)|^2 + r^*\alpha^*(t)\left[ta(t) + rv(t)\right] \right.$$
$$\left. + r\alpha(t)\left[t^* a^\dagger(t) + r^* v^\dagger(t)\right]\right\} \cos(\omega t). \tag{8.4.3}$$

Here we assume a single frequency local oscillator, i.e.

$$\alpha(t) = Ae^{-i\omega_L t - i\phi}, \tag{8.4.4}$$

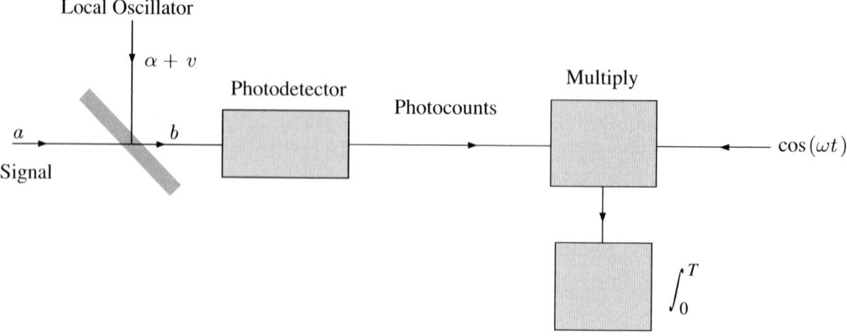

Fig. 8.6 Arrangements for homodyne or heterodyne detection

and let us choose, for convenience, r and t to be *real*. We define a Fourier variable

$$\tilde{a}(\omega) = \frac{1}{T}\int_0^T e^{i\omega t} a(t)\, dt \tag{8.4.5}$$

with a similar notation for $\tilde{v}(\omega)$. We then find

$$\begin{aligned} N(\omega, T) &= \int_0^T r^2 A^2 \cos\omega t\, dt \\ &+ \frac{rAT}{2} e^{i\phi}\left\{t\left[\tilde{a}(\omega_L - \omega) + \tilde{a}(\omega_L + \omega)\right] + r\left[\tilde{v}(\omega_L - \omega) + \tilde{v}(\omega_L + \omega)\right]\right\} \\ &+ \frac{rAT}{2} e^{-i\phi}\left\{t\left[\tilde{a}(\omega_L - \omega)^\dagger + \tilde{a}(\omega_L + \omega)^\dagger\right] + r\left[\tilde{v}(\omega_L - \omega)^\dagger + \tilde{v}(\omega_L + \omega)^\dagger\right]\right\}. \end{aligned} \tag{8.4.6}$$

a) Mean Detected Signal: We now define the mean field,

$$\tilde{\mathcal{E}}(\omega) = \langle \tilde{a}(\omega) \rangle \tag{8.4.7}$$

so that the mean signal detected is, on the assumption that $T \gg 1/\omega$, and terms of order ω^{-1} are neglected in comparison with terms of order T,

$$\boxed{\langle N(\omega, T)\rangle = r^2 A^2 T \delta_{\omega,0} + rtAT\, \mathrm{Re}\left\{e^{i\phi}\left[\tilde{\mathcal{E}}(\omega_L - \omega) + \tilde{\mathcal{E}}(\omega_L + \omega)\right]\right\}.} \tag{8.4.8}$$

b) Variance of Detected Signal: We are also interested in the variance of $N(\omega, T)$. We need to assume a variance for $v(t)$, and we shall assume that of the vacuum, i.e.

$$\langle v^\dagger(t) v(t')\rangle = 0, \qquad \langle v(t) v^\dagger(t')\rangle = \delta(t - t'), \tag{8.4.9}$$

so that the variance can be computed to be

$$\left\langle \left[N(\omega,T) - \langle N(\omega,T)\rangle\right]^2 \right\rangle = \left(\frac{rAT}{2}\right)^2 \left\{ \frac{2r^2}{T}(1+\delta_{\omega,0}) \right. \\ +t^2 \left\langle \left[e^{i\phi} \{\delta\tilde{a}(\omega_L - \omega) + \delta\tilde{a}(\omega_L + \omega)\} \right. \right. \\ \left. \left. \left. + e^{-i\phi} \left\{ \delta\tilde{a}(\omega_L - \omega)^\dagger + \delta\tilde{a}(\omega_L + \omega)^\dagger \right\} \right]^2 \right\rangle \right\} \tag{8.4.10}$$

(where δ means the difference from the mean value). There are two terms in this expression; that proportional to r^2 is *local oscillator noise*, and arises from the fluctuations in the local oscillator part; the other is the signal noise.

8.4.3 Coherent Signal Detection

If the signal is itself coherent, the signal noise takes on exactly the same form as the vacuum, i.e.

$$\langle a(t)a^\dagger(t')\rangle = \delta(t-t'), \quad \langle a^\dagger(t)a(t')\rangle = 0 \tag{8.4.11}$$

and

$$\langle [N(\omega,T) - \langle N(\omega,T)\rangle]^2 \rangle = \frac{T(rA)^2}{2}\{1+\delta_{\omega,0}\} \tag{8.4.12}$$

(where use has been made of $r^2+t^2=1$). We can compute the signal to noise ratio in some practical cases.

a) Homodyne Detection: Here we set $\omega = 0$. We take the *signal* to be that part by which the $\langle N(\omega,T)\rangle$ differs from the "background", r^2A^2T.

$$\text{Sig (Ho)} = 2rtAT\,\mathrm{Re}\left\{e^{i\phi}\tilde{\mathcal{E}}(\omega_L)\right\} \tag{8.4.13}$$

$$\text{Noise (Ho)} = T(rA)^2 \tag{8.4.14}$$

$$\text{S/N(Ho)} = 4t^2T\left[\mathrm{Re}\left\{e^{i\phi}\tilde{\mathcal{E}}(\omega_L)\right\}\right]^2. \tag{8.4.15}$$

b) Heterodyne Detection: Here we set $\omega \neq 0$, and we assume that $\omega_L - \omega = \omega_{\text{Sig}}$, where ω_{Sig} is the frequency around which the signal has significant power. The background term now vanishes, which is an advantage for this method. It is in practice not common to have significant power in both $\omega_L + \omega$ and $\omega_L - \omega$, but it is of course possible. We then find

$$\text{Sig (Het)} = rtAT\,\mathrm{Re}\left\{e^{i\phi}\tilde{\mathcal{E}}(\omega_{\text{Sig}})\right\} \tag{8.4.16}$$

$$\text{Noise (Het)} = \tfrac{1}{2}T(rA)^2 \tag{8.4.17}$$

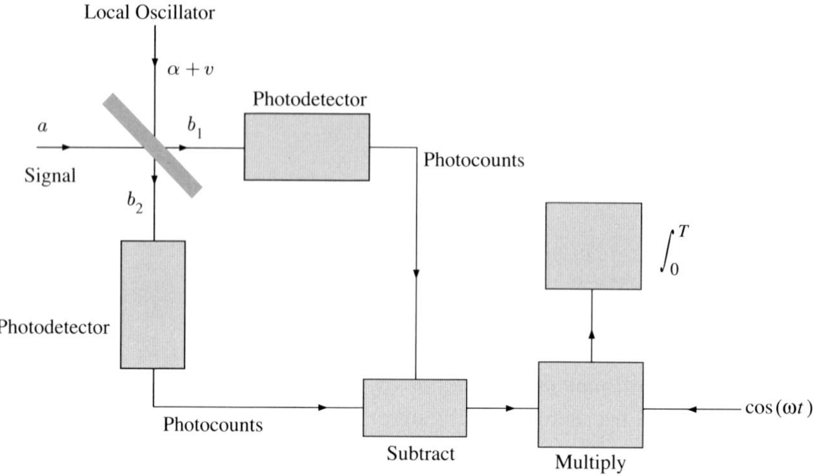

Fig. 8.7 Arrangements for balanced detection

$$\mathbf{S/N}(\text{Het}) = 2t^2 T \left[\text{Re}\left\{e^{i\phi}\tilde{\mathcal{E}}(\omega_{\text{Sig}})\right\}\right]^2. \tag{8.4.18}$$

c) Comment on Signal to Noise Ratios: Note that the signal to noise ratio in heterodyne detection is only half of that for homodyne detection. On the other hand, there is no constant level, as in homodyne detection.

d) Imperfect Photodetectors: If the photodetector has an efficiency of less than 100%, then there is an additional noise source. This can be seen by using the model of the inefficient detector as a perfect detector behind a beam splitter, see Sect. 8.3.7. Thus the light that reaches the photodetector is

$$c(t) = \sqrt{e}\; b(t) + \sqrt{1-e}\; w(t), \tag{8.4.19}$$

where $w(t)$ is a vacuum white noise. In the limit of $\alpha(t)$ large, we find that

$$\begin{aligned} N(\omega, T) = \int_0^T dt \,\cos\omega t \Big\{ & e r^2 |\alpha(t)|^2 \\ & + \sqrt{e}\; r\alpha^*(t)\left[\sqrt{e}\; ta(t) + \sqrt{e}\; rv(t) + \sqrt{1-e}\; w(t)\right] \\ & + \sqrt{e}\; r\alpha(t)\left[\sqrt{e}\; ta(t) + \sqrt{e}\; rv(t) + \sqrt{1-e}\; w(t)\right]^\dagger \Big\}. \end{aligned} \tag{8.4.20}$$

By making $\alpha(t)$ arbitrarily large, and r correspondingly small (to the degree that this may be technically achievable), the term with $v(t)$ can be made as small as desired, but the $w(t)$ term cannot—it can only be eliminated by making the efficiency e as close to 100% as possible.

An arbitrarily large signal to noise ratio is achievable, however, by counting for an arbitrarily long time. But if one is interested in the *noise* itself, which is the case in the study of squeezed light, then great efforts must be made to increase e. Efficiencies of about 90% are in practice achievable. [8.7]

The results for signal to noise ratios are similar to (8.4.15,18), and are obtained by the replacement $t \to t\sqrt{e}$ in these formulae—as might be expected.

8.4.4 Balanced Homodyne/Heterodyne Detection

In balanced detection, the beam splitter is chosen to have equal reflectivity and transmissivity, i.e., $|r| = |t| = 1/\sqrt{2}$, and two detectors are used, as in Fig.8.7. The photocurrents are then subtracted, and processed as before. In the case that the detectors are 100% efficient, the signal to be measured is

$$N_B(\omega, T) = \int_0^T dt \left\{ b_1^\dagger(t) b_1(t) - b_2^\dagger(t) b_2(t) \right\} \cos\omega t \tag{8.4.21}$$

with

$$\begin{aligned} b_1(t) &= \frac{1}{\sqrt{2}} \left[\alpha(t) + v(t) + a(t) \right] \\ b_2(t) &= \frac{1}{\sqrt{2}} \left[-\alpha(t) - v(t) + a(t) \right] . \end{aligned} \tag{8.4.22}$$

Many terms cancel, and we find

$$N_B(\omega, T) = \int_0^T dt \left\{ \alpha(t)^* a(t) + \alpha(t) a^\dagger(t) + v^\dagger(t) a(t) + v(t) a^\dagger(t) \right\} \cos\omega t. \tag{8.4.23}$$

Again, one assumes that $\alpha(t)$ is very large, so that the final two terms can be neglected. However, $\alpha(t)$ need not be nearly so strong as in unbalanced detection since it does not appear multiplied by $r \ll 1$. Thus balanced detection provides a more satisfactory technique, particularly as the constant term is no longer present. The formulae are essentially the same as those for ordinary detection, with the replacement $r \to 1, t \to 1$ (*not* $r \to 1/\sqrt{2}\,, t \to 1/\sqrt{2}$, since the differencing in (8.4.21) provides another factor of 2.)

Of course to get the full advantage of the technique, one must have a perfectly balanced beam splitter, which can never be exactly so.

a) Imperfect Detectors: If the photodetectors are not perfect, we have two additional independent noise sources at each detector. Thus, the inputs measured will be

$$\begin{aligned} c_1(t) &= \sqrt{e_1}\, b_1(t) + \sqrt{1 - e_1}\, w_1(t) \\ c_2(t) &= \sqrt{e_2}\, b_2(t) + \sqrt{1 - e_2}\, w_2(t). \end{aligned} \tag{8.4.24}$$

Here $w_1(t)$ and $w_2(t)$ may be taken as vacuum quantum white noise. In the limit that $\alpha(t)$ is large (and with $e_1 = e_2 \equiv e$ for simplicity) we find

$$N_B(\omega, T) \to \int_0^T dt \left\{ e\left[\alpha(t)^* a(t) + \alpha(t) a^\dagger(t)\right] + \right.$$
$$+ \sqrt{\frac{e(1-e)}{2}}\, \alpha(t)^* \left[w_1(t) + w_2(t)\right]$$
$$\left. + \sqrt{\frac{e(1-e)}{2}}\, \alpha(t) \left[w_1(t) + w_2(t)\right]^\dagger \right\} \cos \omega t. \qquad (8.4.25)$$

If e is very close to 1, the final terms are negligible, but otherwise they can be significant, as in the case of unbalanced detection.

Exercise. In fact, because there are two *independent* noise sources in balanced detection, the signal to noise ratio is half that of unbalanced detection. Show this.

8.5 Input-Output Formulations of Photodetection

The formulation of photodetection given in Sect. 8.3 is based on what really must be regarded as a phenomenological theory, for that is all the Srinivas-Davies theory can claim to be. The formulation is manifestly only an approximation, obtained by assuming a narrow bandwidth detector, and by averaging over many cycles of the relevant optical frequency. Of course both of these are in almost all current situations very good approximations, but their introduction is necessary not only because they are reasonable, but also because ambiguities arise when they are not made.

I want to introduce here a different view of the problem, based on the input-output formalism. The major new idea is to treat the photodetection process as the counting of electrons, represented by an appropriate fully quantized field, rather than merely the counting of photoionizations, which lies at the heart of the Srinivas-Davies formalism. The QSDEs developed in Sect. 8.3.2 have noise terms, which were not interpreted there. But the whole formalism developed there suggests that they must be related to the individual absorptions of photons with the consequent emissions of electrons. Thus, it is tempting to interpret the "out" field corresponding to $\mathbf{b}(\mathbf{r}, t)$ as the electron field operator. However this must be justified in more detail before being put into application.

In this section I start with a very simple model, essentially of one atom and a one dimensional electromagnetic field. The electrons are treated as fully quantized *bosons*, because this allows for great simplicity and clarity. It is able to be shown that, under the conditions which normally apply for a practical photodetector, the "out" electron field has the same statistics as the "in" photon field—and as a natural consequence we can derive Mandel's formula. The next section shows that the results still hold, in a degree of approximation, even with Fermi electrons. This involves the adaptation of the input-output formalism to handle fermions.

The final section reverts to Bose electrons, but treats the fully quantized propagation absorption and photoproduction problem in some detail, without making the rotating wave approximation. In this final model, we see a physical interpretation of the formalism of Sect. 8.3.2 in which the "out" noises are indeed the electron field operators.

8.5.1 A One Atom Model

In order to see why the photoelectron statistics are those of the incident light, let us consider the simplest possible model of a photodetector—a single atom (idealized as a harmonic oscillator), localized at some point in a one-dimensional space, and tuned to the frequency of the incoming light. In photodetection the electrons are never produced with any significant occupation of the available states so the distinction between bosons and fermions is immaterial; let us therefore consider Bose electrons.

We consider a simple Hamiltonian, in which a single atom interacts with the operators of both the electromagnetic field and of the electrons.

$$\begin{aligned} H &= \hbar\Omega a^\dagger a + \int_{-\infty}^{\infty} d\omega\hbar\omega b^\dagger(\omega)b(\omega) + \int_{-\infty}^{\infty} d\omega\hbar\omega d^\dagger(\omega)d(\omega) \\ &+ \mathrm{i}\hbar \int_{-\infty}^{\infty} \bar{\gamma}(\omega)\left[b^\dagger(\omega)a - b(\omega)a^\dagger\right] d\omega \\ &+ \mathrm{i}\hbar \int_{-\infty}^{\infty} \bar{\kappa}(\omega)\left[d^\dagger(\omega)a - d(\omega)a^\dagger\right] d\omega. \end{aligned} \tag{8.5.1}$$

Here a is a destruction operator for the atom, $b(\omega)$ is a destruction operator for an electromagnetic field mode, $d(\omega)$ is a destruction operator for an electron field mode. The coupling between electrons, photons and the atom is of the kind that allows an atom to be excited by the electromagnetic field, and to de-excite by emission of either an electron or a photon—the processes are all completely reversible.

The quantum Langevin equation corresponding to (8.5.1) is

$$\dot{a}(t) = -\mathrm{i}\Omega a(t) - \tfrac{1}{2}(\kappa+\gamma)a(t) - \sqrt{\gamma}\, b_{\mathrm{in}}(t) - \sqrt{\kappa}\, d_{\mathrm{in}}(t). \tag{8.5.2}$$

Here the "first Markov approximation of Sect. 5.3.2 has been made, so that $\bar{\gamma}(\omega) = \sqrt{\gamma/2\pi}$, $\bar{\kappa}(\omega) = \sqrt{\kappa/2\pi}$. The "in" fields are essentially the inputs into the oscillator from the electromagnetic and electron fields; thus $b_{\mathrm{in}}(t)$ represents the incoming electromagnetic field, whose statistics we wish to measure. The process of measurement is to count emitted electrons, so we wish to know not the "in" field d_{in}, but a corresponding "out" field,

$$d_{\mathrm{out}}(t) = d_{\mathrm{in}}(t) + \sqrt{\kappa}\, a(t). \tag{8.5.3}$$

To get the relationship between $d_{\text{out}}(t)$ and $b_{\text{in}}(t)$ now requires us to solve the equations. We go to a frame rotating at angular frequency Ω by writing $a(t) = e^{-i\Omega t}A(t)$ with a similar notation for $B_{\text{in}}(t)$, $D_{\text{in}}(t)$, $D_{\text{out}}(t)$, etc. Now assume that $\kappa + \gamma$ is sufficiently large that we can solve the equation resulting from (8.5.2) by adiabatic elimination, thus

$$A(t) = -\frac{2\sqrt{\gamma}\, B_{\text{in}}(t) + 2\sqrt{\kappa}\, D_{\text{in}}(t)}{\kappa + \gamma} \tag{8.5.4}$$

and using (8.5.3), we get

$$D_{\text{out}}(t) = \frac{\gamma - \kappa}{\gamma + \kappa} D_{\text{in}}(t) - \frac{2\sqrt{\gamma\kappa}}{\gamma + \kappa} B_{\text{in}}(t). \tag{8.5.5}$$

It is clear that the normally ordered moments of $D_{\text{out}}(t)$ are equal to those of $2\sqrt{\gamma\kappa}\, A_{\text{in}}(t)/(\gamma + \kappa)$ provided that the state of $D_{\text{in}}(t)$ corresponds to the vacuum, that is, provided we are not directing electrons at the atom, which is of course the case. If $\gamma = \kappa$, we get the ideal result that $D_{\text{out}}(t) = A_{\text{in}}(t)$, so that the measurement of $D_{\text{out}}(t)$ is equivalent to the measurement of $A_{\text{in}}(t)$. As was shown by *Ghielmetti* [8.8], this is exactly equivalent to Mandel's formula. In the case of lower efficiency, when $\gamma \neq \kappa$, the effect is similar to that of a beam splitter; that is, the output electron field is a mixture of the input photon field and the vacuum fluctuations arising from the input electron field. This is similar to a view introduced by *Yuen* and *Shapiro* [8.9], that such a photodetector is equivalent to an ideal photodetector preceded by a beam splitter.

Mandel's formula is not exactly equivalent to (8.5.5), which would suggest that phase information is also supplied to the electron field. Mandel's formula only requires that the normally ordered moments of the *electron number operator*

$$N(t_a, t_b) = \int_{t_a}^{t_b} D_{\text{out}}^{\dagger}(t) D_{\text{out}}(t) dt \tag{8.5.6}$$

are given by substituting for $D_{\text{out}}(t)$, as mentioned, thus we can say that measuring $N(t_a, t_b)$ counts the number of electrons arriving in the interval (t_a, t_b), and this is equal to

$$\frac{4\gamma\kappa}{(\gamma + \kappa)^2} \int_{t_a}^{t_b} A_{\text{in}}^{\dagger}(t) A_{\text{in}}(t) dt. \tag{8.5.7}$$

The equality of *normally ordered* moments, with the factor $\epsilon = 4\gamma\kappa/(\gamma+\kappa)^2 \leq 1$ as the detector efficiency is equivalent to Mandel's formula. Notice also that the detector efficiency is strictly no greater than 100%, and reaches 100% when $\gamma = \kappa$—a fact which can be regarded as impedance matching. This is not a relevant result to practical photodetection, which is a distributed process, in which attenuation takes place over a finite distance. A more appropriate model is presented in Sect. 8.5.3.

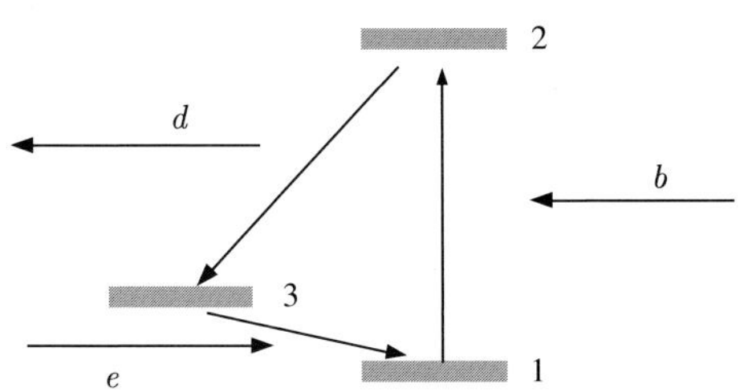

Fig. 8.8 Three level atom model of a photodetector

8.5.2 A Model Using Fermi Electrons

The use of Fermi electrons is much more difficult because the harmonic oscillator can no longer be used as a model of an atom. It is simply not possible to set up a physics in which a photon raises the atom to an excited state, which emits an electron and returns to the *same* state. Instead, a three level model is the bare minimum, as in Fig.8.8. The emitted electron leaves the atom in an intermediate state, which then makes a transition to the ground state by absorption of another electron from a bath of electrons. This is of course the essence of what happens in practice.

a) Input-Output Method with Fermions: To use fermions, an appropriate formulation of the input-output methods is essential. Suppose we consider ionization of an atom, in which a fermion can be attached or detached. The Hamiltonian is

$$H = H_{\text{sys}} + H_{\text{int}} + H_B \tag{8.5.8}$$

where H_{sys} is the free atom Hamiltonian, and

$$H_B = \int d\omega\, \hbar\omega\, d^\dagger(\omega) d(\omega) \tag{8.5.9}$$

$$H_{\text{int}} = \mathrm{i}\hbar \int d\omega\, \bar{\kappa}(\omega) \left\{ d^\dagger(\omega) c - c^\dagger d(\omega) \right\}, \tag{8.5.10}$$

which is exactly the same as in the boson formulation. However, in this case, $d(\omega), d^\dagger(\omega)$, are fermion operators so that

$$[d(\omega), d^\dagger(\omega')]_+ = \delta(\omega - \omega'). \tag{8.5.11}$$

A complication is now introduced by the concept of fermionic and bosonic system operators. At equal times $d(\omega), d^\dagger(\omega)$ commute with all bosonic system operators and anticommute with all fermionic operators. Clearly the operator c which occurs in the interaction Hamiltonian must be Fermionic.

b) Quantum Langevin Equation: The same procedure as for the boson bath case can now be applied, to yield the Fermionic quantum Langevin equation,

$$\begin{aligned} \dot{a} = -\frac{\mathrm{i}}{\hbar}[a, H_{\text{sys}}] - \Big\{ & [a, c^\dagger]_\mp \left(\tfrac{1}{2}\kappa c + \sqrt{\kappa}\, d_{\text{in}}(t) \right) \\ & \mp \left(\tfrac{1}{2}\kappa c^\dagger + \sqrt{\kappa}\, d_{\text{in}}^\dagger(t) \right) [a, c]_\mp \Big\} \end{aligned} \tag{8.5.12}$$

where + corresponds to a fermionic a and − corresponds to a bosonic a. The bosonic equation is the same as in the usual formalism.

c) Inputs, Outputs and Causality: As in the usual formalism, we find

$$d_{\text{out}}(t) = d_{\text{in}}(t) + \sqrt{\kappa}\, c(t) \tag{8.5.13}$$

and

$$\begin{aligned} [a(t), d_{\text{in}}(t')]_\pm &= -u(t-t')\sqrt{\kappa}\,[a(t), c(t')]_\pm \\ [a(t), d_{\text{out}}(t')]_\pm &= u(t'-t)\sqrt{\kappa}\,[a(t), c(t')]_\pm . \end{aligned} \tag{8.5.14}$$

d) Fermionic Quantum White Noise: In this case we define

$$\langle d_0^\dagger(\omega) d_0(\omega')\rangle = \bar{N}\delta(\omega-\omega'), \qquad \langle d_0(\omega) d_0^\dagger(\omega')\rangle = (1-\bar{N})\delta(\omega-\omega') \tag{8.5.15}$$

where $\bar{N} = \{\exp(\hbar\omega/kT)+1\}^{-1} \le 1$.

e) The Detector Model: Define

$$\left.\begin{aligned} |1\rangle\langle 1| &= n \\ |2\rangle\langle 2| &= m \\ |3\rangle\langle 3| &= l = 1-m-n \\ |1\rangle\langle 2| &= p \end{aligned}\right\} \quad \text{bosonic,}$$
$$\left.\begin{aligned} |3\rangle\langle 2| &= q \\ |1\rangle\langle 3| &= r \end{aligned}\right\} \quad \text{fermionic.} \tag{8.5.16}$$

There is a field b for the photons, and two separate electron fields d and e. The first represents ejected electrons, and the second the returning electrons. The interaction Hamiltonian is

$$\begin{aligned} H_{\text{int}} = \mathrm{i}\hbar \int d\omega\,\bar{\kappa}(\omega)\left\{d^\dagger(\omega)q - q^\dagger d(\omega)\right\} \\ +\mathrm{i}\hbar \int d\omega\,\bar{\lambda}(\omega)\left\{e^\dagger(\omega)r^\dagger - re(\omega)\right\} \\ +\mathrm{i}\hbar \int d\omega\,\bar{\gamma}(\omega)\left\{b^\dagger(\omega)p - p^\dagger b(\omega)\right\}, \end{aligned} \tag{8.5.17}$$

and this leads to the equations of motion (in the interaction picture)

$$\begin{aligned} \dot{n} &= \gamma m + \lambda l + \sqrt{\gamma}\left\{p^\dagger b_{\text{in}} + b_{\text{in}}^\dagger p\right\} + \sqrt{\lambda}\left\{r^\dagger e_{\text{in}}^\dagger + e_{\text{in}} r\right\} \\ \dot{m} &= -(\kappa+\gamma)m - \sqrt{\gamma}\left\{p^\dagger b_{\text{in}} + b_{\text{in}}^\dagger p\right\} - \sqrt{\kappa}\left\{q^\dagger d_{\text{in}} + d_{\text{in}}^\dagger q\right\} \\ \dot{l} &= -\dot{m} - \dot{n} \\ \dot{p} &= -\frac{\kappa}{2}p - \sqrt{\kappa}\, r d_{\text{in}} + \sqrt{\lambda}\, q e_{\text{in}}^\dagger - \frac{\gamma}{2}p - \sqrt{\gamma}\,(n-m) b_{\text{in}} \\ \dot{q} &= -\tfrac{1}{2}(\gamma+\kappa+\lambda)q - \sqrt{\gamma}\, r^\dagger b_{\text{in}} - \sqrt{\kappa}\,(1-n) d_{\text{in}} - \sqrt{\lambda}\, e_{\text{in}} p . \\ \dot{r} &= -\frac{\lambda}{2}r - \sqrt{\gamma}\, q^\dagger b_{\text{in}} - \sqrt{\kappa}\, d_{\text{in}}^\dagger p - \sqrt{\lambda}\,(1-m) e_{\text{in}}^\dagger . \end{aligned} \tag{8.5.18}$$

f) Input-Output Boundary Conditions: We want to specify $b_{\rm in}$, the incoming photon field, and to compute $d_{\rm out}$, the ejected electron field. We also need to specify that :

> $d_{\rm in}$ corresponds to a vacuum—no input electrons are present in the channel to which the ejected electron couples.
>
> $e_{\rm in}$ corresponds to a fully filled bath—there are numerous electrons available in the channels required to return the atom to the ground state.

We know that

$$\begin{aligned} b_{\rm out} &= b_{\rm in} + \sqrt{\gamma}\, p \\ d_{\rm out} &= d_{\rm in} + \sqrt{\kappa}\, q \\ e^\dagger_{\rm out} &= e^\dagger_{\rm in} + \sqrt{\lambda}\, r. \end{aligned} \tag{8.5.19}$$

g) Adiabatic Elimination Limit: As before, assume that γ and κ are large and we can solve the equations for q and p by adiabatic elimination.

$$p = \frac{-\sqrt{\gamma}\,(n-m)b_{\rm in} - \sqrt{\kappa}\, r d_{\rm in} + \sqrt{\lambda}\, q e^\dagger_{\rm in}}{\frac{1}{2}(\gamma+\kappa)} \tag{8.5.20}$$

$$q = \frac{-\sqrt{\gamma}\, r^\dagger b_{\rm in} - \sqrt{\kappa}\,(1-n)d_{\rm in} - \sqrt{\lambda}\, e_{\rm in}p}{\frac{1}{2}(\gamma+\kappa+\lambda)}. \tag{8.5.21}$$

The r equation cannot be solved by adiabatic elimination, since we do not put any fast requirement on λ. We can approximate the r equation by noting that the excitation amplitude will be small in practice, for linear operation of the detector, so that we can say $n \sim 1, m \sim l \sim r \sim p \sim q \sim 0$, so that

$$r^\dagger \approx -\sqrt{\lambda} \int_{-\infty}^{t} dt' {\rm e}^{-\frac{1}{2}\lambda(t-t')} e_{\rm in}(t') \tag{8.5.22}$$

and

$$\begin{aligned} q \approx{}& \frac{2\sqrt{\gamma\lambda}}{\gamma+\kappa+\lambda} \int_{-\infty}^{t} dt' {\rm e}^{-\frac{1}{2}\lambda(t-t')} e_{\rm in}(t') b_{\rm in}(t) - \frac{2\sqrt{\kappa}}{\gamma+\kappa+\lambda} d_{\rm in}(t) \\ &+ \frac{4\sqrt{\lambda\gamma}\; e_{\rm in}(t) b_{\rm in}(t)}{(\gamma+\kappa+\lambda)(\gamma+\kappa)}. \end{aligned} \tag{8.5.23}$$

We will only be interested in normally ordered products, and since $d_{\rm in}$ is the vacuum, it will contribute nothing. It is not difficult to see that although the other two terms are similar, the first one is much bigger than the other if $\gamma, \kappa \gg \lambda$.

Thus we find

$$d_{\rm out} \simeq \frac{\gamma-\kappa+\lambda}{\gamma+\kappa+\lambda} d_{\rm in} + \frac{2\sqrt{\gamma\kappa\lambda}}{\gamma+\kappa+\lambda} \int_{-\infty}^{t} dt' {\rm e}^{-\frac{1}{2}\lambda(t-t')} e_{\rm in}(t') b_{\rm in}(t). \tag{8.5.24}$$

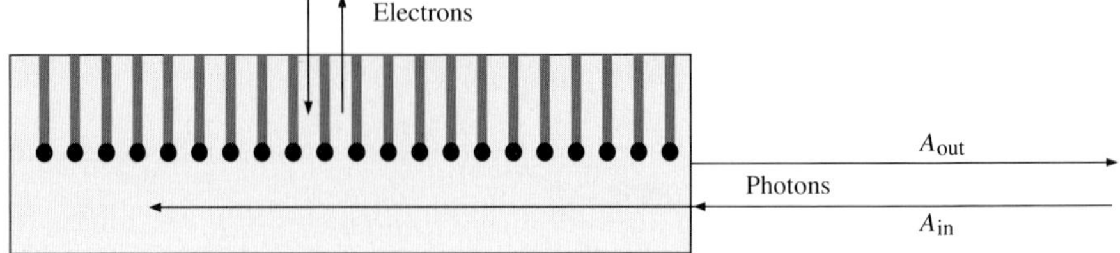

Fig. 8.9 Spatially distributed model of a photodetector

The normally ordered correlation functions of of d_{out} are then given by those of the second term. We take, for a filled $e_{\mathrm{in}}(t)$ that

$$\langle e_{\mathrm{in}}^\dagger(t)e_{\mathrm{in}}(t')\rangle = \delta(t-t') \qquad \langle e_{\mathrm{in}}(t)e_{\mathrm{in}}^\dagger(t')\rangle = 0 \tag{8.5.25}$$

so that, for example

$$\langle d_{\mathrm{out}}^\dagger(t)d_{\mathrm{out}}(t')\rangle = \left(\frac{4\gamma\kappa}{\gamma+\kappa+\lambda}\right) \mathrm{e}^{-\frac{1}{2}\lambda|t-t'|}\langle b_{\mathrm{in}}^\dagger(t)b_{\mathrm{in}}(t')\rangle. \tag{8.5.26}$$

Provided λ is much *smaller* than the time constant of the correlation function $\langle b_{\mathrm{in}}^\dagger(t)b_{\mathrm{in}}(t')\rangle$, this says that the electron correlation function is that of the input photons, up to an efficiency factor of $4\gamma\kappa/(\gamma+\kappa)$, since $\gamma \ll \gamma+\kappa$. This is the same result that we found in the Bose model. The higher order normally ordered correlation functions follow similarly for *even* order correlations, but the odd order correlations will vanish, as should indeed be the case.

Thus the inclusion of Fermi electrons does yield exactly the kind of result predicted by Mandel's formula, without the spurious equality of odd order moments which happens with a Bose electron model. Nevertheless the result for even order moments in the degree of approximation used is exactly the same as that of the Bose model. The reason for this is very simple. The Bose electron model is completely linear, in that the output electron fields are linear functions of the input photon fields. The Fermi electron model cannot be linearized in this sense, since Fermi fields can never be equal to Bose fields, but it is linearized in the only sense that is allowable; namely there is a linear relation between the d_{out} and a bilinear function of the input Bose field and another Fermi field, which arises from the returning electrons. The equivalence of correlation functions is true only if λ, the relaxation time constant of the re-absorption of the electrons, is small compared to those of the photon absorption and electron production. This is appropriate, since it means that the virtual level to which the atom returns after emitting the photoelectron is almost stable, and correlations between photoelectrons and reabsorbed electrons cannot arise.

8.5.3 A Spatially Distributed Detector Model

The fact that we can get essentially the same results from both Bose and Fermi electrons is encouraging, but both of these models are far too simplified. For example, the result of perfect photodetection if $\gamma = \kappa$ is not what one expects from a single atom—rather, it is the property of a lumped oscillator model of a detector. Most treatments of photodetection assume that the medium acts locally as a very weak inefficient detector, with high efficiency obtained by the eventual absorption in a sufficiently thick medium. Let us therefore model a detector of this kind, but simplify matters by assuming again Bose electrons. The model is one dimensional, as in Fig.8.9. An appropriate Lagrangian is

$$
\begin{aligned}
L = \tfrac{1}{2} \int_{-\infty}^{\infty} dx \left\{ \dot{A}(x,t)^2 - c^2[\partial_x A(x,t)]^2 \right\} \\
+ \int_{-\infty}^{0} dx \left\{ \tfrac{1}{2}\dot{u}(x,t)^2 - \tfrac{1}{2}\omega^2 u(x,t)^2 + u(x,t)\left[2\sqrt{\kappa v}\, \dot{D}(x,t,0) + \gamma \dot{A}(x,t)\right] \right\} \\
+ \tfrac{1}{2} \int_{-\infty}^{0} dx \int_{0}^{\infty} ds \left\{ \dot{D}(x,t,s)^2 - v^2[\partial_s D(x,t,s)]^2 \right\}.
\end{aligned}
\tag{8.5.27}
$$

This Lagrangian corresponds to a one dimensional electrodynamics on the range $(-\infty, \infty)$, coupled in the left half of space to a distribution of atoms represented by the operators $u(x,t)$. These atoms are coupled *locally* to a distribution of electron baths, represented by the field $D(x,t,s)$. For the electron baths and the atoms the co-ordinate x does not represent any dynamics, but merely acts as a label to distinguish the different electron baths and atoms at each point—the dynamics takes place in another variable s. Writing the bath Lagrangian in the form given in (8.5.27) has two main advantages over the more conventional Hamiltonian method.

—The input-output formalism is easily developed.

—The identification of outputs corresponding to electrons produced is straightforward.

The physics represented is something like that of a solid state photodetector, in which the electrons produced are immediately localized by thermal and other interactions in the solid, so that the photoelectric processes at two points depend on each other only to the extent enforced by the coherence of the electromagnetic field. The basic concept is thus the same as in the previous part, and it is merely a matter of solving for a D_{out} field in terms of the electromagnetic field being input from the right.

a) Equations of Motion: For the three fields, we find, for $x < 0$

$$
\begin{aligned}
\ddot{A} - c^2\partial_x^2 A &= -\gamma\dot{u} \\
\ddot{u} + \omega^2 u &= \gamma\dot{A} + 2\sqrt{\kappa v}\, \dot{D}(x,t,0) \\
\ddot{D} - v^2\partial_s^2 D &= -2\sqrt{\kappa v}\, \delta(s)\dot{u}.
\end{aligned}
\tag{8.5.28}
$$

Since $D(x,t,s)$ represents the electron bath, we have to solve for it in terms of the electron inputs and outputs. The solution of the equation can be written in two equivalent forms;

$$\begin{aligned} D(x,t,s) &= D_{\rm in}(x,t+s/v) + D_{\rm in}(x,t-s/v) \\ &\qquad - \sqrt{\frac{\kappa}{v}}\, u(x,t-s/v) \\ &= D_{\rm out}(x,t-s/v) + D_{\rm out}(x,t+s/v) \\ &\qquad + \sqrt{\frac{\kappa}{v}}\, u(x,t+s/v). \end{aligned} \tag{8.5.29}$$

We can substitute the first of equations (8.5.29) into the equation for u given in (8.5.28) to get as our basic pair of equations, for $x < 0$, (that is, inside the photodetection medium).

$$\begin{aligned} \ddot{A} - c^2\partial_x^2 A &= -\gamma\dot{u} \\ \ddot{u} + \omega^2 u &= \gamma\dot{A} - \kappa\dot{u} + 2\sqrt{\kappa v}\,\dot{D}_{\rm in}(x,t). \end{aligned} \tag{8.5.30}$$

To the right, $x > 0$, where there is no medium, we simply have the free wave equation $\ddot{A} - c^2\partial_x^2 A = 0$. The other variables do not exist for $x > 0$.

b) Solution of the Equations of Motion: If we define a Fourier transform variable by

$$u(t,x) = \frac{1}{\sqrt{2\pi}} \int_{-\infty}^{\infty} d\Omega e^{-i\Omega t}\tilde{u}(\Omega,x) \tag{8.5.31}$$

with a similar definition for $\tilde{A}(\Omega,x)$ and $\tilde{D}(\Omega,x,s)$, we find that the u equation can be reduced to

$$\tilde{u} = \frac{(-i\Omega)\left[2\sqrt{\kappa v}\,\tilde{D}_{\rm in}(\Omega,x) + \gamma\tilde{A}\right]}{\omega^2 - \Omega^2 - i\kappa\Omega} \tag{8.5.32}$$

which, when substituted into the equation for $\tilde{A}$, yields

$$-\tilde{A}\left\{\Omega^2 + \frac{\gamma^2\Omega^2}{\omega^2 - \Omega^2 - i\Omega\kappa}\right\} - c^2\partial_x^2\tilde{A} = \frac{2\gamma\Omega^2\sqrt{\kappa v}}{\omega^2 - \Omega^2 - i\Omega\kappa}\tilde{D}_{\rm in}(\Omega,x). \tag{8.5.33}$$

This is a wave equation with both dispersion and absorption, as expected for such a medium. Notice, however, the appearance of a "quantum noise" term on the right-hand side, which arises from the vacuum fluctuations in the electron baths. This acts to ensure the preservation of canonical commutation relations.

Let us define a quantity ν by

$$\nu^2 = \Omega^2 + \frac{\gamma^2\Omega^2}{\omega^2 - \Omega^2 - i\Omega\kappa}. \tag{8.5.34}$$

In particular, we define ν so that it is the root of (8.5.34) with a *negative imaginary part*. (Notice that this means for large Ω that $\nu \sim -\Omega$). In terms of ν, (8.5.33) can be written

$$-\nu^2 \tilde{A} - c^2 \partial_x^2 \tilde{A} = (\nu^2 - \Omega^2) \frac{2\sqrt{\kappa v}}{\gamma} \tilde{D}_{\text{in}}(\Omega, x). \tag{8.5.35}$$

A general solution of this equation which satisfies the boundary condition that $\tilde{A}(\Omega, x)$ be finite as $x \to -\infty$ is (noting that $\text{Im}(\nu) < 0$ by definition)

$$\tilde{A}(\Omega, x) = Q\mathrm{e}^{i\nu x/c} + \frac{2\sqrt{\kappa v}}{\gamma} \int_{-\infty}^{\infty} dx'(\nu^2 - \Omega^2)\tilde{D}_{\text{in}}(\Omega, x') \frac{\mathrm{e}^{-\mathrm{i}\nu|x-x'|/c}}{2\mathrm{i}\nu c} \tag{8.5.36}$$

where Q is to be determined from boundary conditions.

c) Boundary Conditions of the Interface: The conditions of continuity of electric and magnetic fields must be met at $x = 0$, the edge of the medium. The solution of the wave equation to the right of $x = 0$ is of the form

$$A(x, t) = a_{\text{in}}(t + x/c) + a_{\text{out}}(t - x/c), \tag{8.5.37}$$

so that the boundary conditions are

$$\dot{A}(x, t)|_{x\to 0-} = \dot{a}_{\text{in}}(t + x/c) + \dot{a}_{\text{out}}(t - x/c) \tag{8.5.38}$$

and

$$\partial_x A(x, t)|_{x\to 0-} = \frac{1}{c} \left\{\dot{a}_{\text{in}}(t + x/c) - \dot{a}_{\text{out}}(t - x/c)\right\}. \tag{8.5.39}$$

Taking the Ω frequency component of these boundary conditions and combining with (8.5.36) gives

$$Q = \frac{2\Omega\tilde{a}_{\text{in}}}{\Omega - \nu} + \frac{2\sqrt{\kappa v}}{\gamma}(\Omega + \nu)^2 \int_{-\infty}^{\infty} dx \tilde{D}_{\text{in}}(\Omega, x) \frac{\mathrm{e}^{\mathrm{i}\nu x/c}}{2\mathrm{i}\nu c} \tag{8.5.40}$$

$$\tilde{a}_{\text{out}} = \left(\frac{\Omega + \nu}{\Omega - \nu}\right) \tilde{a}_{\text{in}} + \frac{4\sqrt{\kappa v}}{\gamma}(\Omega + \nu)\nu \int_{-\infty}^{\infty} dx \tilde{D}_{\text{in}}(\Omega, x) \frac{\mathrm{e}^{\mathrm{i}\nu x/c}}{2\mathrm{i}\nu c}. \tag{8.5.41}$$

The equation for $\tilde{a}_{\text{out}}$ shows a reflected term, with reflection coefficient $\left|\frac{\Omega+\nu}{\Omega-\nu}\right|^2$, and a term which ar s from the quantum fluctuations of the electron bath.

d) Attenuation Length: Since ν has an imaginary part, the factor $\mathrm{e}^{\mathrm{i}\nu x/c}$ in (8.5.38) represents an attenuation of the input beam with depth x. The characteristic absorption depth is $l_a = \frac{1}{2}\{|\text{Im}(\nu/c)|\}^{-1}$. An exact expression is very complicated, but if we assume that the second term in (8.5.36) is always smaller than Ω^2, then

$$l_a \approx \frac{(\omega^2 - \Omega^2)^2 + \Omega^2\kappa^2}{\Omega^2\kappa\gamma^2}. \tag{8.5.42}$$

This shows a strong resonant behaviour—the maximum absorption obviously takes place when $\omega \simeq \Omega$. The ratio of the wavelength to the minimum attenuation length

9. Interaction of Light with Atoms

Systems composed of a number of free atoms interacting with a few modes of the electromagnetic field are particularly well studied both theoretically and experimentally. Such systems are much simpler than those consisting of solids or liquids, where a much wider variety of kinds of behaviour is possible. Theoretical treatments often simplify matters even more by assuming that the atoms only have two levels, which experimentally can only be achieved by rather complicated optical pumping techniques. However, such experiments as have been done with two level atoms show impressive agreement with theory [9.1].

This chapter consists of an introductory part, describing the algebra of two level systems, and then treats the behaviour of a single two level atom in thermal and coherent light fields. These results are then used in two major applications, the theory of a gas laser, and the theory of optical bistability induced by an atomic vapour inside an optical cavity.

9.1 Two Level Systems

A fundamental building block in quantum theory is the two level system. This is an idealization of an atom, or molecule, which has only two energy levels. Although it is in principle possible for an atomic system to have only two *bound* states, there would also be a continuum of unbound or scattering states. In the ideal two level atom, these continuum states do not exist. The basic idea is that we can consider situations in which transitions to the continuum are negligible, and similarly, any transitions to other energy levels are negligible. If the energy differences between the energy levels are all very different this makes sense, since radiation which might induce a transition between a given pair of levels will have an energy which will be completely detuned from any other transition. The only situation for which this does not make sense is the harmonic oscillator, since the energy levels are evenly spaced, so that it is impossible to consider incident radiation as only exciting one transition.

9.1.1 Pauli Matrix Description

The two level atom is described by a wavefunction which is a two component matrix

$$u = \begin{pmatrix} u^{(+)} \\ u^{(-)} \end{pmatrix} \qquad (9.1.1)$$

and all operators can be expressed in terms of the Pauli matrices

$$\sigma_x = \begin{pmatrix} 0 & 1 \\ 1 & 0 \end{pmatrix} \qquad \sigma_y = \begin{pmatrix} 0 & -\mathrm{i} \\ \mathrm{i} & 0 \end{pmatrix} \qquad \sigma_z = \begin{pmatrix} 1 & 0 \\ 0 & -1 \end{pmatrix} \tag{9.1.2}$$

and the identity matrix, **1**. The Hamiltonian for such a two level system then can always be written

$$H = \tfrac{1}{2}\hbar\Omega\boldsymbol{\sigma}\cdot\hat{\mathbf{n}} \tag{9.1.3}$$

where $\hat{\mathbf{n}}$ is a unit vector—if this vector is along the z axis, then

$$H = \tfrac{1}{2}\hbar\Omega\sigma_z = \begin{pmatrix} \frac{1}{2}\hbar\Omega & 0 \\ 0 & -\frac{1}{2}\hbar\Omega \end{pmatrix}. \tag{9.1.4}$$

9.1.2 Pauli Matrix Properties

The proofs are left as an exercise :

i)
$$\sigma_i\sigma_j = \delta_{ij} + \mathrm{i}\sum_k \epsilon_{ijk}\sigma_k. \tag{9.1.5}$$

ii)
$$(\boldsymbol{\sigma}\cdot\mathbf{A})(\boldsymbol{\sigma}\cdot\mathbf{B}) = \mathbf{A}\cdot\mathbf{B} + \mathrm{i}\boldsymbol{\sigma}\cdot\mathbf{A}\times\mathbf{B}; \tag{9.1.6}$$

A and **B** are c-number or operator vectors.

iii)
$$[\sigma_i, \sigma_j] = 2\mathrm{i}\sum_k \epsilon_{ijk}\sigma_k. \tag{9.1.7}$$

iv)
$$[\sigma_i, \sigma_j]_+ = 2\delta_{ij}. \tag{9.1.8}$$

v) If G is any two dimensional matrix

$$G = \tfrac{1}{2}\left\{\mathbf{1}\mathrm{Tr}\{G\} + \sum_i \sigma_i \mathrm{Tr}\{\sigma_i G\}\right\}. \tag{9.1.9}$$

vi)
$$\exp\{\mathrm{i}\theta\boldsymbol{\sigma}\cdot\mathbf{n}\} = \cos\theta + \mathrm{i}\boldsymbol{\sigma}\cdot\mathbf{n}\sin\theta. \tag{9.1.10}$$

vii) If

$$\sigma_+ = \tfrac{1}{2}(\sigma_x + \mathrm{i}\sigma_y) = \begin{pmatrix} 0 & 1 \\ 0 & 0 \end{pmatrix} \tag{9.1.11}$$

$$\sigma_- = \tfrac{1}{2}(\sigma_x - \mathrm{i}\sigma_y) = \begin{pmatrix} 0 & 0 \\ 1 & 0 \end{pmatrix} \tag{9.1.12}$$

then

$$\sigma_+^2 = \sigma_-^2 = 0, \quad \sigma_z^2 = 1 \tag{9.1.13}$$

$$\sigma_+\sigma_- = \tfrac{1}{2}(1 + \sigma_z) \tag{9.1.14}$$

$$\sigma_-\sigma_+ = \tfrac{1}{2}(1 - \sigma_z) \tag{9.1.15}$$

$$\sigma_z\sigma_+ = -\sigma_+\sigma_z = \sigma_+ \tag{9.1.16}$$

$$\sigma_z\sigma_- = -\sigma_-\sigma_z = -\sigma_-. \tag{9.1.17}$$

viii) For any matrix G

$$G = \tfrac{1}{2}\mathbf{1}\mathrm{Tr}\{G\} + \tfrac{1}{2}\sigma_z \mathrm{Tr}\{\sigma_z G\} + \sigma_+ \mathrm{Tr}\{\sigma_- G\} + \sigma_- \mathrm{Tr}\{\sigma_+ G\}. \quad (9.1.18)$$

ix)

$$\exp(a + \mathbf{b}\cdot\boldsymbol{\sigma}) = \mathrm{e}^a \left\{\cosh|\mathbf{b}| + \frac{\boldsymbol{\sigma}\cdot\mathbf{b}}{|\mathbf{b}|}\sinh|\mathbf{b}|\right\}. \quad (9.1.19)$$

9.1.3 Atoms with More Than Two Levels

Notice that atoms with more than two levels can be handled by similar methods. If the atom has eigenstates $|i\rangle$, one defines matrices by

$$\lambda_{ij} = |i\rangle\langle j|. \quad (9.1.20)$$

These matrices are of course linearly complete, and also have a closed algebra corresponding to (9.1.6)

$$\lambda_{ij}\lambda_{kl} = \delta_{jk}\lambda_{il}. \quad (9.1.21)$$

With care, almost all two level problems in this chapter can be generalized.

9.2 Two Level Atom in the Electromagnetic Field

The quantum Langevin equation for the two level atom was derived in Sect. 3.4.4, where it was noted that exact solutions were not possible. Using the same notation as in that section, it is straightforward to substitute into the master equation, and to show that one obtains essentially the same results as in Sect. 1.5.1

$$X \to \sqrt{\frac{\hbar}{\Omega}}\,\sigma_x = \sqrt{\frac{\hbar}{\Omega}}\,(\sigma^+ + \sigma^-) \quad (9.2.1)$$

$$H_{\mathrm{sys}} \to \tfrac{1}{2}\hbar\Omega\sigma_z. \quad (9.2.2)$$

We see that the process of splitting the operator X into eigenoperators as in (3.6.61) corresponds exactly to that in (9.2.1) so that m takes on only one value, and

$$X^{\pm} \to \hbar\sigma^{\pm}. \quad (9.2.3)$$

Using the Pauli matrix algebra, the master equation can be reduced to

$$\begin{aligned}\dot{\rho} = &-\tfrac{1}{2}\mathrm{i}\Omega[\sigma_z, \rho] \\ &+\frac{\pi}{2}\left(\bar{N}(\Omega)+1\right)\kappa(\Omega)^2\left\{2\sigma^-\rho\sigma^+ - \rho\sigma^+\sigma^- - \sigma^+\sigma^-\rho + \sigma^-\rho\sigma^- + \sigma^+\rho\sigma^+\right\} \\ &+\frac{\pi}{2}\bar{N}(\Omega)\kappa(\Omega)^2\left\{2\sigma^+\rho\sigma^- - \rho\sigma^-\sigma^+ - \sigma^+\sigma^-\rho + \sigma^-\rho\sigma^- + \sigma^+\rho\sigma^+\right\} \\ &+\frac{\mathrm{i}}{2\Omega}\mathrm{P}\int_{-\infty}^{\infty}\frac{\omega\kappa(\omega)^2 d\omega}{\Omega-\omega}\left(\bar{N}(\omega)+\tfrac{1}{2}\right)\left\{-[\sigma_z,\rho] - 2\sigma^+\rho\sigma^+ + 2\sigma^-\rho\sigma^-\right\}. \quad (9.2.4)\end{aligned}$$

Exercise. Using the algebra of Pauli matrices, including in particular the formulae in (9.1.13–18) derive the Master equation in the form (9.2.4).

Exercise. Using the cyclic property of the trace the Pauli matrix algebra, and

$$\frac{d}{dt}\langle\sigma^+\rangle = \mathrm{Tr}\{\dot{\rho}\sigma^+\}, \quad \text{etc.,} \tag{9.2.5}$$

show that the equations of motion for the means of $\sigma^{\pm}, \sigma_z$ are

$$\frac{d}{dt}\langle\boldsymbol{\sigma}\rangle = \boldsymbol{\Omega}\langle\boldsymbol{\sigma}\rangle - \boldsymbol{\gamma} \tag{9.2.6}$$

in which

$$\boldsymbol{\sigma} = \begin{pmatrix}\sigma^+\\ \sigma^-\\ \sigma_z\end{pmatrix}, \qquad \boldsymbol{\gamma} = \begin{pmatrix}0\\ 0\\ \gamma\end{pmatrix} \tag{9.2.7}$$

and

$$\boldsymbol{\Omega} = \begin{pmatrix} \mathrm{i}(\Omega+\Delta\Omega) - \gamma(\bar{N}+\frac{1}{2}) & \gamma(\bar{N}+\frac{1}{2}) + 2\mathrm{i}\Delta\Omega & 0 \\ \gamma(\bar{N}+\frac{1}{2}) - 2\mathrm{i}\Delta\Omega & -\mathrm{i}(\Omega+\Delta\Omega) - \gamma(\bar{N}+\frac{1}{2}) & 0 \\ 0 & 0 & -\gamma(2\bar{N}+1) \end{pmatrix}. \tag{9.2.8}$$

Quantities in (9.2.7,8) are defined by

$$\bar{N} = \bar{N}(\Omega) \tag{9.2.9}$$

$$\gamma = \pi\kappa(\Omega)^2 \tag{9.2.10}$$

$$\Delta\Omega = \frac{1}{\Omega}\mathrm{P}\int_{-\infty}^{\infty} \frac{\omega\kappa(\omega)^2 d\omega}{\Omega - \omega}\left(\bar{N}(\omega) + \tfrac{1}{2}\right). \tag{9.2.11}$$

9.2.1 Lamb and Stark Shifts

From the structure of (9.2.8), $\Delta\Omega$ represents a frequency shift, in the sense that it occurs in (9.2.4) as the coefficient of a term $[\sigma_z, \rho]$ which is of the same kind as the original system Hamiltonian. Thus, in interaction with the bath (or radiation field in this case) this frequency observed will be $\Omega + \Delta\Omega$.

i) If $\kappa(\omega)$ is constant, $\Delta\Omega$ is divergent—this means that for a point atom this kind of formulation does not make sense. The original model is unrealistic, in that there is no kinetic term in the atomic Hamiltonian, and thus recoil is impossible. In fact, it is only in the case of a fully relativistic Hamiltonian that the Lamb shift can be made finite, and even then after QED renormalization, which involves infinities.

ii) For the purposes of this book it is best to regard the model of an atom fixed in space as having a small but finite extent, so that $\kappa(\omega)$ drops off sufficiently rapidly at high ω to make the integral (9.2.11) convergent.

iii) Of course the $\bar{N}(\omega)$ term is finite at all $T > 0$, and it vanishes when $T = 0$. This part is a shift caused by the thermal radiation which the atom experiences. This is called the Stark shift; the Lamb shift is that part proportional to the $\frac{1}{2}$, and has its source in the vacuum fluctuations.

iv) Notice that one can modify the density of states, perhaps by use of a cavity, so that $\kappa(\omega)^2$ varies strongly around $\omega \approx \Omega$. If $\kappa(\omega)^2$ is smooth, the principal value integral is almost zero, but if it changes rapidly in this region the cancellation will not occur, and the Lamb and Stark shifts can both be very large.

v) One should keep in mind that these frequency shifts are small in comparison to Ω, but are nevertheless of the same order of magnitude as the damping γ—indeed they have a common origin. The damping is more noticeable only because it gives a decay which is otherwise absent. The frequency shifts can normally be neglected; or at least absorbed into Ω.

9.2.2 Rotating Wave Approximation

But it is not quite fair to say this, since there are a larger number of other terms. However these do not have a large effect. The simplest way to argue is via the method of the rotating wave approximation of Sect. 3.6.3a. Applied directly to (9.2.6), this says that $\Omega + \Delta\Omega$ is much larger than all other coefficients, so to a first approximation the time dependence of $\langle\sigma^{\pm}\rangle$ is $e^{\pm i(\Omega + \Delta\Omega t)}$. If we substitute

$$\begin{aligned}\langle\sigma^{\pm}(t)\rangle &= S^{\pm}(t)e^{\pm(i\Omega+\Delta\Omega)t}\\ \langle\sigma_z(t)\rangle &= S_z(t),\end{aligned} \tag{9.2.12}$$

the equations of motion become

$$\begin{aligned}\dot{S^{\pm}} &= -\tfrac{1}{2}\gamma(2\bar{N}+1)S^{\pm} + \left\{\tfrac{1}{2}\gamma(2\bar{N}+1) \pm 2i\Delta\Omega\right\}S^{\pm}e^{\pm 2i(\Omega+\Delta\Omega)t}\\ \dot{S}_z &= -\gamma(2\bar{N}+1)S_z - \gamma.\end{aligned} \tag{9.2.13}$$

The very rapidly varying terms $e^{\pm 2i(\Omega+\Delta\Omega)t}$ average out to zero on the time scale of the other time constants, $\gamma(2\bar{N}+1)$ and are thus neglected. This means the solutions can be approximated by

$$\langle\sigma^{\pm}(t)\rangle = \langle\sigma^{\pm}(0)\rangle \exp\left\{\pm(i\Omega + \Delta\Omega)t - \tfrac{1}{2}\gamma(2\bar{N}+1)t\right\} \tag{9.2.14}$$

$$\langle\sigma_z(t)\rangle = \langle\sigma_z(0)\rangle \exp\left\{-\gamma(2\bar{N}+1)t\right\} - \frac{1}{2\bar{N}+1}\left\{1 - \exp(2\bar{N}+1)t\right\} \tag{9.2.15}$$

Compared to the equations of motion for a free spin system, we see exponential damping factors, which cause $\langle\sigma^{\pm}\rangle$ to approach zero, and $\langle\sigma_z\rangle$ to approach the value $-1/(2\bar{N}+1)$. Using $\bar{N} = \left(\exp(\hbar\Omega/\kappa T) - 1\right)^{-1}$ gives the correct result for the Boltzmann distribution over the energy levels.

Exercise. Solve (9.2.6) by diagonalizing the coefficient matrix. In particular, show that the two complex eigenvalues are, to first order in Ω^{-1}

$$\pm i(\Omega + \Delta\Omega) - \tfrac{1}{2}\gamma(2\bar{N}+1) \pm \frac{i}{\Omega}\left\{\frac{3(\Delta\Omega)^2}{2} + \frac{\gamma^2(2\bar{N}+1)^2}{4}\right\} \tag{9.2.16}$$

which agree with the rotating wave approximation, (9.2.14), to order Ω^0.

9.2.3 Master Equation and QSDE

The practical investigation of atomic spectra is carried out by shining light onto atoms, and observing light which is emitted. Because the world is three dimensional, the mode which represents the incoming light is not usually the same as that which represents the outgoing light. As a simplified representation of the situation, let us consider an atom coupled to two inputs, as illustrated in Fig.9.1, and write down the Ito QSDEs using terms (5.3.36) and (5.3.50). We use the interaction picture, and thus write

$$\begin{aligned} &H_{\mathrm{sys}} = 0 \\ &b_{\mathrm{in}}(1,t) = b_{\mathrm{in}}(t); \quad b_{\mathrm{in}}(2,t) = f_{\mathrm{in}}(t); \quad c_1 = c_2 = \sigma^- \end{aligned} \tag{9.2.17}$$

and

$$\langle b_{\mathrm{in}}^\dagger(t) b_{\mathrm{in}}(t')\rangle = \bar{N}_0 \delta(t-t'); \quad \langle f_{\mathrm{in}}^\dagger(t) f_{\mathrm{in}}(t')\rangle = 0. \tag{9.2.18}$$

Using the Pauli matrix formulae of Sect. 9.1.2, we find the three QSDE's

$$\dot{\sigma}^+ = -\left\{\frac{\gamma_1}{2}(2\bar{N}_0+1)+\frac{\gamma_2}{2}\right\}\sigma^+ + \sigma_z\left\{\sqrt{\gamma_1}\, b_{\mathrm{in}}^\dagger(t) + \sqrt{\gamma_2}\, f_{\mathrm{in}}^\dagger(t)\right\} \tag{9.2.19}$$

$$\dot{\sigma}^- = -\left\{\frac{\gamma_1}{2}(2\bar{N}_0+1)+\frac{\gamma_2}{2}\right\}\sigma^- + \sigma_z\left\{\sqrt{\gamma_1}\, b_{\mathrm{in}}(t) + \sqrt{\gamma_2}\, f_{\mathrm{in}}(t)\right\} \tag{9.2.20}$$

$$\begin{aligned} \dot{\sigma}_z = &-\left\{\gamma_1(2\bar{N}_0+1)+\gamma_2\right\}\sigma_z - (\gamma_1+\gamma_2) \\ &-2\sigma^+\left(\sqrt{\gamma_1}\, b_{\mathrm{in}}(t) + \sqrt{\gamma_2}\, f_{\mathrm{in}}(t)\right) - 2\sigma^-\left(\sqrt{\gamma_1}\, b_{\mathrm{in}}^\dagger(t) + \sqrt{\gamma_2}\, f_{\mathrm{in}}^\dagger(t)\right). \end{aligned} \tag{9.2.21}$$

a) Input and Output Fields: We assume that the atom is illuminated with the field $b_{\mathrm{in}}(t)$, and observed with the field $f_{\mathrm{out}}(t)$, where, of course

$$f_{\mathrm{out}}(t) = f_{\mathrm{in}}(t) + \sqrt{\gamma_2}\,\sigma^-(t). \tag{9.2.22}$$

We will want to compute the 2nd and 4th order correlation functions, which correspond to the amplitude and intensity correlation functions, and are experimentally accessible. These are best done using the master equation, and the formulae of Sect. 5.2.1.

b) Master Equation Evolution Operator: For simplicity, we define

$$\begin{aligned} \gamma &= \gamma_1 + \gamma_2 \\ \bar{N} &= \frac{\gamma_1 \bar{N}_0}{\gamma_1+\gamma_2} \\ \kappa &= \tfrac{1}{2}\left[\gamma_1(2\bar{N}_0+1)+\gamma_2\right] \equiv \tfrac{1}{2}\gamma(2\bar{N}+1). \end{aligned} \tag{9.2.23}$$

The equations of motion for the atomic averages are thus

$$\begin{aligned} \langle\dot{\sigma}^+\rangle &= -\kappa\langle\sigma^+\rangle \\ \langle\dot{\sigma}^-\rangle &= -\kappa\langle\sigma^-\rangle \\ \langle\dot{\sigma}_z\rangle &= -2\kappa\langle\sigma_z\rangle - \gamma \end{aligned} \tag{9.2.24}$$

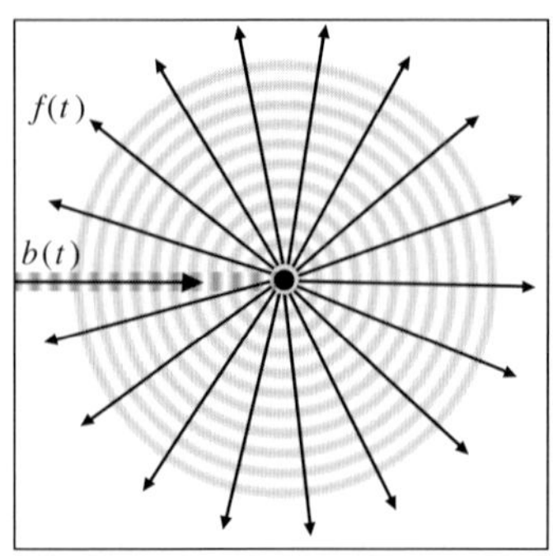

Fig. 9.1 Atom coupled to an almost plane wave mode and to a spherical wave mode

with the solutions

$$\begin{aligned} \langle\sigma^+(t)\rangle &= \mathrm{e}^{-\kappa t}\langle\sigma^+(0)\rangle \\ \langle\sigma^-(t)\rangle &= \mathrm{e}^{-\kappa t}\langle\sigma^-(0)\rangle \\ \langle\sigma_z(t)\rangle &= \mathrm{e}^{-2\kappa t}\langle\sigma_z(0)\rangle - \frac{1}{2\bar{N}+1}\left(1-\mathrm{e}^{-2\kappa t}\right). \end{aligned} \tag{9.2.25}$$

Using the Pauli matrix properties, we can write the density matrix $\rho(t)$ as

$$\rho(t) = \tfrac{1}{2}\left\{1 + \langle\sigma_z(t)\rangle\sigma_z\right\} + \langle\sigma^+(t)\rangle\sigma^- + \langle\sigma^-(t)\rangle\sigma^+ \tag{9.2.26}$$

which gives us the general solution to the master equation. Indeed, it gives a description of the evolution operator $V(t)$ acting on any matrix *with unit trace*.

In general if $\mathrm{Tr}\{\rho(t)\} \neq 1$ one writes

$$\rho(t) = \tfrac{1}{2}\mathrm{Tr}\left\{\rho(0)\right\} + \tfrac{1}{2}\langle\sigma_z(t)\rangle\sigma_z + \langle\sigma^+(t)\rangle\sigma^- + \langle\sigma^-(t)\rangle\sigma^+ \tag{9.2.27}$$

in which

$$\langle\sigma_z(t)\rangle = \mathrm{Tr}\left\{\sigma_z\rho(t)\right\} \tag{9.2.28}$$

i.e. we ignore the fact that $\rho(t)$ does not have unit trace in this definition.

c) Stationary Solutions—Boltzmann Formula:

As $t \to \infty$, we find

$$\langle\sigma^+(t)\rangle, \quad \langle\sigma^-(t)\rangle \to 0 \tag{9.2.29}$$

$$\langle\sigma_z(t)\rangle \to -1/(2\bar{N}+1) \tag{9.2.30}$$

$$\rho(t) \to \rho_s = \tfrac{1}{2}\left[1 - \sigma_z/(2\bar{N}+1)\right] = \begin{pmatrix} \frac{N}{2N+1} & 0 \\ 0 & \frac{N+1}{2N+1} \end{pmatrix}. \tag{9.2.31}$$

The relations (9.2.23) show that the equations of motion are the same as if we had only one noise, characterized by $\bar{N}$. If this is a thermal field, approximated by white noise, then $\bar{N} = 1/[\exp(\hbar\Omega/kT) - 1]$, and the stationary density operator shows that the ratio of the population in the upper level to that in the lower level is $\mathrm{e}^{-\hbar\Omega/kT}$, the Boltzmann formula.

d) Amplitude Correlation Function: Using the formulae of Sect. 5.4.6 and (9.2.22), we note that the input $f_{\text{in}}(t)$ is in the vacuum state, so that (5.4.21) may be used to get

$$\langle f_{\text{out}}^{\dagger}(t) f_{\text{out}}(t') \rangle = \gamma_2 \langle \sigma^+(t) \sigma^-(t') \rangle \tag{9.2.32}$$

and using (5.2.23), the stationary output amplitude correlation function can be written in terms of

$$\langle \sigma^+(t) \sigma^-(0) \rangle_s = \text{Tr}\left\{ \sigma^+ V(t,0) \sigma^- \rho_s \right\}. \tag{9.2.33}$$

Notice now that

i) Using the Pauli matrix formulae and (9.2.31)

$$\sigma^- \rho_s = \frac{\bar{N}}{2\bar{N}+1} \sigma^-. \tag{9.2.34}$$

ii) σ^- does not have unit trace: hence we can use the solutions (9.2.25) where, for example, we now have

$$\begin{aligned} \langle \sigma^+(0) \rangle &= \text{Tr}\left\{ \sigma^+ \sigma^- \right\} = 1 \\ \langle \sigma^-(0) \rangle &= \text{Tr}\left\{ \sigma^- \sigma^- \right\} = 0 \\ \langle \sigma_z(0) \rangle &= \text{Tr}\left\{ \sigma_z \sigma^- \right\} = 0. \end{aligned} \tag{9.2.35}$$

Thus

$$V(t)\{\sigma^- \rho_s\} = \frac{\bar{N}}{2\bar{N}+1} \sigma^- \mathrm{e}^{-\kappa t}. \tag{9.2.36}$$

iii) Hence

$$\langle \sigma^+(t) \sigma^-(0) \rangle_s = \text{Tr}\left\{ \sigma^+ V(t)\{\sigma^- \rho_s\} \right\} = \frac{\bar{N}}{2\bar{N}+1} \mathrm{e}^{-\kappa t}. \tag{9.2.37}$$

Similarly,

$$\langle \sigma^-(t) \sigma^+(0) \rangle_s = \frac{\bar{N}+1}{2\bar{N}+1} \mathrm{e}^{-\kappa t}. \tag{9.2.38}$$

and thus

$$\langle f_{\text{out}}^{\dagger}(t) f_{\text{out}}(0) \rangle_s = \frac{\gamma_2 \bar{N}}{2\bar{N}+1} \mathrm{e}^{-\gamma(\bar{N}+\frac{1}{2})|t|}. \tag{9.2.39}$$

e) Intensity Correlation Function: In this case, the correlation function related to photon counting, as explained in Sect. 8.3.4, is

$$\langle f_{\text{out}}^\dagger(0) f_{\text{out}}^\dagger(t)\, f_{\text{out}}(t) f_{\text{out}}(0)\rangle_s = \gamma_2^2 \langle \sigma^+(0)\sigma^+(t)\sigma^-(t)\sigma^-(0)\rangle_s \tag{9.2.40}$$

$$= \text{Tr}\left\{\sigma^+\sigma^- V(t)\{\sigma^-\rho_s\sigma^+\}\right\}. \tag{9.2.41}$$

Here we note that

$$\sigma^-\rho_s\sigma^+ = \frac{\bar{N}}{2\bar{N}+1}\left\{\tfrac{1}{2}(1-\sigma_z)\right\} \tag{9.2.42}$$

and the term inside the curly brackets has unit trace. Thus, proceeding as before, we find

$$V(t)(\sigma^-\rho_s\sigma^+) = \frac{\bar{N}}{2\bar{N}+1}\left\{\tfrac{1}{2} - \tfrac{1}{2}\sigma_z\left[\mathrm{e}^{-2\kappa t} + \frac{1-\mathrm{e}^{-2\kappa t}}{2\bar{N}+1}\right]\right\} \tag{9.2.43}$$

so that (9.2.41) gives

$$\langle f_{\text{out}}^\dagger(0) f_{\text{out}}^\dagger(t) f_{\text{out}}(t) f_{\text{out}}(0)\rangle = \gamma_2^2\left(\frac{\bar{N}}{2\bar{N}+1}\right)^2\left(1-\mathrm{e}^{-2\kappa t}\right). \tag{9.2.44}$$

Comments:

i) In terms of the notation of Sect. 8.3.4, we find that

$$g^{(2)}(t)_s = (1-\mathrm{e}^{-2\kappa t}) \tag{9.2.45}$$

$$g^{(1)}(t)_s = \mathrm{e}^{-\kappa t}. \tag{9.2.46}$$

Thus, $g^{(2)}(0) = 0$, which is completely non-classical behaviour, called antibunching. From the formula (9.2.40), we see that this vanishing at $t = 0$ is not really a property of the solutions of the equations, but rather of the fact that $\sigma^+(0)\sigma^+(0) = \sigma^-(0)\sigma^-(0) = 0$.

ii) The antibunching says that there is no probability of counting a photon immediately after one has just been counted. This can be viewed as the requirement that the atom will be in its ground state after emitting a photon, and it will take a time of order $(2\kappa)^{-1}$ before it is again significantly excited. The property of antibunching was first pointed out by *Carmichael* and *Walls* [9.2] in the related problem of *coherent* resonance fluorescence, in which it is assumed that $b_{\text{in}}(t)$ is a coherent field, with mean $\beta(t)$, and thus $\bar{N}_0 = 0$. The resonance fluorescence problem can be solved by exactly the same methods as given here, but is more complicated because the equations for the averages do not decouple, as they do in (9.2.24).

9.2.4 Two Level Atom in a Coherent Driving Field

For further application to the laser, and for its own intrinsic interest, we are led to the situation in which we drive the atom with a coherent driving field. To do this we follow the methods of Chap.5, and modify the QSDEs for the atom in a thermal field by adding a constant term to the incoming noise, thus

$$\sqrt{\gamma}\, b_{\rm in}(t) \to \sqrt{\gamma}\, b_{\rm in}(t) + \mathcal{E}\mathrm{e}^{\mathrm{i}\phi} \tag{9.2.47}$$

where $\mathcal{E}\mathrm{e}^{\mathrm{i}\phi}$ represents the coherent field by the amplitude $\mathcal{E} > 0$ and the phase ϕ. (Actually, $\mathcal{E}$ is the product of the field amplitude and an appropriate coupling constant, which has been absorbed into one constant for simplicity). If we now define

$$\langle\sigma^+\rangle = -\frac{\mathrm{i}}{2}\mathrm{e}^{-\mathrm{i}\phi}(S_x + \mathrm{i}S_y) = \langle\sigma^-\rangle^*, \qquad \langle\sigma_z\rangle = S_z, \tag{9.2.48}$$

then the mean value equations corresponding to (9.2.2) become

$$\begin{aligned}\dot{S}_x &= -\kappa S_x \\ \dot{S}_y &= -\kappa S_y + 2\mathcal{E} S_z \\ \dot{S}_z &= -2\kappa S_z - 2\mathcal{E} S_y - \gamma.\end{aligned} \tag{9.2.49}$$

a) Stationary Solutions: Setting the time derivatives to zero, we get the stationary solutions, $\bar{\mathbf{S}}$,

$$\begin{aligned}\bar{S}_x &= 0 \\ \bar{S}_y &= -\frac{2\gamma\mathcal{E}}{2\kappa^2 + 4\mathcal{E}^2} \\ \bar{S}_z &= -\frac{\gamma\kappa}{2\kappa^2 + 4\mathcal{E}^2}.\end{aligned} \tag{9.2.50}$$

The stationary density matrix is then

$$\rho_s = \tfrac{1}{2}\left(1 + \sigma_z\bar{S}_z\right) + \sigma^+\langle\sigma^-\rangle + \sigma^-\langle\sigma^+\rangle \tag{9.2.51}$$

$$= \tfrac{1}{2}\left(1 + \bar{\mathbf{S}}\cdot\boldsymbol{\sigma}'\right) \tag{9.2.52}$$

where

$$\sigma' = \left\{\mathrm{i}\left(\mathrm{e}^{\mathrm{i}\phi}\sigma^+ - \mathrm{e}^{-\mathrm{i}\phi}\sigma^-\right), \left(\mathrm{e}^{\mathrm{i}\phi}\sigma^+ + \mathrm{e}^{-\mathrm{i}\phi}\sigma^-\right), \sigma_z\right\} \tag{9.2.53}$$

and

$$(\boldsymbol{\sigma})'^{\pm} = \pm\mathrm{i}\mathrm{e}^{\pm\mathrm{i}\phi}\sigma^{\pm}. \tag{9.2.54}$$

From the above, we can see that we may simply make the replacement $\boldsymbol{\sigma}' \to \boldsymbol{\sigma}$, since the algebra is the same. In what follows we will not explicitly write $\boldsymbol{\sigma}'$, though it should be understood.

b) Time Dependent Solutions: The equations of motion (9.2.49) can be written symbolically as

$$\dot{\mathbf{S}}(t) = -A(\mathbf{S}(t) - \bar{\mathbf{S}}) \tag{9.2.55}$$

with

$$A = \begin{pmatrix} -\kappa & 0 & 0 \\ 0 & -\kappa & 2\mathcal{E} \\ 0 & -2\mathcal{E} & -2\kappa \end{pmatrix} \tag{9.2.56}$$

and solutions given by

$$\mathbf{S}(t) = \mathrm{e}^{-At}\left(\mathbf{S}(0) - \bar{\mathbf{S}}\right) + \bar{\mathbf{S}}. \tag{9.2.57}$$

Using the property (9.1.19) for exponentiating the 2×2 submatrix, one can write

$$\mathrm{e}^{-At} = \begin{pmatrix} \mathrm{e}^{-\kappa t} & 0 & 0 \\ 0 & f_1(t) & f_3(t) \\ 0 & -f_3(t) & f_2(t) \end{pmatrix} \tag{9.2.58}$$

where

$$\begin{aligned} f_1(t) &= \mathrm{e}^{-\frac{3}{2}\kappa t}\left(\cosh \Lambda t + \frac{\kappa}{2\Lambda}\sinh \Lambda t\right) \\ f_2(t) &= \mathrm{e}^{-\frac{3}{2}\kappa t}\left(\cosh \Lambda t - \frac{\kappa}{2\Lambda}\sinh \Lambda t\right) \\ f_3(t) &= \mathrm{e}^{-\frac{3}{2}\kappa t}\frac{2\mathcal{E}}{\Lambda}\sinh \Lambda t \end{aligned} \tag{9.2.59}$$

and

$$\Lambda = \sqrt{\frac{\kappa^2}{4} - 4\mathcal{E}^2}\,. \tag{9.2.60}$$

The solutions change from hyperbolic to circular functions when $\mathcal{E} > \kappa/4$; for larger $\mathcal{E}$ one makes the replacement

$$\Lambda \to \mathrm{i}\Omega = \mathrm{i}\sqrt{4\mathcal{E}^2 - \frac{\kappa^2}{4}} \tag{9.2.61}$$

where the frequency Ω is known as the *Rabi frequency*.

c) The Time Dependent Density Matrix: This is of course given (using (9.2.27)) by

$$\rho(t) = \tfrac{1}{2}\operatorname{Tr}\{\rho(0)\}\{\mathbf{1} + \boldsymbol{\sigma}\cdot\mathbf{S}(t)\}\,, \tag{9.2.62}$$

where for a genuine density matrix, $\operatorname{Tr}\{\rho(0)\}$ should be equal to 1, but for generality, we will see that it is best to not make this restriction.

d) Correlation Functions: All the two-time correlation functions can be written in terms of the expression

$$\mathrm{Cor}(t) = \mathrm{Tr}\left\{(a + \mathbf{b}\cdot\boldsymbol{\sigma})V(t,0)\tfrac{1}{2}(1 + \mathbf{S}(0)\cdot\boldsymbol{\sigma})\right\} \tag{9.2.63}$$

where the evolution operator $V(t,0)$ is the notation of Sect. 6.2. Using the time dependent solutions,

$$\mathrm{Cor}(t) = a + \mathbf{b}\cdot\mathbf{S}(t), \tag{9.2.64}$$

and this can be written, using the solution (9.2.57), as

$$\mathrm{Cor}(t) = a + \mathbf{b}\cdot\bar{\mathbf{S}} + \mathbf{b}\cdot \mathrm{e}^{-At}\left(\mathbf{S}(0) - \bar{\mathbf{S}}\right). \tag{9.2.65}$$

e) Amplitude Correlation Function: Using the formulae of (5.2.21–5.2.23), we want to evaluate the stationary correlation function

$$\langle\sigma^+(t)\sigma^-(0)\rangle_s = \begin{cases} \mathrm{Tr}\left\{\sigma^+V(t,0)\sigma^-\rho_s\right\} & t>0 \\ \mathrm{Tr}\left\{\sigma^-V(-t,0)\rho_s\sigma^+\right\} & t<0. \end{cases} \tag{9.2.66}$$

By comparing with (9.2.63), one sees that here $a = 0$ and

$$\mathbf{b} = \begin{cases} \left(\tfrac{1}{2},\ \tfrac{1}{2}\mathrm{i},\ 0\right) & t>0 \\ \left(\tfrac{1}{2},\ -\tfrac{1}{2}\mathrm{i},\ 0\right) & t<0. \end{cases} \tag{9.2.67}$$

Using the formula (9.1.6), one finds

$$\sigma^-\rho_s = -\frac{\mathrm{i}\bar{S}_y}{4}\left\{1 + \boldsymbol{\sigma}\cdot(\mathrm{i}g, g, -1)\right\} \tag{9.2.68}$$

where

$$g = \frac{1+\bar{S}_z}{2\mathrm{i}\bar{S}^-} = \frac{1+\bar{S}_z}{\bar{S}_y}. \tag{9.2.69}$$

Thus using (9.2.65) (multiplied by a factor $-\mathrm{i}\bar{S}_y/2$)

$$\langle\sigma^+(t)\sigma^-(0)\rangle_s = \frac{\bar{S}_y^2}{4} + \frac{(1+\bar{S}_z)}{4}\mathrm{e}^{-\kappa|t|} + \frac{1+\bar{S}_z-\bar{S}_y^2}{4}f_1(|t|) - \frac{\bar{S}_y(1+\bar{S}_z)}{4}f_3(|t|) \tag{9.2.70}$$

where we have implicitly included the similarly derivable result for $t < 0$, by writing $|t|$ instead of t. Again, we can derive in the same way the reverse correlation function

$$\langle\sigma^-(t)\sigma^+(0)\rangle_s = \frac{\bar{S}_y^2}{4} + \frac{(1-\bar{S}_z)}{4}\mathrm{e}^{-\kappa|t|} + \frac{1-\bar{S}_z-\bar{S}_y^2}{4}f_1(|t|) + \frac{\bar{S}_y(1-\bar{S}_z)}{4}f_3(|t|). \tag{9.2.71}$$

Reverting to the transformation $\sigma \to \sigma'$, we see that indeed this does not change the result.

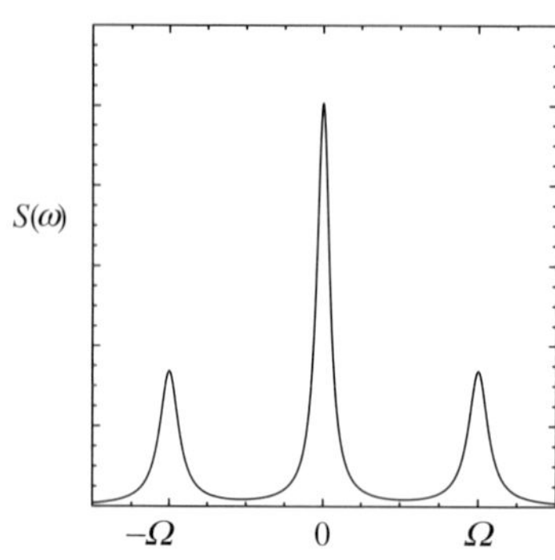

Fig. 9.2 The spectrum of resonance fluorescence as given by (9.2.75)

f) Intensity Correlation Function: We follow much the same procedure to evaluate

$$\langle\sigma^+(0)\sigma^+(t)\sigma^-(t)\sigma^-(0)\rangle = \mathrm{Tr}\left\{\sigma^+\sigma^- V(t,0)\sigma^-\rho_s\sigma^+\right\}$$

$$= \frac{(1+\bar{S}_z)^2}{4}(1 - f_2(t)) + \frac{\bar{S}_y(1+\bar{S}_z)}{4} f_3(t) \qquad t > 0. \tag{9.2.72}$$

g) Spectrum of Fluorescent Light: The spectrum of the fluorescent light is given by the Fourier transform of $\langle b^\dagger_{\text{out}}(t) b_{\text{out}}(0)$, which, by using the formula (5.2.5)b, relating the output correlation function to the internal correlation function, is given by the Fourier transform of $\langle\sigma^+(t)\sigma^-(0)\rangle$, i.e.

$$S(\omega) = \frac{\gamma_2}{2\pi}\int_{-\infty}^{\infty} \mathrm{e}^{\mathrm{i}\omega t}\langle\sigma^+(t)\sigma^-(0)\rangle_s\, dt. \tag{9.2.73}$$

Noting that $\langle\sigma^+(-t)\sigma^-(0)\rangle = \langle\sigma^+(0)\sigma^-(t)\rangle = \langle\sigma^+(t)\sigma^-(0)\rangle^*$, we see that

$$S(\omega) = \frac{\gamma_2}{\pi}\mathrm{Re}\left\{\int_0^{\infty}\langle\sigma^+(t)\sigma^-(0)\rangle \mathrm{e}^{-\mathrm{i}\omega t}\, dt\right\}. \tag{9.2.74}$$

The Fourier transforms are straightforward to evaluate, but the spectrum which results is an unsightly formula. In the very strong field limit, $\Lambda \to \mathrm{i}\Omega$, $\bar{S}_y = \bar{S}_z = 0$, and

$$S(\omega) = \frac{\gamma_2}{4\pi}\frac{\kappa}{\kappa^2+\omega^2} + \frac{\gamma_2}{4\pi}\left\{\frac{\frac{3}{4}\kappa}{\left(\frac{3}{2}\kappa\right)^2 + (\omega+\Omega)^2} + \frac{\frac{3}{4}\kappa}{\left(\frac{3}{2}\kappa\right)^2 + (\omega-\Omega)^2}\right\}. \tag{9.2.75}$$

We see three peaks; a central peak, and two sidebands at the Rabi frequency, Ω, as shown in Fig.9.2.

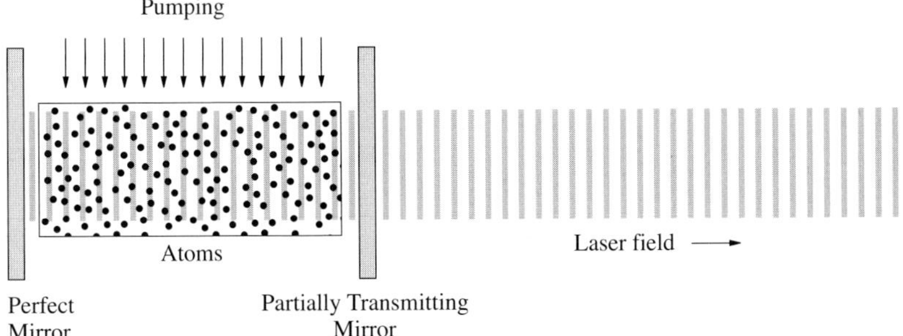

Fig. 9.3 Configuration of the essential elements of a gas laser

9.3 The Laser

The impetus to develop a useful form of quantum noise theory as a computational tool came from the invention of the laser, and the consequent need to understand the properties of laser light. The basis of laser action is stimulated emission of radiation. As we have seen in Sect. 7.2.2, if a bath of atoms is coupled to a mode of the electromagnetic field and can be maintained in a situation in which there is a net inversion, the electromagnetic field will be amplified.

A practical laser must include two more features. The first is the fact that the mode of the electromagnetic field will most often be a standing wave mode between two mirrors, which are not perfect—indeed for practical use, some light must get through the mirrors to be available to the user, so that one mirror will be deliberately made partially reflecting.

The second feature is that of depletion of the bath of inverted atoms. As lasing takes place, atoms lose their inversion. Some mechanism must be provided to pump them back up again, and this cannot be absolutely perfect.

A simple model of a laser is found by considering a system composed of:

i) A single mode of the light field, confined between mirrors with some small losses through them. This is basically a damped harmonic oscillator, represented by a system Hamiltonian

$$H_F = \hbar\omega a^\dagger a \tag{9.3.1}$$

coupled to a heat bath via a coupling

$$H_{FB} = a^\dagger \Gamma_F + a\Gamma_F^\dagger . \tag{9.3.2}$$

The precise nature of the heat bath Hamiltonian is not very important unless we are concerned with the output field—in that case it consists of the appropriate modes of the free electromagnetic field, and the output field can be determined via the input-output methods of Sect. 3.2.

ii) A large number, N, of two level atoms, which are held in an inverted state by some appropriate pumping. These are represented as in Sect. 5.1.4, but are not treated as a heat bath. Instead, they are coupled to another bath, which is the source of the pumping. The Hamiltonian for the atoms is

$$H_A = \sum_{\mu=1}^{N} \tfrac{1}{2}\hbar\omega\sigma_z^{\mu} \tag{9.3.3}$$

where the σ^{μ} are different representations of the Pauli matrices, one for each atom, and the transition frequency ω is (for simplicity) chosen exactly equal to that of the mode of the light field. The coupling to the pumping bath is given by the Hamiltonian

$$H_{AB} = \sum_{\mu=1}^{N} \left(\Gamma_{A,\mu}\sigma_{\mu}^{+} + \Gamma_{A,\mu}^{\dagger}\sigma_{\mu}^{-} \right) . \tag{9.3.4}$$

iii) The atoms and the field are now coupled by a RWA coupling

$$H_{AF} = \mathrm{i}g\hbar \sum_{\mu} \left(a^{\dagger}\sigma_{\mu}^{-} - a\sigma_{\mu}^{+} \right) . \tag{9.3.5}$$

This model is in the form given by *Haken* [9.3], and is also very similar to those of *Lax*, *Louisell* and *Gordon* [9.4, 9.5].

The laser is most naturally treated with master equation techniques, which allow the application of techniques of adiabatic elimination of fast variables, like those used in Sect. 5.1.2. There are two natural time scales; the lifetime of a photon in the laser cavity and the natural lifetime of an excited atom. By assuming that the lifetime of the atoms is very small, their motion can be assumed to follow that of the much slower laser field, and thus a quite simple equation can be derived.

9.3.1 Quantum Langevin Equations for the Laser

We can use the techniques used in Sect. 6.3.1 to write quantum stochastic differential equations for the variables $a, \sigma_{\mu}^{\pm}, \sigma_z^{\mu}$. The heat bath coupling to the atomic variables has a negative temperature, so the damping parts of equations take the form given in (7.2.36). We use the notation

$$c \to \sigma_{\mu}^{-} \tag{9.3.6}$$

$$\gamma(\bar{N}+1) \to W_{12} \tag{9.3.7}$$

$$\gamma\bar{N} \to W_{21} \tag{9.3.8}$$

in this equation. Similarly, the damping part field mode a is taken from the Ito QSDE form (9.5.44), with

$$\bar{N} \to n \tag{9.3.9}$$

$$\frac{\gamma}{2} \to \kappa . \tag{9.3.10}$$

The equations of motion are then straightforwardly worked out as (in the interaction picture)

$$(\mathbf{I})da = \left(-\kappa a + g\sum_{\mu}\sigma_{\mu}^{-}\right)dt - \sqrt{2\kappa}\,dB(t) \tag{9.3.11}$$

$$(\mathbf{I})d\sigma_{\mu}^{+} = \left(ga^{\dagger}\sigma_{\mu}^{z} - \tfrac{1}{2}(W_{12}+W_{21})\sigma_{\mu}^{+}\right)dt + \sigma_{\mu}^{z}dH_{\mu}^{\dagger}(t) \tag{9.3.12}$$

$$(\mathbf{I})d\sigma_{\mu}^{z} = \left\{-2g\left(a^{\dagger}\sigma_{\mu}^{-} + a\sigma_{\mu}^{+}\right) - (W_{12}+W_{21})\sigma_{\mu}^{z} + (W_{12}-W_{21})\right\}dt$$
$$-2\sigma_{\mu}^{+}dH_{\mu}(t) - 2\sigma_{\mu}^{-}dH_{\mu}^{\dagger}(t). \tag{9.3.13}$$

Here, the Ito noise terms satisfy the relations

$$dB^{\dagger}(t)dB(t) = \bar{n}dt, \quad dB(t)dB^{\dagger}(t) = (\bar{n}+1)dt \tag{9.3.14}$$

$$dH_{\mu}^{\dagger}(t)dH_{\nu}(t) = \delta_{\mu\nu}W_{12}dt, \quad dH_{\mu}(t)dH_{\nu}^{\dagger}(t) = \delta_{\mu\nu}W_{21}dt \tag{9.3.15}$$

and all other products vanish.

Since the equations are linear, we can simply add them up, and derive equations of motion for the "total spin operator" $\mathbf{S} = \sum_{\mu}\boldsymbol{\sigma}_{\mu}$; namely

$$(\mathbf{I})da = (-\kappa a + gS^{-})dt - \sqrt{2\kappa}\,dB(t) \tag{9.3.16}$$

$$(\mathbf{I})dS^{+} = \left(ga^{+}S_{z} - \tfrac{1}{2}(W_{12}+W_{21})S^{+}\right)dt + dH^{\dagger}(t) \tag{9.3.17}$$

$$(\mathbf{I})dS_{z} = \left\{-2g(a^{+}S^{-} + aS^{+}) - (W_{12}+W_{21})S_{z} + N(W_{12}-W_{21})\right\}dt$$
$$-2dH_{z}(t). \tag{9.3.18}$$

Here, $dH(t)$, $dH^{\dagger}(t)$, and $dH_{z}(t)$ are defined as the sums of the noises occurring in (9.3.12,13), and satisfy rather simple product formulae, which are easily evaluated using the fact that the Ito increment commutes with and is independent of the variables $\boldsymbol{\sigma}_{\mu}$. Thus, for example

$$dH^{\dagger}(t)dH(t) = \sum_{\mu\nu}\sigma_{\mu}^{z}dH_{\mu}^{\dagger}(t)\sigma_{\nu}^{z}dH_{\nu}(t) \tag{9.3.19}$$

$$= \sum_{\mu\nu}W_{12}\delta_{\mu\nu}dt = NW_{12}dt. \tag{9.3.20}$$

We can devise all the rules similarly; the complete set of nonvanishing products is

$$\begin{aligned}
dH^{\dagger}(t)dH(t) &= NW_{12}dt\\
dH(t)dH^{\dagger}(t) &= NW_{21}dt\\
dH^{\dagger}(t)dH_{z}(t) &= W_{12}S^{+}dt\\
dH_{z}(t)dH^{\dagger}(t) &= -W_{21}S^{+}dt\\
dH_{z}(t)dH(t) &= W_{12}S^{-}dt\\
dH(t)dH_{z}(t) &= -W_{21}S^{-}dt\\
dH_{z}(t)^{2} &= \tfrac{1}{2}\left\{(W_{21}-W_{12})S_{z} + (W_{12}+W_{21})N\right\}dt.
\end{aligned} \tag{9.3.21}$$

a) Asymptotic Expansion in the Number of Atoms: In any normal laser the number, N, of atoms is very large. Noting that $\mathbf{S}$ is the sum of N parts, we can define appropriate scaled variables by

$$\begin{aligned} \tilde{a} &= a/\sqrt{N} \\ \tilde{g} &= g\sqrt{N} \\ \tilde{\mathbf{S}} &= \mathbf{S}/N \\ d\tilde{H}(t) &= dH(t)/\sqrt{N} \\ d\tilde{H}_z(t) &= dH_z(t)/\sqrt{N}\ . \end{aligned} \tag{9.3.22}$$

This can be viewed as a kind of system size expansion in N, the number of atoms, and is a quantum analogue of the system size expansion introduced by *van Kampen* [9.6] in classical stochastic systems.

The scaled equations now are of the same appearance as the original equations, (9.3.16–18), except that the deterministic terms are of order N^0, while the noises are all of order $N^{-\frac{1}{2}}$: explicitly

$$\begin{aligned} d\tilde{a} &= \left(-\kappa\tilde{a} + \tilde{g}\tilde{S}^-\right) dt - \sqrt{\frac{2\kappa}{N}}\, dB(t) \\ d\tilde{S}^+ &= \left(\tilde{g}\tilde{a}^\dagger\tilde{S}_z - \tfrac{1}{2}(W_{12}+W_{21})\tilde{S}^+\right) dt + \sqrt{\frac{1}{N}}\, d\tilde{H}^\dagger(t) \\ d\tilde{S}_z &= \Big\{ -2\tilde{g}\left(\tilde{a}^\dagger\tilde{S}^- + \tilde{a}\tilde{S}^+\right) - (W_{12}+W_{21})\tilde{S}_z \\ &\quad +(W_{12}-W_{21})\Big\} dt - \frac{2}{\sqrt{N}} d\tilde{H}_z(t) \end{aligned} \tag{9.3.23}$$

with

$$\begin{aligned} &d\tilde{H}^\dagger(t)d\tilde{H}(t) = W_{12}dt; &&d\tilde{H}(t)d\tilde{H}^\dagger(t) = W_{21}dt \\ &d\tilde{H}^\dagger(t)d\tilde{H}_z(t) = W_{12}\tilde{S}^+dt; &&d\tilde{H}_z(t)d\tilde{H}^\dagger(t) = -W_{21}\tilde{S}^+dt \\ &d\tilde{H}_z(t)^2 = \tfrac{1}{2}\left\{(W_{21}-W_{12})\tilde{S}_z + W_{12} + W_{21}\right\} dt \end{aligned} \tag{9.3.24}$$

and the corresponding Hermitian conjugate equations.

b) Ito Form of QSDE is Appropriate for Asymptotic Expansions: It is implicit in the formulation (9.3.22), that the noise terms are small compared with the deterministic terms. In the Ito form, the mean values of the noise terms are zero, and their mean squares in (9.3.22) are small. If we use Stratonovich equations, the mean values of the noises are proportional to the mean values of the W_{12} and W_{21} terms in (9.3.22)—this is how the damping terms arise in the averages of the Stratonovich equations, even though they are not in the equations themselves. Thus the averages of the Stratonovich noise terms are proportional to N^0, and thus no analogous expansion in $1/\sqrt{N}$ is possible.

c) Lowest Order—Macroscopic Laser Equations: To zeroth order in N we can neglect the quantum nature of the variables since the commutators are all of

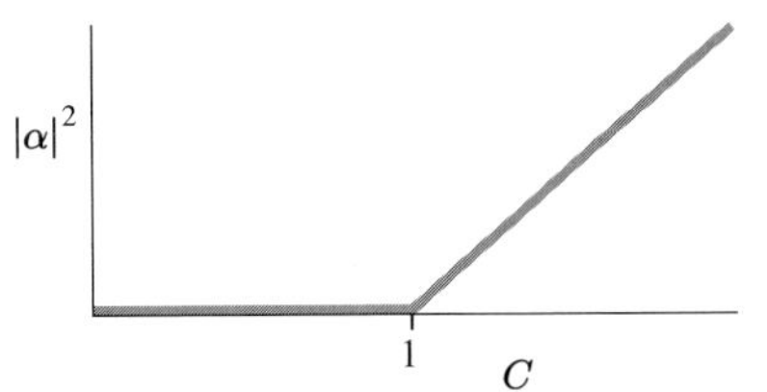

Fig. 9.4 Macroscopic stationary solutions of the laser equations

order $1/N$, and hence arise from higher order contributions. We are then led to the deterministic laser equations for the variables,

$$\begin{aligned}\tilde{\alpha} &= \tilde{a}\\ \tilde{v} &= \tilde{S}^{-}\\ \tilde{D} &= \tilde{S}_z\end{aligned} \tag{9.3.25}$$

in the form

$$\boxed{\begin{aligned}\dot{\tilde{\alpha}} &= -\kappa\tilde{\alpha} + \tilde{g}\tilde{v}\\ \dot{\tilde{v}} &= \tilde{g}\tilde{\alpha}\tilde{D} - \bar{\gamma}\tilde{v}\\ \dot{\tilde{D}} &= -2\tilde{g}\left(\tilde{\alpha}^*\tilde{v} + \tilde{\alpha}\tilde{v}^*\right) - 2\bar{\gamma}(\tilde{D} - d)\end{aligned}} \tag{9.3.26}$$

with

$$\bar{d} = \frac{W_{12} - W_{21}}{W_{12} + W_{21}}, \qquad \bar{\gamma} = \tfrac{1}{2}(W_{12} + W_{21}). \tag{9.3.27}$$

d) Stationary Solutions of the Macroscopic Laser Equations: The three equations (9.3.26) have within them a rather rich range of behaviours. It is simplest to start first with the stationary solution, obtained by setting $\dot{\tilde{\alpha}} = \dot{\tilde{v}} = \dot{\tilde{D}} = 0$. Elementary algebra shows that the stationary solution for the field variable $\tilde{\alpha}$ is given by

$$\tilde{\alpha}\left[1 - \frac{C}{1 + |\tilde{\alpha}|^2/\tilde{n}_0}\right] = 0 \tag{9.3.28}$$

where we define

$$C \equiv \frac{\bar{d}\tilde{g}^2}{\bar{\gamma}\kappa} \tag{9.3.29}$$

which is known as the *pump* or *co-operativity parameter*, and

$$n_0 \equiv \tilde{n}_0 N = \frac{\bar{\gamma}^2}{2g^2} \tag{9.3.30}$$

which is called the *saturation photon number*. There are two solutions for (9.3.28), as illustrated in Fig.9.4, and these are

$$\tilde{\alpha} = 0 \tag{9.3.31}$$

and

$$|\tilde{\alpha}|^2 = \tilde{n}_0(C - 1). \tag{9.3.32}$$

Clearly, the second is only possible if $C > 1$. We can see that C may be varied by changing $W_{12} - W_{21}$, that is, by changing the pumping of the atoms. The behaviour of the system is completely different, depending on whether C is greater than or less than 1, and there is a threshold at $C = 1$. Above threshold, when $C > 1$, a mean coherent field arises—this is the lasing action. The intensity, $|\alpha|^2$, is proportional to the pumping, and to the number of atoms, if we consider that $g \approx 1/\sqrt{N}$. For fixed g, the lasing intensity is proportional to the square of the number of atoms, which is the hallmark of a coherent process. Below threshold, when $C < 1$, there is no mean coherent field—in fact in this macroscopic field, $\tilde{\alpha} = 0$. We cannot discuss what field is actually present without considering the fluctuations.

e) Time Dependent Solutions: There are three time constants in this problem, γ^{-1}, κ^{-1}, and $\tilde{g}^{-1}$, and the full analysis of behaviour depends on their relative sizes. In a practical laser the largest of these is γ, which determines the rate at which the atoms radiate. We can then use the procedure known as adiabatic elimination, which means that the variables whose equations have a time development determined by γ rapidly approach quasi-stationary values obtained by setting equal to zero the left-hand side of the appropriate equations. Thus we transform the equations (9.3.26) by setting $\dot{\tilde{v}} = \dot{\tilde{D}} = 0$, and substitute into the equation for α—the resulting equation is

$$\dot{\tilde{\alpha}} = -\kappa\tilde{\alpha}\left(1 - \frac{C}{1 + |\tilde{\alpha}|^2/\tilde{n}_0}\right). \tag{9.3.33}$$

If we use real and imaginary parts of $\tilde{\alpha} = x + iy$, the equation (9.3.33) can be written in terms of the gradient of a potential

$$\dot{x} = -\frac{\partial}{\partial x}V\left(\sqrt{x^2 + y^2}\right) \tag{9.3.34}$$

$$\dot{y} = -\frac{\partial}{\partial y}V\left(\sqrt{x^2 + y^2}\right) \tag{9.3.35}$$

where

$$V(r) = \tfrac{1}{2}\kappa r^2 - \tfrac{1}{2}\tilde{n}_0\kappa C\ln(1 + r^2/\tilde{n}_0). \tag{9.3.36}$$

This potential (see Fig.9.5) clearly shows that for $C < 1, x = y = 0$ is a stable stationary solution, while for $C > 1$ the stable solution is given by $x^2 + y^2 = \tilde{n}_0(C - 1)$, which means that there is neutral stability for all solutions of this equation. This will mean that above the threshold at $C = 1$ the field variable develops an amplitude which is determined by C, but the phase of the field is arbitrary.

f) Solution to Next Order: We can expand all the variables to first order in

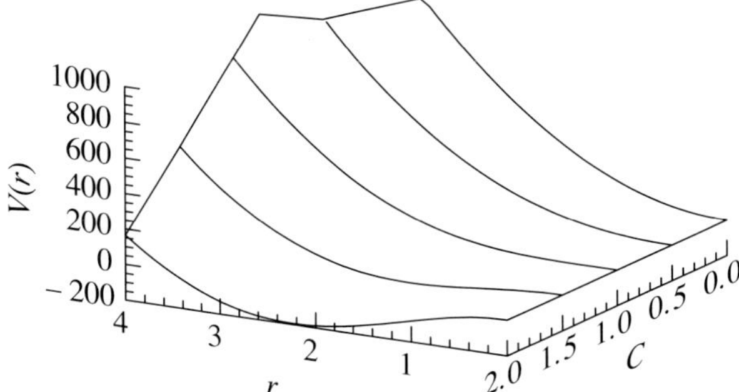

Fig. 9.5 The laser potential $V(r)$ for different values of the pump parameter C

$1/\sqrt{N}$, to derive equations to the next order. Writing

$$\begin{aligned} \tilde{a} &= \tilde{\alpha} + N^{-\frac{1}{2}}\alpha_1 \\ \tilde{S}^- &= \tilde{v} + N^{-\frac{1}{2}}v_1 \\ \tilde{S}_z &= \tilde{D} + N^{-\frac{1}{2}}D_1, \end{aligned} \tag{9.3.37}$$

we can derive equations for the operators α_1, v_1, D_1, which are linear, with coefficients which are functions of $\tilde{\alpha}, \tilde{v}, \tilde{D}$, the deterministic solutions, which are c-numbers. Thus, the equations can be written as

$$\begin{bmatrix} d\alpha_1 \\ d\alpha_1^\dagger \\ dv_1 \\ dv_1^\dagger \\ dD_1 \end{bmatrix} = \begin{bmatrix} -\kappa & 0 & g & 0 & 0 \\ 0 & -\kappa & 0 & g & 0 \\ g\tilde{D} & 0 & -\bar{\gamma} & 0 & g\tilde{\alpha} \\ 0 & g\tilde{D} & 0 & -\bar{\gamma} & g\tilde{\alpha}^* \\ -2\tilde{g}\tilde{v}^* & 2\tilde{g}\tilde{v} & 2\tilde{g}\tilde{\alpha}^* & 2\tilde{g}\tilde{\alpha} & 2\bar{\gamma} \end{bmatrix} \begin{bmatrix} \alpha_1 \\ \alpha_1^\dagger \\ v_1 \\ v_1^\dagger \\ D_1 \end{bmatrix} dt + \begin{bmatrix} -\sqrt{2\kappa}\; dB(t) \\ -\sqrt{2\kappa}\; dB^\dagger(t) \\ dH(t) \\ dH^\dagger(t) \\ 2dH_z(t) \end{bmatrix}. \tag{9.3.38}$$

It must be remembered that these are operator equations, though this does not make them difficult to solve, especially if the zeroth order solution is a stationary state, so that the 5×5 matrix has constant matrix elements.

Exercise. Write (9.3.38) as

$$d\mathbf{v} = -A\mathbf{v}dt + d\mathbf{B}(t). \tag{9.3.39}$$

Show that, in a similar manner to that used for classical SDEs, the stationary solution is

$$\mathbf{v}(t) = \int_{-\infty}^{t} \mathrm{e}^{-A(t-t')} d\mathbf{B}(t'). \tag{9.3.40}$$

The matrix of products (9.3.18) can be written as

$$d\mathbf{B}(t)d\mathbf{B}^T(t) = Gdt. \tag{9.3.41}$$

Using this, follow the reasoning in S.M. Sect. 4.4.6 to show that the stationary variance matrix can be written

$$\langle \mathbf{v}(t)\mathbf{v}^T(t)\rangle = \sigma \tag{9.3.42}$$

and that it satisfies the equation

$$A\sigma + \sigma A^T = G. \tag{9.3.43}$$

The autocorrelation function is

$$\langle \mathbf{v}(t)\mathbf{v}^T(s)\rangle_s = \begin{cases} \mathrm{e}^{-A(t-s)}\sigma, & t \geq s \\ \sigma\,\mathrm{e}^{-A^T(s-t)}, & t \leq s \end{cases} \tag{9.3.44}$$

and the spectrum matrix is

$$S(\omega) = \frac{1}{2\pi}\int_{-\infty}^{\infty} \mathrm{e}^{-\mathrm{i}\omega\tau}\langle \mathbf{v}(t+\tau)\mathbf{v}^T(t)\rangle_s\, d\tau \tag{9.3.45}$$

$$= \frac{1}{2\pi}(A+\mathrm{i}\omega^{-1})G(A^T-\mathrm{i}\omega)^{-1}. \tag{9.3.46}$$

Hence work out the laser spectrum, i.e. the spectrum of fluctuations of $a(t)$ to first order in $1/\sqrt{N}$.

Exercise. Of course the previous exercise assumes that A is nonsingular. Show that above threshold A is singular. After reading Sect. 9.3.3 you will see that this problem is related to the existence of phase diffusion.

9.3.2 Derivation of Laser Equations

a) Master Equation in the Interaction Picture: The QSDE approach to the laser has the advantage of possessing simple equations, and of dealing directly with the quantities of interest, the operators $\mathbf{S}$ and a. But it does have the disadvantage that any practical calculations must be carried out in the perturbation expansion in $N^{-\frac{1}{2}}$, which assumes that N is large.

The QSDEs (9.3.11–13) are equivalent to a master equation. From these, the standard techniques give a master equation which can be written (in the interaction picture of the atom and field Hamiltonians)

$$\frac{\partial \rho}{\partial t} = -\frac{\mathrm{i}}{\hbar}[H_{AF},\rho] + \left.\frac{\partial \rho}{\partial t}\right|_A + \left.\frac{\partial \rho}{\partial t}\right|_F . \tag{9.3.47}$$

The last two terms are the damping terms arising from the two bath-system couplings given in (9.3.2,4), namely

$$\left.\frac{\partial \rho}{\partial t}\right|_A = \frac{W_{21}}{2}\sum_{\mu=1}^{N}\left(2\sigma_\mu^-\rho\sigma_\mu^+ - \sigma_\mu^+\sigma_\mu^-\rho - \rho\sigma_\mu^+\sigma_\mu^-\right)$$

$$+\frac{W_{12}}{2}\sum_{\mu=1}^{N}\left(2\sigma_\mu^+\rho\sigma_\mu^- - \sigma_\mu^-\sigma_\mu^+\rho - \rho\sigma_\mu^-\sigma_\mu^+\right) \tag{9.3.48}$$

and

$$\left.\frac{\partial\rho}{\partial t}\right|_F = \kappa n\left(2a^\dagger\rho a - aa^\dagger\rho - \rho aa^\dagger\right) + \kappa(n+1)\left(2a\rho a^\dagger - a^\dagger a\rho - \rho a^\dagger a\right). \tag{9.3.49}$$

We can use the techniques of adiabatic elimination directly on this master equation, and derive in the end some laser equations which do *not* depend on the large N limit.

b) Evolution Operators in the P-representation: We represent the joint density operator of the atoms and the light field by a P-representation thus

$$\rho = \int d^2\alpha|\alpha\rangle\langle\alpha|\boldsymbol{\rho}(\alpha), \tag{9.3.50}$$

where $\boldsymbol{\rho}(\alpha)$ is an operator in the many atom space. The methods of Sect. 3.5 and Sect. 6.2.1 can be used in the interaction picture to give the equation of motion for $\boldsymbol{\rho}(\alpha)$:

$$\begin{aligned}\frac{\partial\boldsymbol{\rho}}{\partial t} &= \left\{\kappa\left(\frac{\partial}{\partial\alpha}\alpha + \frac{\partial}{\partial\alpha^*}\alpha^*\right) + 2\kappa n\frac{\partial^2}{\partial\alpha\partial\alpha^*}\right\}\boldsymbol{\rho} + \left.\frac{\partial\boldsymbol{\rho}}{\partial t}\right|_A \\ &+g\sum_\mu\left\{\alpha^*[\sigma_\mu^-,\boldsymbol{\rho}] - \alpha[\sigma_\mu^+,\boldsymbol{\rho}] - \frac{\partial}{\partial\alpha}\sigma_\mu^-\boldsymbol{\rho} - \frac{\partial}{\partial\alpha^*}\boldsymbol{\rho}\sigma_\mu^+\right\}.\end{aligned} \tag{9.3.51}$$

The operator on the RHS of (9.3.51) divides into three parts, in a way that is not immediately obvious. Since we are considering the situation in which the atoms relax much faster than the field, we can assume that they are in a quasi-stationary state, which depends however on the electric field α. The relevant equation of motion for the atoms is thus given by

$$\begin{aligned}\left.\frac{\partial\boldsymbol{\rho}}{\partial t}\right|_A = L_1\boldsymbol{\rho} &= \tfrac{1}{2}W_{21}\sum_{\mu=1}^N\left(2\sigma_\mu^-\boldsymbol{\rho}\sigma_\mu^+ - \sigma_\mu^+\sigma_\mu^-\boldsymbol{\rho} - \boldsymbol{\rho}\sigma_\mu^+\sigma_\mu^-\right) \\ &+\tfrac{1}{2}W_{12}\sum_{\mu=1}^N\left(2\sigma_\mu^+\boldsymbol{\rho}\sigma_\mu^- - \sigma_\mu^-\sigma_\mu^+\boldsymbol{\rho} - \boldsymbol{\rho}\sigma_\mu^-\sigma_\mu^+\right) \\ &+g\sum_{\mu=1}^N\left(\alpha^*[\sigma_\mu^-,\boldsymbol{\rho}] - \alpha[\sigma_\mu^+,\boldsymbol{\rho}]\right)\end{aligned} \tag{9.3.52}$$

which is obtained by including the first part of the summation in (9.3.51) together with the purely atomic part of the master equation.

c) Stationary Means of Atomic Operators: Equation (9.3.52) is simply the equation of motion for N driven two level atoms, as in Sect. 9.2.3, Sect. 9.2.4. The only connection between the atoms is the common external field, α. As in the previous treatment, we define some parameters

$$\bar{d} = \frac{W_{12} - W_{21}}{W_{12} + W_{21}} \tag{9.3.53}$$

$$\bar{\gamma} = \tfrac{1}{2}(W_{12} + W_{21}) \tag{9.3.54}$$

so that the parameters of Sect. 9.2.3, Sect. 9.2.4 are given by (remembering that the bath temperature is now negative, so that γ is negative)

$$\begin{aligned} \kappa &\to \bar{\gamma} \\ \gamma &\to -2\bar{d}\bar{\gamma} \ , \\ \mathcal{E} &\to g\alpha \end{aligned} \tag{9.3.55}$$

and thus the stationary means are (from (9.2.48,50)

$$\langle \sigma_\mu^- \rangle = \langle \sigma_\mu^+ \rangle^* = \frac{\bar{d} g\alpha/\bar{\gamma}}{1 + |\alpha|^2/n_0} \tag{9.3.56}$$

where

$$n_0 = \bar{\gamma}^2/2g^2 \tag{9.3.57}$$

is a parameter known as the *saturation photon number*, which will be explained soon.

d) Adiabatic Elimination: The master equation can now be written as

$$\frac{\partial \rho}{\partial t} = L_1\rho + L_2\rho + L_3\rho \tag{9.3.58}$$

with L_1 defined in (9.3.52), and

$$L_3\rho = \left\{ \frac{\partial}{\partial \alpha}\left(\kappa\alpha - g\sum_\mu \langle \sigma_\mu^- \rangle\right) + \frac{\partial}{\partial \alpha^*}\left(\kappa\alpha^* - g\sum \langle \sigma_\mu^+ \rangle\right) + 2\kappa n \frac{\partial^2}{\partial \alpha \partial \alpha^*} \right\}\rho \tag{9.3.59}$$

$$L_2\rho = -g\sum_\mu \left\{ \frac{\partial}{\partial \alpha}\left(\sigma_\mu^- - \langle \sigma_\mu^- \rangle\right)\rho + \frac{\partial}{\partial \alpha^*}\rho\left(\sigma_\mu^+ - \langle \sigma_\mu^+ \rangle\right) \right\}. \tag{9.3.60}$$

The adiabatic elimination is done by the projector techniques of Sect. 5.1.2 and S.M. Chap.6. The projector onto the stationary atomic state, ρ_A^s, is defined by

$$\begin{aligned} \mathcal{P}L_1 &= L_1\mathcal{P} = 0 \\ \mathcal{P}L_2\mathcal{P} &= 0 \\ \mathcal{P}L_3 &= L_3\mathcal{P}. \end{aligned} \tag{9.3.61}$$

There is no parametric dependence explicitly written for L_1, L_2, L_3, but it is clear that

$$L_1 \sim \gamma, \quad L_3 \sim \kappa, \quad L_2 \sim g \tag{9.3.62}$$

so that if γ is much larger than κ and g, the adiabatic elimination will be justified and we will be able to derive the equation of motion

$$\frac{\partial P(\alpha)}{\partial t} = L_3 P(\alpha) - \mathrm{Tr}_A\left\{L_2 L_1^{-1} L_2 \rho_A^s\right\} P(\alpha) \tag{9.3.63}$$

in which

$$P(\alpha) = \mathrm{Tr}_{\mathrm{A}}\left\{\rho(\alpha)\right\}. \tag{9.3.64}$$

e) Evaluation of Noise Terms: The second term on the RHS of (9.3.63) can, as explained in S.M. (6.5.33), be put in the form

$$\mathrm{Tr}_{\mathrm{A}}\left\{L_2 \int_0^\infty \mathrm{e}^{L_1 t}(L_2 \rho_A^s)dt\right\} \tag{9.3.65}$$

of which a typical term is

$$g^2 \frac{\partial}{\partial\alpha} \int_0^\infty \mathrm{Tr}_{\mathrm{A}}\left\{\left(\sigma_\mu^- - \langle\sigma_\mu^-\rangle\right) \mathrm{e}^{L_1 t} \frac{\partial}{\partial\alpha^*} \rho_A^s \left(\sigma_\mu^+ - \langle\sigma_\mu^+\rangle\right)\right\} dt. \tag{9.3.66}$$

We can show that this term is small, so we will be able to move the $\partial/\partial\alpha^*$ to the left, with the addition of a small term to L_3, which will be neglected. In this case

$$(9.3.66) \rightarrow \quad g^2 \frac{\partial^2}{\partial\alpha\partial\alpha^*} \int_0^\infty \left\{\langle\sigma_\mu^+(0)\sigma_\mu^-(t)\rangle - \langle\sigma_\mu^+\rangle\langle\sigma_\mu^-\rangle\right\} dt. \tag{9.3.67}$$

One can now look up the correlation functions for the driven atom and substitute them. In the case of not too large a field, we can in fact simply use the correlation function for the atom without a driving field, for which $\langle\sigma_\mu^+\rangle = \langle\sigma_\mu^-\rangle = 0$. Thus from (9.2.38)

$$\langle\sigma_\mu^+(t)\sigma_\mu^-(0)\rangle = \langle\sigma_\mu^+(0)\sigma_\mu^-(t)\rangle = \tfrac{1}{2}(1+\bar{d})\mathrm{e}^{-\bar{\gamma}t} \tag{9.3.68}$$

and substituting into (9.3.67), and similar terms

$$(9.3.65) \rightarrow \quad (1+\bar{d})\frac{g^2}{\bar{\gamma}} \frac{\partial^2}{\partial\alpha\partial\alpha^*}. \tag{9.3.69}$$

f) Laser Equation: Substituting (9.3.69) and the values for $\langle\sigma_\mu^\pm\rangle$, (9.3.56) into (9.3.63), we finally derive

$$\begin{aligned}\frac{\partial P(\alpha)}{\partial t} = \Bigg\{&\kappa \frac{\partial}{\partial\alpha}\left[\alpha\left(1 - \frac{C}{1+|\alpha|^2/n_0}\right)\right] + \kappa\frac{\partial}{\partial\alpha^*}\left[\alpha^*\left(1 - \frac{C}{1+|\alpha|^2/n_0}\right)\right] \\ &+ \left[2\kappa n + N(1+\bar{d})\frac{g^2}{\bar{\gamma}}\right] \frac{\partial^2}{\partial\alpha\partial\alpha^*}\Bigg\} P(\alpha)\end{aligned} \tag{9.3.70}$$

where it should be noted that there are N terms exactly the same as (9.3.69), and the pump parameter C is defined by

$$C = \frac{N\bar{d}g^2}{\bar{\gamma}\kappa}. \tag{9.3.71}$$

g) Behaviour of Parameters: The situation of most interest occurs when the noise terms are small, but the terms proportional to C, which depend on the pumping, are significant. In order to derive the adiabatically eliminated equation (9.3.70), it was necessary to assume large $\bar{\gamma}$, which does seem to indicate that the noise term must be small, since it is divided by $\bar{\gamma}$. However, one must not forget that N, the number of atoms, is normally very large. Bearing in mind that $1+\bar{d}$ is of order of magnitude 1, we can see that in order to be confident of a finite numerator, we must have g^2 of the same order of magnitude as $1/N$, which can be formalized by defining a scaled quantity $\tilde{g}$, by

$$g = \tilde{g}/\sqrt{N} \tag{9.3.72}$$

where it is understood that $\tilde{g}$ is a quantity that is of order of magnitude 1.

This means that we can write a reduced saturation photon number by use of the equation

$$n_0 = \tilde{n}_0 N = \frac{\bar{\gamma}^2}{2\tilde{g}^2} N \tag{9.3.73}$$

and this shows that n_0 is expected to scale with N. The cooperativity parameter C, can be written

$$C = \frac{\bar{d}\,\tilde{g}^2}{\bar{\gamma}\kappa} \tag{9.3.74}$$

and does not scale with N—indeed it can have a variety of values, and is the parameter which determines the behaviour of the laser.

We can make the behaviour of the solutions with respect to N appear simplest by also defining a reduced variable $\tilde{\alpha}$ by

$$\alpha = \tilde{\alpha}\sqrt{N}\,, \tag{9.3.75}$$

in terms of which the Fokker-Planck equation (9.3.70) takes the form

$$\begin{aligned}\frac{\partial P}{\partial t} = \Bigg\{ \frac{\partial}{\partial \tilde{\alpha}} \left[\kappa\tilde{\alpha} \left(1 - \frac{C}{1 + |\tilde{\alpha}|^2/\tilde{n}_0} \right) \right] + \frac{\partial}{\partial \tilde{\alpha}^*} \left[\kappa\tilde{\alpha}^* \left(1 - \frac{C}{1 + |\tilde{\alpha}|^2/\tilde{n}_0} \right) \right] \\ + \frac{2}{N} \frac{\partial^2}{\partial\tilde{\alpha}\partial\tilde{\alpha}^*} Q(\tilde{\alpha}) \Bigg\} P \end{aligned} \tag{9.3.76}$$

where in this case $Q(\tilde{\alpha})$ is actually independent of $\tilde{\alpha}$, and is given by

$$Q(\tilde{\alpha}) = \kappa n + \frac{\tilde{g}^2(1+\bar{d})}{2\bar{\gamma}}. \tag{9.3.77}$$

The form (9.3.77) for $Q(\tilde{\alpha})$ arose because we assumed that $|\tilde{\alpha}|^2 \ll \tilde{n}_0$. If this is not so, the correlation function expressions (9.3.68) have to be replaced by appropriate expressions derived from (9.2.70,71), which can be done with sufficient labour. For complete generality, we can adopt the form (9.3.76).

9.3.3 Solutions of the Laser Equations

a) Stochastic Differential Equations: The Fokker-Planck equation (9.3.76) is equivalent to the stochastic differential equation

$$\frac{d\tilde{\alpha}}{dt} = -\kappa\tilde{\alpha}\left(1 - \frac{C}{1 + |\tilde{\alpha}|^2/\tilde{n}_0}\right) + \frac{1}{\sqrt{N}}\tilde{\Gamma}(t) \tag{9.3.78}$$

and its complex conjugate. The correlation function of the noise term $\tilde{\Gamma}(t)$ is

$$\langle \tilde{\Gamma}^*(t)\tilde{\Gamma}(t')\rangle = 2Q(\tilde{\alpha})\delta(t - t') \tag{9.3.79}$$

with other correlation functions equal to zero.

b) Macroscopic Laser Equations: If we take the limit $N \to \infty$ we obtain deterministic equations in which the noise on the RHS of (9.3.78) is simply dropped. One obtains the same equations as in Sect. 9.3.1c–e, and the analysis is exactly the same.

c) Stationary Solution of the Fokker Planck Equation: Although it is possible to solve for the stationary solution of Fokker-Planck equation (9.3.76) exactly (in terms of integrals), the result is complicated and uninformative. It is simpler to consider the situation in which the $|\tilde{\alpha}^2|$ dependence is neglected, and take Q to be a constant. In that case, the solution is quite straightforward, and we find

$$P(\alpha, \alpha^*) = \mathcal{N}\exp\left(-\frac{N}{Q}V(|\tilde{\alpha}|)\right) \tag{9.3.80}$$

where $V(|\alpha|)$ is the classical potential given in (9.3.36). Using the explicit form of this potential, the solution is

$$P(\alpha, \alpha^*) = \mathcal{N}(1 + |\tilde{\alpha}|^2/\tilde{n}_0)^{\tilde{n}_0\kappa CN/2Q}\exp\left(-\frac{N\kappa|\tilde{\alpha}|^2}{2Q}\right). \tag{9.3.81}$$

This P-function is plotted in Fig.9.6.

This distribution takes on two characteristic shapes, depending on whether the laser is above or below threshold. Below threshold, the distribution has the form of a single peak centred on the origin. Above threshold, the distribution develops a maximum at the classical stationary value, given by $|\tilde{\alpha}| = \sqrt{\tilde{n}_0(C-1)}$. This is therefore an annular distribution in the variables $x = \text{Re}(\tilde{\alpha}), y = \text{Im}(\tilde{\alpha})$, showing that the amplitude $|\tilde{\alpha}|$ has a well defined most probable value, while the phase of $\tilde{\alpha}$ is indeterminate. The lasing action determines the amplitude quite precisely, while not determining the phase.

d) Linearized Solution of the Langevin Equation: Let us write the Langevin equation in the form

$$\frac{d\tilde{\alpha}}{dt} = -\kappa\tilde{\alpha}f(|\tilde{\alpha}|^2) + \frac{1}{\sqrt{N}}\tilde{\Gamma}(t) \tag{9.3.82}$$

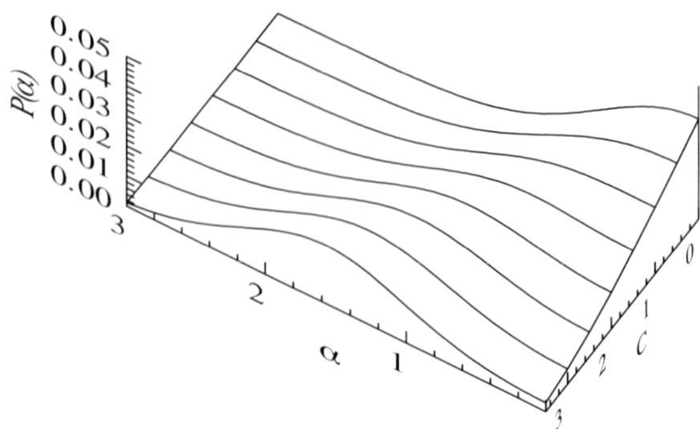

Fig. 9.6 Plot of the P-function $P(\alpha)$ for the laser

and assume that N is large, so that we may set $\tilde{\alpha} = \tilde{\alpha}_0 + \epsilon$, where $\tilde{\alpha}_0$ is a stationary solution of the equation without the fluctuating term. Thus, we can approximate (9.3.82) by

$$\frac{d\tilde{\alpha}_0}{dt} + \frac{d\epsilon}{dt} = -\kappa(\tilde{\alpha}_0 + \epsilon)\left\{ f(|\tilde{\alpha}|^2) + \epsilon\frac{\partial f}{\partial \tilde{\alpha}_0} + \epsilon^*\frac{\partial f}{\partial \tilde{\alpha}_0^*} \right\} + \frac{1}{\sqrt{N}}\tilde{\Gamma}(t). \tag{9.3.83}$$

e) Below Threshold: $\tilde{\alpha}_0 = 0$, so that to lowest order in ϵ, the equation is

$$\dot{\epsilon} = -\kappa\epsilon + \frac{1}{\sqrt{N}}\tilde{\Gamma}(t) \tag{9.3.84}$$

(where we note that $f(0) = 1$.) This is a simple Ornstein-Uhlenbeck equation (S.M. Sect. 3.8.4, 4.4.4. In the stationary state the time correlation function is

$$\langle \epsilon^*(t)\epsilon(t')\rangle = \frac{Q}{\kappa N}e^{-\kappa|t-t'|}. \tag{9.3.85}$$

In this case, the correlation function (and in fact the total dynamical behaviour) of the light field is determined by the properties of the cavity, whose decay constant is κ, and the number of atoms N.

f) Above Threshold: The annular symmetry complicates matters here, and it is best to transform to polar co-ordinates by setting

$$\tilde{\alpha} = Re^{i\phi}. \tag{9.3.86}$$

We follow the procedure of S.M. Sect. 4.4.5. Firstly, set $R = e^{\mu}$ so that

$$\mu + i\phi = \ln(\tilde{\alpha}). \tag{9.3.87}$$

We write the Langevin equations in SDE form as

$$d\tilde{\alpha} = -\kappa\tilde{\alpha} f(|\tilde{\alpha}|^2)dt + \frac{1}{\sqrt{N}}d\tilde{\Gamma}(t) \tag{9.3.88}$$

where

$$\begin{aligned} d\tilde{\Gamma}(t)^2 &= d\tilde{\Gamma}^*(t)^2 = 0 \\ d\tilde{\Gamma}(t)d\tilde{\Gamma}^*(t) &= 2Q dt \end{aligned} \tag{9.3.89}$$

and interpret them as Ito equations, since the Fokker-Planck equation always comes out in this form. Then, in changing variables, we have to expand the infinitesimals to second order, so that

$$d\mu + \mathrm{i}d\phi = \frac{d\tilde{\alpha}}{\tilde{\alpha}} - \frac{(d\tilde{\alpha})^2}{\tilde{\alpha}^2} \tag{9.3.90}$$

but in this case, using (9.3.89), we see that the second order term is zero. Hence

$$d\mu + \mathrm{i}d\phi = -\kappa f(\mathrm{e}^{2\mu})dt + \frac{\mathrm{e}^{-\mu-\mathrm{i}\phi}}{\sqrt{N}} d\tilde{\Gamma}(t) \tag{9.3.91}$$

so that

$$\begin{aligned} d\mu &= -\kappa f(\mathrm{e}^{2\mu})dt + \frac{\mathrm{e}^{-\mu}}{\sqrt{N}} d\Gamma_R(t) \\ d\phi &= \frac{\mathrm{e}^{-\mu}}{\sqrt{N}} d\Gamma_\phi(t). \end{aligned} \tag{9.3.92}$$

As discussed in S.M. Sect. 4.4.5, $d\Gamma_R(t)$ and $d\Gamma_\phi(t)$ are given by

$$\begin{aligned} d\Gamma_R(t) &= \mathrm{Re}\left(\mathrm{e}^{-\mathrm{i}\phi} d\tilde{\Gamma}(t)\right) \\ d\Gamma_\phi(t) &= \mathrm{Im}\left(\mathrm{e}^{-\mathrm{i}\phi} d\tilde{\Gamma}(t)\right) \end{aligned} \tag{9.3.93}$$

and are independent stochastic sources, i.e.

$$\begin{aligned} d\Gamma_R(t)^2 = d\Gamma_\phi(t)^2 &= Q dt \\ d\Gamma_R(t) d\Gamma_\phi(t) &= 0 \end{aligned} \tag{9.3.94}$$

We now let $R = \exp(\mu)$, so that

$$dR = \exp(\mu)d\mu + \tfrac{1}{2}\exp(\mu)d\mu^2 \tag{9.3.95}$$

and find the equivalent SDEs :

$$\boxed{\begin{aligned} dR &= \left\{-\kappa R f(R^2) + \frac{Q}{2NR}\right\} dt + \frac{1}{\sqrt{N}} d\Gamma_R(t) \\ d\phi &= \frac{1}{R\sqrt{N}} d\Gamma_\phi(t). \end{aligned}} \tag{9.3.96}$$

The correlation function of interest is $\langle\tilde{\alpha}^*(t)\tilde{\alpha}(t')\rangle$, since this is proportional to $\langle a^\dagger(t)a(t')\rangle$, which is the quantity whose Fourier transform is the optical spectrum. Using the polar variables, we find

$$\langle\tilde{\alpha}^*(t)\tilde{\alpha}(t')\rangle = \langle R(t)R(t')\exp[\mathrm{i}\phi(t') - \mathrm{i}\phi(t)]\rangle. \tag{9.3.97}$$

From (9.3.96), it is clear that $R(t)$ fluctuates about

$$R = |\tilde{\alpha}_0| = \sqrt{\tilde{n}_0(C-1)} \tag{9.3.98}$$

with fluctuations of order $1/N$. Thus, to order $1/N$, we can write

$$\langle \tilde{\alpha}^*(t)\tilde{\alpha}(t')\rangle = \tilde{n}_0(C-1)\langle \exp \mathrm{i}\left(\phi(t') - \phi(t)\right)\rangle. \tag{9.3.99}$$

The equation for ϕ is also accurately represented to order $1/N$ by substituting for R by (9.3.98). We then find

$$\phi(t) - \phi(t') = \frac{1}{\sqrt{N\tilde{n}_0(C-1)}} \int_{t'}^{t} d\Gamma_\phi(t'') \tag{9.3.100}$$

so that

$$\langle \left(\phi(t) - \phi(t')\right)^2\rangle = \frac{Q|t-t'|}{N\tilde{n}_0(C-1)}. \tag{9.3.101}$$

Since Γ_ϕ is Gaussian, $\phi(t) - \phi(t')$ is also, so we can use

$$\langle \exp x\rangle = \exp \tfrac{1}{2}\langle x^2\rangle \tag{9.3.102}$$

which is valid for Gaussian x with zero mean, to get

$$\langle \tilde{\alpha}^*(t)\tilde{\alpha}(t')\rangle = \tilde{n}_0(C-1)\exp\left(-\frac{Q|t-t'|}{2N\tilde{n}_0(C-1)}\right) \tag{9.3.103}$$

or

$$\boxed{\langle \alpha^*(t)\alpha(t')\rangle = \langle a^\dagger(t)a(t')\rangle = n_0(C-1)\exp\left(-\frac{Q|t-t'|}{2n_0(C-1)}\right).} \tag{9.3.104}$$

9.3.4 The Nature of Laser Light

a) Below Threshold: Below and above threshold we find two different kinds of behaviour. To the extent that the linearization (9.3.84) is valid, it is clear that $\alpha(t) = \sqrt{N}\,\epsilon(t)$ is a Gaussian function with zero mean, and intensity

$$\langle a^\dagger a\rangle_s = \frac{Q}{\kappa} \tag{9.3.105}$$

and

$$\langle a^\dagger(t)a(0)\rangle_s = \frac{Q}{\kappa}\mathrm{e}^{-\kappa|t|}. \tag{9.3.106}$$

This light is thus essentially the same as thermal light, as described in Sect. 8.3.4. Thus, below threshold, the intensity of the light is essentially independent of the number of atoms, N, though careful inspection of the definition (9.3.77) of Q shows that, there is a dependence through $\tilde{g} = g\sqrt{N}$.

Since the light is represented by a Gaussian function, we can use the results of Sect. 8.3.4, to find that

$$g^{(1)}(t) = \mathrm{e}^{-\kappa|t|} \tag{9.3.107}$$

$$g^{(2)}(t) = 1 + \mathrm{e}^{-2\kappa|t|}. \tag{9.3.108}$$

Thus, below threshold, the laser behaves like an ordinary lamp; in no way is the light distinguishable from that emitted by a number of atoms.

b) At Threshold: We can get an estimate of $g^{(1)}(0)$ and $g^{(2)}(0)$ at threshold by using the method of steepest descents to evaluate the moments using the P-function (9.3.81).

Exercise. Write $1+|\tilde{\alpha}^2|/n_0 = \exp\left[\log(1 + |\tilde{\alpha}|^2/n_0)\right]$, and expand the logarithm to the second order. Set $C = 1$, and show that at threshold, this approximation yields

$$\langle : (a^\dagger a)^n : \rangle = \frac{2^{n/2}}{\sqrt{\pi}} n_0 \Gamma\left(\frac{n+1}{2}\right) \tag{9.3.109}$$

and hence

$$g^{(2)}(0) = \pi/2 \approx 1.57 \tag{9.3.110}$$

which is intermediate between the Gaussian value of 2, and the Poissonian value of 1.

c) Above Threshold: To a first approximation, above threshold one can ignore the fluctuations, and $\alpha(t)$ is simply a constant, say α_0. This means that all correlations of $a^\dagger(t)$ will, to this accuracy, factorize—the laser produces essentially coherent light above threshold.

A more careful interpretation of the fluctuations follows by still ignoring the amplitude fluctuations, but taking note of the phase fluctuations, which gives the correlation formula (9.3.104). The phase correlation time is

$$\tau_c = N\frac{2\tilde{n}_0(C-1)}{Q} \tag{9.3.111}$$

which grows with N, and can be very large. Thus if the phase is measured, and then remeasured within a time much less than τ_c, almost the same result will be found. The laser field can be represented by a slowly drifting $\alpha(t)$. Taking the Fourier transform of (9.3.104) will give a spectrum with bandwidth τ_c^{-1}; i.e., a very narrow bandwidth.

The intensity correlation functions can be deduced from the Langevin equations (9.3.96).

Exercise. Linearize the R equation (9.3.96) about $\sqrt{\tilde{n}_0(C-1)}$, to obtain

$$\langle R(t)R(0)\rangle \approx \tilde{n}_0^2(C-1) + \frac{Q^2}{2\kappa n_0(C-1)} \exp\left[-2\kappa\tilde{n}_0(C-1)|t|\right] \tag{9.3.112}$$

and thus

$$g^{(2)}(t) \approx 1 + \frac{1}{N}\frac{Q^2}{2\kappa\tilde{n}_0^3(C-1)^2} \exp\left[-2\kappa\tilde{n}_0(C-1)|t|\right]. \tag{9.3.113}$$

For large numbers of atoms N, this is barely distinguishable from the Poissonian value, $g^{(2)}(t) = 1$. Notice however that the time constant is quite different from that of the amplitude correlation function. Thus, the amplitude fluctuations are small ($\sim 1/\sqrt{N}$ times the amplitude), and relatively fast (i.e., the time constant does not increase with N). The phase fluctuations are very slow, and large, and give the principal contribution to the observed laser linewidth.

9.4 Optical Bistability

Optical bistability can be generated by a number of mechanisms, all of which involve a nonlinearity in optical response. A system which does exhibit optical bistability, and has been well studied experimentally, is modelled very similarly to the laser model of the previous section. The major differences are

i) The atomic bath is not pumped to a negative temperature—indeed its temperature may be zero.

ii) An external driving field is added; thus the mode of the light field is coupled to an external field $F(t)$.

iii) The atomic transition frequency, the resonant frequency of the cavity, and the frequency of the driving field are not all exactly the same. This makes it necessary to introduce detuning parameters to describe the situation.

Optical bistability was first correctly treated theoretically by *Bonifacio* and *Lugiato* [9.7] using a different method from that presented here.

a) Detunings: The driving field frequency is taken to be ω, the resonant frequency of the cavity is $\omega + \Delta$, and the resonant frequency of the atoms is $\omega + \theta$. This means that we now work in a frame rotating at frequency ω, in which we find

$$H_F = \hbar\Delta a^\dagger a + \mathrm{i}\hbar F a^\dagger - \mathrm{i}\hbar F^* a \tag{9.4.1}$$

$$H_A = \sum_{\mu=1}^{N} \tfrac{1}{2}\hbar\theta\sigma_\mu^z . \tag{9.4.2}$$

In this frame, the master equation can be written as in (9.3.47), in which however,

$$\begin{aligned}\left.\frac{\partial\rho}{\partial t}\right|_A &= \tfrac{1}{2}W_{21}\sum_{\mu=1}^{N}\left(2\sigma_\mu^-\rho\sigma_\mu^+ - \sigma_\mu^+\sigma_\mu^-\rho - \rho\sigma_\mu^+\sigma_\mu^-\right)\\ &\quad+\tfrac{1}{2}W_{12}\sum_{\mu=1}^{N}\left(2\sigma_\mu^+\rho\sigma_\mu^- - \sigma_\mu^-\sigma_\mu^+\rho - \rho\sigma_\mu^-\sigma_\mu^+\right) - \tfrac{1}{2}\mathrm{i}\theta\sum_{\mu=1}^{N}\left[\sigma_\mu^z,\rho\right]\end{aligned} \tag{9.4.3}$$

$$\begin{aligned}\left.\frac{\partial\rho}{\partial t}\right|_F &= \kappa n\left(2a^\dagger\rho a - aa^\dagger\rho - \rho aa^\dagger\right) + \kappa(n+1)\left(2a\rho a^\dagger - a^\dagger a\rho - \rho a^\dagger a\right)\\ &\quad-\mathrm{i}\Delta\left[a^\dagger a,\rho\right] + F\left[a^\dagger,\rho\right] - F^*\left[a,\rho\right].\end{aligned} \tag{9.4.4}$$

Otherwise, the equations are the same.

b) Evolution Equations in the P-Representation: We find that, analogously to Sect. 9.3.2d

$$\frac{\partial \rho}{\partial t} = (L_1 + L_2 + L_3)\rho, \tag{9.4.5}$$

in which

$$\begin{aligned} L_1\rho = {} & \tfrac{1}{2}W_{21}\sum_{\mu=1}^{N}\left(2\sigma_\mu^-\rho\sigma_\mu^+ - \sigma_\mu^+\sigma_\mu^-\rho - \rho\sigma_\mu^+\sigma_\mu^-\right) \\ & + \tfrac{1}{2}W_{12}\sum_{\mu=1}^{N}\left(2\sigma_\mu^+\rho\sigma_\mu^- - \sigma_\mu^-\sigma_\mu^+\rho - \rho\sigma_\mu^-\sigma_\mu^+\right) \\ & - \tfrac{1}{2}\mathrm{i}\theta\sum_{\mu=1}^{N}\left[\sigma_\mu^z, \rho\right] + g\sum_{\mu=1}^{N}\left(\alpha^*\left[\sigma_\mu^-, \rho\right] - \alpha\left[\sigma_\mu^+, \rho\right]\right) \end{aligned} \tag{9.4.6}$$

$$\begin{aligned} L_3\rho = \Bigg\{ & \frac{\partial}{\partial\alpha}\left[-F + (\kappa + \mathrm{i}\Delta)\alpha + g\sum_{\mu=1}^{N}\langle\sigma_\mu^-\rangle\right] \\ & + \frac{\partial}{\partial\alpha^*}\left[-F^* + (\kappa - \mathrm{i}\Delta)\alpha^* - g\sum_{\mu=1}^{N}\langle\sigma_\mu^+\rangle\right] + 2\kappa n\frac{\partial^2}{\partial\alpha\partial\alpha^*}\Bigg\}\rho \end{aligned} \tag{9.4.7}$$

$$L_2\rho = -g\sum_{\mu=1}^{N}\left\{\frac{\partial}{\partial\alpha}\left(\sigma_\mu^- - \langle\sigma_\mu^-\rangle\right)\rho + \frac{\partial}{\partial\alpha^*}\rho\left(\sigma_\mu^+ - \langle\sigma_\mu^+\rangle\right)\right\}. \tag{9.4.8}$$

Exercise. Show that the equations of motion for the averages of σ_μ are

$$\left.\begin{aligned} \langle\dot\sigma_\mu^+\rangle &= -(\bar\gamma - \mathrm{i}\theta)\langle\sigma_\mu^+\rangle + g\alpha^*\langle\sigma_\mu^z\rangle \\ \langle\dot\sigma_\mu^-\rangle &= -(\bar\gamma + \mathrm{i}\theta)\langle\sigma_\mu^-\rangle + g\alpha\langle\sigma_\mu^z\rangle \\ \langle\dot\sigma_\mu^z\rangle &= -2\bar\gamma\langle\sigma_\mu^z\rangle - 2g\alpha\langle\sigma_\mu^+\rangle - 2g\alpha^*\langle\sigma_\mu^-\rangle - 2d\bar\gamma \end{aligned}\right\} \tag{9.4.9}$$

and that the stationary solutions are given by

$$\begin{aligned} \langle\sigma_\mu^+\rangle_s = \langle\sigma_\mu^-\rangle_s^* &= -\frac{g(\bar\gamma + \mathrm{i}\theta)\alpha^*}{\bar\gamma^2 + \theta^2 + 2g^2|\alpha|^2} \\ \langle\sigma_\mu^z\rangle &= -\frac{\bar\gamma^2 + \theta^2}{\bar\gamma^2 + \theta^2 + 2g^2|\alpha|^2}. \end{aligned} \tag{9.4.10}$$

These solutions have been computed with $\bar d = 1$, which, according to (9.3.53), corresponds to $W_{21} = 0$, implying that the atoms are not pumped, and are treated as being at zero temperature.

c) Adiabatic Elimination of Atoms: This can be carried out in the same way as for the laser; the Fokker-Planck equation is

$$\begin{aligned}\frac{\partial P}{\partial t} = \Bigg\{ &\frac{\partial}{\partial \alpha}\left[-F + \alpha\left(\kappa + \mathrm{i}\Delta + \frac{Ng^2(\bar{\gamma} - \mathrm{i}\theta)}{\bar{\gamma}^2 + \theta^2 + 2g^2|\alpha|^2}\right)\right] \\ &+ \frac{\partial}{\partial \alpha^*}\left[-F^* + \alpha^*\left(\kappa - \mathrm{i}\Delta + \frac{Ng^2(\bar{\gamma} + \mathrm{i}\theta)}{\bar{\gamma}^2 + \theta^2 + 2g^2|\alpha|^2}\right)\right] \\ &+ 2\kappa n \frac{\partial^2}{\partial \alpha \partial \alpha^*} + N\left[\frac{\partial^2}{\partial \alpha \partial \alpha^*} Q + \frac{\partial^2}{\partial \alpha^2} R + \frac{\partial^2}{\partial \alpha^{*2}} R^*\right]\Bigg\} P. \end{aligned} \tag{9.4.11}$$

The coefficients Q and R are given by the same reasoning as for the laser, and are

$$Q = g^2 \int_0^\infty dt \left\{ \langle \sigma_\mu^+(0) \sigma_\mu^-(t) \rangle_s + \langle \sigma_\mu^-(t) \sigma_\mu^+(0) \rangle_s - 2 \langle \sigma_\mu^+ \rangle_s \langle \sigma_\mu^- \rangle_s \right\} \tag{9.4.12}$$

$$R = g^2 \int_0^\infty \left\{ \langle \sigma_\mu^-(t) \sigma_\mu^-(0) \rangle_s - \langle \sigma_\mu^- \rangle_s^2 \right\} dt. \tag{9.4.13}$$

To evaluate these is not particularly pleasant in the general case, though the procedure which must be used is quite clear.

i) Write the equation of motion for the averages, (9.4.9), as

$$\langle \dot{\boldsymbol{\sigma}} \rangle = -A \langle \boldsymbol{\sigma} \rangle + \mathbf{b}. \tag{9.4.14}$$

ii) The quantum regression theorem can then be used to say

$$\begin{aligned}\int_0^\infty dt \left\{ \langle \boldsymbol{\sigma}(t) \sigma^\alpha(0) \rangle_s - \langle \boldsymbol{\sigma} \rangle_s \langle \sigma^\alpha \rangle_s \right\} &= \int_0^\infty dt\, e^{-At} \left\{ \langle \boldsymbol{\sigma} \sigma^\alpha \rangle_s - \langle \boldsymbol{\sigma} \rangle_s \langle \sigma^\alpha \rangle_s \right\} \\ &= A^{-1} \left\{ \langle \boldsymbol{\sigma} \sigma^\alpha \rangle_s - \langle \boldsymbol{\sigma} \rangle_s \langle \sigma^\alpha \rangle_s \right\}. \end{aligned} \tag{9.4.15}$$

iii) In these last two equations, $\boldsymbol{\sigma} = \begin{pmatrix} \sigma^+ \\ \sigma^- \\ \sigma_z \end{pmatrix}$ and σ^α is one of the components of $\boldsymbol{\sigma}$.

iv) The terms $\langle \boldsymbol{\sigma} \sigma^\alpha \rangle_s$ can all be evaluated by using the algebra of Pauli matrices, e.g., $\langle \sigma^+ \sigma^- \rangle_s = \frac{1}{2} \langle \sigma_z + 1 \rangle_s$, etc.

v) Terms like $\langle \sigma^-(0) \sigma^+(t) \rangle_s$ can be evaluated by using

$$\langle \sigma^-(0) \sigma^+(t) \rangle_s = \langle \sigma^-(t) \sigma^+(0) \rangle_s^*. \tag{9.4.16}$$

Exercise. Show that if

$$\begin{pmatrix} \int_0^\infty dt \langle \sigma^+(t), \sigma^+(0) \rangle_s & \int_0^\infty dt \langle \sigma^+(t), \sigma^-(0) \rangle_s \\ \int_0^\infty dt \langle \sigma^-(t), \sigma^+(0) \rangle_s & \int_0^\infty dt \langle \sigma^-(t), \sigma^-(0) \rangle_s \\ \int_0^\infty dt \langle \sigma_z(t), \sigma^+(0) \rangle_s & \int_0^\infty dt \langle \sigma_z(t), \sigma^-(0) \rangle_s \end{pmatrix} \equiv \mathbf{S} \tag{9.4.17}$$

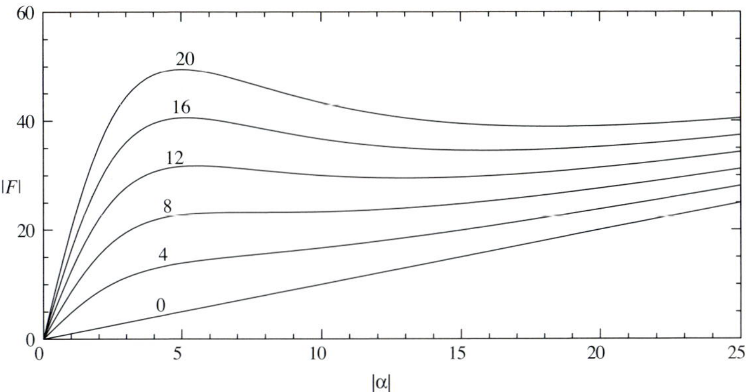

Fig. 9.7 Plots of bistability curves for the equation (9.4.21)

where $\langle A(t), B(0)\rangle_s \equiv \langle A(t)B(0)\rangle_s - \langle A(t)\rangle_s \langle B(0)\rangle_s$, then

$$\mathbf{S} = A^{-1} \begin{pmatrix} -\langle\sigma^+\rangle_s^2 & \frac{1}{2}\langle 1+\sigma_z\rangle_s - |\langle\sigma^+\rangle_s|^2 \\ \frac{1}{2}\langle 1-\sigma_z\rangle_s - |\langle\sigma^-\rangle_s|^2 & -\langle\sigma^-\rangle_s^2 \\ \langle\sigma^+\rangle_s\langle 1-\sigma_z\rangle_s & -\langle\sigma^-\rangle_s\langle 1+\sigma_z\rangle_s \end{pmatrix}. \tag{9.4.18}$$

From the formula the noise coefficients Q and R can be computed. The explicit results are a little complicated.

d) Absorptive Limit: Existence of Bistability: In the case that θ and Δ are zero, i.e. there is no atomic or cavity detuning, the stationary solution of the deterministic equation is given by

$$F = \alpha \left(\kappa + \frac{Ng^2\bar{\gamma}}{\bar{\gamma}^2 + 2g^2|\alpha|^2} \right). \tag{9.4.19}$$

We can use the fact that the solutions have F an α in phase to write

$$|F| = |\alpha| \left(\kappa + \frac{Ng^2\bar{\gamma}}{\bar{\gamma}^2 + 2g^2|\alpha|^2} \right). \tag{9.4.20}$$

Using the same notations n_s and C as in the laser, we can write the equation more simply as

$$|F| = \kappa|\alpha| \left(1 + \frac{C}{1 + |\alpha|^2/n_0} \right). \tag{9.4.21}$$

If $C > 4$, there are the solutions for this equation for any n_0, κ, and a range of $|F|$. Graphing the function as in Fig.9.7 demonstrates this clearly.

Exercise. Show that (9.4.21) exhibits bistability if $C > 4$. The curve has two branches;

i) Lower branch; $|\alpha|^2/n_0 \ll 1$,

$$\Rightarrow |F| \approx \kappa|\alpha|(1 + C). \tag{9.4.22}$$

In this branch we see the extra effect of atomic absorption in the factor $(1 + C)$.

The different contributions are defined as follows:

$$
\begin{aligned}
L_A = \frac{W_{12}}{2} \Bigg\{ & N\left(e^{-2\frac{\partial}{\partial D}} - 1\right) + N\, e^{2\frac{\partial}{\partial D}} \frac{\partial^4}{\partial v^2 \partial v^{*2}} + 2N \frac{\partial^2}{\partial v \partial v^*} \\
& + \frac{\partial}{\partial v}\left[2\frac{\partial^2}{\partial v \partial v^*} + 2e^{-2\frac{\partial}{\partial D}} - 1 \right] v \\
& + \frac{\partial}{\partial v^*}\left[2\frac{\partial^2}{\partial v \partial v^*} + 2e^{-2\frac{\partial}{\partial D}} - 1 \right] v^* \\
& -2\left[\left(e^{-2\frac{\partial}{\partial D}} - 1\right) - e^{2\frac{\partial}{\partial D}} \frac{\partial^4}{\partial^2 v \partial^2 v^*} \right] \frac{D}{2} \Bigg\} \\
& + \frac{W_{21}}{2} \left\{ N\left(e^{2\frac{\partial}{\partial D}} - 1\right) + \frac{\partial}{\partial v} v + \frac{\partial}{\partial v^*} v^* + 2\left(e^{2\frac{\partial}{\partial D}} - 1\right)\frac{D}{2} \right\}
\end{aligned}
\tag{9.5.7}
$$

$$
\begin{aligned}
L_{AL} = -g \Bigg\{ & \left[e^{-2\frac{\partial}{\partial D}} v^* - \frac{\partial^2}{\partial v^2} v + \frac{\partial}{\partial v} D \right] \alpha \\
& - \left[-\frac{\partial^2}{\partial v^{*2}} v^* + e^{-2\frac{\partial}{\partial D}} v + \frac{\partial}{\partial v^*} D \right] \alpha^* \\
& + \left[-\frac{\partial}{\partial \alpha} + \alpha^* \right] v - \left[-\frac{\partial}{\partial \alpha^*} + \alpha \right] v^* \Bigg\}
\end{aligned}
\tag{9.5.8}
$$

$$
L_L = \kappa \left[\frac{\partial}{\partial \alpha}\alpha + \frac{\partial}{\partial \alpha^*}\alpha^* \right] + 2\kappa\, n_{\text{th}} \frac{\partial^2}{\partial \alpha \partial \alpha^*}. \tag{9.5.9}
$$

a) Scaling Arguments: The equation as presented is almost impossible to use—recourse must be made to scaling arguments. As before, one assumes $g \approx 1/\sqrt{N}$, and one considers scaling with

$$
\begin{aligned}
D_0 &\equiv N \frac{W_{12} - W_{21}}{W_{12} + W_{21}} \\
D &\sim N \\
|\alpha|^2 &\sim N^x \\
v &\sim N^{(x+1)/2}.
\end{aligned}
\tag{9.5.10}
$$

With this scaling, all derivatives of order higher than the second are negligible. One can see that the adiabatic elimination results, and the exercise of Sect. 9.3.4, show that $x = 1/2$ is appropriate at threshold, while $x = 1$ is appropriate above threshold. In the threshold region, one finds the Fokker-Planck equation simplifies to

$$
\begin{aligned}
\frac{\partial P}{\partial t} = \Bigg\{ & -\frac{\partial}{\partial \alpha}(-\kappa\alpha + gv) - \frac{\partial}{\partial \alpha^*}(-\kappa\alpha^* + gv^*) \\
& - \frac{\partial}{\partial v}(-\bar{\gamma}\, v + g\alpha D) - \frac{\partial}{\partial v^*}(-\bar{\gamma}\, v^* + g\alpha^* D) \\
& - \frac{\partial}{\partial D}\left[-2\bar{\gamma}\,(D - D_0) - 2g(v^*\alpha + v\alpha^*) \right] \\
& + \frac{\partial^2}{\partial v \partial v^*}(NW_{12}) \Bigg\} P.
\end{aligned}
\tag{9.5.11}
$$

Here $\bar{\gamma} = \frac{1}{2}(W_{12} + W_{21})$ as in (9.3.54), and the zero temperature case n_{th} has been assumed. Adiabatic elimination can then be carried out, to yield essentially the same results as have already been derived. The same methods have been used for optical bistability by *Reid*, *Walls*, *Drummond* and their co-workers [9.11].

In the limit of large N, this method is quite simple and straightforward to apply, unlike its derivation. Of course it is only applicable in the limit that N is large, and since this is exactly the same limit in which one can linearize the quantum Langevin equations (9.3.16,24), the results obtained by both must be in agreement.

One can use the method of adiabatic elimination of the atoms on the Fokker-Planck equation (9.5.11), and rather surprisingly, one finds essentially the same equations as obtained without any small noise assumption. See Haken's book [9.3] for details.

9.5.2 A More Direct Phase Space Method

The equations of motion in the previous section are of infinite order in D because the characteristic function chosen involves exponentials. Although this characteristic function has an ordering which readily allows the finding of averages when the atomic operators are normally ordered, it is still quite singular. In fact, O^A can be written explicitly as

$$O^A = \prod_{\mu=1}^{N} (1 + \mathrm{i}\xi^* \sigma_\mu^+) \left[\cos(\varsigma \sigma_{\mu,z}) + \mathrm{i}\ \sin(\varsigma \sigma_{\mu,z})\right] (1 + \mathrm{i}\xi \sigma_\mu^-) \tag{9.5.12}$$

whose Fourier transform (to be defined in (9.5.34)) will contain terms of the form

$$\delta^m(v)\delta(D - m')\delta^n(v^*) \tag{9.5.13}$$

for all m, n such that $0 < m, n < N$ and for all m' such that $-N \leq m' \leq N$.

The P-function conventionally obtained by the truncation procedure of the previous section amounts to a smoothed version of this quite singular distribution. It is an approximation in the sense that physical averages evaluated with a smooth distribution can be quite close to the exact result obtained with the exact, nonsmooth distribution, although the smoothed distribution is not a good pointwise approximation to the exact distribution.

It is clear, however, that O^A is, in fact, simply a linear combination of the Pauli spin matrices and the identity matrix, so a simpler way of looking at the characteristic function is to change to a new set of variables and define

$$\chi = \mathrm{Tr}\left\{O\rho\right\} \tag{9.5.14}$$

where

$$O = O^A O^F \tag{9.5.15}$$

and, as before,

$$O^F = \mathrm{e}^{\mathrm{i}\beta^* a^\dagger} \mathrm{e}^{\mathrm{i}\beta a} \tag{9.5.16}$$

but

$$O^A = \prod_{\mu=1}^{N} \left(b + c^+\sigma_\mu^+ + c^-\sigma_\mu^- + c\sigma_{\mu,z}\right) = \prod_{\mu=1}^{N} \left(b + \mathbf{c}.\boldsymbol{\sigma}_\mu\right). \tag{9.5.17}$$

This characteristic function has six variables, though one of them is redundant, since, by construction, O^A is a homogeneous polynomial in b and $\mathbf{c}$ of degree N, and the physics is thus contained only in the *ratios* of b and $\mathbf{c}$.

We can derive an equation of motion for the characteristic function as follows.

a) Field Contribution: For the terms involving $a, a^\dagger$, we find

$$\mathrm{Tr}\left\{ O \left.\frac{\partial\rho}{\partial t}\right|_F \right\} = \kappa\left[-\beta\frac{\partial}{\partial\beta} - \beta^*\frac{\partial}{\partial\beta^*} - 2n\beta\beta^* \right]\chi. \tag{9.5.18}$$

b) Atomic Terms: For the atomic contribution, write O as

$$O = Q^{A,\mu}O^{A,\mu} \tag{9.5.19}$$

where

$$Q^{A,\mu} = O^F \prod_{i\neq\mu} O^{A,i} \tag{9.5.20}$$

and

$$O^{A,\mu} = b + \mathbf{c}.\boldsymbol{\sigma}_\mu. \tag{9.5.21}$$

Hence

$$\mathrm{Tr}\left\{ O \left.\frac{\partial\rho}{\partial t}\right|_A \right\} = \mathrm{Tr}\Big\{ \sum_\mu Q^{A,\mu}O^{A,\mu} \left[\tfrac{1}{2}W_{21}\left(2\sigma_\mu^-\rho\sigma_\mu^+ - \sigma_\mu^+\sigma_\mu^-\rho - \rho\sigma_\mu^+\sigma_\mu^-\right) \right.$$

$$\left. + \tfrac{1}{2}W_{12}\left(2\sigma_\mu^+\rho\sigma_\mu^- - \sigma_\mu^-\sigma_\mu^+\rho - \rho\sigma_\mu^-\sigma_\mu^+\right)\right] \Big\}. \tag{9.5.22}$$

Using the properties of the trace and of the Pauli matrices we deal with each term individually e.g.,

$$\mathrm{Tr}\left\{ \sum_\mu Q^{A,\mu}\left(b + \mathbf{c}.\boldsymbol{\sigma}_\mu\right) W_{21}\sigma_\mu^-\rho\sigma_\mu^+ \right\}$$

$$= \mathrm{Tr}\left\{ \sum_\mu Q^{A,\mu}W_{21}\sigma_\mu^+\left(b + \mathbf{c}.\boldsymbol{\sigma}_\mu\right)\sigma_\mu^-\rho \right\} \tag{9.5.23}$$

$$= \mathrm{Tr}\left\{ \sum_\mu Q^{A,\mu}\tfrac{1}{2}W_{21}(b - c)\left(\sigma_{z,\mu} + 1\right)\rho \right\} \tag{9.5.24}$$

$$= \mathrm{Tr}\left\{ \sum_\mu Q^{A,\mu}\tfrac{1}{2}W_{21}(b - c)\left(\frac{\partial}{\partial c} + \frac{\partial}{\partial b}\right)O^{A,\mu}\rho \right\}. \tag{9.5.25}$$

Following a similar method for all the other terms we then substitute into (9.5.22) to obtain

$$\mathrm{Tr}\left\{O\left.\frac{\partial\rho}{\partial t}\right|_A\right\}=\mathrm{Tr}\left\{\sum_\mu Q^{A,\mu}\left[-\tfrac{1}{2}(W_{21}+W_{12})\left(c^-\frac{\partial}{\partial c^-}+c^+\frac{\partial}{\partial c^+}+2c\frac{\partial}{\partial c}\right)\right.\right.$$
$$\left.\left.+(W_{12}-W_{21})c\frac{\partial}{\partial b}\right]O^{A,\mu}\rho\right\}. \tag{9.5.26}$$

Now using the product rule for differentiation we can turn the sum of derivatives of $O^{A,\mu}$ into one derivative of the product O, so that

$$\mathrm{Tr}\left\{O\left.\frac{\partial\rho}{\partial t}\right|_A\right\}$$
$$=\left[-\tfrac{1}{2}(W_{21}+W_{12})\left(c^-\frac{\partial}{\partial c^-}+c^+\frac{\partial}{\partial c^+}+2c\frac{\partial}{\partial c}\right)+(W_{12}-W_{21})c\frac{\partial}{\partial b}\right]\chi. \tag{9.5.27}$$

c) Interaction Terms: Expanding out the commutator, the interaction term can be evaluated as

$$g\mathrm{Tr}\left\{O\left[a^\dagger S^-\rho-\rho a^\dagger S^-+\rho aS^+-aS^+\rho\right]\right\} \tag{9.5.28}$$

where

$$S^-=\sum_\mu\sigma_\mu^-\qquad S^+=\sum_\mu\sigma_\mu^+, \tag{9.5.29}$$

and inserting the definition (9.5.15), and following a procedure similar to (9.5.23–25) for the atomic operators (9.5.28) becomes

$$=g\mathrm{Tr}\left\{(b-c)\left[\mathrm{i}\beta\frac{\partial}{\partial c^-}+\mathrm{i}\beta^*\frac{\partial}{\partial c^+}\right]+\tfrac{1}{2}(\mathrm{i}\beta c^++\mathrm{i}\beta^*c^-)\left[\frac{\partial}{\partial c}+\frac{\partial}{\partial b}\right]\right.$$
$$\left.+\frac{\partial}{\partial\mathrm{i}\beta^*}\left[-2c\frac{\partial}{\partial c^-}+c^+\frac{\partial}{\partial c}\right]+\frac{\partial}{\partial\mathrm{i}\beta}\left[-2c\frac{\partial}{\partial c^+}+c^-\frac{\partial}{\partial c}\right]\right\}\chi. \tag{9.5.30}$$

d) Characteristic Function Equation: substituting (9.5.18,27,30) into the characteristic function equation we obtain the equation

$$
\begin{aligned}
\frac{\partial \chi}{\partial t} = \Bigg\{ & \kappa \left(-\beta \frac{\partial}{\partial \beta} - \beta^* \frac{\partial}{\partial \beta^*} - 2n\beta^*\beta \right) \\
& - \frac{(W_{12} + W_{21})}{2} \left(c^+ \frac{\partial}{\partial c^+} + c^- \frac{\partial}{\partial c^-} + 2c \frac{\partial}{\partial c} \right) + (W_{12} - W_{21}) c \frac{\partial}{\partial b} \\
& + g \Bigg\{ -2c \frac{\partial^2}{\partial c^+ \partial \mathrm{i}\beta} + c^- \frac{\partial^2}{\partial c \partial \mathrm{i}\beta} \\
& + \mathrm{i}\beta \left[(b - c) \frac{\partial}{\partial c^-} + \tfrac{1}{2} c^+ \left(\frac{\partial}{\partial b} + \frac{\partial}{\partial c} \right) \right] \\
& - 2c \frac{\partial^2}{\partial c^- \partial \mathrm{i}\beta^*} + c^+ \frac{\partial^2}{\partial c \partial \mathrm{i}\beta^*} \\
& + \mathrm{i}\beta^* \left[(b - c) \frac{\partial}{\partial c^+} + \tfrac{1}{2} c^- \left(\frac{\partial}{\partial b} + \frac{\partial}{\partial c} \right) \right] \Bigg\} \Bigg\} \chi .
\end{aligned} \tag{9.5.31}
$$

e) Generalized Fokker-Planck Equation: For any given number, N, of atoms the characteristic function is a homogeneous polynomial on b and c of degree N. A Fourier transform does not exist in the ordinary sense, but can exist in the sense of a distribution. If, however, there is a distribution of numbers of atoms, (e.g., a Gaussian distribution) the characteristic function can be well behaved and possess a well behaved Fourier transform. This is sufficient for our needs, and this Fourier transform will be a quasiprobability. Because the means of the variables are defined by

$$
\begin{aligned}
\langle \sum_\mu \sigma_\mu \rangle &= \left. \frac{\partial \chi}{\partial c} \right|_{b=1, c=0, \beta=\beta^*=0} \\
N &= \left. \frac{\partial \chi}{\partial b} \right|_{b=1, c=0, \beta=\beta^*=0}
\end{aligned} \tag{9.5.32}
$$

we make the change of variables

$$
\begin{aligned}
b &\to 1 + b' \\
\frac{\partial}{\partial b} &\to \frac{\partial}{\partial b'},
\end{aligned} \tag{9.5.33}
$$

which means that the expectation values are now evaluated at $b' = 0$. In analogy to the method of the previous section, we could expect to define a quasiprobability function $P(x) = P(\alpha^*, \alpha, v^*, D, v)$, where $x = (\alpha^*, \alpha, v^*, D, v)$, via

$$
\chi(\beta, \beta^*, \xi^*, \varsigma, \xi) = \int d\alpha^* \int d\alpha \int dv^* \int dv \int dD \, \mathrm{e}^{\mathrm{i}\beta^*\alpha^*} \mathrm{e}^{\mathrm{i}\beta\alpha} \mathrm{e}^{\mathrm{i}\xi^* v^*} \mathrm{e}^{\mathrm{i}\varsigma D/2} \mathrm{e}^{\mathrm{i}\xi v} P(x). \tag{9.5.34}
$$

In this case we have two independent complex variables and one real variable in both χ and P, so that (9.5.34) defines $P(x)$ to be the five-dimensional Fourier transform of the characteristic function. We may then define $P(x)$ via the inverse Fourier transform

$$P(x) = \int d\beta^* \int d\beta \int d\xi^* \int d\xi \int d\varsigma \\ e^{-i\beta^*\alpha^*} e^{i\beta\alpha} e^{-i\xi^* v^*} e^{-i\varsigma D/2} e^{-i\xi v} \chi(\beta^*, \beta, \xi^*, \varsigma, \xi). \tag{9.5.35}$$

However, a major problem arises with the P-representation defined in this way for the laser theory, in that the diffusion matrix of the Fokker-Planck equation obtained is not positive definite in the five-dimensional physical space. To overcome this problem we carry out a procedure analogous to that followed for the positive P-representation and consider all the variables in both the characteristic function and the P-function to be independent complex variables. Thus we no longer have $\alpha^* = (\alpha)^*$ but must treat α and α^* as independent complex variables. To make this clear in the following we shall adopt the notation $\alpha^* = \alpha^+$ to explicitly display the lack of conjugacy. So making all the notational changes to the variables, we define the characteristic function χ in a ten dimension phase space by

$$\chi = \int d^2\alpha^+ \int d^2\alpha \int d^2v^+ \int d^2D \int d^2v \exp\left[i(\beta^+, \beta, \xi^+, \varsigma/2, \xi) \cdot x\right] P(x). \tag{9.5.36}$$

It is now no longer possible to uniquely define the quasiprobability function $P(x)$ by the inverse Fourier transform, as was done between (9.5.34) and (9.5.35). However it is this nonuniqueness which makes the guarantee of positive-definite diffusion possible. We arrive at a generalized Fokker-Planck equation via the correspondences :

$$\left.\begin{array}{ll} i\beta \to -\dfrac{\partial}{\partial\alpha} & \dfrac{\partial}{\partial i\beta} \to \alpha \\ i\beta^* \to -\dfrac{\partial}{\partial\alpha^*} & \dfrac{\partial}{\partial i\beta^*} \to \alpha^* \end{array}\right\} \tag{9.5.37}$$

$$\left.\begin{array}{ll} c^- \to -\dfrac{\partial}{\partial v} & \dfrac{\partial}{\partial c^-} \to v \\ c^+ \to -\dfrac{\partial}{\partial v^*} & \dfrac{\partial}{\partial c^+} \to v^* \\ c \to -\dfrac{\partial}{\partial D} & \dfrac{\partial}{\partial c} \to D \end{array}\right\} \tag{9.5.38}$$

$$\left.\begin{array}{ll} b' \to -\dfrac{\partial}{\partial B} & \dfrac{\partial}{\partial b'} \to B. \end{array}\right\} \tag{9.5.39}$$

10. Squeezing

Although quantum noise cannot be eliminated, it is not immune to technical manipulation, a result which was first noted by *Caves et al.* [10.1] in a paper on the possibility of detecting gravitational radiation. The *standard quantum limit* is a term introduced to describe the limitations on the precision of specification of two canonically conjugate variables as given by the following naive argument. If $a, a^\dagger$ are destruction and creation operators for a harmonic oscillator, and we define the quadrature phase operators, X, Y, by

$$a = \frac{X + \mathrm{i}Y}{\sqrt{2\hbar}}, \qquad a^\dagger = \frac{X - \mathrm{i}Y}{\sqrt{2\hbar}},$$

then Heisenberg's uncertainty principle says that

$$\Delta X\, \Delta Y \geq \tfrac{1}{2}\hbar.$$

The standard quantum limit assumes that in the minimum uncertainty states of interest, $\Delta X = \Delta Y$, and hence $\Delta X^2 = \Delta Y^2 = \hbar/2$. Such is certainly the case in the lowest energy state of the harmonic oscillator, which corresponds to a zero temperature thermal state. Intuitively one would believe that the uncertainty, i.e., the fluctuations, at absolute zero would surely represent the irreducible minimum.

However, the uncertainty principle itself does not require this; either ΔX or ΔY can be as small as one likes, as long as the product satisfies the uncertainty principle. This chapter is devoted to developing that theme. It will establish the basic method of generating squeezing, with the degenerate parametric amplifier, and show how the input-output formalism, developed by *Collett* and *Gardiner* [10.2] must be used to study the production of squeezed beams of light.

A simple theory of squeezed white noise is then developed, and the corresponding master equations, etc., developed and applied. Finally, the concept of multi-beam squeezing, in which different beams of light are quantum mechanically correlated, is developed.

10.1 Squeezed States of the Harmonic Oscillator

Squeezing is defined precisely only in the case of the operators of a harmonic oscillator. We have already met squeezed states in the study of Gaussian density operators, undertaken in Sect. 4.4.5. The *ideal* squeezed state of a harmonic oscillator is taken to be a quantum Gaussian state with minimum uncertainty product. This is most straightforwardly defined by means of the linear transformation which will transform a vacuum state into one with prescribed means and covariance matrix. Because of the Gaussian requirement, this is sufficient to specify the squeezed state.

10.1.1 Definition of an Ideal Squeezed State

An ideal squeezed state, $|\alpha, \xi\rangle$ is defined as

$$|\alpha, \xi\rangle = D(\alpha)S(\xi)|0\rangle, \tag{10.1.1}$$

where $D(\alpha)$ is the displacement operator

$$D(\alpha) = \exp(\alpha a^\dagger - \alpha^* a) \tag{10.1.2}$$

used in the definition of coherent states in Sect. 4.3.1b, $S(\xi)$ is the operator

$$S(\xi) = \exp\left(\tfrac{1}{2}\xi^* a^2 - \tfrac{1}{2}\xi a^{\dagger 2}\right) \tag{10.1.3}$$

and

$$\xi = r\mathrm{e}^{\mathrm{i}\theta}. \tag{10.1.4}$$

a) Means and Variances of a Squeezed State: For a squeezed state, $|\alpha, \xi\rangle$, the means are given by

$$\langle a\rangle = \alpha, \qquad \langle a^\dagger\rangle = \alpha^*. \tag{10.1.5}$$

The variances have to be defined in a rotated frame; if variables $\bar{X}$ and $\bar{Y}$ are defined by

$$\sqrt{2\hbar}\; a = X + \mathrm{i}Y = (\bar{X} + \mathrm{i}\bar{Y})\mathrm{e}^{\mathrm{i}\theta/2} \tag{10.1.6}$$

where θ is as defined in (10.1.4), then the variables $\bar{X}, \bar{Y}$ are the independent squeezed variables, i.e.

$$\begin{aligned}
\langle \Delta\bar{X}^2\rangle &= \tfrac{1}{4}\mathrm{e}^{-2r}\\
\langle \Delta\bar{Y}^2\rangle &= \tfrac{1}{4}\mathrm{e}^{2r}\\
\langle \Delta\bar{X}\Delta\bar{Y}\rangle &= 0.
\end{aligned} \tag{10.1.7}$$

Graphically, this can be represented as in Fig. 10.1.

The principal axes of the error ellipse are rotated an angle $\theta/2$ with respect to those of X and Y.

Exercise. Show that the squeezed state *never* has a Glauber-Sudarshan P-function for any non-zero ξ. (It does have of course, a positive P-function.)

Exercise—Complex P-function for a Squeezed State. From the definition (10.1.1) show that the density operator for a squeezed vacuum state (i.e., $\alpha = 0$) with real ξ satisfies the equation

$$\frac{\partial\rho}{\partial\xi} = -2\mathrm{i}[\hat{H}_2, \rho] \tag{10.1.8}$$

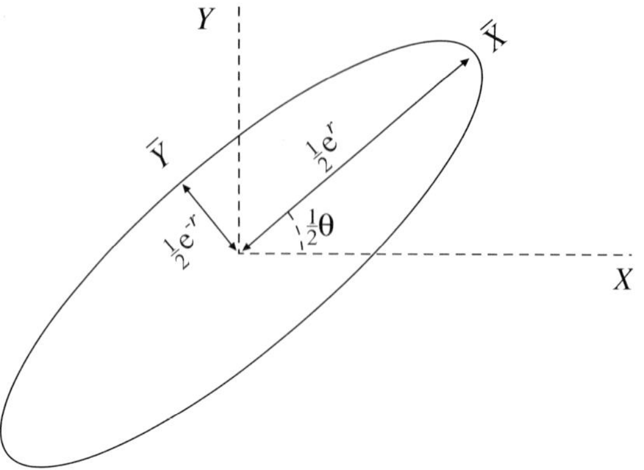

Fig. 10.1 Error ellipse for a squeezed state

with

$$H_2 = \frac{\mathrm{i}}{4}(a^{\dagger 2} - a^2). \tag{10.1.9}$$

Using the operator correspondences for the complex P-function, show that the complex P-function obeys the equation

$$2\frac{\partial P(\alpha,\beta;\xi)}{\partial \xi} = \left(2\beta\frac{\partial}{\partial\alpha} + 2\alpha\frac{\partial}{\partial\beta} - \frac{\partial^2}{\partial\alpha^2} - \frac{\partial^2}{\partial\beta^2}\right) P(\alpha,\beta;\xi). \tag{10.1.10}$$

This corresponds to an Ornstein-Uhlenbeck process in the variables $\mathrm{i}\alpha$ and $\mathrm{i}\beta$. Hence the solution of this equation is

$$P(\alpha,\beta;\xi) = \exp\left\{\alpha\beta + \frac{\coth\xi}{2}(\alpha^2+\beta^2)\right\}. \tag{10.1.11}$$

The contours of integration for this complex P-function are the imaginary axes of both variables.

Using an appropriate transformation of variables, find the complex P-function for an arbitrary squeezed state.

10.2 The Degenerate Parametric Amplifier

From the results of Sect. 6.3.3 and Sect. 7.2.9, the equation (10.1.8) is seen to correspond exactly to that of the degenerate parametric amplifier (DPA) in which the coupling to the heat bath (or modes outside the cavity) is zero. Thus t and ξ are proportional, and the DPA inside a perfect cavity produces an increasingly squeezed state as time increases. In the case that the damping produced by the heat bath is not negligible, a stationary state occurs when losses into the bath balance the gain of the amplifier. Using (7.2.63) of the exercise of Sect. 7.2.9c, it can be seen that the internal field in this stationary state does not become perfectly squeezed. In fact the minimum uncertainty achievable is $\langle \Delta Y^2\rangle = \hbar/8$, not zero. How, therefore, to achieve arbitrary squeezing is not a problem that will be solved this way. However,

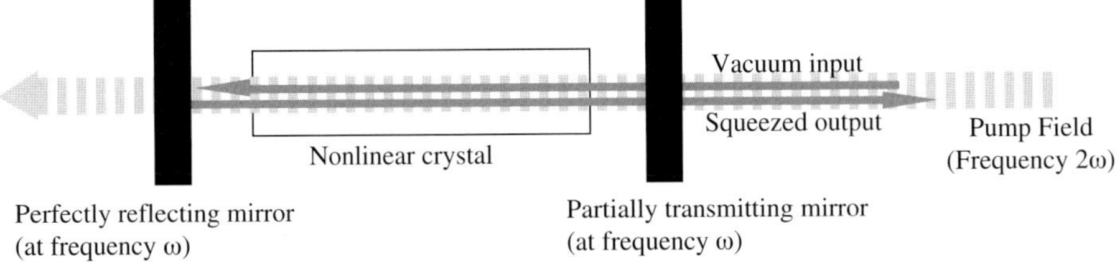

Fig. 10.2 Optical arrangement for a degenerate parametric amplifier

in Sect. 7.2.9c we showed that the *output* from a degenerate parametric amplifier is arbitrarily squeezed, even though the internal field is not. Since, in fact, squeezed light is used in beams, rather than in internal cavity modes, this is just the result that is needed.

10.2.1 Squeezing in the Degenerate Parametric Amplifier

Although this has already been treated in Sect. 7.2.9, from the point of view of its properties as an amplifier, a treatment will be given here which is more appropriate for the study of squeezing.

The systematic Hamiltonian for degenerate parametric amplification with a classical pump can be written as

$$H_{\text{sys}} = \hbar\omega_0 a^\dagger a + \tfrac{1}{2}\mathrm{i}\hbar\left[\epsilon \mathrm{e}^{-\mathrm{i}\omega_p t}(a^\dagger)^2 - \epsilon^* \mathrm{e}^{\mathrm{i}\omega_p t} a^2\right], \tag{10.2.1}$$

where ω_p is the frequency of the pump beam and ϵ a measure of the effective pump intensity. For now the pump and cavity will be considered to be tuned so that $\omega_p = 2\omega_0$. In practice the amplifying medium will be a crystal inside an optical cavity, as in Fig. 10.3, which of necessity has two ends. The mirrors can be chosen to be almost non-reflecting at ω_p, and quite highly reflecting at ω_0. We can use the formalism of quantum Langevin equations as developed in Sect. 5.3.2 to write an equation of motion for the field mode a inside the cavity in terms of inputs $a_{\text{in}}(t)$, $b_{\text{in}}(t)$, coming from either side of the cavity; the resulting equation of motion is

$$\frac{da}{dt} = -\mathrm{i}\omega_0 a + \epsilon \mathrm{e}^{-\mathrm{i}\omega_{pt} t} a^\dagger - \tfrac{1}{2}(\gamma_1 + \gamma_2)a - \sqrt{\gamma_1}\, a_{\text{in}}(t) - \sqrt{\gamma_2}\, b_{\text{in}}(t). \tag{10.2.2}$$

We now transform to a rotating frame with $a \to \mathrm{e}^{\mathrm{i}\omega_p t/2} a$ and similarly for the input operators—for simplicity, we now use only these operators in the rotating frame, without any distinctive notation. In matrix notation, the equations become

$$\frac{d\mathbf{a}}{dt} = \left[\mathbf{A} - \tfrac{1}{2}(\gamma_1 + \gamma_2)\right]\mathbf{a} - \sqrt{\gamma_1}\,\mathbf{a}_{\text{in}} - \sqrt{\gamma_2}\,\mathbf{b}_{\text{in}} \tag{10.2.3}$$

where

$$\mathbf{A} = \begin{pmatrix} 0 & \epsilon \\ \epsilon^* & 0 \end{pmatrix} = \begin{pmatrix} 0 & |\epsilon|\mathrm{e}^{\mathrm{i}\theta} \\ |\epsilon|\mathrm{e}^{-\mathrm{i}\theta} & 0 \end{pmatrix}, \tag{10.2.4}$$

and the notation

$$\mathbf{a}_{\text{in}} = \begin{pmatrix} a_{\text{in}} \\ a_{\text{in}}^{\dagger} \end{pmatrix} \tag{10.2.5}$$

is used for $\mathbf{a}, \mathbf{a}_{\text{in}}, \mathbf{b}_{\text{in}}$, etc.

a) Fourier Transform Solution: Define a Fourier transform variable by

$$\tilde{a}(\omega) = \frac{1}{\sqrt{2\pi}} \int dt\, e^{i\omega t} a(t), \tag{10.2.6}$$

with similar notation for other Fourier transforms. In frequency space (10.2.3) transforms to

$$-i\omega\tilde{\mathbf{a}}(\omega) = \left[\mathbf{A} - \tfrac{1}{2}(\gamma_1 + \gamma_2)\right] \tilde{\mathbf{a}}(\omega) - \sqrt{\gamma_1}\, \tilde{\mathbf{a}}_{\text{in}}(\omega) - \sqrt{\gamma_2}\, \tilde{\mathbf{b}}_{\text{in}}(\omega) \tag{10.2.7}$$

where now, to allow for the rotating frame,

$$\tilde{\mathbf{a}}(\omega) = \begin{pmatrix} \tilde{a}(\omega_s + \omega) \\ \tilde{a}^{\dagger}(\omega_s - \omega) \end{pmatrix}, \tag{10.2.8}$$

with $\omega_s = \omega_p/2$, and similarly for the input operators.

After performing the matrix inversion we find

$$\begin{aligned} \tilde{a}(\omega_s + \omega) = &- \frac{\left(\frac{1}{2}\gamma_1 + \frac{1}{2}\gamma_2 - i\omega\right)\left[\sqrt{\gamma_1}\, \tilde{a}_{\text{in}}(\omega_s + \omega) + \sqrt{\gamma_2}\, \tilde{b}_{\text{in}}(\omega_s + \omega)\right]}{\left(\frac{1}{2}\gamma_1 + \frac{1}{2}\gamma_2 - i\omega\right)^2 - |\epsilon|^2} \\ &- \frac{\epsilon\left[\sqrt{\gamma_1}\, \tilde{a}_{\text{in}}^{\dagger}(\omega_s - \omega) + \sqrt{\gamma_2}\, \tilde{b}_{\text{in}}^{\dagger}(\omega_s - \omega)\right]}{\left(\frac{1}{2}\gamma_1 + \frac{1}{2}\gamma_2 - i\omega\right)^2 - |\epsilon|^2}. \end{aligned} \tag{10.2.9}$$

If both the input fields are vacuum or coherent, they will have zero normally ordered variance, that is,

$$\mathbf{C}_N(a_{\text{in}}, a_{\text{in}}^{\dagger}) = \mathbf{C}_N(b_{\text{in}}, b_{\text{in}}^{\dagger}) = \mathbf{0} \tag{10.2.10}$$

where we use the notation, for any operator x

$$\mathbf{C}_N(x, x^{\dagger}) = \begin{pmatrix} \langle x, x\rangle & \langle x^{\dagger}, x\rangle \\ \langle x^{\dagger}, x\rangle & \langle x^{\dagger}, x^{\dagger}\rangle \end{pmatrix} \tag{10.2.11}$$

and have used the notation

$$\langle a, b\rangle = \langle ab\rangle - \langle a\rangle\langle b\rangle. \tag{10.2.12}$$

b) Internal Mode Properties: We can compute the normally ordered second moments of the Fourier transform variables directly from the solution (10.2.9). Because the input fields are vacuum fields, the only contribution to the normally ordered variance of the internal field will be from the terms involving antinormally ordered products of the input field operators. The result is easily computed to be

$$\langle \tilde{a}(\omega_s+\omega), \tilde{a}(\omega_s+\omega')\rangle = \frac{(\frac{1}{2}\gamma_1+\frac{1}{2}\gamma_2-\mathrm{i}\omega)\mathrm{e}^{\mathrm{i}\theta}}{2} \times\left[\frac{1}{\left(\frac{1}{2}\gamma_1+\frac{1}{2}\gamma_2-|\epsilon|\right)^2+\omega^2}-\frac{1}{\left(\frac{1}{2}\gamma_1+\frac{1}{2}\gamma_2+|\epsilon|\right)^2+\omega^2}\right]\delta(\omega+\omega') \quad (10.2.13)$$

$$\langle \tilde{a}^\dagger(\omega_s+\omega), \tilde{a}(\omega_s+\omega')\rangle = \frac{|\epsilon|}{2}\left[\frac{1}{\left(\frac{1}{2}\gamma_1+\frac{1}{2}\gamma_2-|\epsilon|\right)^2+\omega^2}-\frac{1}{\left(\frac{1}{2}\gamma_1+\frac{1}{2}\gamma_2+|\epsilon|\right)^2+\omega^2}\right]\delta(\omega-\omega'). \quad (10.2.14)$$

Inverting the Fourier transform, and noting the presence of delta functions in (10.2.13,14), the variances of the internal mode are

$$\langle a^\dagger(t), a(t)\rangle = \frac{1}{2\pi}\int d\omega d\omega' \langle \tilde{a}^\dagger(\omega_s+\omega), \tilde{a}(\omega_s+\omega')\rangle = \frac{1}{2}\frac{|\epsilon|^2}{(\frac{1}{2}\gamma_1+\frac{1}{2}\gamma_2)^2-|\epsilon|^2}, \quad (10.2.15)$$

$$\langle a(t), a(t)\rangle = \frac{1}{2\pi}\int d\omega d\omega' \langle \tilde{a}(\omega_s+\omega), \tilde{a}(\omega_s+\omega')\rangle = \frac{1}{2}\frac{\epsilon\,(\frac{1}{2}\gamma_1+\frac{1}{2}\gamma_2)}{(\frac{1}{2}\gamma_1+\frac{1}{2}\gamma_2)^2-|\epsilon|^2}. \quad (10.2.16)$$

c) Internal Mode Quadrature Phases: The internal mode quadrature phase operators will be defined by

$$a = \mathrm{e}^{\mathrm{i}\theta/2}(X_1+\mathrm{i}X_2). \quad (10.2.17)$$

It is convenient to use the *normally ordered variances* to characterize the squeezing, since, in the case of vacuum input, these tend to give simpler formulae. The normally ordered variances of these operators are

$$\begin{aligned} \langle : X_1, X_1 : \rangle &= \frac{\frac{1}{4}|\epsilon|}{\frac{1}{2}\gamma_1+\frac{1}{2}\gamma_2-|\epsilon|} \\ \langle : X_2, X_2 : \rangle &= -\frac{\frac{1}{4}|\epsilon|}{\frac{1}{2}\gamma_1+\frac{1}{2}\gamma_2+|\epsilon|} \\ \langle : X_1, X_2 : \rangle &= 0. \end{aligned} \quad (10.2.18)$$

Perfect squeezing in one quadrature, corresponding to an eigenstate of the quadrature phase operator, is achieved with a normally ordered variance of $-1/4$. The best that can be achieved in the case of the parametric amplifier in a cavity is when on oscillation threshold, $|\epsilon| = (\gamma_1+\gamma_2)/2$, giving

$$\langle : X_2, X_2 : \rangle = -1/8. \quad (10.2.19)$$

Squeezing by a factor of one-half can thus be obtained in the X_2 quadrature of the generating cavity with the X_1 quadrature infinitely unsqueezed. Note that the properties of the internal mode considered in this section depend only on the total damping, not on the damping through each mirror separately.

d) The Output Field: The output fields can be calculated as the boundary condition (5.3.24) (in Fourier transform form)

$$\tilde{a}_{\text{out}}(\omega) = \tilde{a}_{\text{in}}(\omega) + \sqrt{\gamma_1}\,\tilde{a}(\omega). \tag{10.2.20}$$

It is a matter of straightforward substitution to arrive at the result

$$\begin{aligned}\tilde{a}_{\text{out}}\,(\omega_s+\omega) &= -\frac{\left[(\frac{1}{2}\gamma_1)^2 - (\frac{1}{2}\gamma_2 - \mathrm{i}\omega)^2 + |\epsilon|^2\right]\tilde{a}_{\text{in}}(\omega_s+\omega) + \epsilon\gamma_1\tilde{a}^\dagger_{\text{in}}(\omega_s-\omega)}{(\frac{1}{2}\gamma_1 + \frac{1}{2}\gamma_2 - \mathrm{i}\omega)^2 - |\epsilon|^2} \\ &\quad - \frac{\sqrt{\gamma_1\gamma_2}\,(\frac{1}{2}\gamma_1 + \frac{1}{2}\gamma_2 - \mathrm{i}\omega)\tilde{b}_{\text{in}}(\omega_s+\omega) + \epsilon\sqrt{\gamma_1\gamma_2}\,b^\dagger_{\text{in}}(\omega_s-\omega)}{(\frac{1}{2}\gamma_1 + \frac{1}{2}\gamma_2 - \mathrm{i}\omega)^2 - |\epsilon|^2}.\end{aligned} \tag{10.2.21}$$

As with the internal field, only antinormally ordered terms contribute to the variances so that

$$\begin{aligned}&\langle \tilde{a}^\dagger_{\text{out}}(\omega_s+\omega), \tilde{a}_{\text{out}}(\omega_s+\omega')\rangle \\ &\quad = \frac{|\epsilon|\gamma_1}{2}\left[\frac{1}{\left(\frac{1}{2}\gamma_1 + \frac{1}{2}\gamma_2 - |\epsilon|\right)^2 + \omega^2} - \frac{1}{\left(\frac{1}{2}\gamma_1 + \frac{1}{2}\gamma_2 + |\epsilon|\right)^2 + \omega^2}\right]\delta(\omega-\omega'),\end{aligned} \tag{10.2.22}$$

$$\begin{aligned}&\langle \tilde{a}_{\text{out}}(\omega_s+\omega), \tilde{a}_{\text{out}}(\omega_s+\omega')\rangle \\ &\quad = \frac{|\epsilon|\gamma_1}{2}\left[\frac{1}{\left(\frac{1}{2}\gamma_1 + \frac{1}{2}\gamma_2 - |\epsilon|\right)^2 + \omega^2} + \frac{1}{\left(\frac{1}{2}\gamma_1 + \frac{1}{2}\gamma_2 + |\epsilon|\right)^2 + \omega^2}\right]\delta(\omega+\omega').\end{aligned} \tag{10.2.23}$$

Once again it is the variances in the quadrature phases which are of most interest. Defining the output quadrature phases in the same fashion as the internal ones, from (10.2.22)

$$\begin{aligned}&\langle : \tilde{X}_{1,\text{out}}(\omega_s+\omega), \tilde{X}_{1,\text{out}}(\omega_s+\omega') :\rangle \\ &\qquad = \frac{|\epsilon|(\gamma_1/2)}{\left(\frac{1}{2}\gamma_1 + \frac{1}{2}\gamma_2 - |\epsilon|\right)^2 + \omega^2}\delta(\omega+\omega')\end{aligned} \tag{10.2.24}$$

$$\begin{aligned}&\langle : \tilde{X}_{2,\text{out}}(\omega_s+\omega), \tilde{X}_{2,\text{out}}(\omega_s+\omega') :\rangle \\ &\qquad = -\frac{|\epsilon|(\gamma_1/2)}{\left(\frac{1}{2}\gamma_1 + \frac{1}{2}\gamma_2 + |\epsilon|\right)^2 + \omega^2}\delta(\omega+\omega').\end{aligned} \tag{10.2.25}$$

e) Maximum Squeezing: The maximum squeezing is still attained at threshold, $|\epsilon| = \frac{1}{2}(\gamma_1 + \gamma_2)$, giving

$$\langle : \tilde{X}_{2,\text{out}}(\omega_s + \omega), \tilde{X}_{2,\text{out}}(\omega_s + \omega') : \rangle_{\max} = -\frac{\gamma_1}{4} \frac{\gamma_1 + \gamma_2}{(\gamma_1 + \gamma_2)^2 + \omega^2} \delta(\omega + \omega'). \tag{10.2.26}$$

The most interesting thing about these results is that while the squeezing in the total field is independent of γ_2, this is not the case for the individual frequency components. The δ function in (10.2.26) can be removed by integrating over ω' to give the normally ordered spectrum of the operator $X_{2,\text{out}}$:

$$: S_{2,\text{out}}(\omega_s + \omega) :_{\max} = -\frac{\gamma_1}{4} \frac{\gamma_1 + \gamma_2}{(\gamma_1 + \gamma_2)^2 + \omega^2}, \tag{10.2.27}$$

which is a convenient way of describing the squeezing in the output field. It may be thought of loosely as the squeezing at a particular frequency, although in this case it results from the coupling of pairs of frequencies on either side of resonance: This spectrum is, ignoring the sign, a Lorentzian with peak height $\frac{1}{4}\left[\gamma_1/(\gamma_1 + \gamma_2)\right]$ and width $\gamma_1 + \gamma_2$. Thus for a symmetric double-ended cavity with $\gamma = \gamma_1 = \gamma_2$

$$: S_{2,\text{out}}(0) :_{\max} = -1/8. \tag{10.2.28}$$

The resonant mode of the output field is squeezed by a factor of one-half, the same as the internal field. If, however, the cavity is single-ended, with $\gamma_2 = 0$,

$$: S_{2,\text{out}}(0) :_{\max} = -1/4 \tag{10.2.29}$$

and this is the maximum squeezing available.

We get degradation of squeezing with a two-sided cavity because there are two inputs. With a one-sided cavity, the radiated field from the cavity (the term $\sqrt{\gamma}\,\tilde{a}(\omega)$ in (10.2.20)) and the reflected field $\tilde{a}_{\text{in}}(\omega)$ interfere in one quadrature to cancel out all fluctuations at $\omega = 0$. With a two-sided cavity, there is an additional component from $b_{\text{in}}(\omega)$ which is transmitted, and, being independent of $a_{\text{in}}(\omega)$, cannot interfere with the reflection.

f) Correlation Functions of the Output Field: By Fourier transforming (10.2.22) we can easily deduce the correlation functions. Using the definition (10.2.6), we find that the correlation functions can be compactly represented as

$$\langle a^\dagger_{\text{out}}(t)\, a_{\text{out}}(t') \rangle = \frac{\gamma_1}{\gamma_1 + \gamma_2} \left\{ \frac{\lambda^2 - \mu^2}{4} \left[\frac{e^{-\mu(t-t')}}{2\mu} - \frac{e^{-\lambda(t-t')}}{2\lambda} \right] \right\} \tag{10.2.30}$$

$$\langle a_{\text{out}}(t)\, a_{\text{out}}(t') \rangle = \frac{\gamma_1}{\gamma_1 + \gamma_2} \left\{ \frac{\lambda^2 - \mu^2}{4} \left[\frac{e^{-\mu(t-t')}}{2\mu} + \frac{e^{-\lambda(t-t')}}{2\lambda} \right] \right\} \tag{10.2.31}$$

where

$$\begin{aligned} \lambda &= \tfrac{1}{2}(\gamma_1 + \gamma_2) + |\epsilon| \\ \mu &= \tfrac{1}{2}(\gamma_1 + \gamma_2) - |\epsilon|. \end{aligned} \tag{10.2.32}$$

g) Squeezed White Noise Limit: In the case that the cavity is single-ended, $\gamma_2 \to 0$, and γ_1 and $|\epsilon|$ are both very large, so that $\mu, \lambda \to \infty$, we can set

$$e^{-\mu|t-t'|} \to \frac{2}{\mu}\delta(t-t') \tag{10.2.33}$$

and so on, to get

$$\langle a_{\rm out}^\dagger(t)a_{\rm out}(t')\rangle = N\delta(t-t') \tag{10.2.34}$$

$$\langle a_{\rm out}(t)a_{\rm out}(t')\rangle = M\delta(t-t') \tag{10.2.35}$$

with

$$N = \frac{(\lambda^2-\mu^2)^2}{4\mu^2\lambda^2}, \qquad M = \frac{\lambda^4-\mu^4}{4\mu^2\lambda^2}. \tag{10.2.36}$$

Notice that in fact here

$$|M|^2 = N(N+1). \tag{10.2.37}$$

Exercise. Use the squeezed beam $a_{\rm out}(t)$ as produced by the DPA, as the input to a single-ended passive cavity, and show that the internal mode is an ideal squeezed state. Thus ideal squeezing can be produced inside a cavity, but not by an active medium inside that cavity.

10.2.2 Squeezed White Noise

We can define ideal squeezed white noise on the basis of the previous section; abstractly, we can say that $b(t)$ represents squeezed white noise if

$$\begin{aligned} dB(t)^2 &= M\,dt, \\ dB^\dagger(t)^2 &= M^*dt, \\ dB(t)dB^\dagger(t) &= (N+1)dt, \\ dB^\dagger(t)dB(t) &= N\,dt. \end{aligned} \tag{10.2.38}$$

Quantum stochastic calculus can be derived in the same way as for ordinary white noise, and the resulting rules are

a) Ito QSDE: This takes the form

$$\begin{aligned} da = &-\frac{i}{\hbar}[a,H_{\rm sys}] + \frac{\gamma}{2}(N+1)(2c^\dagger ac - ac^\dagger c - c^\dagger ca)dt \\ &+\frac{\gamma}{2}N(2cac^\dagger - acc^\dagger - cc^\dagger a)dt - \frac{\gamma}{2}M(2c^\dagger ac^\dagger - ac^\dagger c^\dagger - c^\dagger c^\dagger a)dt \\ &-\frac{\gamma}{2}M^*(2cac - acc - cca)dt - \sqrt{\gamma}\,[a,c^\dagger]dB(t) + \sqrt{\gamma}\,[a,c]dB^\dagger(t). \end{aligned} \tag{10.2.39}$$

b) Relation between Ito and Stratonovich Integrals: This now takes the form

$$(\mathbf{S})\int_{t_0}^{t} dB^\dagger(t')g(t') = (\mathbf{I})\int_{t_0}^{t} g(t')dB^\dagger(t')$$
$$-\tfrac{1}{2}\sqrt{\gamma}\,N\int_{t_0}^{t}[g(t'),c^\dagger(t')]dt' + \tfrac{1}{2}\sqrt{\gamma}\,M^*\int_{t_0}^{t}[g(t'),c(t')]dt' \tag{10.2.40}$$

$$(\mathbf{S})\int_{t_0}^{t} g(t')dB(t') = (\mathbf{I})\int_{t_0}^{t} g(t')dB(t')$$
$$+\tfrac{1}{2}\sqrt{\gamma}\,N\int_{t_0}^{t}[g(t'),c(t')]dt' - \tfrac{1}{2}\sqrt{\gamma}\,M\int_{t_0}^{t}[g(t'),c^\dagger(t')]dt. \tag{10.2.41}$$

(The Stratonovich integrals with $g(t')$ and increment permuted differ from these by the replacement $\bar{N} \to \bar{N}+1$, corresponding to the commutation relation (5.3.55) still being true.) The Stratonovich QSDE corresponding to (10.2.39) is exactly (5.3.64) as is expected. As previously noted, the Stratonovich QSDE is independent of the statistics of the incoming field, and in this sense is more general than the Ito QSDE.

c) Master Equation: The master equation can be derived as in Sect. 5.4.2; it is

$$\begin{aligned}\frac{\partial\rho}{\partial t} &= \frac{\mathrm{i}}{\hbar}[\rho, H_{\mathrm{sys}}] + \frac{\gamma}{2}(N+1)(2c\rho c^\dagger - c^\dagger c\rho - \rho c^\dagger c)\\ &+\frac{\gamma}{2}N(2c^\dagger\rho c - cc^\dagger\rho - \rho cc^\dagger) - \frac{\gamma}{2}M(2c^\dagger\rho c^\dagger - c^\dagger c^\dagger\rho - \rho c^\dagger c^\dagger)\\ &-\frac{\gamma}{2}M^*(2c\rho c - cc\rho - \rho cc).\end{aligned} \tag{10.2.42}$$

This equation can be expected to give the correct description of a system driven by a squeezed noise (which arises from a squeezed vacuum) provided the squeezing is reasonably constant over a bandwidth significantly larger than that expected from the system.

d) Squeezed White Noise in a Non-Rotating Frame: When we have ordinary white noise, there is no difference between $dB(t)$, and $dB(t)\mathrm{e}^{-\mathrm{i}\Omega t} \equiv d\bar{B}(t)$; both $dB(t)$ and $d\bar{B}(t)$ have the same stochastic and commutator properties. However, with squeezing,

$$d\bar{B}(t)^2 = M\mathrm{e}^{-2\mathrm{i}\Omega t}dt, \quad d\bar{B}^\dagger(t)^2 = M^*\mathrm{e}^{2\mathrm{i}\Omega t}dt. \tag{10.2.43}$$

Looking at the equation (10.2.39), one can see that if H_{sys} is the full system Hamiltonian, the operator c will have, to lowest order, a time dependence of the form $\mathrm{e}^{\mathrm{i}\Omega t}$ for some Ω, and this would produce rapidly changing coefficients of M, M^*, which would be negligible when time averaged. If however $d\bar{B}(t)$ is understood, then $M \to M\mathrm{e}^{-2\mathrm{i}\Omega t}$, and cancels this time dependence out. What this means is that in practice, squeezed white noise does have a characteristic frequency Ω, and the form (10.2.38), which may be regarded as the fundamental form, has $\Omega = 0$. For the master equation and QSDE to be true, the squeezing frequency must be matched to the transition frequency of interest.

For an experimental laser physicist this is not a major point—in practice one has real squeezed noise, which has an obvious central frequency, and is squeezed only over a rather narrow bandwidth, which nevertheless may be sufficiently wide to enable the white noise limit to be used sometimes.

The equations (10.2.39–42) are best interpreted as being valid in an interaction picture rotating at frequency Ω, which makes the squeezed white noise representation in the form (10.2.38) valid—$H_{\rm sys}$ is then only that part of the Hamiltonian not removed by this process.

e) Quadrature Phases of Squeezed White Noise: Writing in terms of input fields, and not in the interaction picture, the correlation functions of the squeezed white noise are

$$\begin{aligned}\langle b(t)b(t')\rangle &= M\,\mathrm{e}^{-\mathrm{i}\Omega(t+t')}\delta(t-t')\\ \langle b^\dagger(t)b^\dagger(t')\rangle &= M^*\,\mathrm{e}^{\Omega(t-t')}\delta(t-t')\\ \langle b^\dagger(t)b(t')\rangle &= N\,\delta(t-t')\\ \langle b(t)b^\dagger(t')\rangle &= (N+1)\delta(t-t').\end{aligned} \tag{10.2.44}$$

We can define quadrature phases of squeezed white noise by setting $M = |M|\mathrm{e}^{-2\mathrm{i}\theta}$, and

$$\begin{aligned}X(t) &= \tfrac{1}{2}\left[b^\dagger(t)\mathrm{e}^{\mathrm{i}(\theta-\Omega t)} + b(t)\mathrm{e}^{-\mathrm{i}(\theta-\Omega t)}\right]\\ Y(t) &= \tfrac{1}{2}\mathrm{i}\left[b^\dagger(t)\mathrm{e}^{\mathrm{i}(\theta-\Omega t)} - b(t)\mathrm{e}^{-\mathrm{i}(\theta-\Omega t)}\right]\end{aligned} \tag{10.2.45}$$

and we then find

$$\begin{aligned}\langle X(t)X(t')\rangle &= \tfrac{1}{2}(N + |M| + \tfrac{1}{2})\delta(t-t')\\ &\sim (N+\tfrac{1}{2})\delta(t-t') \qquad \text{if } N\to\infty \text{ and } |M| = \sqrt{N(N+1)}\end{aligned} \tag{10.2.46}$$

$$\begin{aligned}\langle Y(t)Y(t')\rangle &= \tfrac{1}{2}(N + \tfrac{1}{2} - |M|)\delta(t-t')\\ &\sim \delta(t-t')/16N \qquad \text{if } N\to\infty \text{ and } |M| = \sqrt{N(N+1)}\,.\end{aligned} \tag{10.2.47}$$

Thus, in this notation, the Y quadrature is squeezed, and the X quadrature unsqueezed.

10.3 Squeezed Light on a Single Atom

As a simple application, consider a two-level atom interacting with squeezed light, whose frequency is tuned to that of the atomic transition. We use the QSDE (10.2.39) with

$$c \to \sigma^-, \qquad c^\dagger \to \sigma^+ \tag{10.3.1}$$

and choose M, M^* real, since this is merely a phase choice. In the interaction picture $H_{\text{sys}} \to 0$, and we get the Ito QSDEs for the variables $\sigma_x, \sigma_y, \sigma_z$

$$\begin{aligned}
\dot\sigma_x &= -\gamma(N+M+\tfrac{1}{2})\sigma_x + \tfrac{1}{2}\sqrt{\gamma}\,\sigma_z\left\{b_{\text{in}}^\dagger(t)+b_{\text{in}}(t)\right\}\\
\dot\sigma_y &= -\gamma(N-M+\tfrac{1}{2})\sigma_y - \tfrac{1}{2}\mathrm{i}\sqrt{\gamma}\,\sigma_z\left\{b_{\text{in}}^\dagger(t)-b_{\text{in}}(t)\right\}\\
\dot\sigma_z &= -\gamma(2N+1)\sigma_z - \gamma\\
&\quad -\sqrt{\gamma}\,\sigma_x\left\{b_{\text{in}}^\dagger(t)+b_{\text{in}}(t)\right\} + \mathrm{i}\sqrt{\gamma}\,\sigma_y\left\{b_{\text{in}}^\dagger(t)-b_{\text{in}}(t)\right\}.
\end{aligned} \tag{10.3.2}$$

a) Mean Value Solutions: The equations of motion for the mean values take the form

$$\begin{aligned}
\langle\dot\sigma_x\rangle &= -\gamma(N+M+\tfrac{1}{2})\langle\sigma_x\rangle \equiv -\gamma_x\langle\sigma_x\rangle\\
\langle\dot\sigma_y\rangle &= -\gamma(N-M+\tfrac{1}{2})\langle\sigma_y\rangle \equiv -\gamma_y\langle\sigma_y\rangle\\
\langle\dot\sigma_z\rangle &= -\gamma(2N+1)\langle\sigma_z\rangle - \gamma = -\gamma_z\langle\sigma_z\rangle - \gamma.
\end{aligned} \tag{10.3.3}$$

If we have pure squeezed state, in which $M = \sqrt{N(N+1)}$, for large N,

$$\left.\begin{aligned}
N+M+\tfrac{1}{2} &\quad\to\quad 2N+1\\
N-M+\tfrac{1}{2} &\quad\to\quad 1/8N
\end{aligned}\right\}\text{ as } N\to\infty. \tag{10.3.4}$$

We can see therefore that for large squeezing both γ_z and γ_x become large. In fact γ_x and γ_y are proportional to the coefficients of intensity of the respective quadrature phases of the input, as defined in (10.2.46,47).

Since this noise can approach zero in the Y quadrature phase, it is clear that the decay constant γ_y can be as small as we please, while at the same time γ_x becomes extremely large. Thus on a time scale short compared with γ_y^{-1}, but substantially larger than γ_x^{-1} and γ_z^{-1}, we find

$$\langle\sigma_x\rangle \to 0, \qquad \langle\sigma_z\rangle \to -1/(2N+1), \tag{10.3.5}$$

while $\langle\sigma_y\rangle$ is essentially unchanged. Thus the projection of the original orientation of the Bloch vector on the direction of the low-noise quadrature phase is preserved—nevertheless the inversion $\langle\sigma_z\rangle$ decays rapidly to its stationary value.

b) Correlation Functions of the Output: To detect this effect one can look at the spectrum of the fluorescent light. According to the quantum regression theorem and standard methods connected with it (see Sect. 5.2.3), the stationary correlation functions of the atom are

$$\begin{aligned}
\langle\sigma^+(t)\sigma^-(0)\rangle &= \tfrac{1}{2}[N/(2N+1)][\exp(-\gamma_x t)+\exp(-\gamma_y t)]\\
\langle\sigma^-(t)\sigma^+(0)\rangle &= \tfrac{1}{2}[(N+1)/(2N+1)][\exp(-\gamma_x t)+\exp(-\gamma_y t)]\\
\langle\sigma^+(t)\sigma^+(0)\rangle &= \tfrac{1}{2}[(N+1)/(2N+1)][\exp(-\gamma_x t)-\exp(-\gamma_y t)]\\
\langle\sigma^-(t)\sigma^-(0)\rangle &= \tfrac{1}{2}[N/(2N+1)][\exp(-\gamma_x t)-\exp(-\gamma_y t)].
\end{aligned} \tag{10.3.6}$$

The above has been formulated in the case that there is one input and one output channel. In the case of interaction of light with an atom, this is realistic if we view

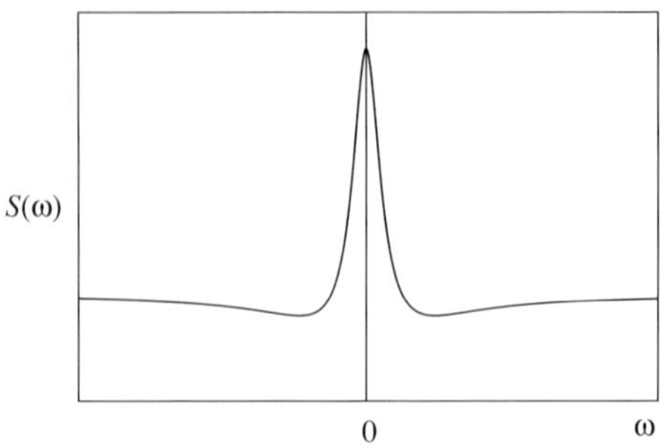

Fig. 10.3 Spectrum corresponding to the correlation function (10.3.9)

the channels as corresponding to the various partial waves, since only the electric dipole partial wave (corresponding to the electric dipole approximation) interacts appreciably with the atom. If the experiment can be carried out with a squeezed electric dipole wave, then the output would be of the same kind.

In this case the amplitude correlation function of the fluorescent light (whose Fourier transform is the spectrum) is computed from the boundary condition

$$b_{\text{out}}(t) = \sqrt{\gamma}\,\sigma^-(t) + b_{\text{in}}(t). \tag{10.3.7}$$

We find

$$\begin{aligned}\langle b^\dagger_{\text{out}}(t) b_{\text{out}}(0)\rangle &= N\delta(t) + \gamma u(t)\langle[\sigma^+(t), N\sigma^-(0) - M\sigma^+(0)]\rangle \\ &+\gamma u(-t)\langle[N\sigma^+(t) - M\sigma^-(t), \sigma^-(0)]\rangle + \gamma\langle\sigma^+(t)\sigma^-(0)\rangle\end{aligned} \tag{10.3.8}$$

(where $u(t) = 0$ for $t < 0; u(t) = 1$ for $t > 0$). This correlation function contains elements arising from the input correlation function $N\delta(t)$ and elements from the atom itself—arising because the fluorescent light consists of a radiated part and a reflected part. The correlation function can be evaluated as

$$\langle b^\dagger_{\text{out}}(t) b_{\text{out}}(0)\rangle = N\delta(t) + \tfrac{1}{2}[\gamma M/(2N+1)][\exp(-\gamma_y t) - \exp(-\gamma_x t)] \tag{10.3.9}$$

which displays the characteristic decay constants.

The spectrum obtained by the Fourier transform of (10.3.9) consists of a flat background, from $\delta(t)$, plus a negative peak of width γ_x (which will be very broad), and finally a positive peak of width γ_y, which can be as narrow as we please. Such narrowing is an effect which can only be produced by some reduction of noise, i.e., squeezing. The narrowed width is in fact directly proportional to the squeezing, and thus provides a measurement of squeezing.

10.4 Simulation Methods Based on the Adjoint Equation

The squeezed white noise results are intriguing, and elegant, but one must ask to what extent these results will be preserved when real squeezed light is used, since it will almost certainly not be exactly white. In fact, the correlation functions (10.2.30,31) show two time constants λ^{-1}, and μ^{-1}. According to (10.2.24), the maximum squeezing is attained at threshold, when $\mu \to 0$, and this means that one of the time constants becomes *infinite*, which of course means that the squeezed white noise approximation cannot be used. Consequently, neither the master equation nor QSDEs can be used. However, the adjoint equation introduced in Sect. 3.5 can be used here, and has been fruitfully applied to computations of these effects by *Parkins* and *Gardiner*, using simulation methods [10.3] and also by *Ritsch* and *Zoller* [10.4], using eigenfunction methods.

10.4.1 Adjoint Equation for Squeezed Light

We can take the quantum Langevin equations (3.4.48–3.4.50), with the resulting equations

$$\begin{aligned}\dot{\sigma}_x &= -\Omega\sigma_y \\ \dot{\sigma}_y &= \Omega\sigma_x + (\hbar\Omega)^{-\frac{1}{2}}\left[\xi(t),\sigma_z\right]_+ \\ \dot{\sigma}_z &= -\gamma - (\hbar\Omega)^{-\frac{1}{2}}\left[\xi(t),\sigma_y\right]_+\end{aligned} \qquad (10.4.1)$$

where $\xi(t) = \sqrt{2\gamma c}\,\dot{A}_{\rm in}(0,t)$ is the incoming electric field evaluated at the atom. This light beam is assumed to have all excitation in a rather narrow bandwidth around Ω, and we can represent this by defining the quadrature phase operators $X_{\rm in}(t)$, $Y_{\rm in}(t)$, by

$$\xi(t) = -\sqrt{2\gamma c}\,\left[Y_{\rm in}(t)\cos(\Omega t) + X_{\rm in}(t)\sin(\Omega t)\right]. \qquad (10.4.2)$$

One now moves to an interaction picture, defined by

$$\begin{aligned}S_x(t) \pm {\rm i}S_y(t) &= {\rm e}^{\mp{\rm i}\Omega t}(\sigma_x(t) \pm {\rm i}\sigma_y(t)) \\ S_z(t) &= \sigma_z(t),\end{aligned} \qquad (10.4.3)$$

and dropping rapidly rotating terms like ${\rm e}^{\pm{\rm i}\Omega t}$, one obtains

$$\begin{aligned}\dot{S}_x &= -\tfrac{1}{2}c_0\,[X_{\rm in}(t), S_z]_+ \\ \dot{S}_y &= -\tfrac{1}{2}c_0\,[Y_{\rm in}(t), S_z]_+ \\ \dot{S}_z &= -\gamma + \tfrac{1}{2}c_0\left\{[X_{\rm in}(t), S_x]_+ + \left[Y_{\rm in}(t), S_y\right]_+\right\}\end{aligned} \qquad (10.4.4)$$

where $c_0 = \sqrt{2\gamma c/\hbar\Omega}$

a) Adjoint Equation: Using the same techniques as in Sect. 3.5 the adjoint equation can be derived in this interaction picture. If $\mu(t)$ is the adjoint equation density operator, then it is defined by

$${\rm Tr}_{\rm s}\left\{\rho S_i(t)\right\} = {\rm Tr}_{\rm s}\left\{\mu(t)\sigma_i\right\}\rho_B \qquad (10.4.5)$$

where σ_i are the Pauli matrices. For ease of working, it is simplest to work in terms of $\bar{\mathbf{S}} \equiv {\rm Tr}_{\rm s}\left\{\mathbf{S}\mu(t)\right\}$, for which the equations of motion are

$$\begin{aligned}\dot{\bar{S}}_x &= -\alpha_X(t)\bar{S}_z \\ \dot{\bar{S}}_y &= -\alpha_Y(t)\bar{S}_z \\ \dot{\bar{S}}_z &= -\gamma + \left[\alpha_X(t)\bar{S}_x + \alpha_Y(t)\bar{S}_y\right].\end{aligned} \qquad (10.4.6)$$

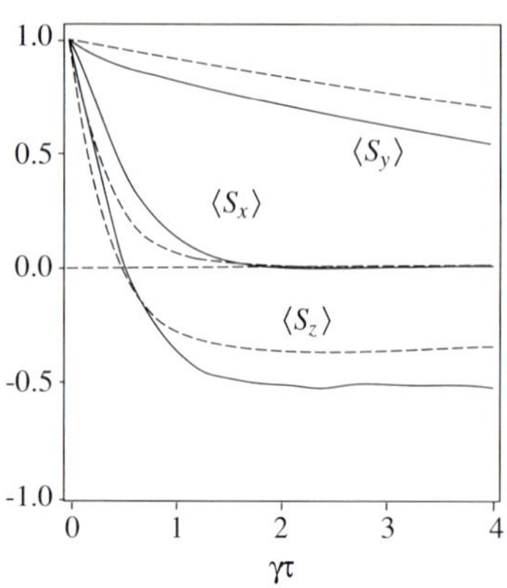

Fig. 10.4 Comparison of the spin averages for simulations (solid line); white noise theory (dashed line)

Since there are both real and imaginary terms in the equations, the values of $\bar{\mathbf{S}}(t)$ obtained will have an imaginary part. However, on average, provided the initial conditions are real, the solutions will be real—that is, in practical situations, one takes simply the real parts of the computed averages over $\alpha_x^c(t)$ and $\alpha_y^c(t)$.

Parkins and *Gardiner* carried out simulations using this kind of algorithm, and found that convergence was surprisingly rapid—for details see [10.6] The major effect is a destruction of the inhibition effect when the squeezing becomes too strong, caused largely by the very slow fluctuations of the unsqueezed quadrature, as seen in Fig.10.4.

11. The Stochastic Schrödinger Equation

One of the earliest concepts in quantum mechanics was the idea of a "quantum jump", in which an atomic system manages to go from one energy level to another "instantaneously", while emitting a photon in the process. Such a picture was mandated by *Bohr's* original formulation of the nuclear atom [11.1] in 1913. Later developments in quantum mechanics gave a more detailed picture of the atom-light interaction in terms of the continuous evolution of wave functions or operators, from which probabilities for all physical processes could be computed. The probabilities of Chap.8 result from this methodology. Yet these probabilities necessarily imply that photons will be counted at discrete times, and the measurement interpretation of quantum mechanics then requires that there is some sense in which the observation of a photon at a definite time must be associated with a sudden change in the quantum state of the atom which emitted the photon.

In this chapter we will present a formulation of quantum stochastic processes in which the "quantum jumps" are made explicit, and indeed become a powerful computational tool. This formulation became important when in the 1980s it became possible to trap individual ions, and perform experiments [11.2] in which one could monitor the quantum jumps of a single ion. A formulation of quantum mechanics appropriate to this situation was necessary. The first comprehensive description of the quantum jump formulation was in fact presented by *Mollow* in 1975 [11.3], in a very perceptive paper that was ahead of its time. A practical algorithmic formulation based on Mollow's work was first achieved by *Zoller et al.* [11.4] in 1987, as a simulation of the probability structure inherent in quantum mechanics, in which quantum jumps were *simulated*, and the overall measured quantities obtained, as averages over many runs. However, much of the development of the relevant structure was done independently by *Barchielli et al.* [11.5], as a formulation of quantum stochastic methods in terms of the state vector of the combined system of atom plus the radiation field (see also [11.6]). The use of *quantum trajectories* as a *calculational tool* to simulate the time evolution of dissipative quantum optical systems in terms of system wave functions was first emphasized by *Castin*, *Mølmer* and *Dalibard* [11.7]. This quantum trajectory picture was also developed independently by *Carmichael* [11.8], and extended to describe the diffusive evolution of the system wave function in homodyne detection. A *practical implementation* of the quantum trajectory technique was established in the early 1990s by three main groups,

Castin, Mølmer and *Dalibard* [11.7], *Carmichael* [11.8] and *Zoller et al.* [11.6].

The techniques used depend on two principal formulations of the Schrödinger equation for a system in interaction with a light field:

i) The quantum stochastic Schrödinger equation, which is formally equivalent to a Schrödinger picture version of the QSDE for Heisenberg operators of Chap.5.
ii) The stochastic Schrödinger equation, which uses a wave function of the atom alone, in which the detections of the photons generate quantum jumps in the atom wave function.

In essence, the stochastic Schrödinger equation is equivalent to the master equation used in conjunction with the quantum regression theorem for multitime correlation functions. This is a non-unitary evolution, which is appropriate to the situation in which only the atomic properties are described.

The relationship between the QSSE and the SSE is established by the continuous measurement formalism for counting processes (see Sect. 8.3) which allows an interpretation as a probabilistic description in terms of *quantum jumps* in a system. Thus the continuous measurement theory can be *simulated* probabilistically, and this simulation yields a sequence of system wave functions with jumps at times $t_1, t_2, \ldots, t_n$ by a rule which is directly related to the structure of the appropriate QSDE. The trajectories of countsgenerated in these simulations have the same statistics as the photon statistics derived within the standard photodetection theory, and in this sense the individual counting sequences generated in a single computer run can, with some caution, be interpreted as "what an observation would yield in a single run of the experiment". These simulations can thus illustrate the dynamics of single quantum systems, for example in the quantum jumps in ion traps [11.4] or squeezing dynamics [11.8].

Instead of direct counting of photons, it is possible to use homodyne measurement techniques, in which the observed light is mixed with a coherent laser field, as shown in Sect. 8.4—this can give a very efficient method of measurement. However one now detects many more photons because of the coherent field, and the quantum jumps become so closely spaced in time that a description in terms of the diffusive evolution, described by a classical stochastic differential equation for the wavefunction now emerges in the limit of very strong coherent laser field.

The practical utility of quantum stochastic simulation methods lies in the fact that an n dimensional wavefunction gives an n^2 dimensional density matrix, whose storage can be quite impractical. Accuracy is obtained in stochastic simulations by making repeated runs, which can be performed in parallel distributed over many computers. Using these techniques, simulations have been able to give quite remarkable quantitative results, and this chapter ends with a summary of some practical examples of the application of stochastic simulation which convincingly demonstrate this point. An up to date review of the status of the field in a form more comprehensive than is appropriate here has been given by *Plenio and Knight* [11.10].

In conclusion, we should point out that the techniques based on the stochastic Schrödinger equation have also been studied as a general description of decoherence associated with phenomenological description of measurement, [11.9] rather than as a consequence of quantum optics.

11.1 Quantum Stochastic Schrödinger Equation

The quantum stochastic Schrödinger equation can be derived as a quantum stochastic differential equation for the quantum state vector representing a system interacting with a noise source. We will formulate this in terms of the idealized model of an atom interacting with a model electromagnetic field as has been used in Chap.5. The emphasis in this chapter is not on the equations of motion for operators as in Chap.5, but rather on equations of motion for the wavefunctions, and the associated evolution operators. Since the methodology needed is rather different from that used in Chap.5, it seems wisest to present the whole subject in a self-contained way.

11.1.1 The Model

The standard model of quantum optics considers a system interacting with a heat bath consisting of many harmonic oscillators representing the electromagnetic field. The Hamiltonian for the combined system is

$$H_{\text{tot}} = H_{\text{sys}} + H_B + H_{\text{Int}} \tag{11.1.1}$$

with

$$H_B = \int_0^\infty d\omega\, \hbar\omega\, b^\dagger(\omega) b(\omega) \tag{11.1.2}$$

the Hamiltonian for the heat bath, and $b(\omega)$ the annihilation operator satisfying the canonical commutation relations

$$[b(\omega), b(\omega')^\dagger] = \delta(\omega - \omega') \,. \tag{11.1.3}$$

We do not explicitly specify H_{sys}, but the interaction part of the Hamiltonian is

$$H_{\text{Int}} = \mathrm{i}\hbar \int_0^\infty d\omega\, \kappa(\omega) \left[c - c^\dagger\right] \left[b^\dagger(\omega) + b(\omega)\right] \,. \tag{11.1.4}$$

For simplicity, we consider here only a single heat bath. Generalization to many reservoirs represents no conceptual difficulty.

Notice that in this section we have started from a more realistic model of the interaction of atom with a light field, rather than the highly idealized version in Sect. 5.3.1, in which the frequency spectrum was written in a form which was unbounded both above and below.

11.1.2 Validity of the Model

The development of quantum noise theory in Sect. 5.3 is based on

i) The rotating wave approximation with a smooth system-bath coupling;

ii) The Markov, or white noise, approximation.

a) Rotating Wave Approximation: We assume that for the bare (that is, interaction free) system the system dipole c evolves as $c(t) = ce^{-i\omega_0 t}$ where ω_0 is the resonance frequency of the system. Thus the system will be dominantly coupled to a band of frequencies centred on ω_0. The validity of the rotating wave approximation requires that the frequency integration in H_{Int} be restricted to a range of frequencies

$$\int_{\omega_0-\vartheta}^{\omega_0+\vartheta} d\omega \ldots \tag{11.1.5}$$

with cutoff $\vartheta \ll \omega_0$. This assumes a separation of time scales—the optical frequency ω_0 is much larger than the cutoff ϑ which again is much larger than the typical frequencies of the system dynamics, and of the frequency scale induced by the system-bath couplings, i.e., the decay rates. Furthermore, we assume a smooth system-bath coupling $\kappa(\omega)$ in the frequency interval (11.1.5), that is we set $\kappa(\omega) \to \sqrt{\gamma/2\pi}$ a constant within this interval.

We transform to the interaction picture with respect to the free dynamics of the system and field. This means we make a unitary transformation

$$U_I(t) = \exp\left(-\frac{i}{\hbar}[H_B + H_{\text{sys}}]t\right) \tag{11.1.6}$$

from the Schrödinger picture, so that the interaction picture operators are

$$O_I(t) = U_I^\dagger(t) O_S U_I(t), \tag{11.1.7}$$

and the interaction picture state vectors are

$$|\Psi_I(t)\rangle = U_I^\dagger(t)|\Psi_S(t)\rangle. \tag{11.1.8}$$

This amounts to the replacements $c \to ce^{-i\omega_0 t}$, $b(\omega) \to b(\omega)e^{-i\omega t}$ in the interaction Hamiltonian (11.1.4), which thus in the interaction picture takes the form

$$H_{\text{Int}}^{(\vartheta)}(t) = i\hbar \int_{\omega_0-\vartheta}^{\omega_0+\vartheta} d\omega \kappa(\omega) \left[b(\omega)e^{-i\omega t} + b^\dagger(\omega)e^{i\omega t}\right] \left[ce^{-i\omega_0 t} - c^\dagger e^{i\omega_0 t}\right]. \tag{11.1.9}$$

The rapidly oscillating terms, containing $e^{\pm i(\omega+\omega_0)t}$, will average to almost zero compared to those containing $e^{\pm i(\omega-\omega_0)t}$, and are therefore neglected—this is the rotating wave approximation. Now using $\kappa(\omega) \to \sqrt{\gamma/2\pi}$, the interaction Hamiltonian in the interaction picture can be written as

$$H_{\text{Int}}^{(\vartheta)}(t) = i\hbar\sqrt{\gamma}\left[b^{(\vartheta)}(t)^\dagger c - b^{(\vartheta)}(t)c^\dagger\right] \tag{11.1.10}$$

with

$$b^{(\vartheta)}(t) = \frac{1}{\sqrt{2\pi}} \int_{\omega_0-\vartheta}^{\omega_0+\vartheta} d\omega\, b(\omega)e^{-i(\omega-\omega_0)t}. \tag{11.1.11}$$

The time evolution operator $U^{(\vartheta)}(t)$ from time 0 to t in the interaction picture obeys the Schrödinger equation

$$\frac{d}{dt}U^{(\vartheta)}(t) = -\frac{i}{\hbar}H_{\text{Int}}^{(\vartheta)}(t)U^{(\vartheta)}(t) \quad (U^{(\vartheta)}(0) = 1). \tag{11.1.12}$$

The time evolution operator from time $t = t_0$ to t is given by

$$U^{(\vartheta)}(t, t_0) = U^{(\vartheta)}(t) U^{(\vartheta)}(t_0)^\dagger \quad (U^{(\vartheta)}(t_0, t_0) = 1) \ . \tag{11.1.13}$$

Note that by going to the interaction picture fast optical frequencies ω_0 have disappeared—this is commonly known as *transformation to a rotating frame.*

b) The Markov Approximation or White Noise Approximation: In the weak coupling limit, where γ is rather small, the time scale implied by (11.1.12) will be relatively slow. Assuming that this time scale is much slower than $1/\vartheta$, we can consider ϑ to be practically infinite. Thus, taking the limit $\vartheta \to \infty$ in (11.1.12),

$$b^{(\vartheta)}(t) \to b(t) \equiv \frac{1}{\sqrt{2\pi}} \int_{-\infty}^{\infty} d\omega \, b(\omega) \mathrm{e}^{-\mathrm{i}(\omega - \omega_0)t} \ . \tag{11.1.14}$$

so that the commutator for the fields $b(t)$ in the time domain acquires a δ-function form,

$$[b(t), b^\dagger(t')] = \delta(t - t') \ . \tag{11.1.15}$$

The operator $b(t)$ is a driving field for the equation of motion (11.1.12) at time t. In the following we should interpret the parameter t to mean *the time at which the initial incoming field will interact with the system*, rather than specifying that $b(t)$ is a time-dependent operator at time t.

In typical quantum optical problems one has one (or more) external laser fields of well defined frequencies, which are in the vicinity of ω_0. These give rise to a time-dependent Hamiltonian in the Schrödinger picture. It is then convenient to go to a reference frame where this explicit time-dependence has been transformed away. This leads to an interaction picture in which there is a residue of H_{sys}, which we will simply call H in what follows, and the resulting equation we will be considering takes the form

$$\frac{d}{dt} U^{(\vartheta)}(t) = \left\{ -\frac{\mathrm{i}}{\hbar} H + \sqrt{\gamma}\, b^{(\vartheta)}(t)^\dagger c - \sqrt{\gamma}\, c^\dagger b^{(\vartheta)}(t) \right\} U^{(\vartheta)}(t) \ . \tag{11.1.16}$$

11.2 QSDE for Time Evolution Operators and State Vectors

11.2.1 QSDE in Stratonovich Form

The definition of Ito and Stratonovich integration allows us to give meaning to a *Quantum Stochastic Schrödinger equation* for the time evolution operator $U(t)$ of the system interacting with Bose fields $b(t)$, $b^\dagger(t)$. In the limit $\vartheta \to \infty$ the Schrödinger equation (11.1.16) can be interpreted as a Stratonovich QSDE

$$(\mathbf{S})\; dU(t) = \left\{ -\frac{\mathrm{i}}{\hbar} H\, dt + \sqrt{\gamma}\, dB^\dagger(t) c - \sqrt{\gamma}\, dB(t) c^\dagger \right\} U(t) \tag{11.2.1}$$

with initial condition $U(0) = 1$.

For simplicity we consider the situation in which the field is in a vacuum state, so that $b(t)|\text{vac}\rangle = 0$, and thus

$$\left\langle b(t)b^\dagger(t')\right\rangle = \langle \text{vac}|b(t)b^\dagger(t')|\text{vac}\rangle = \delta(t-t') \tag{11.2.2}$$

$$\langle b(t)b(t')\rangle = \left\langle b^\dagger(t)b(t')\right\rangle = \left\langle b^\dagger(t)b^\dagger(t')\right\rangle = 0. \tag{11.2.3}$$

We define

$$B(t) \equiv \int_0^t b(s)\,ds, \quad B^\dagger(t) \equiv \int_0^t b^\dagger(s)\,ds. \tag{11.2.4}$$

A quantum stochastic calculus of the Ito type, based on the increments

$$dB(t) = B(t{+}dt) - B(t), \quad dB^\dagger(t) = B^\dagger(t{+}dt) - B^\dagger(t) \tag{11.2.5}$$

and dt has been developed in Sect. 5.3.6. The pair $B(t)$, $B^\dagger(t)$ are the complex non-commutative analogues of classical Wiener processes as in S.M. Chap.3. The corresponding Ito rules are, from (5.3.52)

$$\begin{aligned} \left[dB(t)\right]^2 &= \left[dB^\dagger(t)\right]^2 = 0, \\ dB(t)\,dB^\dagger(t) &= dt, \\ dB^\dagger(t)\,dB(t) &= 0. \end{aligned} \tag{11.2.6}$$

Comparison of (11.2.1) with the QSDE for the system operator as given by (5.3.64) in Sect. 5.3.10 shows that this form of evolution equation is of a different kind, in that all operators appear on the left, which arises from the fact that this is essentially a Schrödinger equation.

Notice that in contrast to Sect. 5.3 we find it convenient to set the initial time $t_0 = 0$ in the following. Comparison of (11.2.4) with (5.3.41) shows that $B(t,t_0) \equiv B(t) - B(t_0)$, $B^\dagger(t,t_0) = B^\dagger(t) - B^\dagger(t_0)$, and $b(t) \equiv b_{\text{in}}(t)$.

11.2.2 Conversion from Stratonovich to Ito Form

It is advantageous to convert the Stratonovich equation (11.2.1) into the Ito form. In this way, we can get a modified equation in which the increments $dB(t)$, $dB^\dagger(t)$ are independent of $U(t)$. We can show that this Ito form of the quantum stochastic Schrödinger equation is

$$\textbf{(I)}\ dU(t) = \left\{\left(-\frac{\mathrm{i}}{\hbar}H - \tfrac{1}{2}\gamma c^\dagger c\right) dt + \sqrt{\gamma}\,dB^\dagger(t)\,c - \sqrt{\gamma}\,dB(t)\,c^\dagger\right\} U(t) \tag{11.2.7}$$

with initial condition $U(0) = 1$.

From this initial condition it follows that the solution $U(t)$ of (11.2.7) depends only on $B(t')$, $B^\dagger(t')$ for $0 \le t' \le t$. Since the increments $dB(t)$, $dB^\dagger(t)$ point to the future, we have

$$[U(t), dB(t)] = [U(t), dB^\dagger(t)] = 0\,. \tag{11.2.8}$$

Proof: The conversion of the Stratonovich equation (11.2.1) to the Ito form (11.2.7) is analogous to the classical case as presented in S.M. We start with the Ito equation

$$(\mathbf{I})\; dU(t) = \Big\{\alpha(t)\, dt + \beta(t)\, dB^\dagger(t) - \beta^\dagger(t)\, dB(t)\Big\}\, U(t) \tag{11.2.9}$$

and consider an arbitrary Stratonovich integral of $U(t)$ which obeys (11.2.9)

$$\begin{aligned}(\mathbf{S}) \int_{t_0}^{t} dB(s)U(s) &= \lim_{n\to\infty} \sum \Delta B_i \frac{U(t_{i+1}) + U(t_i)}{2} \\ &= \lim_{n\to\infty} \sum \Delta B_i[U(t_i) + \tfrac{1}{2}(\alpha_i \Delta t_i + \beta_i \Delta B_i^\dagger - \beta_i^\dagger \Delta B_i)U(t_i)]\end{aligned} \tag{11.2.10}$$

with n subdivisions of the integration interval. We now make the replacements $\Delta B_i\, \Delta B_i^\dagger \to \Delta t_i$ according to the Ito table (11.2.6), so that

$$\begin{aligned}(\mathbf{S}) \int_{t_0}^{t} dB(s)U(s) &= \lim_{n\to\infty} \sum \left\{\Delta B_i U(t_i) + \tfrac{1}{2}\beta_i \Delta t_i U(t_i)\right\} \\ &= (\mathbf{I}) \int_{t_0}^{t} dB(s)\, U(s) + \tfrac{1}{2} \int_{t_0}^{t} ds\, \beta(s)\, U(s),\end{aligned} \tag{11.2.11}$$

with similar expressions for the other integrals. Thus the corresponding Stratonovich equation is

$$(\mathbf{S})\; dU(t) = \left(\Big\{\alpha(t) + \tfrac{1}{2}\beta^\dagger(t)\beta(t)\Big\}\, dt + \beta(t)\, dB^\dagger(t) - \beta^\dagger(t)\, dB^\dagger(t)\right) U(t). \tag{11.2.12}$$

Remembering that (11.2.9) is a shorthand for an integral equation we see that (11.2.9) is equivalent to (11.2.1) when we set

$$\beta(t) \to \sqrt{\gamma}\, c, \quad \beta^\dagger(t) \to \sqrt{\gamma}\, c^\dagger, \quad \alpha(t) \to -\frac{\mathrm{i}}{\hbar} H - \tfrac{1}{2}\gamma c^\dagger c. \tag{11.2.13}$$

11.2.3 Formal Solution

Since the QSDEs (11.2.1) and (11.2.7) are linear in $U(t)$, there are formally exact solutions. The formal solution of the (11.2.7) can be written as

$$U(t) = T \exp \int_0^t \left(-\frac{\mathrm{i}}{\hbar} H\, ds + \sqrt{\gamma}\, dB^\dagger(s) c - \sqrt{\gamma}\, dB(s)\, c^\dagger\right) \tag{11.2.14}$$

where T denotes the time ordered product. We now use Ito rules to show that this is also a solution of an Ito QSDE as follows:

$$\begin{aligned} dU(t) &\equiv U(t{+}dt) - U(t) \\ &= \left[\exp\left(-\frac{\mathrm{i}}{\hbar} H\, dt + \sqrt{\gamma}\, dB^\dagger(t) c - \sqrt{\gamma}\, dB(t)\, c^\dagger\right) - 1\right] U(t) \\ &= \sum_{n=1}^{\infty} \frac{1}{n!} \left(-\frac{\mathrm{i}}{\hbar} H\, dt + \sqrt{\gamma}\, dB^\dagger(t) c - \sqrt{\gamma}\, dB(t)\, c^\dagger\right)^n U(t). \end{aligned} \tag{11.2.15}$$

According to the Ito rules (11.2.6) terms with $n > 2$ vanish, the term with $n = 1$ gives

$$-\frac{i}{\hbar}H\,dt + \sqrt{\gamma}\,dB^{\dagger}(t)c - \sqrt{\gamma}\,dB(t)\,c^{\dagger}$$

and $n = 2$ gives

$$-\tfrac{1}{2}\gamma c^{\dagger}c\,dt. \tag{11.2.16}$$

We thus recover (11.2.7). Obviously, from (11.2.14) $U(t)$ is unitary, and satisfies the initial condition $U(0) = 1$.

11.2.4 QSDE for the State Vector

The state vector $|\Psi(t)\rangle = U(t)|\Psi(0)\rangle$ of the combined system plus heat bath obeys a stochastic Schrödinger equation

$$(\mathbf{I})\; d|\Psi(t)\rangle = \left\{\left(-\frac{i}{\hbar}H - \tfrac{1}{2}\gamma c^{\dagger}c\right)dt + \sqrt{\gamma}\,dB^{\dagger}(t)\,c - \sqrt{\gamma}\,dB(t)\,c^{\dagger}\right\}|\Psi(t)\rangle \tag{11.2.17}$$

with initial condition $|\Psi(0)\rangle = |\psi\rangle \otimes |\text{vac}\rangle$. Since $|\Psi(0)\rangle$ contains the vacuum of the electromagnetic field, $b(\omega)|\Psi(0)\rangle = 0$ and thus $dB(t)|\Psi(0)\rangle = 0$. Because $U(t)$ commutes with $dB(t)$ (see (11.2.8)) it follows that $dB(t)|\Psi(t)\rangle = 0$. This means, as far as $dB(t)$ is concerned, $|\Psi(t)\rangle$ is in the vacuum state and therefore the Ito equation (11.2.17) can be simplified to

$$(\mathbf{I})\; d|\Psi(t)\rangle = \left\{-\frac{i}{\hbar}H_{\text{eff}}\,dt + \sqrt{\gamma}\,dB^{\dagger}(t)\,c\right\}|\Psi(t)\rangle \tag{11.2.18}$$

with an *effective Hamiltonian* for the system

$$H_{\text{eff}} = H - \frac{i}{2}\hbar\gamma c^{\dagger}c. \tag{11.2.19}$$

Interpretation:

i) The effective Hamiltonian H_{eff} is not Hermitian, and the appellation "Hamiltonian" is therefore merely conventional. The full evolution is of course unitary, as demonstrated in Sect. 11.2.3.

ii) This equation has the following physical interpretation: The term $dB^{\dagger}(t)$ involves the incoming radiation field evaluated in the immediate future of t. Thus this field is not affected by the system. However, the system does create a self-field which causes the radiation damping by reacting back on the system. This is the meaning of the term $-\frac{1}{2}\gamma c^{\dagger}c|\Psi(t)\rangle$. The Stratonovich equation does not have this term because the evaluation of $dB^{\dagger}(t)$, half in the future and half in the past, itself generates the radiation reaction.

11.2.5 QSDE for the Stochastic Density Operator

In the particular case that we start with a pure state, the density operator can be written

$$\hat{\rho}(t) = |\Psi(t)\rangle\langle\Psi(t)|, \tag{11.2.20}$$

and when the state vector, $|\Psi(t)\rangle$, satisfies the quantum stochastic Schrödinger equation (11.2.17), the density operator $\hat{\rho}(t)$ will be called the *stochastic density operator*. This represents a pure state at any time, but one which changes stochastically as time progresses. From the Schrödinger equation (11.2.17), we derive an equation of motion for a *stochastic density operator* for the combined system and heat bath as

$$\begin{aligned} d\hat{\rho}(t) &\equiv \hat{\rho}(t+dt) - \hat{\rho}(t) \\ &= U(t+dt,t)\hat{\rho}(t)U^\dagger(t+dt,t) - \hat{\rho}(t) \qquad (11.2.21) \\ &= -\frac{\mathrm{i}}{\hbar}\left(H_{\mathrm{eff}}\hat{\rho}(t) - \hat{\rho}(t)H_{\mathrm{eff}}^\dagger\right)dt + \gamma\, dB^\dagger(t)\, c\, \hat{\rho}(t)\, c^\dagger\, dB(t) \\ &\quad + \sqrt{\gamma}\, dB^\dagger(t)c\, \hat{\rho}(t) + \hat{\rho}(t)\sqrt{\gamma}\, dB(t)\, c^\dagger \qquad (11.2.22) \end{aligned}$$

with time evolution operator from time t_0 to t, as in (11.1.13),

$$U(t,t_0) \equiv U(t)U^\dagger(t_0)\,. \tag{11.2.23}$$

Equation (11.2.21) is derived by expanding $U(t+dt,t)$ to second order in the increments, and using the Ito rules (11.2.6). If we now trace over the bath variables, this will perform an average over the $dB(t)$, $dB^\dagger(t)$ operators. We use the cyclic properties of the trace, and derive the usual master equation (as given in Chap.5) for the reduced system density operator $\rho(t) = \mathrm{Tr}_B\{\hat{\rho}(t)\}$ (compare (5.4.12)),

$$\begin{aligned} \dot{\rho}(t) &= -\frac{\mathrm{i}}{\hbar}\left(H_{\mathrm{eff}}\,\rho - \rho\, H_{\mathrm{eff}}^\dagger\right) + \mathcal{J}\rho \qquad (11.2.24) \\ &\equiv \mathcal{L}\rho. \qquad (11.2.25) \end{aligned}$$

Notice that the Liouville operator $\mathcal{L}$ is composed of two parts; the term involving $H_{\mathrm{eff}}\,\rho - \rho\, H_{\mathrm{eff}}^\dagger$, which is a generalization of the commutator appropriate to the non-Hermitian H_{eff}, and the *recycling operator*,

$$\mathcal{J}\rho = \gamma c\,\rho\, c^\dagger\,. \tag{11.2.26}$$

Generalization to Many Channels: For reference in later sections we need the master equation for N_c output channels. It is given by

$$\begin{aligned} \dot{\rho}(t) &= -\frac{\mathrm{i}}{\hbar}[H,\rho] + \sum_{j=1}^{N_c} \tfrac{1}{2}\gamma_j(2c_j\rho c_j^\dagger - c_j^\dagger c_j\rho - \rho\, c_j^\dagger c_j) \qquad (11.2.27) \\ &\equiv -\frac{\mathrm{i}}{\hbar}(H_{\mathrm{eff}}\rho - \rho H_{\mathrm{eff}}^\dagger) + \sum_{j=1}^{N_c} \mathcal{J}_j\rho \qquad (11.2.28) \\ &\equiv \mathcal{L}\rho \qquad (11.2.29) \end{aligned}$$

The number process arises by noting that the the operator for the total number of photons counted between time 0 and t is according to (8.3.76)

$$\Lambda(t) \equiv \int_0^t ds\, b^\dagger(s) b(s)\ . \tag{11.3.1}$$

This assumes a perfect photodetector with unit efficiency and instantaneous δ-function response. The operator $\Lambda(t)$ leads to the construction of the quantum analogue of a Poisson process.

Physically, $d\Lambda(t)$ is an operator whose eigenvalues are the number of photons counted in the time interval $(t, t+dt]$. It is only possible to make sense of $d\Lambda(t)$ in situations in which the initial state is a vacuum. The reason for this is physically quite understandable as a white noise field would lead to an infinite number of counts in any finite time interval.

a) Ito Rules for Number Processes: As well as the vacuum Ito rules (11.2.6), we will also need Ito rules for $d\Lambda(t)$,

$$d\Lambda(t)\, d\Lambda(t) = d\Lambda(t) \tag{11.3.2}$$
$$dB(t)\, d\Lambda(t) = dB(t) \tag{11.3.3}$$
$$d\Lambda(t)\, dB^\dagger(t) = dB^\dagger(t). \tag{11.3.4}$$

All the other products involving $dB(t)$, $dB^\dagger(t)$, $d\Lambda(t)$ and dt vanish. Notice that the Ito rule (11.3.2) implies that $d\Lambda(t)^n = d\Lambda(t)$ for any positive integer n, and that this means that the corresponding Ito calculus requires us to expand power series to all orders. Doing this we find that, for any function $f(z)$,

$$f\big(a\, d\Lambda(t)\big) = f(0) + \big(f(a) - f(0)\big)\, d\Lambda(t). \tag{11.3.5}$$

b) Outline of the Proof of $d\Lambda(t)^2 = d\Lambda(t)$: We can deduce these Ito rules by noting that for any optical field with no white noise component, the probability of finding more than one photon in the time interval dt will go to zero at least as fast as dt^2. We can use this fact to show that if we discretize the integral

$$|A\rangle = \int [d\Lambda(t)^2 - d\Lambda(t)]|\phi(t)\rangle, \tag{11.3.6}$$

then in the limit of infinitely fine discretization the mean of the norm of $|A\rangle$ goes to zero. This means that in a mean-norm topology, $|A\rangle \rightarrow 0$, and thus we can formally write inside integrals $d\Lambda(t)^2 = d\Lambda(t)$. This equation ensures that the only eigenvalues of $d\Lambda(t)$ are 0 and 1, and that we can only count either one or no photons in a time interval dt.

c) Mnemonic Interpretation: The rules for manipulating $d\Lambda(t)$ can be easily remembered by formally setting

$$d\Lambda(t) \equiv \frac{dB^\dagger(t)\, dB(t)}{dt} \tag{11.3.7}$$

and using the vacuum Ito rules for $dB(t)$, $dB^\dagger(t)$. For example

$$d\Lambda(t)\, d\Lambda(t) = \frac{dB^\dagger(t)\, dB(t)\, dB^\dagger(t)\, dB(t)}{dt^2} \tag{11.3.8}$$

$$= \frac{dB^\dagger(t)\,dt\,dB(t)}{dt^2} \tag{11.3.9}$$

$$= d\Lambda(t). \tag{11.3.10}$$

The other rules follow similarly.

11.3.2 Input and Output

In quantum optics the Bose fields (11.1.14) correspond to the various modes of the electromagnetic field. According to Sect. 5.3 these fields represent the input and output channels through which the system interacts with the outside world. In particular, $b(t) \equiv b_{\rm in}(t)$ is the *input* field which is *the field before the interaction with the system.* In (5.3.15) of Sect. 5.3.2 the Heisenberg equations for system operators were expressed as operator Langevin equations with driving terms $b(t) \equiv b_{\rm in}(t)$. In a similar way *output* operators for the fields after the interaction with the system were defined in (5.3.20) as

$$b_{\rm out}(t) = \frac{1}{\sqrt{2\pi}} \int_{-\infty}^{\infty} d\omega\, b(\omega, t_1) e^{-i(\omega-\omega_0)(t-t_1)} \tag{11.3.11}$$

with $b(\omega, t_1)$ the Heisenberg operator of $b(\omega)$ evaluated at some t_1 in the distant future (Note: (11.3.11) differs slightly from the definition (5.3.20) in Sect. 5.3 since we work here in the interaction picture). According to (5.3.24) we have furthermore

$$b_{\rm out}(t) = b_{\rm in}(t) + \sqrt{\gamma}\, c(t) \tag{11.3.12}$$

which expresses the output field as the sum of the input field and the field radiated by the source.

The fundamental problem of quantum optics consists in calculating the statistics of the "out" field for a system of interest, in particular the normally ordered field correlation functions

$$\left\langle b^\dagger_{\rm out}(t_1) \dots b^\dagger_{\rm out}(t_n) b_{\rm out}(t_{n+1}) \dots b_{\rm out}(t_{n+m}) \right\rangle , \tag{11.3.13}$$

given all correlation functions for the "in" field.

a) Definition of "out" Processes: In the present context we introduce the *output processes* via the definitions

$$B_{\rm out}(t) \equiv U^\dagger(t)\, B(t)\, U(t) \tag{11.3.14}$$

$$\Lambda_{\rm out}(t) \equiv U^\dagger(t) \Lambda(t)\, U(t). \tag{11.3.15}$$

b) Physical Identification: The identification between the formal physical $b_{\rm out}(t)$ and the output process comes from

$$B_{\rm out}(t, t_0) = B_{\rm out}(t) - B_{\rm out}(t_0) = \int_{t_0}^{t} b_{\rm out}(s)\, ds. \tag{11.3.16}$$

Formally, in all formulae which follow in this section, we can make the identifications

$$b_{\rm out}(t)\, dt \equiv dB_{\rm out}(t), \tag{11.3.17}$$

$$b^\dagger_{\rm out}(t) b_{\rm out}(t)\, dt \equiv d\Lambda(t), \tag{11.3.18}$$

where it is understood that the increments are all *Ito* increments, i.e., they point to the future.

c) Alternate Form: We can also write, for any $t_1 \geq t$

$$B_{\text{out}}(t) = U^\dagger(t_1)B(t)U(t_1). \tag{11.3.19}$$

This can be seen from (11.2.7), which shows that the time evolution operator $U(t, s)$ from time s to time t depends only on the fields $dB(s')$, $dB^\dagger(s')$ with s' between s and t. The commutation relations then show that

$$[U(t, s), B(\tau)] = 0 \quad \text{for } \tau \leq s. \tag{11.3.20}$$

The right hand side of (11.3.19) then gives

$$U^\dagger(t)U^\dagger(t_1, t)B(t)U(t_1, t)U(t) = U^\dagger(t)B(t)U(t) \tag{11.3.21}$$

$$= B_{\text{out}}(t) \quad (t_1 \geq t). \tag{11.3.22}$$

d) QSDEs for the Output Processes: These can be derived using quantum stochastic calculus as follows:

$$\begin{aligned} dB_{\text{out}}(t) &\equiv B_{\text{out}}(t+dt) - B_{\text{out}}(t) \\ &= U^\dagger(t+dt)B(t+dt)U(t+dt) - U^\dagger(t)B(t)U(t) \\ &= U^\dagger(t)U^\dagger(t+dt, t)\, dB(t)U(t+dt, t)U(t) \\ &= dB(t) + \sqrt{\gamma}\, c(t)\, dt \end{aligned} \tag{11.3.23}$$

where the second line follows from the definition of the "out" field (11.3.14), the third line is a consequence of (11.3.20), and the last line is obtained from the stochastic Schrödinger equation in Ito form (11.2.7) together with the Ito rules (11.2.6). The result agrees with (11.3.12).

In a similar way one finds

$$d\Lambda_{\text{out}}(t) = d\Lambda(t) + \sqrt{\gamma}\, c(t)\, dB^\dagger(t) + \sqrt{\gamma}\, c^\dagger(t)\, dB(t) + \gamma c^\dagger(t)c(t)\, dt, \tag{11.3.24}$$

which expresses the operator for the count rate for the "out" field as the sum of the "in" operator, an interference term between the source and "in" field, and the direct system term.

11.3.3 Photon Counting as a Measurement of the $\Lambda(t)$ Operator

The Hilbert space in which a QSDE such as (11.2.33) functions has a structure which reflects the Markov nature of the operators $dB^\dagger(t)$, $dB(t)$ and $d\Lambda(t)$, which all commute with all similar operators evaluated at a different time. This means that the photon field can be described—in a discretized approximation—as the direct product of separate Hilbert spaces for each discretized time t_i. For each time t_i, each Hilbert space behaves like a harmonic oscillator, with creation, destruction, and number operators given by

$$a^\dagger = dB^\dagger(t)/\sqrt{dt}\,, \tag{11.3.25}$$

$$a = dB(t)/\sqrt{dt}\,, \tag{11.3.26}$$

$$N = a^\dagger a = d\Lambda(t). \tag{11.3.27}$$

The harmonic oscillator levels are only infinitesimally occupied, since n photons can only appear as a result of the application of $\{dB^\dagger(t)\}^n$ to the vacuum, so that the occupation will be of order of magnitude $dt^{n/2}$. This is reflected in the fact that $d\Lambda(t)d\Lambda(t) = d\Lambda(t)$, so that the only eigenvalues of photon number are 0 and 1. The time t_i can be regarded as the time at which the operator $dB(t_i)$ will interact with the atom, and the QSDE (11.2.18) operates on the wavefunction $|\Psi(t_i)\rangle$ as

$$|\Psi(t_i + dt_i)\rangle = \left\{1 - \frac{\mathrm{i}}{\hbar}H_{\mathrm{eff}}\,dt_i + \sqrt{\gamma}\;dB^\dagger(t_i)c\right\}|\Psi(t_i)\rangle. \tag{11.3.28}$$

The result of one time step is to produce, to first order in dt_i, a state which is a linear combination two states, one with no t_i-photon in it, and one with one t_i-photon in it. The atomic parts of these states are quite different from each other because of the operator c in the last term on the right hand side of (11.3.28)—this state is then what is known as an *entangled state*. As time evolves, this process will be repeated at successive times t_i, so that at the end of a finite time, the atom-field system is in a highly entangled state, representing the superposition of all possible ways in which the atom could have emitted a photon in the given time interval.

a) Probability of Counting No Photons: Let us now consider measuring the operator $d\Lambda(t_i)$ immediately after the time t_i. The only possible results will be 0 and 1, and according to the quantum theory of measurement, there will be two projectors, which project onto the available eigenstates of definite photon number. These projectors will be

$$P(1,t) \equiv d\Lambda(t), \qquad P(0,t) \equiv 1 - d\Lambda(t). \tag{11.3.29}$$

Consider first the situation in which we measure $d\Lambda(t_i)$ at all time intervals in the time range (t_0, t_f), and the measured value of $d\Lambda(t)_i$ is zero. It is clear that the resulting wavefunction is given by

$$|\Psi(t_f)\rangle = \prod_{i=0}^{f}\left\{\big(1 - d\Lambda(t_i)\big)\Big(1 - \frac{\mathrm{i}}{\hbar}H_{\mathrm{eff}}\,dt_i + \sqrt{\gamma}\;dB^\dagger(t_i)c\Big)\right\}|\Psi(t_0)\rangle \tag{11.3.30}$$

$$= \prod_{i=0}^{f}\left\{\big(1 - d\Lambda(t_i)\big)\Big(1 - \frac{\mathrm{i}}{\hbar}H_{\mathrm{eff}}\,dt_i\Big)\right\}|\Psi(t_0)\rangle. \tag{11.3.31}$$

In going from the first line to the second, we have used the property (11.3.4) to remove the $dB^\dagger(t)$ term. In the resulting expression, all of the projectors commute with each other and H_{eff}, so they may be all taken to the right. The product

$$\mathbf{P}(t_0, t_f) = \prod_{i=0}^{f}\big(1 - d\Lambda(t_i)\big) \tag{11.3.32}$$

is the projector onto the subspace in which there were no photons produced during the time interval, and the resulting wavefunction is then

$$|\Psi(t_f)\rangle = U_{\mathrm{eff}}(t_f, t_0)\mathbf{P}(t_0, t_f)|\Psi(t_0)\rangle. \tag{11.3.33}$$

where we define the *non-unitary* evolution operator

$$U_{\text{eff}}(t_f, t_0) = \exp\left(-\frac{\mathrm{i}}{\hbar} H_{\text{eff}}(t_f - t_0)\right), \tag{11.3.34}$$

The probability that no photons are detected in the time interval (t_0, t_f) is the norm of the final state vector,

$$\langle \Psi(t_f)|\Psi(t_f)\rangle = \langle \Psi(t_0)|U^\dagger_{\text{eff}}(t_f, t_0) U_{\text{eff}}(t_f, t_0)|\Psi(t_0)\rangle. \tag{11.3.35}$$

b) Probability of Counting One or More Photons: Suppose we carry out the sequence of measurements, and count one photon in each of the time intervals dt_1 and dt_2, and none at any other time. When we count the first photon, we find the wavefunction just after the measurement is given by

$$|\Psi(t_1 + dt_1)\rangle = \left\{ d\Lambda(t_1)\big(1 - \frac{\mathrm{i}}{\hbar} H_{\text{eff}}\, dt_1 + \sqrt{\gamma}\, dB^\dagger(t_1) c\big) \right\} |\Psi(t_1)\rangle \tag{11.3.36}$$

$$= \sqrt{\gamma}\, dB^\dagger(t_1) c |\Psi(t_1)\rangle. \tag{11.3.37}$$

where we have used the Ito rule that $d\Lambda(t)\, dt = 0$, as well as (11.3.4). Combining this with the evolution from t_2 to $t + dt_2$, when another photon is counted, and the evolutions in the intervals (t_0, t_1), $(t_1 + dt_1, t_2)$, $(t_2 + dt_2, t_f)$, where no photon is counted, we find that the wavefunction at the end of the interval is

$$\begin{aligned} |\Psi(t_f)\rangle = {} & U_{\text{eff}}(t_f, t_2)\mathbf{P}(t_f, t_2 + dt_2)\sqrt{\gamma}\, dB^\dagger(t_2) c \\ & \times U_{\text{eff}}(t_2, t_1)\mathbf{P}(t_2, t_1 + dt_1) \\ & \times \sqrt{\gamma}\, dB^\dagger(t_1) c U_{\text{eff}}(t_1, t_0)\mathbf{P}(t_1, t_0)|\Psi(t_0)\rangle. \end{aligned} \tag{11.3.38}$$

The projectors, $dB^\dagger(t_1)$ and $dB^\dagger(t_2)$ all commute with each other and the system operators, so we can write

$$\begin{aligned} |\Psi(t_f)\rangle = {} & \gamma\, dB^\dagger(t_2)\, dB^\dagger(t_1) U_{\text{eff}}(t_f, t_2) c U_{\text{eff}}(t_2, t_1) c U_{\text{eff}}(t_1, t_0) \\ & \times \mathbf{P}(t_f, t_2 + dt_2)\mathbf{P}(t_2, t_1 + dt_1)\mathbf{P}(t_1, t_0)|\Psi(t_0)\rangle. \end{aligned} \tag{11.3.39}$$

The initial state corresponds to having no photons corresponding to the time interval (t_0, t_f), so the projectors simply give 1 when acting on it. Using the Ito rules we find that the required probability, of detecting one photon at time interval dt_1, one at dt_2 and no others in the time interval (t_0, t_f), is the norm

$$\begin{aligned} \langle \Psi(t_f)|\Psi(t_f)\rangle = {} & dt_1 dt_2\, \langle \Psi(t_0)|U^\dagger_{\text{eff}}(t_1, t_0) c^\dagger U^\dagger_{\text{eff}}(t_2, t_1) c^\dagger U^\dagger_{\text{eff}}(t_f, t_2) \\ & \times U_{\text{eff}}(t_f, t_2) c U_{\text{eff}}(t_2, t_1) c U_{\text{eff}}(t_1, t_0)|\Psi(t_0)\rangle. \end{aligned} \tag{11.3.40}$$

This form can obviously be generalized to arbitrary numbers of detections in the time interval (t_0, t_f), but we shall leave this to Sect. 11.3.6, in which Barchielli's elegant characteristic functional method handles the general case very simply.

c) Relationship to "out" Operators: The formula (11.3.38) can be written in a form which does not explicitly evaluate terms involving $dB^\dagger(t)$ in the evolution operators as

$$|\Psi(t_f)\rangle = \prod_{i=0}^{f} P(C_i, t_i) U(t_{i+1}, t_i)|\Psi(t_0)\rangle. \tag{11.3.41}$$

Where $C_i = 0$, for all i except when $i = 1$ or $i = 2$, when it is 1. Using the same methods as in the previous section, it is not difficult to show that

$$|\Psi(t_f)\rangle = U(t_f, t_0) \prod_{i=1}^{f} P_{\text{out}}(C_i, t_i)|\Psi(t_0)\rangle \tag{11.3.42}$$

The probability is then obtained from the norm, using the fact that the projectors all commute with each other, as

$$\langle\Psi(t_f)|\Psi(t_f)\rangle = \langle\Psi(t_0)| \prod_{i=1}^{f} P_{\text{out}}(C_i, t_i)|\Psi(t_0)\rangle. \tag{11.3.43}$$

We can now use the definitions (11.3.29), and two normally ordering identities, which are of course equally valid for $\Lambda_{\text{out}}(t)$:

$$\text{i)} \quad : \exp(d\Lambda(t)) : \quad = 1 - d\Lambda(t) \tag{11.3.44}$$

$$\text{ii)} \quad : d\Lambda(t)\big(1 - d\Lambda(t)\big) : = d\Lambda(t). \tag{11.3.45}$$

These can both be proved by explicitly writing $d\Lambda(t)$ in terms of $dB^\dagger$ and $dB(t)$ as in (11.3.7), normally ordering the result, and finally using the Ito rules (11.2.6). These rules apply with equal validity to the "out" operators, which leads to the expression

$$\langle\Psi(t_f)|\Psi(t_f)\rangle = \langle\Psi(t_0)| : d\Lambda_{\text{out}}(t_1)d\Lambda_{\text{out}}(t_2)\exp\left(-\int_{t_0}^{t_f} d\Lambda_{\text{out}}(t)\right) : |\Psi(t_0)\rangle. \tag{11.3.46}$$

This is now equivalent to the case of 100% efficiency two-photon case of the photon counting formula (8.3.75).

11.3.4 Photon Counting and Exclusive Probability Densities

The starting point of our discussion is photon counting of the output field as realized by a photoelectron counter, as in Fig. 11.1.

We consider a system which is coupled to $j = 1, \ldots, N_c$ output channels with counters which act continuously to register arrival of photons. A complete description of counting process is given by the family of *exclusive* (or *elementary*) *probability densities* (EPDs) (compare Sect. 8.3.3):

i) $P^t_{t_0}(0|\rho)$ = probability of having no count in the time interval $(t_0, t]$ when the system is prepared in the state ρ at time t_0.

ii) $p^t_{t_0}(j_m, t_m; \ldots; j_1, t_1|\rho)$ = multitime probability density of having a count in detector j_1 at time t_1, a count in detector j_2 at time t_2 *etc.*, and no other counts in the rest of the time interval $(t_0, t]$ (with $j_k = 1, \ldots, N_c$; $t_0 < t_1 < \ldots < t_m \le t$).

Knowledge of EPDs allows the reconstruction of the whole counting statistics—this is shown in Sect. 8.3.3, which uses a different notation from that of this chapter.

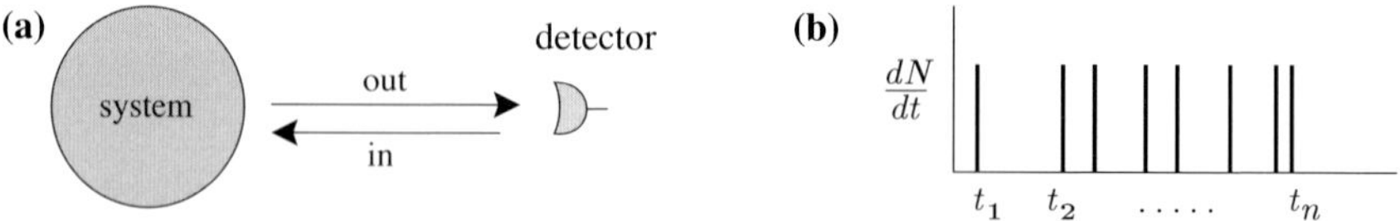

Fig. 11.1 **(a)** Photon counting arrangement; **(b)** Typical trajectory of photocounts

11.3.5 Mandel's Counting Formula

Physically, the operator for the output current of a photodetector is proportional to the integral of photon flux over the surface of the detector. To simplify notation we restrict ourselves in the following to the case of a single output channel $N_c = 1$. Furthermore, since our goal is the derivation of a stochastic Schrödinger equation, we will concentrate on the case of ideal photodetection with unit efficiency. The operator associated with the photon count rate is according to (11.3.1)

$$d\Lambda_{\text{out}}(t) \equiv b^{\dagger}_{\text{out}}(t) b_{\text{out}}(t)\, dt \equiv \hat{I}_c(t)\; dt. \tag{11.3.47}$$

Here $\Lambda_{\text{out}}(t)$ is the operator corresponding to the number of photons up to time t in the "out" field. We note that the operators $\Lambda_{\text{out}}(t)$ are a family of commuting selfadjoint operators. Photoelectric detection realizes a measurement of the compatible observables $\Lambda_{\text{out}}(t)$.

In quantum optics the usual starting point is the photon counting formula given in (8.3.75) of Sect. 8.3.5:

$$P_m(t, t_0) = \left\langle : \frac{1}{m!} \left(\int_{t_0}^{t} ds\, \hat{I}_c(s) \right)^m \exp\left(- \int_{t_0}^{t} ds\, \hat{I}_c(s) \right) : \right\rangle . \tag{11.3.48}$$

This equation gives the probability for m photocounts in the time interval (t_0, t) where $: \ldots :$ indicates normal ordering, and $\langle \ldots \rangle = \text{Tr}_{\text{sys}} \text{Tr}_B \{ \ldots \hat{\rho} \}$ with $\hat{\rho} = \rho \otimes |\text{vac}\rangle\langle\text{vac}|$ is the quantum expectation value with respect to the initial state of the system. Equation (11.3.48) is valid for a detector with unit efficiency. In particular, the probability that there is no count in a time interval $(t_0 = 0, t]$ is

$$P^t_{t_0=0}(0|\rho) = \left\langle : \exp\left(- \int_0^t ds\, \hat{I}_c(s) \right) : \right\rangle , \tag{11.3.49}$$

and the multitime probability density for detecting exactly m photons at times $t_1, \ldots, t_m$ in the time interval $(t_0 = 0, t]$ is

$$p^t_{t_0=0}(t_m, \ldots, t_1|\rho) = \left\langle : \hat{I}_c(t_1) \ldots \hat{I}_c(t_m) \exp\left(- \int_0^t ds \hat{I}_c(s) \right) : \right\rangle . \tag{11.3.50}$$

This establishes the relation to the EPDs introduced above.

On the other hand, the *non-exclusive multitime probability density* (or *coincidence probability density* in the terminology of Chap.8) for detecting a first photon at time t_1, a second (not necessarily the next) photon at time t_2 *etc.*, and the m-th photon at t_m is given by the intensity correlation function (see Sect. 8.3.3)

$$\left\langle : \hat{I}_c(t_1) \dots \hat{I}_c(t_m) : \right\rangle . \tag{11.3.51}$$

11.3.6 The Characteristic Functional and System Averages

As a next step we wish to express the output field correlation functions in terms of system averages. In Sect. 5.4.6, equation (5.4.20), we expressed the output field in (11.3.49) and (11.3.50) in terms of the input field plus the source contribution, $b_{\text{out}}(t) = b_{\text{in}}(t) + \sqrt{\gamma}\, c(t)$, and applied (5.2.20) of Sect. 5.2, to evaluate the normally- and time ordered multitime correlation function for the source operator $c(t)$. Here we will follow a different procedure, devised by Barchielli [11.5] who defined a characteristic functional from which he could evaluate the EPDs.

a) Characteristic Functional: The characteristic functional $\Phi_t[k]$ is defined as the expectation value

$$\Phi_t[k] = \left\langle \hat{V}_t^{\text{out}}[k] \right\rangle \tag{11.3.52}$$

of the *characteristic operator* $\hat{V}_t^{\text{out}}[k]$, which is defined by the equation

$$\hat{V}_t^{\text{out}}[k] \equiv \exp\left(\mathrm{i} \int_0^t k(s)\, \hat{I}_c(s)\, ds \right) , \tag{11.3.53}$$

where $k(s)$ is an arbitrary function of s. We now use the normal ordering relation for bosonic fields, which is essentially a many variable version of (8.3.94),

$$\exp\left(\mathrm{i} \int_0^t k(s)\, \hat{I}_c(s)\, ds \right) =: \exp\left(\int_0^t (\mathrm{e}^{\mathrm{i}k(s)} - 1)\, \hat{I}_c(s)\, ds \right) : , \tag{11.3.54}$$

expand the exponential on the right hand side, and use expressions (11.3.49) and (11.3.50). We obtain

$$\Phi_t[k] = \\ P_0^t(0|\rho) + \sum_{m=1}^{\infty} \int_0^t dt_m \int_0^{t_m} dt_{m-1} \dots \int_0^{t_2} dt_1 \exp\left(\mathrm{i} \sum_{r=1}^{m} k(t_r) \right) p_0^t(t_m, \dots, t_1|\rho) . \tag{11.3.55}$$

This equation relates the characteristic functional to the exclusive probability densities. Thus $\Phi_t[k]$ uniquely determines the whole counting process up to time t. Functional differentiation of $\Phi_t[k]$ with respect to the test functions $k(t)$ gives the moments, correlation functions and counting distributions.

b) QSDE for the Output Characteristic Operator: According to (11.3.15) the output and input processes are related by $\Lambda_{\text{out}}(t) = U^\dagger(t)\Lambda(t)U(t)$, and the characteristic operator satisfies the QSDE

$$
\begin{aligned}
d\hat{V}_t^{\text{out}}[k] &= \hat{V}_{t+dt}^{\text{out}}[k] - \hat{V}_t^{\text{out}}[k] \\
&= \hat{V}_t^{\text{out}}[k]\left(\mathrm{e}^{\mathrm{i}k(t)d\Lambda_{\text{out}}(t)} - 1\right) \\
&= \hat{V}_t^{\text{out}}[k]\left(\mathrm{e}^{\mathrm{i}k(t)} - 1\right) d\Lambda_{\text{out}}(t) \quad (\hat{V}_0^{\text{out}}[k] = 1),
\end{aligned}
\tag{11.3.56}
$$

where the transition from the second to the third line involves the use of the formula $d\Lambda_{\text{out}}(t)^n = d\Lambda_{\text{out}}(t)$, which follows from the Ito rules (11.3.2). This equation, together with the equation of motion (11.3.24) for $\Lambda_{\text{out}}(t)$, and the QSDE (11.2.41) for the system operators, form the full set of equations which when integrated specify the process completely.

c) Reduced Characteristic Operator: We can express the characteristic operator for the "out" field in terms of the "in" field:

$$
\begin{aligned}
\hat{V}_t^{\text{out}}[k] &= U^\dagger(t)\exp\left(\mathrm{i}\int_0^t k(s)\, d\Lambda(s)\right) U(t) \\
&\equiv U^\dagger(t)\hat{V}_t[k]U(t),
\end{aligned}
\tag{11.3.57}
$$

where $\hat{V}_t[k]$ is the characteristic operator for the "in" field. This allows us to write the characteristic functional of the output field in terms of a *reduced characteristic operator*, $\chi_t[k]$ according to

$$
\begin{aligned}
\Phi_t[k] &= \mathrm{Tr_{sys}Tr}_B\left\{\hat{V}_t^{\text{out}}[k]\hat{\rho}(0)\right\} \\
&= \mathrm{Tr_{sys}Tr}_B\left\{\hat{V}_t[k]\,U(t)\hat{\rho}(0)U(t)^\dagger\right\} \\
&= \mathrm{Tr_{sys}}\left\{\chi_t[k]\right\}
\end{aligned}
\tag{11.3.58}
$$

with

$$
\chi_t[k] \equiv \mathrm{Tr}_B\{\hat{V}_t[k]\,\hat{\rho}(t)\}. \tag{11.3.59}
$$

Equation (11.3.58) expresses the characteristic functional as the system trace over the reduced characteristic operator $\chi_t[k]$.

Using quantum stochastic calculus we can derive an equation for $\hat{V}_t[k]\,\hat{\rho}(t)$. According to the Ito product rule (5.3.51) we have

$$
d[\hat{V}_t[k]\,\hat{\rho}(t)] = d\hat{V}_t[k]\,\hat{\rho}(t) + \hat{V}_t[k]\,d\hat{\rho}(t) + d\hat{V}_t[k]\,d\hat{\rho}(t) \tag{11.3.60}
$$

where the stochastic density operator $\hat{\rho}(t)$ obeys equation (11.2.21), and $\hat{V}_t[k]$ obeys an equation similar to (11.3.56):

$$
d\hat{V}_t[k] = \hat{V}_t[k]\left(\mathrm{e}^{\mathrm{i}k(t)} - 1\right) d\Lambda(t) \quad (\hat{V}_0[k] = 1). \tag{11.3.61}
$$

Taking the trace over the bath, $\mathrm{Tr}_B\{\ldots\}$, as in the derivation of the reduced density matrix (11.2.24), we obtain the following equation for the reduced characteristic operator

$$
\frac{d}{dt}\chi_t[k] = \left(\mathcal{L} + (\mathrm{e}^{\mathrm{i}k(t)} - 1)\mathcal{J}\right)\chi_t[k] \quad (\chi_0[k] = \rho)\,. \tag{11.3.62}
$$

Here $\mathcal{L}$ is the Liouville operator as introduced in the master equation (11.2.24), and $\mathcal{J}$ the recycling operator defined in (11.2.26). Eq. (11.3.62) is one of the key results in this section. It allows us to calculate the complete count statistics of the output field from a "master equation-like" equation for $\chi_t[k]$.

Note that for $k(t) = 0$ this characteristic density operator coincides with the density operator,

$$\chi_t[k = 0] = \rho(t), \tag{11.3.63}$$

and (11.3.62) reduces to the master equation (11.2.24).

d) Relation to Exclusive Probability Densities: We solve (11.3.62) by iteration

$$\chi_t[k] = \mathcal{S}(t, 0)\rho + \sum_{m=1}^{\infty} \int_0^t dt_m \int_0^{t_m} dt_{m-1} \ldots \int_0^{t_2} dt_1 \exp\left(\mathrm{i}\sum_{r=1}^{m} k(t_r)\right)$$
$$\times \mathcal{S}(t, t_m)\mathcal{J}(t_m)\mathcal{S}(t_m, t_{m-1}) \ldots \mathcal{S}(t_2, t_1)\mathcal{J}(t_1)\mathcal{S}(t_1, 0)\rho \tag{11.3.64}$$

with propagators

$$\mathcal{S}(t, t_0)\rho = U_{\mathrm{eff}}(t, t_0)\rho U^{\dagger}_{\mathrm{eff}}(t, t_0) \quad (t \geq t_0), \tag{11.3.65}$$
$$\dot{U}_{\mathrm{eff}}(t, t_0) = -\frac{\mathrm{i}}{\hbar} H_{\mathrm{eff}}(t) U_{\mathrm{eff}}(t, t_0) \quad (U_{\mathrm{eff}}(t_0, t_0) = 1)\,. \tag{11.3.66}$$

Taking a trace over the system degrees of freedom, and comparing with (11.3.55) we can express the EPDs in terms of system averages

$$P_0^t(0|\rho) = \mathrm{Tr}_{\mathrm{sys}}\{\mathcal{S}(t, 0)\rho\}, \tag{11.3.67}$$
$$p_0^t(t_1, t_2, \ldots, t_m|\rho) = \mathrm{Tr}_{\mathrm{sys}}\{\mathcal{S}(t, t_m)\mathcal{J}(t_m)\mathcal{S}(t_m, t_{m-1}) \ldots$$
$$\ldots \mathcal{S}(t_2, t_1)\mathcal{J}(t_1)\mathcal{S}(t_1, 0)\rho\}\,. \tag{11.3.68}$$

The structure of these expressions agrees with what is expected from continuous measurement theory of photon counting as discussed in Sect. 8.3.

e) Unravelling of the System Density Operator: Finally, we note that according to (11.3.64) the density matrix can be *unravelled* into a sum of contributions corresponding to $n = 0, 1, 2, \ldots$ photodetections according to

$$\rho(t) = \mathcal{S}(t, 0)\rho + \sum_{n=1}^{\infty} \int_0^t dt_n \int_0^{t_n} dt_{n-1} \ldots \int_0^{t_2} dt_1$$
$$\mathcal{S}(t, t_n)\mathcal{J}(t_n)\mathcal{S}(t_n, t_{n-1}) \ldots \mathcal{S}(t_2, t_1)\mathcal{J}(t_1)\mathcal{S}(t_1, 0)\rho\,. \tag{11.3.69}$$

f) Generalization to Many Channels: As reference for the following sections, we give the generalization of these equations to N_c channels:

$$P^t_{t_0=0}(0|\rho) = \mathrm{Tr}_{\mathrm{sys}}\left\{\mathcal{S}(t, 0)\rho\right\}\,, \tag{11.3.70}$$
$$p^t_{t_0=0}(j_1, t_1; \ldots; j_m, t_m|\rho) = \mathrm{Tr}_{\mathrm{sys}}\{\mathcal{S}(t, t_m)\mathcal{J}_{j_m}(t_m)\mathcal{S}(t_m, t_{m-1}) \ldots$$
$$\ldots \mathcal{S}(t_2, t_1)\mathcal{J}_{j_1}(t_1)\mathcal{S}(t_1, 0)\rho\} \tag{11.3.71}$$

where $t_0 = 0 < t_1 < \ldots t_m \leq t$ and $j_k = 1, \ldots, N_c$ is the channel index, and $k = 1, \ldots, m$.

g) Examples of the Use of the Characteristic Functional: We conclude our discussion with a few examples for the application of the characteristic functional: The mean intensity of the outgoing field is

$$\left\langle \hat{I}_c(t) \right\rangle = -\mathrm{i}\frac{\delta}{\delta k(t)} \Phi_T[k]\big|_{k=0} = \mathrm{Tr}_{\mathrm{sys}}\{\mathcal{J}(t)\rho(t)\}. \tag{11.3.72}$$

The intensity correlation function can be written as the sum of a normally ordered contribution plus a shot noise term,

$$\begin{aligned} \left\langle \hat{I}_c(t_1)\hat{I}_c(t_2) \right\rangle &= \left\langle : \hat{I}_c(t_1)\hat{I}_c(t_2) : \right\rangle + \delta(t_1 - t_2) \left\langle \hat{I}_c(t_1) \right\rangle \\ &= (-\mathrm{i})^2 \frac{\delta^2}{\delta k(t_1)\, \delta k(t_2)} \Phi_T[k]\big|_{k=0} \end{aligned} \tag{11.3.73}$$

$$= \mathrm{Tr}_{\mathrm{sys}}\{\mathcal{J}(t_1)\mathcal{T}(t_1, t_2)\mathcal{J}(t_2)\rho(t_2)\} + \delta(t_1 - t_2)\mathrm{Tr}_{\mathrm{sys}}\{\mathcal{J}(t_1)\rho(t_1)\}. \tag{11.3.74}$$

with $\mathcal{T}(t, t_0) = \exp[\mathcal{L}(t - t_0)]$ the time evolution operator for the density matrix according to the master equation. Notice that the transition from the second to the last two lines requires that $t_1 \geq t_2$.

Similarly, the normally ordered and time ordered m-th order intensity correlation function is

$$\left\langle : \hat{I}_c(t_1) \ldots \hat{I}_{\mathrm{out}}(t_m) : \right\rangle = \mathrm{Tr}_{\mathrm{sys}}\{\mathcal{J}(t_m)\mathcal{T}(t_m, t_{m-1}) \ldots \mathcal{T}(t_2, t_1)\mathcal{J}(t_1)\rho(t_1)\}. \tag{11.3.75}$$

This result is, of course, consistent with (5.4.21) when the time ordered system correlation function is evaluated with the help of the quantum regression theorem (5.2.20).

11.3.7 Conditional Dynamics and *a Posteriori* States

If during a continuous measurement a certain trajectory of a measured observable is registered, the state of the system conditional on this information is called a *conditional state* or *a posteriori state* denoted by $\rho_c(t)$, and the corresponding dynamics is referred to as *conditional dynamics*.

a) *A Posteriori* State: In the present case of photodetection (counting processes) we assume that the counters have registered the sequence $j_1, t_1; \ldots; j_n, t_n$ $(t_1 < \ldots < t_n)$ in the time interval $(t_0 = 0, t]$. To derive an expression for $\rho_c(t)$ we consider, following Barchielli and Belavkin [11.5],

$$P(0, (t, t + \bar{t}\,]|\rho) = \frac{p_0^{t+\bar{t}}(j_1, t_1; \ldots; j_n, t_n|\rho)}{p_0^t(j_1, t_1; \ldots; j_n, t_n|\rho)} \tag{11.3.76}$$

which is the conditional probability of observing no count in $(t, t+\bar{t}]$ given ρ at time $t_0 = 0$ and having observed the sequence $j_1, t_1; \ldots; j_n, t_n$. Obviously, one can write

$$P(0,(t,t+\bar{t}]|\rho) = P_t^{t+\bar{t}}(0|\rho_c(t)) \tag{11.3.77}$$

with

$$\rho_c(t) = \frac{\mathcal{S}(t,t_n)\mathcal{J}_{j_n}(t_n)\mathcal{S}(t_n,t_{n-1})\ldots\mathcal{S}(t_2,t_1)\mathcal{J}_{j_1}(t_1)\mathcal{S}(t_1,0)\rho}{\mathrm{Tr}_{\mathrm{sys}}\{\ldots\}} \tag{11.3.78}$$

where $\mathrm{Tr}_{\mathrm{sys}}\{\ldots\}$ is a normalization factor to ensure $\mathrm{Tr}_{\mathrm{sys}}\rho_c(t) = 1$. Similar arguments can be given for other EPDs. Thus we identify (11.3.78) with the *a posteriori state.*

b) Interpretation: The interpretation of (11.3.78) is as follows: When no count is registered the system evolution is given by $\mathcal{S}(t,t_0)$ and the state of the system between two counts is

$$\rho_c(t) = \mathcal{S}(t,t_r)\rho_c(t_r)/\mathrm{Tr}_{\mathrm{sys}}\{\mathcal{S}(t,t_r)\rho_c(t_r)\} \quad (t \geq t_r) \tag{11.3.79}$$

where t_r is the time of the last count, and $\rho_c(t_r)$ is the state of the system just after this count.

When a count j at time $t = t_r$ is registered the action on the system is given by the operator $\mathcal{J}_j$, and the state of the system immediately after is

$$\rho_c(t_r+dt) = \mathcal{J}_j(t_r)\rho_c(t_r)/\mathrm{Tr}_{\mathrm{sys}}\{\mathcal{J}_j(t_r)\rho_c(t_r)\}\ . \tag{11.3.80}$$

We will call this a *quantum jump*.

c) An Initial Pure State: When the system is initially in a pure state described by the state vector ψ, $\rho = |\psi\rangle\langle\psi|$, it will remain in a pure state $\psi_c(t)$, $\rho_c(t) = |\psi_c(t)\rangle\langle\psi_c(t)|$, for all later times. The time evolution between the counts is

$$\tilde{\psi}_c(t) = U_{\mathrm{eff}}(t,t_r)\tilde{\psi}_c(t_r), \quad \psi_c(t) = \tilde{\psi}_c(t)/||\tilde{\psi}_c(t)|| \tag{11.3.81}$$

i.e. $\tilde{\psi}_c(t)$ obeys the Schrödinger equation

$$\frac{d}{dt}\tilde{\psi}_c(t) = -\frac{\mathrm{i}}{\hbar}H_{\mathrm{eff}}\tilde{\psi}_c(t) \tag{11.3.82}$$

with time evolution governed by the non-Hermitian effective Hamiltonian H_{eff} as given by (11.2.30). Here $\tilde{\psi}_c(t)$ and $\psi_c(t)$ denote the unnormalized and normalized wave functions, respectively. A count j, t_r is associated with the quantum jump

$$\tilde{\psi}_c(t_r+dt) = \lambda_j c_j \tilde{\psi}_c(t_r)\ . \tag{11.3.83}$$

The complex numbers $\lambda_j \neq 0$ are arbitrary and can be chosen, for example, to renormalize the wave function $\tilde{\psi}_c(t)$ after the jump.

d) Jump Probabilities: To prepare for the derivation of a stochastic Schrödinger equation (SSE) for a system wavefunction ("quantum trajectories") in Sect. 11.3.9

we note: The mean number of counts of type j in the time interval $(t, t+dt]$ conditional upon the observed trajectory $j_1, t_1; \ldots; j_n, t_n$ is

$$p_t^{t+dt}(j,t|\rho_c(t))dt = \begin{cases} \mathrm{Tr}_{\mathrm{sys}}\{\mathcal{J}_j(t)\rho_c(t)\}\,dt \\ \gamma_j \|c_j\psi_c(t)\|^2\,dt \end{cases} \tag{11.3.84}$$

and the joint probability that no jump occurred in the time interval $(0, t]$, and a jump of type j occurred in $(t, t+dt]$ given ρ at time $t = 0$ is

$$p_t^{t+dt}(j,t|\ \rho_c(t)\ ) \times P_0^t(0|\rho) = \begin{cases} \mathrm{Tr}_{\mathrm{sys}}\{\mathcal{J}_j(t)\mathcal{S}(t,0)\rho\}\,dt \\ \gamma_j \|c_j\, U_{\mathrm{eff}}(t,0)\psi\|^2\,dt \end{cases} \tag{11.3.85}$$

with

$$\rho_c(t) = \mathcal{S}(t,0)\rho/\mathrm{Tr}_{\mathrm{sys}}\{\mathcal{S}(t,0)\rho\}\,. \tag{11.3.86}$$

11.3.8 Stochastic Schrödinger Equation for Counting Processes

a) Stochastic Schrödinger Equation: We can combine the time evolution for the unnormalized conditional wave function $\tilde{\psi}_c(t)$ according to (11.3.81,82) into a single (c-number) Stochastic Schrödinger Equation (SSE). A typical trajectory $N_j(t)$, where $N_j(t)$ is the number of counts up to time t, $j = 1, \ldots, N_c$ is a sequence of step functions, such that $N_j(t)$ increases by one if there is a count of type j and $N_j(t)$ is constant otherwise. Therefore, the Ito differential

$$dN_j(t) \equiv N_j(t+dt) - N_j(t) = \begin{cases} 1 & \text{if count } j \text{ is in } (t, t+dt] \\ 0 & \text{else} \end{cases} \tag{11.3.87}$$

fulfils $[dN_j(t)]^2 = dN_j(t)$. Moreover, the probability of more than one count in an interval dt vanishes faster than dt. We thus have the Ito table

$$dN_j(t)\,dN_k(t) = \delta_{jk} dN_j(t), \tag{11.3.88}$$

$$dN_j(t)\,dt = 0 \tag{11.3.89}$$

and obtain a *Stochastic Schrödinger Equation for the unnormalized wave function* $\tilde{\psi}_c(t)$,

$$\begin{aligned} d\tilde{\psi}_c(t) &\equiv \tilde{\psi}_c(t+dt) - \tilde{\psi}_c(t) \\ &= \left\{ -\frac{\mathrm{i}}{\hbar} H_{\mathrm{eff}}\,dt + \sum_{j=1}^{N_c} \left(\lambda_j c_j - 1\right) dN_j(t) \right\} \tilde{\psi}_c(t), \end{aligned} \tag{11.3.90}$$

with $\lambda_j \neq 0$.

According to (11.3.84) the mean number of counts of type j in the interval $(t, t+dt]$ conditional on the trajectory $j_1, t_1; \ldots; j_n, t_n$ (denoted by the subscript c) is

$$\langle dN_j(t)\rangle_c = \gamma_j \|c_j\psi_c(t)\|^2\,dt\,. \tag{11.3.91}$$

The SSE (11.3.90) can be converted to an equation for the *normalized* stochastic system wave function

$$\psi_c(t) = \tilde{\psi}_c(t)/||\tilde{\psi}_c(t)||. \tag{11.3.92}$$

$$\begin{aligned} d\psi_c(t) &\equiv \psi_c(t+dt) - \psi_c(t) \\ &= \Big\{ -\frac{\mathrm{i}}{\hbar} H\, dt - \tfrac{1}{2} \sum_{j=1}^{N_c} \gamma_j \left(c_j^\dagger c_j - \langle c_j^\dagger c_j \rangle_c \right) dt \\ &\quad + \sum_{j=1}^{N_c} \left(c_j / \sqrt{\langle c_j^\dagger c_j \rangle_c} - 1 \right) dN_j(t) \Big\} \psi_c(t). \end{aligned} \tag{11.3.93}$$

where

$$\langle \ldots \rangle_c = \langle \psi_c(t) | \ldots | \psi_c(t) \rangle\ . \tag{11.3.94}$$

See c) below for the derivation; equation (11.3.93) is a *nonlinear* Schrödinger equation.

b) Non-uniqueness: Notice that (11.3.93) is non-unique up to a phase factor: $\psi_c(t) \to \mathrm{e}^{\mathrm{i}\phi(t)} \psi_c(t)$ where $\phi(t)$ is a solution of a stochastic differential equation $d\phi(t) = a(t)dt + \sum b_j(t) dN_j(t)$ with arbitrary $a(t)$ and $b_j(t)$. This can lead to different forms of the same equation.

c) Derivation of the Normalized Form of the SSE: As an exercise in Ito calculus we derive (11.3.93) for the normalized wavefunction ψ_c from (11.3.90) for the unnormalized $\tilde{\psi}_c$. To simplify notation consider the case of a single channel $N_c = 1$. We show: If $\tilde{\psi}_c(t)$ obeys

$$d\tilde{\psi}_c(t) = [A\, dt + (B - 1)\, dN(t)]\tilde{\psi}_c(t) \tag{11.3.95}$$

with A and B operators, then the normalized $\psi_c(t)$ obeys

$$d\psi_c(t) = [(A - \tfrac{1}{2} \langle A + A^\dagger \rangle_c) dt + (B/\sqrt{\langle B^\dagger B \rangle_c} - 1) dN(t)]\psi_c(t) \tag{11.3.96}$$

with $\langle \ldots \rangle_c$ defined in (11.3.94). In the present context we can read off A and B from (11.3.90). Substituting these expressions into (11.3.96) gives (11.3.93).

The derivation proceeds as follows: From

$$\tilde{\psi}_c(t+dt) = [1 + A\, dt + (B - 1) dN(t)]\tilde{\psi}_c(t) \tag{11.3.97}$$

the change of the norm in the time step dt is

$$||\tilde{\psi}_c(t+dt)||^2 = ||\tilde{\psi}_c(t)||^2 \left[1 + \langle A + A^\dagger \rangle_c dt + \langle B^\dagger B - 1 \rangle_c dN(t) \right] \tag{11.3.98}$$

and

$$\begin{aligned} ||\tilde{\psi}_c(t+dt)||^{-1} &= ||\tilde{\psi}_c(t)||^{-1} \left[1 + \langle A + A^\dagger \rangle_c\, dt + \langle B^\dagger B - 1 \rangle_c dN(t) \right]^{-1/2} \\ &= ||\tilde{\psi}_c(t)||^{-1} \left[1 - \tfrac{1}{2} \langle A + A^\dagger \rangle_c dt + \left(1/\sqrt{\langle B^\dagger B \rangle_c} - 1 \right) dN(t) \right]. \end{aligned}$$

current $\hat{I}_{\rm hom}(t)$—the quantity which is experimentally measured—by subtracting the local oscillator contribution:

$$\hat{I}_{\rm hom}(t) = \lim_{\mathcal{E}\to\infty} \frac{d\Lambda^h_{\rm out}(t) - \mathcal{E}^2\, dt}{\mathcal{E}\, dt} . \tag{11.4.3}$$

b) Quadrature Operators: We define quadrature operators for the input and output field,

$$d\hat{X}(t) \equiv dB(t) + dB^\dagger(t), \tag{11.4.4}$$

$$d\hat{Y}(t) \equiv -\mathrm{i}dB(t) + \mathrm{i}dB^\dagger(t), \tag{11.4.5}$$

and for the system dipole

$$x \equiv c + c^\dagger, \quad y \equiv -\mathrm{i}c + \mathrm{i}c^\dagger. \tag{11.4.6}$$

This allows us to write

$$d\hat{X}_{\rm out}(t) = \hat{I}_{\rm hom}(t)\, dt \tag{11.4.7}$$

$$= \sqrt{\gamma}\, x(t)\, dt + d\hat{X}(t), \tag{11.4.8}$$

where the second line is the QSDE for the X-quadrature of the "out" field, which follows from (11.3.23). *This equation shows that homodyne detection with $\mathcal{E}$ real corresponds to a measurement of the compatible observables*

$$b_{\rm out}(t) + b^\dagger_{\rm out}(t) = d\hat{X}_{\rm out}(t)/dt \quad (t \geq 0), \tag{11.4.9}$$

the quadrature components of the field.

In homodyne measurement the experimentally accessible quantities are the multitime correlation functions of the homodyne current

$$\left\langle \hat{I}_{\rm hom}(t_1)\hat{I}_{\rm hom}(t_2)\hat{I}_{\rm hom}(t_3)\ldots \right\rangle \tag{11.4.10}$$

which are equivalent to the multitime correlation functions of $d\hat{X}(t)$.

11.4.2 The Characteristic Functional and System Averages

Our first goal is to express the output field correlation functions of the quadrature components in terms of system averages. The procedure parallels our derivations in Sect. 11.3.6 for counting processes.

a) Characteristic Functional: The complete statistics of the homodyne current, including all the correlation functions (11.4.10), is obtained from a *characteristic functional* corresponding to a Gaussian diffusive measurement:

$$\Phi_t[k] = \left\langle \hat{V}^{\rm out}_t[k] \right\rangle \tag{11.4.11}$$

with

$$\hat{V}^{\rm out}_t[k] \equiv \exp\left(\mathrm{i} \int_0^t k(s)\, d\hat{X}_{\rm out}(t) \right) . \tag{11.4.12}$$

If one is interested in the normally ordered correlation functions, these can be obtained using the connection between (11.4.12) and the normally ordered characteristic operator

$$\exp\left(\mathrm{i}\int_0^t k(s)\,d\hat{X}_{\text{out}}(t)\right) = :\exp\left(\mathrm{i}\int_0^t k(s)\,d\hat{X}_{\text{out}}(t) - \tfrac{1}{2}\int_0^t k(s)^2\,ds\right): \tag{11.4.13}$$

b) QSDE for the Output Characteristic Operator: Using Ito rules and (11.4.4), the characteristic operator obeys the QSDEs,

$$d\hat{V}_t^{\text{out}}[k] = \hat{V}_t^{\text{out}}[k]\,\left(\mathrm{i}k(t)\,d\hat{X}_{\text{out}}(t) - \tfrac{1}{2}k(t)^2\,dt\right) \tag{11.4.14}$$

$$d\hat{X}_{\text{out}}(t) = \sqrt{\gamma}\,x(t)\,dt + d\hat{X}(t) \tag{11.4.15}$$

together with the initial condition $\hat{V}_0^{\text{out}}[k] = 1$. The second line reproduces (11.4.8) for $d\hat{X}_{\text{out}}(t)$. These equations are the analogues of (11.3.24,56) for counting processes.

c) Reduced Characteristic Operator: With arguments similar to those presented in the context of (11.3.58–62), we define a *reduced characteristic operator* $\chi_t[k]$ which gives the characteristic functional $\Phi_t[k]$ (11.4.11). As a final result we obtain

$$\frac{d}{dt}\chi_t[k] = \mathcal{L}\chi_t[k] - \tfrac{1}{2}k(t)^2\chi_t[k] + \mathrm{i}k(t)\left(\sqrt{\gamma}\,c\chi_t[k] + \chi_t[k]\sqrt{\gamma}\,c^\dagger\right) \tag{11.4.16}$$

with initial condition $\chi_0[k] = \rho(0)$.

This is a "master-equation-like" equation for $\chi_t[k]$, which allows computation of to complete homodyne statistics from its solutions. An expression for the mean homodyne current and current-current correlation function in terms of $\chi_t[k]$ will be given below in (11.4.37) and (11.4.38), respectively.

11.4.3 Stochastic Schrödinger Equation

a) Derivation of the Stochastic Schrödinger Equation for Diffusive Processes: We now turn to a c-number stochastic description and the derivation of a c-number stochastic Schrödinger equation for diffusion processes. Such an equation was first derived by *Carmichael* [11.8] from an analysis of homodyne detection, and independently in a more formal context by Barchielli and Belavkin [11.5].

From (11.3.84) the mean rate of photon counts is

$$\begin{aligned}\langle dN(t)\rangle_c &= \mathrm{Tr}_{\text{sys}}\{(\sqrt{\gamma}\,c^\dagger + \mathcal{E})(\sqrt{\gamma}\,c + \mathcal{E})\rho_c(t)\}\,dt \\ &= \mathrm{Tr}_{\text{sys}}\{(\mathcal{E}^2 + \mathcal{E}\sqrt{\gamma}\,x + \gamma c^\dagger c)\rho_c(t)\}\,dt\,,\end{aligned} \tag{11.4.17}$$

and registration of a count in homodyne detection at time t_r is associated with a quantum jump

$$\tilde{\psi}_c(t_r + dt) \propto (\sqrt{\gamma}\,c + \mathcal{E})\tilde{\psi}_c(t_r) \tag{11.4.18}$$

of the system wave function. Furthermore, we note that the master equation (11.2.24) is invariant under the transformation

$$\sqrt{\gamma}\, c \to \sqrt{\gamma}\, c + \mathcal{E} \tag{11.4.19}$$

$$H \quad \to H - \frac{\mathrm{i}}{2}\hbar \mathcal{E}\sqrt{\gamma}\,(c - c^\dagger) \tag{11.4.20}$$

which involves a displacement of the jump operators $\sqrt{\gamma}\, c$ by a number $\mathcal{E}$, which we have chosen to be real. With these replacements we obtain from the SSE (11.3.93)

$$\begin{aligned} d\psi_c(t) = &\left(-\frac{\mathrm{i}}{\hbar} H - \tfrac{1}{2}\left(\gamma c^\dagger c + 2\mathcal{E}\sqrt{\gamma}\, c - \gamma \left\langle c^\dagger c\right\rangle_c - \mathcal{E}\sqrt{\gamma}\,\langle x\rangle_c\right)\right)\psi_c(t)\, dt \\ &+ \left(\frac{\sqrt{\gamma}\, c + \mathcal{E}}{\sqrt{\left\langle (\sqrt{\gamma}\, c^\dagger + \mathcal{E})(\sqrt{\gamma}\, c + \mathcal{E})\right\rangle_c}} - 1\right)\psi_c(t)\, dN(t). \end{aligned} \tag{11.4.21}$$

b) Limit of Strong Local Oscillator: We will now compute the limit of this equation as the local oscillator amplitude $\mathcal{E}$ goes to infinity. As pointed out before, in the limit that $\mathcal{E}$ is much larger than the source term $\sqrt{\gamma}\, c$, the count rate (11.4.17) consists of a large constant term (the counting rate from the local oscillator), a term proportional to the x quadrature of the system dipole, and a small term, the direct count rate from the system. Subtracting the contribution of the local oscillator from the counting rate, we write

$$dN(t) = \mathcal{E}^2\, dt + \mathcal{E}\; dX(t)\,. \tag{11.4.22}$$

This defines the stochastic process $X(t)$. For $\mathcal{E} \to \infty$ the stochastic process $X(t)$ has the Gaussian properties

$$\text{i)} \quad \langle dX(t)\rangle_c = \sqrt{\gamma}\;\langle x\rangle_c\; dt \tag{11.4.23}$$

$$\text{ii)} \quad [dX(t)]^2 = dt, \tag{11.4.24}$$

$$\text{iii)} \quad dX(t)\, dt = 0 \tag{11.4.25}$$

$$\Longrightarrow \quad dX(t) - \langle dX(t)\rangle_c = dW(t) \tag{11.4.26}$$

where $dW(t)$ is a Wiener increment $[dW(t)]^2 = dt$, $\langle dW(t)\rangle_c = 0$. Thus the jumps in the SSE (11.4.21) are replaced by a diffusive evolution.

The derivation of the Gaussian properties of $X(t)$ is straightforward: The first line of the above equation, (11.4.23), follows from (11.4.17) for the mean count rate in homodyne detection, while the second line, (11.4.24), is obtained from the Ito table (11.3.88),

$$\begin{aligned} [dX(t)]^2 &= (dN(t) - \mathcal{E}^2\, dt)^2/\mathcal{E}^2 \\ &= dN(t)/\mathcal{E}^2 \\ &= dt + dX(t)/\mathcal{E} \\ &\to dt \quad (\mathcal{E} \to \infty). \end{aligned} \tag{11.4.27}$$

Physically we interpret $dX(t)$ to be related to a c-number stochastic homodyne current

$$I_{\text{hom}}(t) = \frac{dX(t)}{dt} \tag{11.4.28}$$

$$\equiv \lim_{\mathcal{E}\to\infty} \frac{dN(t) - \mathcal{E}^2\,dt}{\mathcal{E}\,dt} \tag{11.4.29}$$

$$= \sqrt{\gamma}\,\langle x\rangle_c(t) + \xi(t) \tag{11.4.30}$$

where $\xi(t) = dW(t)/dt$, the Gaussian white noise function.

c) Stochastic Schrödinger Equation for Diffusion Processes: Taking this limit in (11.4.21), we obtain a SSE for diffusive processes

$$d\psi_c(t) = \left\{ \left(-\frac{\mathrm{i}}{\hbar}H - \tfrac{1}{2}(\gamma c^\dagger c - \gamma\,\langle x\rangle_c\, c + \frac{1}{4}\gamma\,\langle x\rangle_c^2) \right) dt + \sqrt{\gamma}\,\left(c - \langle x/2\rangle_c\right) dW(t) \right\} \psi_c(t)\,. \tag{11.4.31}$$

Equation (11.4.31) is not unique; in particular we can make a phase transformation $\psi_c(t) \to \mathrm{e}^{\mathrm{i}\phi(t)}\psi_c(t)$ which allows us to rewrite this equation as

$$d\psi_c(t) = \left\{ \left(-\frac{\mathrm{i}}{\hbar}H - \tfrac{1}{2}\gamma c^\dagger c + \gamma\,\left\langle c^\dagger\right\rangle_c c + \tfrac{1}{2}\gamma|\,\langle c\rangle_c\,|^2 \right) dt + \sqrt{\gamma}\,\left(c - \langle c\rangle_c\right) dW(t) \right\} \psi_c(t)\,. \tag{11.4.32}$$

d) Alternative Forms: In the literature one finds several versions of the SSE for diffusive processes. Carmichael's derivations [11.8] lead to a SSE for an unnormalized wavefunction $\tilde{\psi}_c(t)$

$$d\tilde{\psi}_c(t) = \left[-\frac{\mathrm{i}}{\hbar}H - \tfrac{1}{2}\gamma c^\dagger c + I_{\text{hom}}(t)\,\sqrt{\gamma}\,c \right] dt\;\tilde{\psi}_c(t)\,. \tag{11.4.33}$$

This form of the equation demonstrates directly how the state is conditioned on the measured photocurrent $I_{\text{hom}}(t)$. Furthermore, there are various versions of the SSE with complex noises: we refer to the literature for details [11.9, 11.14].

e) Derivation from the QSSE: The above c-number SSEs for diffusive processes can also be derived directly from the QSSE (11.2.18) [11.11]. The idea is to use $dB(t)|\Psi(t)\rangle = 0$ to replace $dB^\dagger(t)\,c|\Psi(t)\rangle \to (dB(t) + dB^\dagger(t))\,c|\Psi(t)\rangle \equiv d\hat{X}(t)\,c|\Psi(t)\rangle$ in (11.2.18) and subsequently to project on an eigenbasis of the set of commuting operators $d\hat{X}$. Generalization of the SSEs to a squeezed bath has been given in [11.11, 11.15].

f) Stochastic Density Matrix: A stochastic density matrix equation for diffusive processes is directly obtained from these equations,

$$d\rho_c(t) = \mathcal{L}\rho_c\,dt + \left[\sqrt{\gamma}\,\left(c - \langle c\rangle_c\right)\rho_c + \rho_c\sqrt{\gamma}\,\left(c^\dagger - \left\langle c^\dagger\right\rangle_c\right) \right] dW(t) \tag{11.4.34}$$

and taking stochastic averages we see—as expected—that the a priori dynamics satisfies the master equation (11.2.24).

g) Characteristic Functional for the Stochastic Homodyne Current: The stochastic homodyne current $I_{\rm hom}(t)$ (11.4.28) has the same statistical properties as the operator version $\hat{I}_{\rm hom}(t)$. This becomes evident by defining a characteristic functional

$$\Phi_t[k] = \langle V_t[k] \rangle_{\rm st} \text{ with } \quad V_t[k] \equiv \exp\left(\mathrm{i} \int_0^t k(s)\, dX(t) \right) \tag{11.4.35}$$

where $V_t[k]$ satisfies the SDE

$$dV_t[k] = V_t[k] \; \left(\mathrm{i}k(t)\, dX(t) - \tfrac{1}{2} k(t)^2\, dt \right) \; . \tag{11.4.36}$$

Analogously to (11.3.108) for counting processes, we again define a reduced characteristic operator $\chi_t[k]$. When we use Ito calculus (11.3.110), together with the above equation for $V_t[k]$ and the stochastic density matrix equation for diffusive processes (11.4.34) we find that $\chi_t[k]$ obeys an equation identical to (11.4.16), and with the same initial conditions. Thus the characteristic functionals $\Phi_t[k]$ defined in (11.4.11) and (11.4.35), respectively, are identical.

h) Examples of the Use of the Characteristic Functional: As an example, we readily derive an expression for the mean homodyne current

$$\left\langle \hat{I}_{\rm hom}(t) \right\rangle \equiv \langle I_{\rm hom}(t) \rangle_{\rm st} = -\mathrm{i} \frac{\delta}{\delta k(t)} \Phi_T[k]|_{k=0} = \sqrt{\gamma}\, \mathrm{Tr}_{\rm sys} \left\{ x\rho(t) \right\} . \tag{11.4.37}$$

The stationary stationary homodyne correlation function is given by

$$\begin{aligned} \left\langle \hat{I}_{\rm hom}(t_1) \hat{I}_{\rm hom}(t_2) \right\rangle &\equiv \langle I_{\rm hom}(t_1) I_{\rm hom}(t_2) \rangle_{\rm st} \\ &= (-\mathrm{i})^2 \frac{\delta^2}{\delta k(t_1)\, \delta k(t_2)} \Phi_T[k]|_{k=0} \\ &= \gamma \mathrm{Tr}_{\rm sys} \left\{ x \mathrm{e}^{\mathcal{L}|t_2 - t_1|} \left(c\rho + \rho c^\dagger \right) \right\} + \delta(t_2 - t_1) \end{aligned} \tag{11.4.38}$$

where the last term is the shot noise contribution. The Fourier transform of (11.4.38) gives the spectrum of squeezing according to Sect. 8.4.

i) Dynamical Theories of Wave Function Reduction: As a final remark we note that a completely different interpretation of the stochastic Schrödinger equations of the type (11.4.31) has been proposed by *Gisin* and *Diósi* in [11.9, 11.14] in connection with *dynamical theories of wave function reduction*. The assumption is that the reduction of the wavefunction associated with a measurement is a stochastic process and an equation of the type (11.4.32) is postulated.

11.5 Applications and Illustrations

In the previous sections we have derived quantum stochastic Schrödinger equations and c-number stochastic Schrödinger equations for counting and diffusion processes. The emphasis has been on the basic concepts and the structure of the theory, in particular the relation to continuous measurement theory. These SSEs have found numerous applications in quantum optics. Typically, applications of the SSE are either *illustrations* of the physics of the particular model problem in terms of "typical" trajectories, or they are *numerical simulations* to solve the master equation when a direct solution of the master equation is difficult or not feasible.

In this section we will illustrate application of the SSE for counting processes in the context of specific examples. We will first discuss resonance fluorescence of strongly driven atomic systems. This illustrates the notion of quantum jumps in atomic systems undergoing spontaneous emission, and the simulation of the optical Bloch equations in terms of quantum trajectories. We then extend the theory to include mechanical light effects. The corresponding SSE is the basis of wave function simulations of laser cooling. While it is well outside the scope of the present book to enter a detailed discussion of the physics of laser cooling, our goal is to make a connection with the extensive literature in this subject. Numerical simulation of the master equation of laser cooling has been one of the prime applications of the SSE as a simulation tool.

Applications of the diffusive SSE in quantum optics can be found in Carmichael's book [11.8].

11.5.1 Resonance Fluorescence of Strongly Driven Two Level Systems

a) Resonance Fluorescence from Two Level Atoms: A theory of strongly driven two level atoms coupled to a heat bath of the electromagnetic field modes was developed in Sect. 9.2. Here we will reformulate the problem in a form and notation which allows simple generalization to include mechanical light effects.

We consider the dynamics of a two level atom at position $\mathbf{x} = 0$ with ground state $|g\rangle$ and excited state $|e\rangle$, which is driven by a classical light field and coupled to a reservoir of vacuum modes of the radiation field. The total Hamiltonian of the atomic system and heat bath (11.1.1) has the form

$$H_{\mathrm{tot}} = H_{\mathrm{A}} + H_{\mathrm{B}} + H_{\mathrm{Int}}. \tag{11.5.1}$$

Here

$$H_{\mathrm{A}} = \hbar\omega_{eg}\sigma_{ee} \tag{11.5.2}$$

is the atomic system Hamiltonian with ω_{eg} the atomic transition frequency. For the atomic operators we use the notation

$$\sigma_{ee} = |e\rangle\langle e|, \qquad \sigma_{gg} = |g\rangle\langle g|, \tag{11.5.3}$$

$$\sigma_{+} = |e\rangle\langle g|, \qquad \sigma_{-} = |g\rangle\langle e|. \tag{11.5.4}$$

The Hamiltonian of the free radiation field (the heat bath) is

$$H_B = \sum_{\lambda} \int d^3\mathbf{k}\, \hbar\omega_{\mathbf{k},\lambda} b^{\dagger}_{\mathbf{k},\lambda} b_{\mathbf{k},\lambda} \tag{11.5.5}$$

with $b_{\mathbf{k},\lambda}$ the photon destruction operator for the mode with wave vector $\mathbf{k}$ and polarization $\lambda = 1, 2$.

The interaction Hamiltonian H_{Int} is

$$H_{\text{Int}}(t) = -\boldsymbol{\mu}_{eg}^{*} \cdot [\, \mathbf{E}^{(+)}(0)^{\dagger} + \mathbf{E}_{cl}^{(+)}(0,t)^{*}]\, \sigma_{-} - \sigma_{+}\boldsymbol{\mu}_{eg} \cdot [\mathbf{E}^{(+)}(0) + \mathbf{E}_{cl}^{(+)}(0,t)] \tag{11.5.6}$$

which describes the interaction of the atom $\mathbf{x} = 0$ with the bath of radiation modes and an external classical driving field in the dipole approximation. In (11.5.6) $\boldsymbol{\mu}_{eg}$ is the atomic dipole matrix element,

$$\mathbf{E}^{(+)}(\mathbf{x}) = \mathrm{i} \sum_{\lambda} \int d^3\mathbf{k} \sqrt{\frac{\hbar\omega}{2\epsilon_0(2\pi)^3}}\, \mathbf{e}^{\lambda}(\mathbf{k}) b_{\mathbf{k},\lambda}\, \mathrm{e}^{\mathrm{i}\mathbf{k}\cdot\mathbf{x}} \tag{11.5.7}$$

is the positive frequency part of the quantized electric field operator, and

$$\mathbf{E}_{cl}^{(+)}(0,t) = \mathcal{E}\, \boldsymbol{\epsilon}\, \mathrm{e}^{-\mathrm{i}\omega_L t} \tag{11.5.8}$$

is the positive frequency part of the electric field of the laser at the position of the atom.

The state vector $|\Psi(t)\rangle$ of the combined atom-field system has an expansion of the form

$$\begin{aligned} |\Psi(t)\rangle &= |\psi(t)\rangle \otimes |\text{vac}\rangle \\ &+ \sum_{\lambda} \int d^3\mathbf{k} |\psi_{\mathbf{k},\lambda}(t)\rangle \otimes b^{\dagger}_{\mathbf{k},\lambda} |\text{vac}\rangle \\ &+ \frac{1}{\sqrt{2!}} \sum_{\lambda_1,\lambda_2} \int d^3\mathbf{k}_1 \int d^3\mathbf{k}_2 |\psi_{\mathbf{k}_1,\lambda_1;\mathbf{k}_2,\lambda_2}(t)\rangle \otimes b^{\dagger}_{\mathbf{k}_1,\lambda_1} b^{\dagger}_{\mathbf{k}_2,\lambda_2} |\text{vac}\rangle + \ldots \end{aligned} \tag{11.5.9}$$

which describes the presence of $n = 0, 1, 2, \ldots$ scattered photons in the field. We assume that the radiation field is initially in the vacuum state, $|\Psi(0)\rangle = |\psi\rangle \otimes |\text{vac}\rangle$.

As outlined in Sect. 11.1.1 it is convenient to work in a rotating frame where the optical oscillations of at the laser frequency ω_L are transformed away. In this frame the atomic Hamiltonian H (as defined in (11.1.16)) is

$$H = -\hbar\Delta\sigma_{ee} - \tfrac{1}{2}\hbar\Omega\,(\sigma_{+} + \sigma_{-}) \tag{11.5.10}$$

with $\Delta = \omega_L - \omega_{eg}$ the laser detuning, and $\Omega = 2\boldsymbol{\mu}_{eg} \cdot \boldsymbol{\epsilon}\mathcal{E}/\hbar$ is the Rabi frequency (which we take to be real).

The reduced atomic density matrix, which is obtained by tracing over the modes of the radiation field,

$$\rho(t) = \mathrm{Tr}_B\{|\Psi(t)\rangle\langle\Psi(t)|\} \equiv \mathrm{Tr}_B\hat{\rho}(t) \tag{11.5.11}$$

obeys a master equation of the form (11.2.24) (also known as optical Bloch equation) with effective Hamiltonian (11.2.30)

$$H_{\text{eff}} = \hbar\left(-\Delta - \tfrac{1}{2}\mathrm{i}\gamma\right)\sigma_{ee} - \tfrac{1}{2}\Omega\hbar\,(\sigma_{+} + \sigma_{-}) \tag{11.5.12}$$

and recycling operator (11.2.26)

$$\mathcal{J}\rho(t) = \gamma\sigma_{-}\rho(t)\sigma_{+}\,. \tag{11.5.13}$$

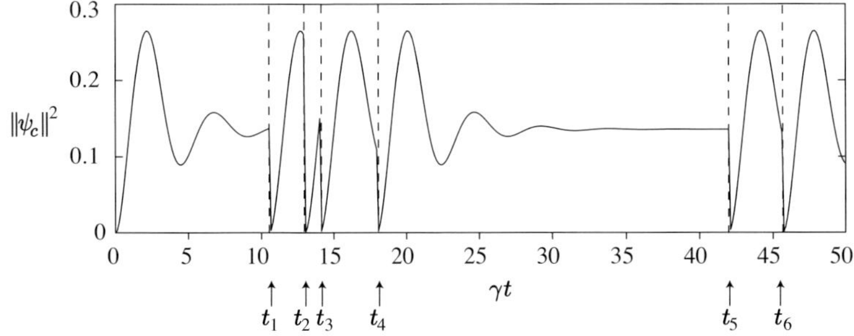

Fig. 11.3 Plot of a realization of the Monte Carlo wavefunction as a function of time: excited state probability $|\langle e|\psi_c(t)\rangle|^2$

H_{eff} is known as the *Wigner-Weisskopf Hamiltonian* for a radiatively damped and driven two level atom, and γ is the radiative decay width of the upper state $|e\rangle$. With these identifications we can readily write down the SSE (11.3.90). Emission of a photon is associated with a quantum jump (11.3.83)

$$|\tilde{\psi}_c(t_r+dt)\rangle \propto \sqrt{\gamma}\,\sigma_-|\tilde{\psi}_c(t_r)\rangle \tag{11.5.14}$$

which returns the atomic electron to the ground state. The probability density for *emission of the next photon* can be read off from (11.3.85) which, as expected, is proportional to the probability of finding the atom in the excited states times the rate of spontaneous emission γ, that is $\gamma|\langle e|\psi_c\rangle(t)|^2$. The time evolution between the photon emissions thus is governed by (11.3.82) with an effective Hamiltonian H_{eff}, which describes the re-excitation of the radiatively damped atom by the laser field.

Fig.11.3 shows the excited state population $|\langle e|\psi_c(t)\rangle|^2$, where the normalized wavefunction is $\psi_c(t) = \tilde{\psi}_c(t)/\|\tilde{\psi}_c(t)\|$, for a "typical run" (quantum trajectory) as a function of time t for a Rabi frequency $\Omega = \gamma$ and detuning $\Delta = -\gamma$. The decay times (indicated by arrows) and quantum jumps, where the atomic electron returns to the ground state, are clearly visible as interruptions of the Rabi oscillations. After a quantum jump the atom is reset to the ground state, and it then is re-excited by the laser. In the photon statistics of the emitted light this leads to antibunching, i.e. two photons will not be emitted at the same time (see Sect. 9.2.3). Averaging over these trajectories, gives the familiar transient solution of the optical Bloch equations. Fig.11.4 gives a comparison between the exact solutions of the optical Bloch equations and the results of such a simulation.

Mollow's Pure State Analysis of Resonant Light Scattering: As a historical remark, we note that an "unravelling" of the master equation in terms of pure state wave functions was first derived by *Mollow* in 1975 [11.3], within his theory of resonance fluorescence. Mollow showed that the atomic density matrix can be decomposed into contributions from subensembles corresponding to $n = 0, 1, 2, \ldots$ photon emissions, where the density matrix of each of these subensembles can be represented in terms of pure state *atomic wave functions*. This formulation anticipated and provided the starting point for some of the recent developments in the

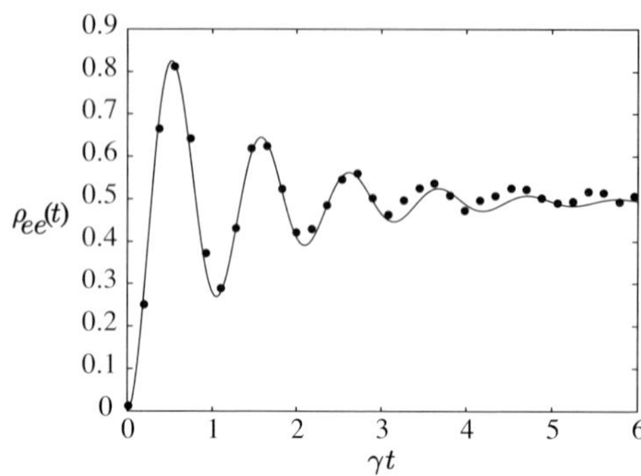

Fig. 11.4 Comparison of the exact solutions of the optical Bloch equations for the excited state population $\rho_{ee}(t)$ with a simulation results for 100 realizations. For 10,000 realizations, the simulation results are indistinguishable from the exact solution. The parameters are for $\Omega = 6\gamma$ and $\Delta = 0$.

application of the SSE to quantum optics. We will briefly summarize below some of Mollow's ideas and relate them to the results of the previous section.

The starting point of Mollow's pure state analysis is the definition of a reduced atomic density operator in the subspace containing exactly n scattered photons according to

$$\rho^{(n)}(t) = \mathrm{Tr}_B\{\hat{P}^{(n)}\hat{\rho}(t)\} \quad (n = 0, 1, 2, \ldots) \tag{11.5.15}$$

where $\hat{\rho}(t) \equiv |\Psi(t)\rangle\langle\Psi(t)|$ is the density operator of the combined atom plus field system, and $\hat{P}^{(n)}$ is the projection operator onto the n-photon subspace. The probability of finding exactly n photons in the field is obtained by tracing over the atomic degrees of freedom

$$P^{(n)}(t) = \mathrm{Tr}_A\{\rho^{(n)}(t)\} \equiv \rho^{(n)}_{gg}(t) + \rho^{(n)}_{ee}(t)\,. \tag{11.5.16}$$

Mollow showed that $\rho^{(n)}(t)$ obeys the equation of motion

$$\frac{d}{dt}\rho^{(0)}(t) = -\frac{\mathrm{i}}{\hbar}\left(H_{\mathrm{eff}}\rho^{(0)}(t) - \rho^{(0)}(t)H^{\dagger}_{\mathrm{eff}}\right) \tag{11.5.17}$$

$$\frac{d}{dt}\rho^{(n)}(t) = -\frac{\mathrm{i}}{\hbar}\left(H_{\mathrm{eff}}\rho^{(n)}(t) - \rho^{(n)}(t)H^{\dagger}_{\mathrm{eff}}\right) + \gamma\sigma_{-}\rho^{(n-1)}(t)\sigma_{+} \qquad (n = 1, 2, \ldots) \tag{11.5.18}$$

with H_{eff} given by (11.5.12). Summing over the n-photon contributions,

$$\rho(t) = \sum_{n=0}^{\infty}\rho^{(n)}(t) \tag{11.5.19}$$

gives, of course, the optical Bloch equations. The n-photon density matrix $\rho^{(n)}(t)$ is thus seen to obey a hierarchy of equations where the $n-1$ photon density matrix provides the feeding term for the n-photon term, $\ldots \to \rho^{(n-1)} \to \rho^{(n)} \ldots$. By formal integration of this hierarchy we obtain equations identical to (11.3.69). Mollow noted that, for initial pure states, this equation can be rewritten in terms of pure atomic states,

$$\rho(t) = |\tilde{\psi}_c(t|)\rangle\langle\tilde{\psi}_c(t|)| + \sum_{n=1}^{\infty}\int_0^t dt_n \int_0^{t_n} dt_{n-1}\ldots\int_0^{t_2} dt_1 \times|\tilde{\psi}_c(t|t_n, t_{n-1}, \ldots, t_1)\rangle\langle\tilde{\psi}_c(t|t_n, t_{n-1}, \ldots, t_1)| \tag{11.5.20}$$

where we have introduce a hierarchy of atomic wave functions

$$|\tilde{\psi}_c(t|)\rangle \tag{11.5.21}$$
$$|\tilde{\psi}_c(t|t_1)\rangle \tag{11.5.22}$$
$$\vdots$$
$$|\tilde{\psi}_c(t|t_n, t_{n-1}, \ldots, t_1)\rangle. \tag{11.5.23}$$

We call (11.5.21) the atomic vacuum wave function, (11.5.22) the 1-photon, and (11.5.23) the n-photon wave functions respectively. The atomic vacuum wave function $|\tilde{\psi}_c(t|)\rangle$ obeys the equation of motion

$$\frac{d}{dt}|\tilde{\psi}_c(t|)\rangle = -\frac{\mathrm{i}}{\hbar} H_{\mathrm{eff}} |\tilde{\psi}_c(t|)\rangle \tag{11.5.24}$$

with $|\tilde{\psi}_c(t = 0|)\rangle = |\psi\rangle$, while the the n-photon wave functions $|\tilde{\psi}_c(t|t_n, t_{n-1} \ldots, t_1)\rangle$ obey

$$\frac{d}{dt}|\tilde{\psi}_c(t|t_n, \ldots, t_1)\rangle = -\frac{\mathrm{i}}{\hbar} H_{\mathrm{eff}} |\tilde{\psi}_c(t|t_n, \ldots, t_1)\rangle \tag{11.5.25}$$

for $t \geq t_n \geq t_{n-1} \geq \ldots \geq t_1$. The initial conditions for these wavefunctions are defined recursively according to the quantum jump condition

$$|\tilde{\psi}_c(t_n|t_n, t_{n-1}, \ldots, t_1)\rangle = \sqrt{\gamma}\,\sigma_- |\tilde{\psi}_c(t_n|t_{n-1}, \ldots, t_1)\rangle \tag{11.5.26}$$

with $\||\tilde{\psi}_c(t = 0)\|| = 1$. Equation (11.5.20) is Mollow's pure state representation of the atomic density matrix.

We interpret $|\tilde{\psi}_c(t|t_n, t_{n-1} \ldots, t_1)\rangle$ as the probability amplitudes which describe the time evolution of the atom in a time interval $(0, t]$ with exactly n photons detected at times $t_1, t_2, \ldots, t_n$. According to (11.5.26) atoms in the ground state with n photons in the scattered field are created at a rate $\gamma\|\sigma_- \tilde{\psi}_c(t|t_{n-1}, \ldots, t_1)\|^2$, which is the decay rate γ times the probability of the atom being in the excited state and having emitted $n-1$ photons. Each photon detection is accompanied by a reduction of the system wave function (quantum jump) according to (11.5.26). According to (11.3.67) the exclusive probability densities for photoemission are

$$P_0^t(0|\rho) = \||\tilde{\psi}_c(t|)\||^2, \tag{11.5.27}$$
$$p_0^t(t_1, t_2, \ldots, t_n|\rho) = \||\tilde{\psi}_c(t|t_n, t_{n-1}, \ldots, t_1)\||^2. \tag{11.5.28}$$

Thus, in the context of the present example of resonance fluorescence, Mollow's construction has the structure expected from the theory of continuous measurement, and the interpretation as a stochastic wave function evolution is equivalent to the one discussed in Sect. 11.3.8.

11.5.2 Quantum Jumps in Three Level Atoms

Ion traps provide a tool to store and observe laser cooled *single ions* for an essentially unlimited time. Experiments with single trapped ions thus represent an

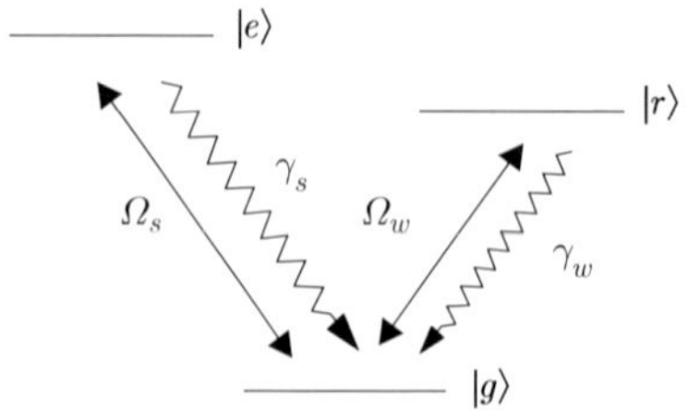

Fig. 11.5 Three level system with $|g\rangle \leftrightarrow |e\rangle$ a strong transition, and $|g\rangle \leftrightarrow |r\rangle$ a weak transition with metastable state $|r\rangle$.

experimental realization of continuous observation of a single quantum system in the context of quantum optics. The observation of quantum jumps in a three level system [11.2] is probably one of the best-known examples in quantum optics where quantum jumps are "seen" in experiments where fluorescence from single ions is observed by continuously monitoring the atom with a photodetector—for reviews and references see [11.10, 11.16].

The system of interest involves a double resonance scheme where two excited states $|e\rangle$ and $|r\rangle$ are connected to a common lower level $|g\rangle$ via a *strong* and *weak* transition, respectively as in Fig.11.5. The fluorescence photons from the strong transition are observed in a photon counting experiment. Excitation of the weak transition with a laser will induce a quantum jump to the metastable state $|r\rangle$, or will temporarily *shelve* the atomic electron in $|r\rangle$. This will cause the fluorescence from the strong transition to be turned off. According to this picture, it is, therefore, possible to monitor the quantum jumps of the weak transition via emission windows in the signal provided by the fluorescence of the strong transition. Experimental observation of this effect has been reported in [11.2].

We are interested in the conditional dynamics of the three level system which is continuously observed on the strong transition with a photodetector. To illustrate the *a posteriori* dynamics of the three level atom we calculate the counting statistics of the photon on the "strong line" (sometimes called the delay function). Simulation of this function [11.4] illustrates the conditional system dynamics. The master equation for a three level atom has the form

$$\dot{\rho} = -\frac{\mathrm{i}}{\hbar}(H_{\mathrm{eff}}\rho - \rho H_{\mathrm{eff}}^{\dagger}) + \mathcal{J}_{\mathrm{s}}\rho + \mathcal{J}_{\mathrm{r}}\rho \quad (\equiv \mathcal{L}\rho)\,. \tag{11.5.29}$$

The effective atomic Hamiltonian is (compare (11.5.3))

$$\begin{aligned} H_{\mathrm{eff}} = {} & \hbar\left(-\Delta_s - \mathrm{i}\frac{\gamma_s}{2}\right)|e\rangle\langle e| + \hbar\left(-\Delta_w - \mathrm{i}\frac{\gamma_w}{2}\right)|r\rangle\langle r| \\ & -\tfrac{1}{2}\hbar\Omega_s\left(|g\rangle\langle e| + |e\rangle\langle g|\right) - \tfrac{1}{2}\hbar\Omega_w\left(|g\rangle\langle r| + |r\rangle\langle g|\right) \end{aligned} \tag{11.5.30}$$

where the first two lines give the bare atomic Hamiltonians including the radiative decay terms with γ_s and γ_w, the decay width of the $|e\rangle$ and $|r\rangle$, respectively ($\gamma_s \gg \gamma_w$). The detunings and Rabi frequencies of the lasers are denoted by $\Delta_{s,w}$ and $\Omega_{s,w}$, respectively as in Fig.11.5. The recycling operators for the atomic electron on the strong and weak transition are respectively

$$\mathcal{J}_{\mathrm{s}}\rho = |g\rangle\langle g|\,\gamma_s\langle e|\rho|e\rangle, \tag{11.5.31}$$

and

$$\mathcal{J}_{\mathrm{w}}\rho = |g\rangle\langle g|\,\gamma_w\langle r|\rho|r\rangle. \tag{11.5.32}$$

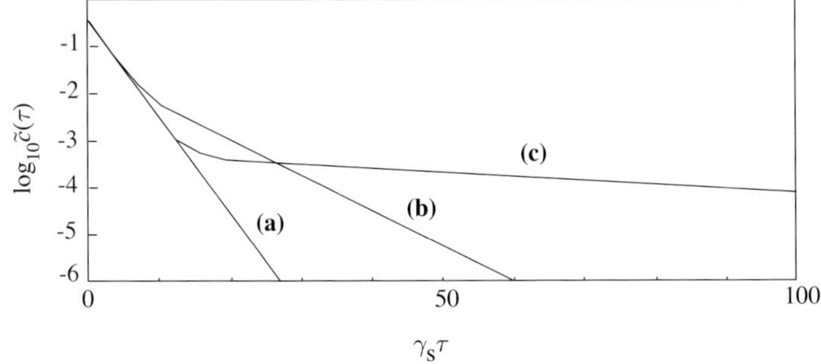

Fig. 11.6 The conditional probability density $\tilde{c}(\tau)$ as a function of $\gamma_s\tau$ according to [11.4]. $\tilde{c}(\tau)$ is the conditional probability density that; given a count has occurred on the strong transition $|g\rangle \leftrightarrow |e\rangle$ at time $\tau = 0$—the *next* count on the strong transition will occur at time τ. The three curves show: **(a)** The metastable state is not excited, $\Omega_w = 0$, and $\tilde{c}(\tau)$ decays essentially exponentially; **(b)** The lifetime of the metastable state $|r\rangle$ is ten times longer than the lifetime of the rapidly decaying state $|e\rangle$, $W_{gr} = \gamma_w = 10^{-1}\gamma_s$. **(c)** The lifetime of the metastable state $|r\rangle$ is one hundred times longer than the lifetime of the rapidly decaying state $|e\rangle$, $W_{gr} = \gamma_w = 10^{-2}\gamma_s$. W_{gr} is an incoherent excitation rate on the weak transition.

The probability density for the emission of a photon on the strong line at time t, given the previous photon on the strong transition was emitted at t_r is, according to (11.3.84),

$$\begin{aligned}\tilde{c}(s,t|s,t_r) &= p_{t_r}^{t_r+dt}(s,t|\rho_c(t_r))/dt \\ &= \frac{\mathrm{Tr}_A\{\mathcal{J}_s e^{(\mathcal{L}-\mathcal{J}_s)(t-t_r)}\mathcal{J}_s\rho\}}{\mathrm{Tr}_A\{\mathcal{J}_s\rho\}}\end{aligned} \tag{11.5.33}$$

where

$$\rho_c(t_r) = |g\rangle\langle g| \tag{11.5.34}$$

is the density matrix at time t_r after emission of a photon on the strong line at time t_r. We note that in the present model this conditional state is independent of the previous history of photon emissions; this is due to the fact that a quantum jump always prepares the atom in the ground state $|g\rangle$. In addition, we have summed over an arbitrary number of possible transitions on the weak line which replaces

$$\mathcal{S}(t,t_r) \to \exp(\mathcal{L}-\mathcal{J}_s)(t-t_r) \tag{11.5.35}$$

in (11.3.79); ρ is the steady state density matrix. An explicit calculation of (11.5.33) shows that it consists of a sum of exponentials with different decay constants, reflecting the different time scales for the excitation and decay on the strong and metastable transition. Fig.11.6 shows a plot of this delay function for a typical set of parameters. We see from this figure that the existence of a weak transition leads to a *long time tail* in the delay function. In a simulation of a photon emission sequence on the strong line this will manifest itself in the *periods of brightness* and *darkness*, as shown in Fig.11.7, in obvious relation to the experimental observation.

(a)

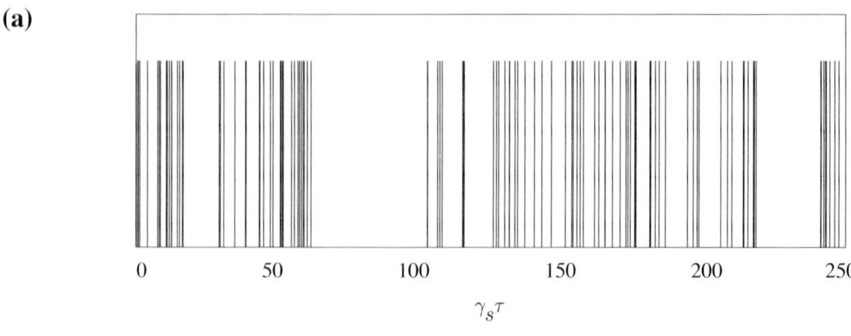

(b)

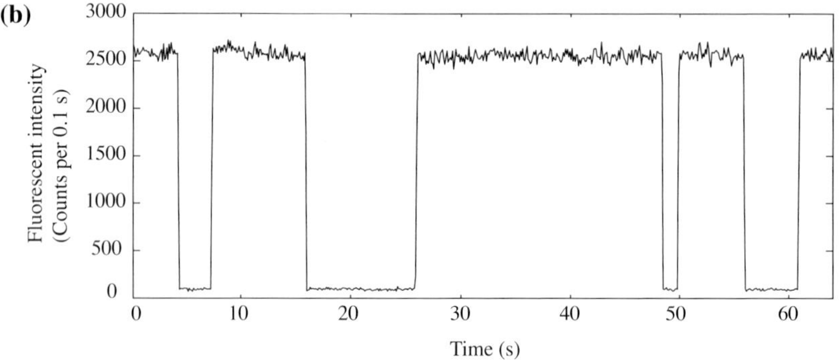

Fig. 11.7 **(a)** Simulation of the photon emission on the strong transition $|g\rangle \leftrightarrow |e\rangle$. A vertical line (of arbitrarily chosen height) corresponds to the emission of one photon. The parameters are: $W_{gr} = \gamma_w = 10^{-1}\gamma_s$, as in Fig.11.6 [11.4]. **(b)** Experimental observation of quantum jumps in the fluorescence light from a single Ba^+ ion at 493nm (data kindly supplied by R. Blatt).

The atomic density matrix conditional on observing an emission window on the strong line when the *last photon on the strong transition* was emitted at time t_n is from (11.3.79)

$$\rho_c(t) = e^{(\mathcal{L}-\mathcal{J}_s)(t-t_r)}\rho/\mathrm{Tr}_A\{\ldots\} \tag{11.5.36}$$

$$\rightarrow |r\rangle\langle r|, \qquad (\gamma_s(t-t_r) \gg 1). \tag{11.5.37}$$

That is, *observation of a window* in a single trajectory of counts corresponds to a *preparation of the electron in the metastable state* $|r\rangle$. This state preparation is what is usually referred to as *shelving of the electron.*

11.5.3 Mechanical Light Effects

One of the prime examples of the application of wave function simulations in recent years has been in the theory of laser cooling of atoms. While it is well outside the scope of the present book to enter into a comprehensive discussion of the physics of laser cooling, for which we refer the reader to review articles and books, [11.17, 11.24, 11.25], our goal is to connect the formal theory developed to in the foregoing sections to the atomic master equations including mechanical light effects, which are the starting point of the theory of laser cooling.

When an atom absorbs laser light and makes a transition to an excited electronic state, the absorption of the laser photon is accompanied by a momentum transfer of the light field to the atomic centre of mass motion. In a similar way, when an atom spontaneously emits a photon the transition to the ground state is associated with a momentum transfer to the atom, while the photon is emitted in a random direction according to the angular distribution of the emitted light. These processes of induced light absorption and emission, and spontaneous emission can be described by a master equation which generalizes the optical Bloch equations of Sect. 11.5.1 to include the centre of mass degrees of freedom. The resulting set of equations for the matrix elements of the density operator is usually called generalized optical Bloch equations, and we will derive these equations for a two level atom. These generalized optical Bloch equations are the basic equations underlying the description of mechanical light effects of strongly driven atoms undergoing spontaneous emission. We will also give the unravelling of the equivalent master equation in terms of quantum trajectories.

Generalized Optical Bloch Equations: We extend the model of Sect. 11.5.1 to include mechanical light effects during photon absorption and emission. The centre of mass position and momentum operator of the atom will be denoted by $\hat{\mathbf{x}}$ and $\hat{\mathbf{p}}$. The atomic system Hamiltonian consists sum of a kinetic and potential energy term of the atomic motion, and an internal atomic Hamiltonian,

$$H_{\mathrm{A}} = \hat{\mathbf{p}}^2/2M + V(\hat{\mathbf{x}}) + \hbar\omega_{eg}\sigma_{ee} \,. \tag{11.5.38}$$

The interaction Hamiltonian is

$$H_{\mathrm{Int}} = -\boldsymbol{\mu}_{eg}^* \cdot [\mathbf{E}^{(+)}(\hat{\mathbf{x}})^\dagger + \mathbf{E}_{\mathrm{cl}}^{(+)}(\hat{\mathbf{x}},t)^*]\,\sigma_- - \sigma_+\boldsymbol{\mu}_{eg} \cdot [\mathbf{E}^{(+)}(\hat{\mathbf{x}}) + \mathbf{E}_{\mathrm{cl}}^{(+)}(\hat{\mathbf{x}},t)] \tag{11.5.39}$$

which describes the interaction of the atom with the bath of radiation modes and an external classical driving field in the dipole approximation. Note that the electric field of the laser and the electric field operator in (11.5.39) is now evaluated at a position given by the atomic position operator $\hat{\mathbf{x}}$. According to (11.5.39) the absorption and emission of photons is associated with a momentum transfer to the centre of mass motion of the atom. For example, for an incident travelling laser wave with wave vector $\mathbf{k}_L$,

$$\mathbf{E}_{\mathrm{cl}}^{(+)}(\mathbf{x},t) = \mathcal{E}\,\boldsymbol{\epsilon}\,\mathrm{e}^{\mathrm{i}(\mathbf{k}_L\cdot\mathbf{x}-\omega_L t)}, \tag{11.5.40}$$

a transition from the atomic ground state $|g\rangle$ to the excited state $|e\rangle$ is accompanied by a momentum transfer $\hbar\mathbf{k}_L$. In a similar way, spontaneous emission of a photon $\mathbf{k}_s$ returns the atomic electron from $|e\rangle$ to $|g\rangle$ and gives a momentum transfer to the atom.

Similarly to (11.5.11) we define a reduced atomic density matrix $\rho(t)$ in the system Hilbert space for the external and internal atomic degrees of freedom. The reduced density matrix of the two level atom obeys the master equation

$$\frac{d}{dt}\rho(t) = -\frac{\mathrm{i}}{\hbar}(H_{\mathrm{eff}}\rho - \rho H_{\mathrm{eff}}^\dagger) + \int d\Omega_{\hat{\mathbf{n}}}\,\mathcal{J}_{\hat{\mathbf{n}}}\,\rho. \tag{11.5.41}$$

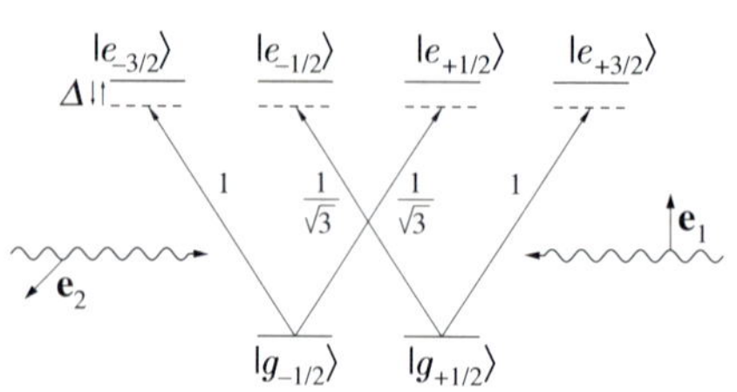

Fig. 11.8 Atomic level scheme and corresponding Clebsch-Gordan coefficients for a $J_g = 1/2$ to $J_e = 3/2$ transition in a lin $\perp$ lin configuration [11.12].

$|g_{m_J=-1/2}\rangle$ and $|g_{m_J=+1/2}\rangle$, and four excited states $|e_{m_J}\rangle$ with $m_J = -3/2, \ldots, +3/2$. In the following we will denote the ground state levels by $|g_\pm\rangle$.
2) *The "lin $\perp$ lin" laser configuration*: For the laser configuration we assume two counterpropagating linearly polarized laser beams with orthogonal polarizations denoted by $\mathbf{e}_1$ and $\mathbf{e}_2$, respectively (see Fig.11.8). Adding these two counterpropagating light waves with propagation axis along z we obtain for the positive frequency part of the electric field of the lasers

$$\mathbf{E}_{\mathrm{cl}}^{(+)}(z,t) = \mathcal{E}(\sin(k_L z)\mathbf{e}_{+1} + \cos(k_L z)\mathbf{e}_{-1})\mathrm{e}^{-\mathrm{i}\omega_L t}. \tag{11.5.49}$$

Here ω_L is the laser frequency, $k_L = 2\pi/\lambda$ is the wave vector with λ the wavelength of the laser light, and $\mathbf{e}_{\pm 1} = \mp(\mathbf{e}_2 \pm \mathrm{i}\mathbf{e}_1)/\sqrt{2}$ are spherical unit vectors corresponding to $\sigma_\pm$ polarized light, respectively. Thus in this laser configuration the electric field (11.5.49) consists of a superposition of a σ_+ polarized standing light wave $\sin(kz)$ and a σ_- polarized standing light wave $\cos(kz)$. Note that the σ_+ component of the light excites the transitions $|g_{m_J=-1/2,+1/2}\rangle \to |e_{m_J=+1/2,+3/2}\rangle$ in Fig.11.8, and similarly the σ_- component will excite $|g_{m_J=-1/2,+1/2}\rangle \to |e_{m_J=-3/2,-1/2}\rangle$.
3) *Laser excitation, optical potentials and optical pumping*: We assume that the exciting laser is detuned towards the red side of the atomic resonance, $\Delta = \omega_L - \omega_{eg} < 0$. Furthermore, we consider the limit of low laser intensities, where the saturation parameter

$$s = \frac{1}{2}\frac{\Omega^2}{(\Delta^2 + \frac{1}{4}\gamma^2)} \ll 1 \tag{11.5.50}$$

is small, with Ω the Rabi frequency and γ the spontaneous decay width (Note: it is convention to normalize the Rabi frequency Ω so that it corresponds to the outermost transition $|g_{1/2}\rangle \to |e_{3/2}\rangle$) and at at the point of maximum intensity of (11.5.49)).

Both ground states $|g_\pm\rangle$ will be shifted by the quadratic AC Stark effect in the laser field (11.5.49). For red laser detunings the atomic ground states will shift downwards. The Stark shifts depend on the position of the atom z in the field, and they provide the *optical potentials* for the motion of the atom in the light field. We will denote these potentials by $U_\pm(z)$ for the two ground states $|g_\pm\rangle$. Due to the large Clebsch-Gordan coefficients for the outer transitions in Fig.11.8, minima (i.e., the points of largest shifts downwards) will occur for the state $|g_+\rangle$ at positions with pure σ^+-light, and for $|g_-\rangle$ at positions with pure σ^--light. For the laser configuration (11.5.49) these optical potentials will thus form a periodic alternating pattern of *optical bipotentials* $U_\pm(z)$, as plotted in the upper part of Fig.11.9. Explicit expressions for these optical potentials for the laser configuration (11.5.49)

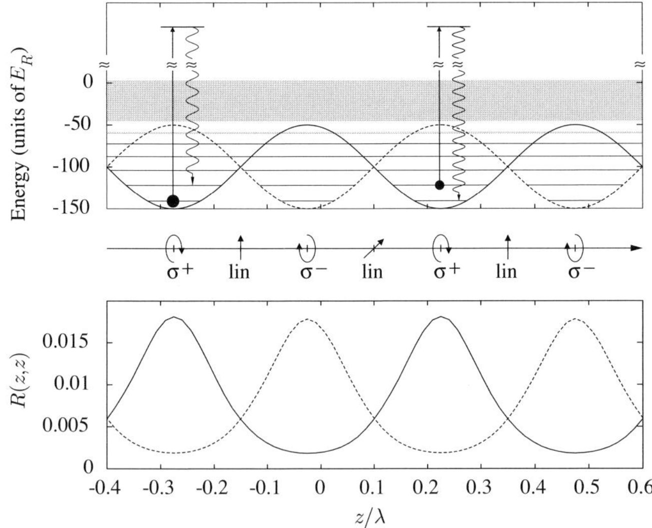

Fig. 11.9 In the upper panel the optical bipotentials and band structure of the atom are plotted as a function of position z for a $1/2$ to $3/2$ transition. The potential depth is $U_0 = 100E_R$ (with $E_R = \hbar^2 k_L^2/2M$ the recoil energy) . The excited states were adiabatically eliminated. We schematically indicate the two Raman processes between the ground and excited states which lead to the red and blue sidebands in resonance fluorescence. In the lower panel we show the spatial distribution of the atoms in the state $|g_+\rangle$ (solid line) and the state $|g_-\rangle$ (dashed line) for $U_0 = 100E_R$ and $\gamma_0 = 5/3E_R$. The atoms are localized in the valleys of the corresponding optical potentials [11.12].

can be found in [11.12]. We will denote by U_0 the depth of the optical potential, which in terms of the saturation parameter and detuning is given by $U_0 = s|\Delta|/2$.

In addition, the ground states are coupled by processes where the atom in one Zeeman ground state absorbs laser light, and makes a transition to the other Zeeman state by spontaneous emission of a photon, $|g_\pm\rangle \to |g_\mp\rangle$. This corresponds to an optical pumping process. We note that these optical pumping processes will be a function of the atomic position z.

4) *Semiclassical laser cooling*: A semiclassical picture of laser cooling for the above atomic and laser configurations has been developed in a seminal paper by Dalibard and Cohen-Tannoudji [11.27]. In semiclassical laser cooling the atomic motion is assumed to be classical while the internal atomic excitations are, of course, described quantum mechanically. The physical picture of laser cooling which emerges from this discussion is one of *Sisyphus cooling*: Consider an atom moving in one of the optical potential curves, say $U_-(z)$ in Fig.11.9. The calculations then show that transitions to the *other* potential $U_+(z)$ then occur preferentially from the tops of $U_-(z)$ down to the valleys of $U_+(z)$, so that on the average the atomic motion is damped.

5) *Quantum picture*: To describe the quantum motion of the atomic centre of mass wave packet in the optical potentials $U_\pm(z)$ we must solve the corresponding Schrödinger equation. In the periodic optical potential Fig.11.9 the spectrum

of energy eigenstates will form a band structure, and the energy eigenstates will be Bloch functions. Depending on the depth of the potential U_0 an optical potential will support several Bloch bands. This is illustrated for a typical example in the upper part of Fig.11.9 where the potential is measured in units of the recoil energy $E_R = \hbar^2 k_L^2/2M$ with M the mass of the atom. For the few lowest levels the optical potential is well approximated by a harmonic oscillator, and the separation between the energy bands is approximately equidistant. The corresponding excitation frequency, ω_{osc}, can be identified with the classical oscillation frequency of atoms in one of the wells.

Laser excitation from the ground state, followed by spontaneous emission will redistribute the atoms between the quantized energy levels. Quantum mechanically, laser cooling can be understood as optical pumping between the quantized energy levels. This picture is valid in the limit when the level separations ω_{osc} are much larger than the optical pumping rate $\gamma_0 \propto s\gamma$, i.e., when that quantized energy levels are spectroscopically well resolved. This condition will be satisfied for large optical detunings. In the time domain this condition corresponds to a situation in which an atomic centre of mass wave packet in the potential undergoes many oscillations with frequency ω_{osc} before an optical pumping process occurs [11.12].

As a result of laser cooling the atom will occupy the lowest energy levels, and will thus be strongly localized on the scale given by the wavelength of the light λ. This is known [11.24] as the *Lamb-Dicke regime*, which is characterized by a small value of the *Lamb-Dicke parameter* $\eta = 2\pi k_L a_0 \ll 1$, where a_0 is the size of the ground state in the optical potential. The lower part of Fig.11.9 shows the corresponding localization of the atom in minima of the $U_\pm(z)$ potentials (obtained from a wave function simulation described below).

6) *Resonance fluorescence and absorption spectrum*: Transitions between the quantized vibrational states will manifest themselves as sidebands around the laser frequency, $\omega \pm \omega_{\text{osc}}$. These motional sidebands can be observed, for example, when resonance fluorescence from an atom is observed [11.20, 11.21, 11.22]. These sidebands can be interpreted as Raman transitions between the vibrational energy levels according to the upper part of Fig.11.9. The sidebands in the energy spectrum have been observed in the NIST experiment by *Jessen et al.* [11.20] for Rb atoms. Our goal below will be to calculate the spectrum of resonance fluorescence including the quantized centre of mass motion using wave function simulations, and compare theory with the experimental result.

b) Generalized Optical Bloch Equations: The basis of a theoretical discussion of laser cooling is the solution of the Generalized Optical Bloch Equations for the atomic density matrix comprising both the internal and external (centre of mass) degrees of freedom. A derivation of the master equation for the atomic configuration of Fig.11.8 can be found in [11.24, 11.25]. The detailed derivation of this master equation is not of direct interest in our present context. We will, therefore, only quote the final result and will confine ourselves to a discussion of the basic structure of this equation and give a physical interpretation of the various terms. The reader interested in the derivation is referred the to [11.12].

The assumption of weak laser excitation allows the excited states of the atom in Fig.11.8 to be adiabatically eliminated. In addition, we consider a one-dimensional model with motion along the z-direction. As a result, one obtains a master equation for the density matrix ρ of the *ground state manifold* [11.12, 11.26]:

$$\dot{\rho} = -\frac{\mathrm{i}}{\hbar}(H_{\mathrm{eff}}(\hat{z})\rho - \rho H_{\mathrm{eff}}(\hat{z})^\dagger) + \gamma_0 \sum_\sigma \int_{-1}^{+1} du\ N_\sigma(u) B_\sigma(\hat{z}) \mathrm{e}^{-iku\hat{z}} \rho B_\sigma(\hat{z})^\dagger e^{iku\hat{z}} . \quad (11.5.51)$$

We now give a physical interpretation of the various terms of the master equations:

i) The non-Hermitian effective atomic Hamiltonian in (11.5.51) is

$$H_{\mathrm{eff}}(\hat{z}) = \frac{\hat{p}^2}{2M} - (U_0 + \tfrac{1}{2}\mathrm{i}\gamma_0) \sum_\sigma B_\sigma(\hat{z})^\dagger B_\sigma(\hat{z}) . \quad (11.5.52)$$

The first term is the kinetic energy of the atom with $\hat{p}$ the atomic momentum operator and M the atomic mass. The potential term has a real and an imaginary part. The real term is the position dependent AC Stark shift of the ground state representing the optical potential , while the imaginary term is associated with the optical pumping process. The height of the optical potential is $U_0 = s|\Delta|/2$, while $\gamma_0 = s\gamma/2$ is the photon scattering rate in the optical pumping process. Note that H_{eff} is an operator on the two-dimensional space of the ground state atoms $\{|g_\pm\rangle\}$. For the atomic and laser configuration described in Fig.(11.9) the optical potential matrix is diagonal with potentials $U_\pm(z)$, i.e., there are no coherent laser induced Raman couplings between the ground state; the ground states are only coupled by optical pumping.

ii) The last term in (11.5.51) is *recycling term*, describing the return of the atomic electron to the ground states after an optical pumping process, i.e. a laser excitation followed by a spontaneous emission.

The operators $B_\sigma(z)$ describe to Raman transitions between the ground state levels by absorption of a laser photon and subsequent emission of a spontaneous photon with polarization $\sigma = \pm 1$. The position dependence of $B_\sigma(z)$ reflects the local intensity which is "seen" by the atom. Explicit expressions for these operators can be found in [11.12].

$N_\sigma(u)$ denotes the angular distribution of spontaneous photons emitted in the direction $u = \cos(\mathbf{k}_s, \mathbf{e}_z)$ $(-1 \le u \le 1)$ with polarization $\sigma = 0, \pm 1$. The distribution is normalized to one, when integrated over u. In the recycling term one integrates over all directions of photon emissions, weighted by the angular distribution $N_\sigma(u)$, and sums over all possible polarizations of the emitted photons.

iii) The master equation (11.5.51) is of the Lindblad form with an infinite number of channels. Thus, solutions of this master equation can be obtained by wave function simulations.

Adapting the formalism of Sect. 11.3.9 a simulation of the master equation (11.5.51) consists of propagation of an atomic wave function $\tilde{\psi}_c(t)$ with the non-Hermitian (damped) atomic Hamiltonian (11.5.52) interrupted at random times by wave function collapses

$$\tilde{\psi}_c(t+dt) \propto \mathrm{e}^{\mathrm{i}k'_z\hat{z}} B_\sigma(\hat{z})\tilde{\psi}_c(t), \quad (11.5.53)$$

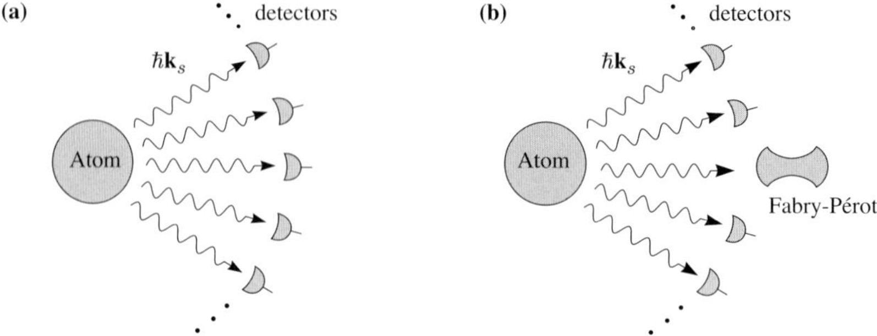

Fig. 11.10 **(a)** The "experiment" simulated in the Monte Carlo wave function description of optical molasses: the atom is surrounded by photo detectors corresponding to an angle resolved detection of the emitted photon $\hbar \mathbf{k}_s$; **(b)** Simulation of the fluorescence spectrum of the emitted light: the atom is "observed" with photodetectors and the spectrum is obtained by filtering one output channel with a Fabry-Pérot interferometer [11.12].

and subsequent wave function renormalization. The Schrödinger equation for $\tilde{\psi}_c(t)$ describes the time evolution of the atomic wave packet in the periodic optical potential, and its coupling to the laser driven internal atomic dynamics. The times of the *quantum jumps* are selected according to the delay function

$$\tilde{c}(u', \sigma, t) = \left\| \sqrt{\gamma_0 \, N_\sigma(u')} \; B_\sigma(\hat{z}) \tilde{\psi}_c(t) \right\|^2 \tag{11.5.54}$$

which gives the probability for emitting a spontaneous photon at time t, with momentum $k'_z = k_{eg} u'$ along the z-axis and polarization σ. The "quantum jump" (11.5.53) corresponds to an optical pumping process between the atomic ground states, including the associated momentum transfer to the atom. Averaging over these wave function realizations gives the density matrix according to (11.3.113).

c) The Spectrum of Resonance Fluorescence: Our goal is to calculate the spectrum of resonance fluorescence of the emitted light. In particular we will compute the spectrum of light emitted along the z-axis, with frequency ω' and polarization σ. This spectrum is proportional to the Fourier transform of the stationary atomic dipole correlation function [11.12] (compare Sect. 5.2.1),

$$c_\sigma(t - t_0) = \left\langle B_\sigma(t)^\dagger \, \mathrm{e}^{\mathrm{i}k'_z \hat{z}(t)} \; B_\sigma(t_0) \, \mathrm{e}^{-\mathrm{i}k'_z \hat{z}(t_0)} \right\rangle . \tag{11.5.55}$$

According to Sect. 11.3.10 and [11.12] we have

$$c_\sigma(t - t_0) = \mathrm{Tr}_S \left\{ B_\sigma^\dagger \; \mathrm{e}^{\mathrm{i}k_z \hat{z}} \; \rho^{(+)}(t) \right\} \quad (t \geq t_0) \quad ; \tag{11.5.56}$$

where the first order perturbed density operator $\rho^{(+)}(t)$ obeys the same master equation as the density matrix but with a different initial condition

$$\rho^{(+)}(t_0) = B_\sigma(\hat{z}) \, \mathrm{e}^{-\mathrm{i}k'_z \hat{z}} \rho(t_0) \, . \tag{11.5.57}$$

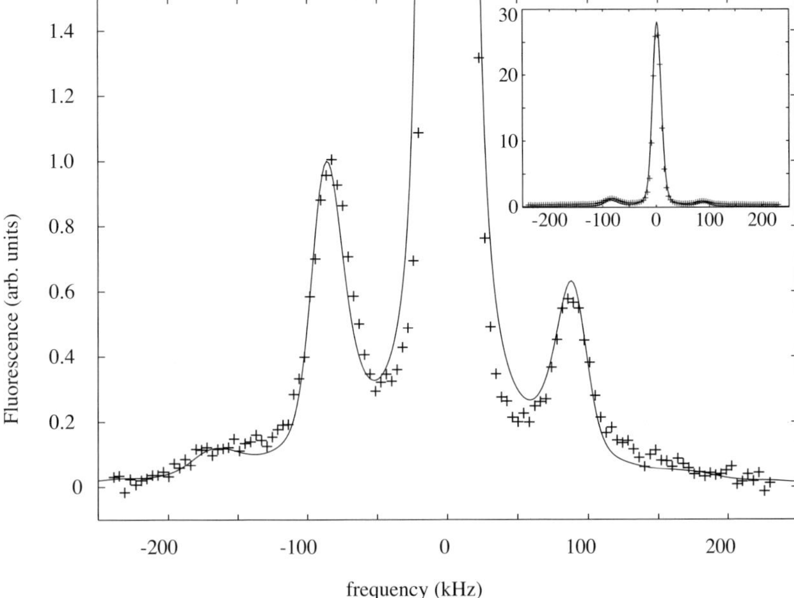

Fig. 11.11 Spectrum of resonance fluorescence as a function of the frequency ν. The solid line is the theoretical spectrum [11.12] convolved with a Lorentzian corresponding to a finite detector width of 3.8 kHz and a Gaussian with width 20 kHz (residual Doppler broadening). Crosses are the experimental result of *Jessen et al.* [11.20]. The inset shows the total spectrum. For parameters we refer [11.12].

d) Physical Interpretation of the Simulation of the Spectrum: Our simulation procedure can be interpreted physically as the following computer experiment [11.6, 11.12] (for a more formal derivation see Sect. 11.3.10). We surround the atom by an (infinite number of) unit efficiency photodetectors covering the 4π solid angle. The simulated "click" of the photodetectors determines the time, polarization and direction of the emitted photon (see Fig.11.10a). This is the simulation of the density matrix of the laser cooled atom. To simulate the fluorescence spectrum, we place a Fabry-Pérot in front of one of these photodetectors and "measure" the transmitted photon flux. This gives the spectrum as a function of the tuning of the resonance frequency of the Fabry-Pérot transmission (see Fig.11.10b).

e) Simulation: In the experiment by *Jessen et al.*, the resonance fluorescence spectrum from Rb^{85} atoms in 1D molasses is observed. The Rb atoms are driven on the $5S_{1/2}\ F_g = 3 \to 5P_{3/2}\ F_e = 4$ transition in a lin $\perp$ lin laser configuration described above. A direct solution of the master equation (11.5.51) to calculate the autocorrelation function (11.5.56) is impractical because the dimensionality of the simulation is $N = 448$. For ^{85}Rb we require a Fourier grid with 64 points corresponding to momenta up to $\pm 32\hbar k$ [11.12]; 10,000 wavefunction realizations are needed for convergence. This algorithm allowed a gain of about a factor 50 in computing time over a density matrix calculation.

The dipole correlation function (11.5.56) can be simulated following the approach outlined in Sect. 11.3.10. The perturbed density matrix $\rho^{(+)}(t)$ in (11.5.56) can be interpreted as a first order response to a "delta-kick" at time $t = t_0$, represented by the initial condition (11.5.57,11.3.128). A simulation is obtained by introducing a "perturbed" wave function $\tilde{\psi}_c^{(+)}(t)$ which obeys the Schrödinger equation for $\tilde{\psi}_c(t)$ but now with initial condition [compare (11.5.57,11.3.128)]

$$\tilde{\psi}_c^{(+)}(t = t_0) = \sqrt{\gamma_0 \, N_\sigma(u)} \; e^{-\mathrm{i}k'_z \hat{z}} B_\sigma(\hat{z}) \tilde{\psi}_c(t_0), \tag{11.5.58}$$

and quantum jumps of $\tilde{\psi}_c^{(+)}(t)$ dictated by the wave function $\tilde{\psi}_c(t)$ according to the delay function (11.5.54). The dipole correlation function is

$$c_\sigma(t - t_0) = \left\langle \langle \tilde{\psi}_c(t) | B_\sigma(\hat{z})^\dagger \mathrm{e}^{\mathrm{i}k'_z \hat{z}} | \tilde{\psi}_c^{(+)}(t) \rangle / \|\tilde{\psi}_c(t)\|^2 \right\rangle_{\mathrm{st}} . \tag{11.5.59}$$

f) Comparison of Simulation with Experiment: Fig.11.11 compares the resonance fluorescence spectrum for σ^+ polarized light obtained by simulation (solid line) with the experimental data of *Jessen et al.* [11.20] (crosses) for laser intensities, detunings etc. taken directly from the experiment with no adjustable parameter. The theoretical spectrum of Fig.11.11 was obtained by convolving the *ab initio* spectrum with the experimental resolution. The central line is scattering at the laser frequency while the first red and blue sidebands correspond to Raman transitions between adjacent vibrational bands in the optical potential.

The asymmetry of the red and blue sideband intensities reflects the populations of the vibrational levels, and from the excellent agreement between theory and experiment we infer that the wave function simulation reproduces the experimental temperature of the atoms. Laser cooling accumulates atoms predominantly in the lowest vibrational states and in the $m_{J_g} = \pm 3$ potentials. This spatial localization of atoms—on a scale small compared with the laser wavelength—suppresses optical pumping transitions between different vibrational levels $n \neq n'$, and is responsible for the narrowing of the lines in the optical spectrum, which is consequently known as *Lamb-Dicke narrowing* [11.24]. A broadening mechanism is present for the sidebands, due to the anharmonicity of the optical potential. This leads to different transition frequencies for $n \to n \pm 1$ (by approximately $E_R/\hbar$). For the present parameters this anharmonicity is not resolved even in the unconvolved spectrum.

11.5.5 Localization by Spontaneous Emission

As outlined in the context of (11.3.114) there is no unique way of decomposing a given master equation (11.2.27) to form quantum trajectories $\tilde{\psi}_c(t)$. This statement is equivalent to noting that (11.2.27) is form invariant under the substitution (11.3.114). Different sets of jump operators $\{c_j\}$ not only lead to a different physical interpretation of trajectories, but an appropriate choice of c_j may be crucial for the formulation of an efficient simulation method for estimating the ensemble distribution [11.23]. We will illustrate these ideas in the context of the quantized motion of an atom moving in optical molasses for the master equation (11.5.51).

a) Localizing Quantum Jumps: Simulation methods for modelling spontaneous emission according to the master equation (11.5.51) have assumed an *angle resolved detection of the photon* as illustrated in Fig.11.12(a). For a one dimen-

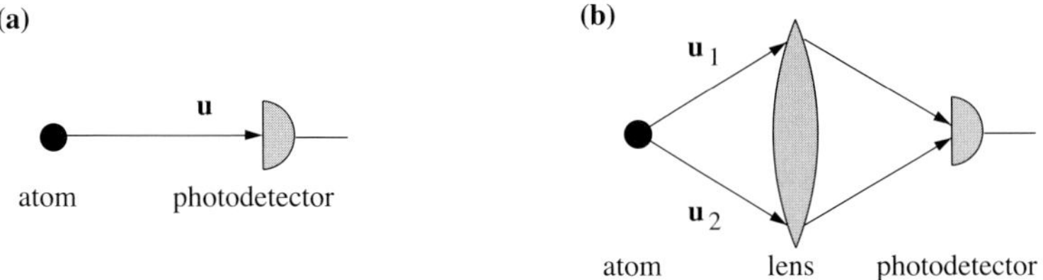

Fig. 11.12 Schematic illustration of the two different measurement bases for the detection of spontaneous emission [11.23].

sional system with adiabatically eliminated excited state this gave the jump operators (11.5.53),

$$c_{u\sigma} = \sqrt{\gamma_0\, N_\sigma(u)}\; \mathrm{e}^{-iku\hat{z}} B_\sigma(\hat{z}) \quad (-1 \leq u \leq +1, \sigma = 0, \pm 1\,)\,. \tag{11.5.60}$$

In a simulation each quantum jump gives information on the direction of the emitted photon.

Alternatively, we could *observe the fluorescence through a lens*, as demonstrated in Fig.11.12(b). The wave function simulation in this case is equivalent to the direct simulation of a Heisenberg microscope: it is not possible to distinguish different paths u_1 and u_2 which may be taken by the photon so the decay operator does not generate a unique recoil. Instead, information about the position coordinate is provided by each emission event, and as a result *the wavefunction is localized in position space*. Applying a Fourier transform to the operators in (11.5.60) to model the action of the ideal lens gives the new decay operators for one dimension

$$c_{\nu\sigma} = \int_{-1}^{1} du\, \sqrt{\gamma_0 N_\sigma(u)}\, B_\sigma(\hat{z}) \mathrm{e}^{-iku(\hat{x}-\nu\lambda/2)} \quad (\nu = 0, \pm 1, \ldots)\,. \tag{11.5.61}$$

For angle resolved detection, u labels a continuous but bounded set of operators, so that in the conjugate basis, ν can be any integer and indexes an infinite set of operators at discrete points. The integral can be evaluated for $c_{\nu=0\sigma}$ to give a localized function centred at the origin and the rest generated by translation by multiples of $\pm\lambda/2$. To prove that both sets of jumps operators (11.5.60) and (11.5.61) give rise to the same a priori dynamics, we consider the following identity for the recycling term in (11.5.51),

$$\begin{aligned} \mathcal{J}_\sigma\, \rho \propto & \int_{-1}^{1} du\, \left(\sqrt{N_\sigma(u)}\, B_\sigma \mathrm{e}^{iku\hat{z}}\right) \rho \left(\sqrt{N_\sigma(u)}\, B_\sigma \mathrm{e}^{iku\hat{z}}\right)^\dagger \\ = & \sum_{\nu=-\infty}^{\infty} \left[\int_{-1}^{1} du \sqrt{\frac{N_\sigma(u)}{2}}\, B_\sigma \mathrm{e}^{iku(\hat{z}-\nu\lambda/2)}\right] \rho \left[\int_{-1}^{1} du' \sqrt{\frac{N_\sigma(u')}{2}}\, B_\sigma \mathrm{e}^{iku'(\hat{z}-\nu\lambda/2)}\right]^\dagger \end{aligned} \tag{11.5.62}$$

where we have used

$$\delta(u-u') = \tfrac{1}{2} \sum_{\nu=-\infty}^{\infty} \mathrm{e}^{i\pi\nu(u-u')} \quad (-1 < u, u' < +1)\,. \tag{11.5.63}$$

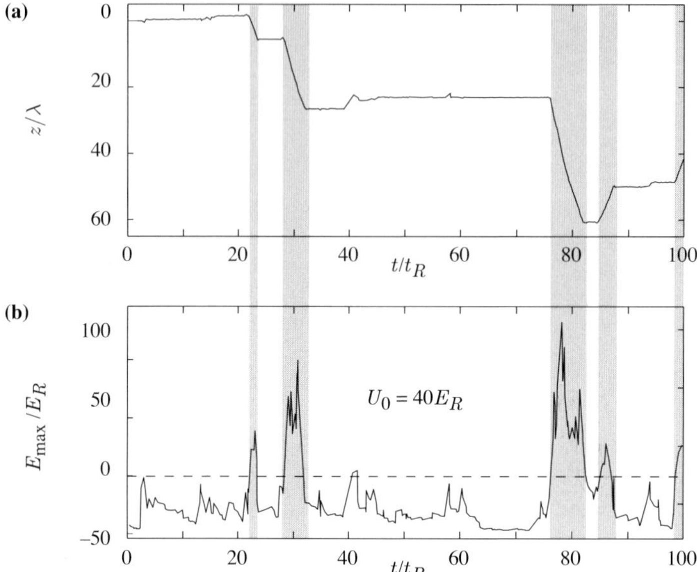

Fig. 11.13 **(a)** Expectation value of the spatial coordinate for a single trajectory versus time. **(b)** Corresponding (kinetic plus potential) energy expectation value. Note the coincidences between above barrier energies and long flight periods over many wavelengths [11.23].

b) Application to Quantum Diffusion of Atoms in Optical Molasses: As an application of the new simulation method, we consider a fully quantum mechanical treatment of atomic diffusion in optical molasses for the lin $\perp$ lin laser configuration described above [11.23]. The calculation of quantum diffusion using the angle resolved detection approach is difficult because the wave function spreads out during the coherent propagation. In contrast to a description of an infinitely extended periodic molasses [11.12], we have here an intrinsically non-periodic problem. Applying the localizing jump operators (11.5.61) allows us to make use of a greatly reduced basis set which we allocate dynamically to follow the atom. One representative trajectory illustrating the random walks is shown in Fig.11.13(a) where we plot the expectation value of the spatial coordinate as a function of time. The long periods when the position does not change appreciably correspond to sub-barrier motion when the total energy of the atom is below the threshold given by the maximum of the optical potential. Energy fluctuations allow eventually the atom to overcome the potential barrier as indicated by the dashed line in Fig.11.13(b). It may then travel over several wavelengths until it is trapped again. For a more complete discussion of the spatial diffusion coefficient we refer to [11.23].

12. Cascaded Quantum Systems

Because it is now possible to produce reliable sources of non-classical light, such as squeezed and antibunched light, it is relevant to consider how one should describe the driving of an optical system by a nonclassical light field. The special situation of incoming squeezed white noise, and methods based on the adjoint equation were used to describe the driving of a single atom with Gaussian squeezed light in Chap.10, but the methods used there do not generalize easily to arbitrary statistics of the incident light. In particular the case of incident antibunched light—the first example of nonclassical light ever produced—is not amenable to this kind of treatment. This is particularly interesting, since antibunching is a property of fourth order correlation functions, while squeezing is a property of second order correlation functions. In any Markovian approximation, only second order correlations are important, so that any fourth order correlation effects will not be noticeable in this degree of approximation.

There are two main issues to be considered.

i) How does one specify a light field with arbitrary statistics? In principle this involves the specification of all possible correlation functions, but this is obviously impractical. The most we should hope for is to find a way of specifying such non-classical statistics as can in practice be produced. The best way of doing this is to model the system which produces the light, and couple the resulting output light into the system under study.

ii) The second issue is then to find a method by which we can model the process of feeding the output of one system into the input of a second system. The major problem here is how one writes a physically acceptable formalism which allows coupling from system 1 to system 2 *without* allowing coupling in the reverse direction. This is achievable in the laboratory with a unidirectional coupler, which utilizes Faraday rotation. Indeed the isolation of the laser used to drive an optical system from reflections off the driven system is a common experimental issue. We can thus be confident that unidirectional coupling can be achieved within a Hamiltonian framework.

The methodology of this chapter was developed independently by *Kolobov* and *Sokolov* [12.1], *Carmichael* [12.2] and *Gardiner* [12.3]. The terminology "cascaded quantum systems" is Carmichael's, and emphasizes that the output of the first system is fed into that of the second system, but that the reverse process does not happen. In principle, there is no limit on the number of systems which may be cascaded.

In this chapter we will develop the required methodology, and apply it to the driving of atoms by non-classical light. We finish the chapter with an application to *quantum information theory*, in which we show how the information contained in quantum state can be transmitted on from place to another.

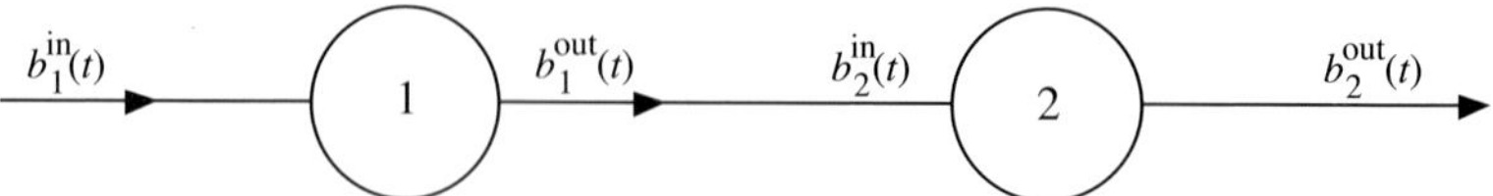

Fig. 12.1 Schematic diagram of system 2 being driven by the output from system 1.

12.1 Coupling Equations

We will develop our coupling equations using an obvious procedure. We consider a quantum system which can be decomposed into two subsystems. Using the notation of Chap.5, a driving field $b_{\rm in}(1,t)$ drives the first system, and gives rise to an output $b_{\rm out}(1,t)$ which, after a propagation delay τ, becomes the input field $b_{\rm in}(2,t)$ to the second system, as illustrated in Fig. 12.1. Now let a_1, a_2 be operators for the two systems, and let $H_{\rm sys} \equiv H_{\rm sys}(1)+H_{\rm sys}(2)$ be the sum of their otherwise independent system Hamiltonians.

The quantum Langevin equations, as in (5.3.15), in this case take the form

$$\begin{aligned}\dot{a}_1 &= -\frac{\mathrm{i}}{\hbar}[a_1, H_{\rm sys}] \\ &\quad -[a_1, c_1^\dagger]\left\{\frac{\gamma_1}{2}c_1+\sqrt{\gamma_1}\, b_{\rm in}(1,t)\right\}+\left\{\frac{\gamma_1}{2}c_1^\dagger+\sqrt{\gamma_1}\, b_{\rm in}^\dagger(1,t)\right\}[a_1, c_1], \quad (12.1.1)\\ \dot{a}_2 &= -\frac{\mathrm{i}}{\hbar}[a_2, H_{\rm sys}] \\ &\quad -[a_2, c_2^\dagger]\left\{\frac{\gamma_2}{2}c_2+\sqrt{\gamma_2}\, b_{\rm in}(2,t)\right\}+\left\{\frac{\gamma_2}{2}c_2^\dagger+\sqrt{\gamma_2}\, b_{\rm in}^\dagger(2,t)\right\}[a_2, c_2]. \quad (12.1.2)\end{aligned}$$

The output field from the first system, as in (5.3.24) is given by

$$b_{\rm out}(1,t) = b_{\rm in}(1,t) + \sqrt{\gamma_1}\, c_1(t). \quad (12.1.3)$$

If it takes a time τ for light to travel from system 1 to system 2, then we can get the effect of feeding the output of system 1 into the input of system 2 by writing

$$b_{\rm in}(2,t) = b_{\rm out}(1,t-\tau) = b_{\rm in}(1,t-\tau) + \sqrt{\gamma_1}\, c_1(t-\tau). \quad (12.1.4)$$

If now a is an operator from either of the systems, the resulting two quantum Langevin equations can be written as one equation for a (using the abbreviated notation $b_{\rm in}(1,t) \to b_{\rm in}(t)$):

$$\begin{aligned}\dot{a} &= -\frac{\mathrm{i}}{\hbar}[a, H_{\rm sys}] \\ &\quad -[a, c_1^\dagger]\left\{\frac{\gamma_1}{2}c_1 + \sqrt{\gamma_1}\, b_{\rm in}(t)\right\} + \left\{\frac{\gamma_1}{2}c_1^\dagger + \sqrt{\gamma_1}\, b_{\rm in}^\dagger(t)\right\}[a, c_1] \\ &\quad -[a, c_2^\dagger]\left\{\frac{\gamma_2}{2}c_2 + \sqrt{\gamma_1\gamma_2}\, c_1(t-\tau) + \sqrt{\gamma_2}\, b_{\rm in}(t-\tau)\right\} \\ &\quad +\left\{\frac{\gamma_2}{2}c_2^\dagger + \sqrt{\gamma_1\gamma_2}\, c_1^\dagger(t-\tau) + \sqrt{\gamma_2}\, b_{\rm in}^\dagger(t-\tau)\right\}[a, c_2]. \quad (12.1.5)\end{aligned}$$

The major technical difficulty is the fact that operators at the two times t and $t-\tau$ both turn up. Because we are considering only the case in which the driving is one way, it is clear that τ must be positive, but otherwise may be chosen arbitrarily—its only effect is to shift the origin of the time axis for the second atom.

To see this explicitly, note that the equations (12.1.5) take on different forms depending on whether a is an operator from the first atom or the second. To put these in the simplest form we proceed as follows:

i) Define an advanced operator $\hat{a}(t) \equiv a(t+\tau)$.

ii) Note that we can write $H_{\text{sys}} = H_1 + H_2$, where the two parts are operators only in the first and second space respectively.

iii) If a_1, a_2 represent arbitrary operators in the first and second systems respectively, the equation (12.1.5) gives the two equations

$$\begin{aligned}\frac{da_1}{dt} &= -\frac{\mathrm{i}}{\hbar}[a_1, H_1] \\ &\quad -[a_1, c_1^\dagger]\left\{\frac{\gamma_1}{2}c_1 + \sqrt{\gamma_1}\, b_{\text{in}}(t)\right\} + \left\{\frac{\gamma_1}{2}c_1^\dagger + \sqrt{\gamma_1}\, b_{\text{in}}^\dagger(t)\right\}[a_1, c_1], \qquad (12.1.6)\\ \frac{d\hat{a}_2}{dt} &= -\frac{\mathrm{i}}{\hbar}[\hat{a}_2, \hat{H}_2] \\ &\quad -[\hat{a}_2, \hat{c}_2^\dagger]\left\{\frac{\gamma_2}{2}\hat{c}_2 + \sqrt{\gamma_1\gamma_2}\, c_1 + \sqrt{\gamma_2}\, b_{\text{in}}(t)\right\} \\ &\quad +\left\{\frac{\gamma_2}{2}\hat{c}_2^\dagger + \sqrt{\gamma_1\gamma_2}\, c_1^\dagger + \sqrt{\gamma_2}\, b_{\text{in}}^\dagger(t)\right\}[\hat{a}_2, \hat{c}_2]. \qquad (12.1.7)\end{aligned}$$

There is now only one time in the resulting equations, and the solutions for these equations provide all information required to describe the system, since they are valid for arbitrary operators in acting on one space or the other, in terms of which any other operator can also be written. The same result is obtained simply by setting $\tau = 0$ in (12.1.5), and omitting the rather cumbersome $\hat{a}_2$ notation, and this is what we shall do from now on.

Thus the final form of the quantum Langevin equation is

$$\begin{aligned}\dot{a} &= -\frac{\mathrm{i}}{\hbar}[a, H_{\text{sys}}] \\ &\quad -[a, c_1^\dagger]\left\{\frac{\gamma_1}{2}c_1 + \sqrt{\gamma_1}\, b_{\text{in}}(t)\right\} + \left\{\frac{\gamma_1}{2}c_1^\dagger + \sqrt{\gamma_1}\, b_{\text{in}}^\dagger(t)\right\}[a, c_1] \\ &\quad -[a, c_2^\dagger]\left\{\frac{\gamma_2}{2}c_2 + \sqrt{\gamma_2}\, b_{\text{in}}(t)\right\} + \left\{\frac{\gamma_2}{2}c_2^\dagger + \sqrt{\gamma_2}\, b_{\text{in}}^\dagger(t)\right\}[a, c_2] \\ &\quad -[a, c_2^\dagger]\sqrt{\gamma_1\gamma_2}\, c_1 + \sqrt{\gamma_1\gamma_2}\, c_1^\dagger[a, c_2]. \qquad (12.1.8)\end{aligned}$$

12.1.1 Relation to Input-Output Formalism

The output of the combined system (in the limit $\tau \to 0+$) is found by using (12.1.3):

$$\begin{aligned}b_{\text{out}}(t) &= b_{\text{out}}(2, t) \\ &= b_{\text{in}}(2, t) + \sqrt{\gamma_2}\, c_2(t) \\ &= b_{\text{in}}(t) + \sqrt{\gamma_1}\, c_1(t) + \sqrt{\gamma_2}\, c_2(t). \qquad (12.1.9)\end{aligned}$$

This form implies that the effective coupling to the input field is by the operator $\sqrt{\gamma_1}\,c_1 + \sqrt{\gamma_2}\,c_2$, and we can rewrite the quantum Langevin equation (12.1.5), after setting $\tau \to 0$, as a quantum Langevin equation of the standard form (5.3.15) plus an extra Hamiltonian term:

$$\begin{aligned}\dot{a} = &-\frac{i}{\hbar}\left[a, H_{sys} + \frac{i\hbar\sqrt{\gamma_1\gamma_2}}{2}\left(c_1^\dagger c_2 - c_2^\dagger c_1\right)\right] \\ &-\left[a, \sqrt{\gamma_1}\,c_1^\dagger + \sqrt{\gamma_2}\,c_2^\dagger\right]\left\{\frac{\sqrt{\gamma_1}\,c_1 + \sqrt{\gamma_2}\,c_2}{2} + b_{in}(t)\right\} \\ &-\left\{\frac{\sqrt{\gamma_1}\,c_1^\dagger + \sqrt{\gamma_2}\,c_2^\dagger}{2} + b_{in}^\dagger(t)\right\}\left[a, \sqrt{\gamma_1}\,c_1 + \sqrt{\gamma_2}\,c_2\right].\end{aligned} \tag{12.1.10}$$

By rewriting the equation in this form, it can be seen that the coupling can be viewed as arising from a *coherent* radiation term from the combined operator $\sqrt{\gamma_1}\,c_1 + \sqrt{\gamma_2}\,c_2$, given by the second two lines, combined with a Hamiltonian term which exchanges the excitation between the two atoms. Under the exchange $1 \leftrightarrow 2$ the radiation term is symmetric, while the Hamiltonian term is antisymmetric, and the asymmetry of the process arises from the interference between these.

12.1.2 Conversion to Quantum Ito Equations

Although the quantum Langevin equations give an elegant description of the physics involved, only an equivalent master equation can provide a tractable way of treating the problems numerically. We will therefore convert the quantum Langevin equations into quantum Ito stochastic differential equations, from which an appropriate master equation can be derived, using the methods of Chap.5.

The most general case which we will consider is that of an input field which can be written in terms of a coherent part and a quantum white noise part

$$b_{in}(t)\,dt = dB(t) + \mathcal{E}_{in}(t)\,dt \tag{12.1.11}$$

with

$$dB(t)^2 = dB^\dagger(t)^2 = 0 \tag{12.1.12}$$

$$dB(t)dB^\dagger(t) = (\bar{N} + 1)dt \tag{12.1.13}$$

$$dB(t)^\dagger dB(t) = \bar{N}\,dt. \tag{12.1.14}$$

The Ito white noise quantum stochastic differential equation is most easily derived using the form (12.1.10) of the quantum Langevin equation, and is

$$\begin{aligned}da = &-\frac{i}{\hbar}[a, H_{sys}]\,dt \\ &-[a, c_1^\dagger]\left\{\tfrac{1}{2}\gamma_1 c_1 + \sqrt{\gamma_1}\,\mathcal{E}_{in}(t)\right\}dt + \left\{\tfrac{1}{2}\gamma_1 c_1^\dagger + \sqrt{\gamma_1}\,\mathcal{E}_{in}^*(t)\right\}[a, c_1]\,dt \\ &-[a, c_2^\dagger]\left\{\tfrac{1}{2}\gamma_2 c_2 + \sqrt{\gamma_2}\,\mathcal{E}_{in}(t)\right\}dt + \left\{\tfrac{1}{2}\gamma_2 c_2^\dagger + \sqrt{\gamma_2}\,\mathcal{E}_{in}^*(t)\right\}[a, c_2]\,dt\end{aligned}$$

$$
\begin{aligned}
&-[a, c_2^\dagger]\sqrt{\gamma_1\gamma_2}\, c_1\, dt + \sqrt{\gamma_1\gamma_2}\, c_1^\dagger [a, c_2]\, dt \\
&-\tfrac{1}{2}\bar{N}\left[\left[a, \sqrt{\gamma_1}\, c_1^\dagger + \sqrt{\gamma_2}\, c_2^\dagger\right], \sqrt{\gamma_1}\, c_1 + \sqrt{\gamma_2}\, c_2\right] dt \\
&-\tfrac{1}{2}\bar{N}\left[\sqrt{\gamma_1}\, c_1^\dagger + \sqrt{\gamma_2}\, c_2^\dagger, \left[\sqrt{\gamma_1}\, c_1 + \sqrt{\gamma_2}\, c_2, a\right]\right] dt \\
&-\sqrt{\gamma_1}\,[a, c_1^\dagger] dB(t) + \sqrt{\gamma_1}\, dB^\dagger(t)[a, c_1] \\
&-\sqrt{\gamma_2}\,[a, c_2^\dagger] dB(t) + \sqrt{\gamma_2}\, dB^\dagger(t)[a, c_2]. \qquad (12.1.15)
\end{aligned}
$$

12.1.3 Master Equation

Using this Ito equation, it follows that a master equation can be derived for the density operator $\rho(t)$ by setting $\langle da(t)\rho\rangle \equiv \langle da\rho(t)\rangle$, and this master equation takes the form

$$
\begin{aligned}
\frac{d\rho}{dt} = {} & \frac{i}{\hbar}[\rho, H_{\text{sys}}] + \frac{\gamma_1}{2}\left\{2c_1\rho c_1^\dagger - \rho c_1^\dagger c_1 - c_1^\dagger c_1 \rho\right\} \\
&+\frac{\gamma_2}{2}\left\{2c_2\rho c_2^\dagger - \rho c_2^\dagger c_2 - c_2^\dagger c_2 \rho\right\} \\
&-\sqrt{\gamma_1\gamma_2}\left\{[c_2^\dagger, c_1\rho] + [\rho c_1^\dagger, c_2]\right\} \\
&+\frac{\bar{N}}{2}\left[[\sqrt{\gamma_1}\, c_1 + \sqrt{\gamma_2}\, c_2, \rho], \sqrt{\gamma_1}\, c_1^\dagger + \sqrt{\gamma_2}\, c_2^\dagger\right] \\
&+\frac{\bar{N}}{2}\left[[\sqrt{\gamma_1}\, c_1^\dagger + \sqrt{\gamma_2}\, c_2^\dagger, \rho], \sqrt{\gamma_1}\, c_1 + \sqrt{\gamma_2}\, c_2\right] \\
&-\left[\mathcal{E}_{\text{in}}(t)(\sqrt{\gamma_1}\, c_1^\dagger + \sqrt{\gamma_2}\, c_2^\dagger) - \mathcal{E}_{\text{in}}^*(t)(\sqrt{\gamma_1}\, c_1 + \sqrt{\gamma_2}\, c_2), \rho\right]. \qquad (12.1.16)
\end{aligned}
$$

12.1.4 The Lindblad Form

Although it is not immediately apparent, the master equation (12.1.16) is of the Lindblad form, which we noted in Sect. 5.2.2 is required to preserve the positivity of the density operator. This is apparent if we derive the master equation from the form (12.1.10), and this is equivalent to using the identity for the coupling terms

$$
\begin{aligned}
2[a\rho, b^\dagger] + 2[b, \rho a^\dagger] = {} & 2(a+b)\rho(a+b)^\dagger - (a+b)^\dagger(a+b)\rho - \rho(a+b)^\dagger(a+b) \\
&+[a^\dagger b - b^\dagger a, \rho]. \qquad (12.1.17)
\end{aligned}
$$

This gives the coupling term explicitly as the sum of a damping term and a commutator, as required by (5.2.29).

12.1.5 Stochastic Schrödinger Equation

Since the master equation (12.1.16) is of the Lindblad form, a stochastic Schrödinger equation formulation can be given for the cascaded systems. Thus, we can write the evolution of a stochastic wavefunction in the form (11.3.90), for the unnormalized wavefunction

$$
d\psi_c(t) = -\frac{i}{\hbar} H_{\text{eff}}\psi_c(t)\, dt
$$

$$+\lambda^{(+)}\sqrt{\bar{N}}\left(\sqrt{\gamma_1}\,c_1^\dagger+\sqrt{\gamma_2}\,c_2^\dagger-1\right)\psi_c(t)\,dN^{(+)}(t)$$
$$+\lambda^{(-)}\sqrt{\bar{N}+1}\left(\sqrt{\gamma_1}\,c_1+\sqrt{\gamma_2}\,c_2-1\right)\psi_c(t)\,dN^{(-)}(t). \qquad (12.1.18)$$

Here, the non-Hermitian effective Hamiltonian can be written

$$\begin{aligned}H_{\text{eff}} = H_{\text{sys}} &+ \mathrm{i}\hbar\left\{\mathcal{E}_{\text{in}}(t)(\sqrt{\gamma_1}\,c_1^\dagger+\sqrt{\gamma_2}\,c_2^\dagger)-\mathcal{E}_{\text{in}}^*(t)(\sqrt{\gamma_1}\,c_1+\sqrt{\gamma_2}\,c_2)\right\}\\ &+\frac{\mathrm{i}\hbar\sqrt{\gamma_1\gamma_2}}{2}\left(c_1^\dagger c_2-c_2^\dagger c_1\right)-\frac{\mathrm{i}\hbar}{2}\left(\sqrt{\gamma_1}\,c_1^\dagger+\sqrt{\gamma_2}\,c_2^\dagger\right)\left(\sqrt{\gamma_1}\,c_1+\sqrt{\gamma_2}\,c_2\right).\\ &-\frac{\mathrm{i}\hbar\bar{N}}{2}\left(\sqrt{\gamma_1}\,c_1+\sqrt{\gamma_2}\,c_2\right)\left(\sqrt{\gamma_1}\,c_1^\dagger+\sqrt{\gamma_2}\,c_2^\dagger\right)\\ &-\frac{\mathrm{i}\hbar\bar{N}}{2}\left(\sqrt{\gamma_1}\,c_1^\dagger+\sqrt{\gamma_2}\,c_2^\dagger\right)\left(\sqrt{\gamma_1}\,c_1+\sqrt{\gamma_2}\,c_2\right). \qquad (12.1.19)\end{aligned}$$

This non-Hermitian Hamiltonian is composed of system and driving terms on the first line, and then the Hamiltonian term from the coupled systems, finally followed by the anti-Hermitian damping terms, including those proportional to $\bar{N}$. In the case that $\bar{N} = 0$, it can be advantageous to simplify the second line to $-\frac{1}{2}\mathrm{i}\hbar\left(\gamma_1 c_1^\dagger c_1+\gamma_2 c_2^\dagger c_2+2\sqrt{\gamma_1\gamma_2}\,c_2^\dagger c_1\right)$, but the form (12.1.19) gives a clearer idea of the structure of H_{eff} in the general case.

12.1.6 Imperfect Coupling

In the previous section we considered the case in which all of the output of the first system was used as the input to the second system. This is usually not realistic, since

i) There must be some losses in transmission.

ii) We may not wish to couple perfectly from the first to the second system.

iii) For example, by (12.1.3), any coherent component of $b_{\text{in}}(1,t)$ will appear in $b_{\text{out}}(1,t)$, and hence in $b_{\text{in}}(2,t)$. We may wish to do this, but we may also wish to investigate the effect of illuminating the second system with only the *fluorescent* light (represented in (12.1.3) by $\sqrt{\gamma_1}\,c_1(t)$) from the first system.

a) Input Channels: This can be done by considering the coupling of the first system to two channels with the same kind of coupling (though of different strength), as illustrated in Fig. 12.2. We thus modify the quantum Langevin equation (12.1.1) by setting

$$b_{\text{in}}(t) = \sqrt{\epsilon_1}\;f_{\text{in}}(t)+\sqrt{(1-\epsilon_1)}\;g_{\text{in}}(t) \qquad (12.1.20)$$

with $0 \le \epsilon_1 \le 1$. Here $f_{\text{in}}(t)$ and $g_{\text{in}}(t)$ together make up all the coupling available to radiated modes, so that the total damping constant is simply γ_1. There are then two output fields given by

$$f_{\text{out}}(t) = f_{\text{in}}(t)+\sqrt{\epsilon_1\gamma_1}\;c_1(t) \qquad (12.1.21)$$
$$g_{\text{out}}(t) = g_{\text{in}}(t)+\sqrt{(1-\epsilon_1)\gamma_1}\;c_1(t). \qquad (12.1.22)$$

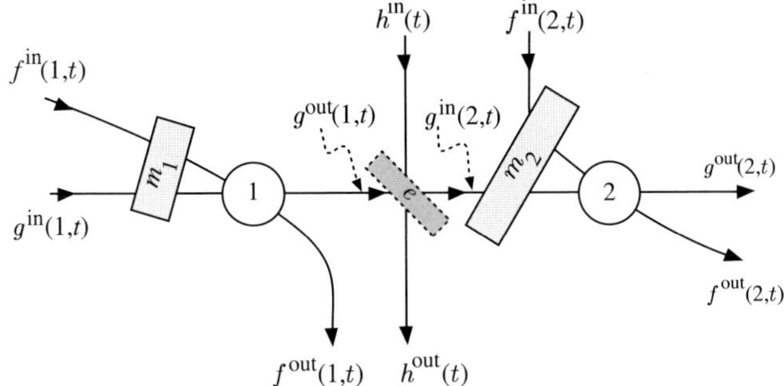

Fig. 12.2 Arrangements required to account for imperfect coupling. The systems are considered to have two input-output channels. The corresponding fields are coupled with strengths $\sqrt{\epsilon_i}$, $\sqrt{1-\epsilon_i}$, so that the *total* coupling is the same independent of ϵ_i. To account for the possibility that the input channel to the second system may not match perfectly with an output channel from the first system, a *unitary mixer* m_1, is inserted between the physical inputs, and those corresponding to the required outputs. The same is done for the second system. The beam splitter e, is necessary to account for the fact the light in the relevant output channel may not all be fed into the relevant input channel in the second system.

If we now consider a situation in which we drive the system with a coherent field in the $f_{\rm in}$ channel, it is clear that in the $g_{\rm out}$ channel we will see no component of the incoming driving field. Physically, we could consider the case where $f_{\rm in}$ represents a laser beam (almost a plane wave) while $g_{\rm out}$ is an outgoing electric dipole wave, observed away from the direction of the laser beam.

b) Coupling to the Next System: In coupling to the next system we must consider that some proportion of $g_{\rm out}$ will be lost in transmission. This can be modelled by inserting a beam splitter in the channel $g_{\rm out}$, so that we have

$$g_{\rm in}(2,t) = \sqrt{e_1}\; g_{\rm out}(1,t) + \sqrt{1-e_1}\; h_{\rm in}(t) \tag{12.1.23}$$

$$= \sqrt{e_1(1-\epsilon_1)\gamma_1}\; c_1(t) + \sqrt{e_1}\; g_{\rm in}(1,t) + \sqrt{1-e_1}\; h_{\rm in}(t) \tag{12.1.24}$$

where $h_{\rm in}(t)$ represents a vacuum field. The total field then enters the second system, with its complementary field $f_{\rm in}(2,t)$. Using (12.1.23) and (12.1.24), we find the quantum Langevin equation

$$\begin{aligned}\dot a = & -\frac{i}{\hbar}[a, H_{\rm sys}] \\ & -\left([a, c_1^\dagger]\left\{\frac{\gamma_1}{2}c_1 + \sqrt{\gamma_1\epsilon_1}\; f_{\rm in}(1,t) + \sqrt{\gamma_1(1-\epsilon_1)}\; g_{\rm in}(1,t)\right\} + \text{conjugate}\right) \\ & -\left([a, c_2^\dagger]\left\{\frac{\gamma_2}{2}c_2 + \sqrt{\gamma_2\epsilon_2}\; f_{\rm in}(2,t) + \sqrt{\gamma_2(1-\epsilon_2)}\left(\sqrt{\gamma_1 e_1(1-\epsilon_1)}\, c_1(t-\tau)\right.\right.\right. \\ & \left.\left.\left. +\sqrt{e_1}\, g_{\rm in}(1,t-\tau) + \sqrt{1-e_1}\; h_{\rm in}(t)\right)\right\} + \text{conjugate}\right).\end{aligned} \tag{12.1.25}$$

In this equation, the term "conjugate" means that one should take the Hermitian conjugate of everything except a.

c) Equivalent Master Equation: In the case that $g_{\rm in}(1,t)$, $h_{\rm in}(t)$ and $f_{\rm in}(2,t)$ represent the vacuum, and we have a coherent driving field in the $f_{\rm in}$ channel of the first system, which we represent by

$$f_{\rm in}(1,t) = E(t) + f^0_{\rm in}(t) \tag{12.1.26}$$

where $f^0_{\rm in}(t)$ is a vacuum field, we can derive the master equation

$$\begin{aligned}\frac{d\rho}{dt} &= \frac{\mathrm{i}}{\hbar}[\rho, H_{\rm sys}] + \frac{\gamma_1}{2}\left\{2c_1\rho c_1^\dagger - \rho c_1^\dagger c_1 - c_1^\dagger c_1 \rho\right\} + \frac{\gamma_2}{2}\left\{2c_2\rho c_2^\dagger - \rho c_2^\dagger c_2 - c_2^\dagger c_2 \rho\right\} \\ &\quad - \sqrt{e_1\gamma_1\gamma_2(1-\epsilon_1)(1-\epsilon_2)}\,\left\{[c_2^\dagger, c_1\rho] + [\rho c_1^\dagger, c_2]\right\} \\ &\quad - \sqrt{\gamma_1\epsilon_1}\,[E(t)c_1^\dagger + E^*(t)c_1, \rho]. \end{aligned} \tag{12.1.27}$$

d) Imperfect Coupling to the Second System: There is the further possibility that $g_{\rm out}$ and $f_{\rm out}$ may not correspond exactly to the $g_{\rm in}$ and $f_{\rm in}$. For example, a fraction of the coherent field in $f_{\rm in}(1,t)$ (see (12.1.26)) may be seen in $g_{\rm out}(1,t)$ due to some misalignment. This is best dealt with by putting a unitary mixer between $f_{\rm in}, g_{\rm in}$ and the system, so that the fields which correspond to $g_{\rm out}, f_{\rm out}$ are in fact

$$g'_{\rm in}(1,t) = \sqrt{m_1}\ g_{\rm in}(1,t) + \sqrt{1-m_1}\ f_{\rm in}(1,t) \tag{12.1.28}$$

$$f'_{\rm in}(1,t) = \sqrt{m_1}\ f_{\rm in}(1,t) - \sqrt{1-m_1}\ g_{\rm in}(1,t) \tag{12.1.29}$$

where $0 \le m_1 \le 1$ is a mixing parameter. (For simplicity we have adjusted arbitrary phases so that no complex coefficients are necessary.)

A similar thing can happen at the input to the second system. From all of these we get the master equation

$$\begin{aligned}\frac{d\rho}{dt} &= \frac{\mathrm{i}}{\hbar}[\rho, H_{\rm sys}] + \frac{\gamma_1}{2}\left\{2c_1\rho c_1^\dagger - \rho c_1^\dagger c_1 - c_1^\dagger c_1 \rho\right\} + \frac{\gamma_2}{2}\left\{2c_2\rho c_2^\dagger - \rho c_2^\dagger c_2 - c_2^\dagger c_2 \rho\right\} \\ &\quad + \frac{\gamma_2}{2}\left\{2c_2\rho c_2^\dagger - \rho c_2^\dagger c_2 - c_2^\dagger c_2 \rho\right\} \\ &\quad - \sqrt{e_1\epsilon_1\gamma_1\gamma_2}\,\left\{\sqrt{m_2\epsilon_2} + \sqrt{(1-m_2)(1-\epsilon_2)}\right\}\left\{[c_2^\dagger, c_1\rho] + [\rho c_1^\dagger, c_2]\right\} \\ &\quad - \sqrt{\gamma_1}\,\left\{\sqrt{m_1\epsilon_1} + \sqrt{(1-m_1)(1-\epsilon_1)}\right\}[E(t)c_1^\dagger + E^*(t)c_1, \rho] \\ &\quad - \sqrt{\gamma_2 e_1(1-m_1)}\,\left\{\sqrt{m_2\epsilon_2} + \sqrt{(1-m_2)(1-\epsilon_2)}\right\}[E(t)c_2^\dagger + E^*(t)c_2, \rho]. \end{aligned} \tag{12.1.30}$$

This equation represents the most general combination of mismatching and inefficient coupling for two coupled systems. For the purposes of solving this master equation, the mixing parameter m_2 is really an unnecessary complication, but in the case of several cascaded systems, it is essential.

12.2 Application to Harmonic Oscillator Systems

The problem of driving a two level atom with squeezed light has already been treated in Sect. 10.4 by using methods based on the adjoint equation. In this section

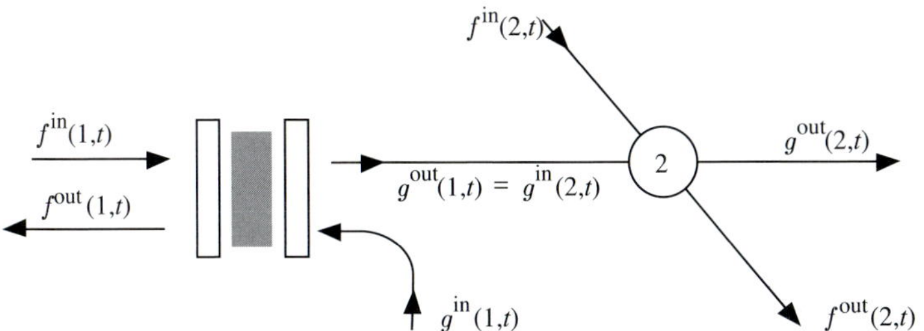

Fig. 12.3 Light from a two sided harmonic oscillator coupled to a second system.

we will show how the coupled systems approach can provide a very much more efficient way of treating this and related problems. The idea is to include the production of the squeezed light in the problem as the first of two coupled systems. This is done by modelling a degenerate parametric oscillator as in Sect. 10.2.1, and using the methods of the previous section to produce a coupled master equation description, which is exactly equivalent to the methods of Sect. 10.4. It proves practicable to solve these equations numerically by truncating the infinite dimensional space of the harmonic oscillator operators used for the parametric oscillator to quite a small size, usually matrices no larger than 10×10 are sufficient for degrees of squeezing up to approximately 90%.

12.2.1 Driving by Squeezed Light

Here we consider the first system to be a one sided cavity driven by a vacuum field, as in Fig. 12.3. The cavity has an appropriate crystal inside which, when pumped appropriately, gives rise to a degenerate parametric amplifier as in Sect. 10.4. There is now only one input-output channel to the first system. The master equation becomes

$$\begin{aligned}\frac{d\rho}{dt} &= \frac{\mathrm{i}}{\hbar}[\rho, H_{\rm sys}] + \frac{\gamma_1}{2}\left\{2a\rho a^\dagger - \rho a^\dagger a - a^\dagger a\rho\right\} + \frac{\gamma_2}{2}\left\{2c_2\rho c_2^\dagger - \rho c_2^\dagger c_2 - c_2^\dagger c_2\rho\right\} \\ &\quad -\sqrt{\eta\gamma_1\gamma_2}\left\{[c_2^\dagger, a\rho] + [\rho a^\dagger, c_2]\right\} \end{aligned} \quad (12.2.1)$$

where,

$$\eta = e_1(1-\epsilon_1)(1-\epsilon_2) \quad (12.2.2)$$

$$\bar{N} = \epsilon_1 \bar{N}_0. \quad (12.2.3)$$

and

$$H_{\rm sys} = \frac{\mathrm{i}\hbar}{2}\left(E a^{\dagger 2} - E^* a^2\right) + H_2. \quad (12.2.4)$$

Here H_2 is an operator only in the second system.

a) P-Representation Method: We can introduce a positive P-representation by a method similar to the Glauber-Sudarshan P-representation used for the laser in Sect. 9.3.2

$$\rho = \int d^2\alpha d^2\beta\, \Lambda(\alpha,\beta)\rho(\alpha,\beta) \quad (12.2.5)$$

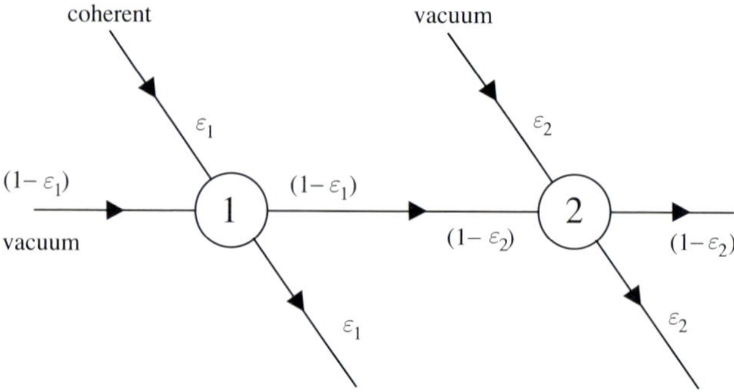

Fig. 12.5 Schematic diagram of the coupling between a coherently-driven atom (1) and a second atom (2). The various input, output, and coupling channels are discussed in the text.

Given that our source of antibunched light is a two level atom, a pumping mechanism must be provided to excite this atom and induce fluorescence from which the source field is derived. The two distinct possibilities that we shall consider are with coherent pumping provided by incident light from a laser, and incoherent pumping provided by finite bandwidth thermal light.

12.3.1 Coherent Excitation of the Source Atom

a) Master Equation: The configuration that we shall consider is shown schematically in Fig. 12.5, where we illustrate the various input, output, and coupling channels for the case of coherent excitation of the source atom. The master equation describing this situation takes the form

$$\begin{aligned}\frac{d\rho}{dt} &= \tfrac{1}{2}\gamma_1\left(2\sigma_1^-\rho\sigma_1^+ - \sigma_1^+\sigma_1^-\rho - \rho\sigma_1^+\sigma_1^-\right) - \sqrt{\epsilon_1\gamma_1}\,[E\sigma_1^+ - E^*\sigma_1^-,\rho] \\ &\quad -\sqrt{(1-\epsilon_1)(1-\epsilon_2)\gamma_1\gamma_2}\left([\sigma_2^+,\sigma_1^-\rho] + [\rho\sigma_1^+,\sigma_2^-]\right) \\ &\quad +\tfrac{1}{2}\gamma_2\left(2\sigma_2^-\rho\sigma_2^+ - \sigma_2^+\sigma_2^-\rho - \rho\sigma_2^+\sigma_2^-\right). \end{aligned} \tag{12.3.1}$$

The coherent electric field E is incident upon atom 1 through the channel labelled ϵ_1. This coherent field is not incident upon atom 2. Antibunched light emitted from atom 1 into channel $(1-\epsilon_1)$ provides the light source for atom 2 with which it is coupled through channel $(1-\epsilon_2)$. The (vacuum input) channel ϵ_2 provides an observation channel for the light emitted from atom 2.

The master equation is solved by direct numerical integration. Application of the quantum regression theorem (Sect. 5.2.3) yields all of the correlation functions and spectra of interest.

b) General Features: The property of antibunching in the light emitted from a two level atom is a consequence of the finite time required for the atom to be re-excited from its ground state following the emission of a photon. This is manifested in the stationary normally ordered intensity correlation function of the emitted light. This function is directly related to photon counting measurements, representing the

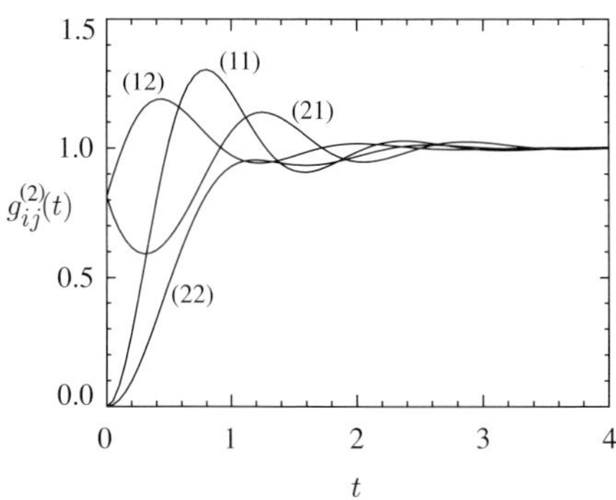

Fig. 12.6 Intensity correlation function $g_{ij}^{(2)}(t)$ for a two level atom driven by light from a coherently driven two level atom, with $\sqrt{\epsilon_1\gamma_1}\,E = 2$, where $\epsilon_1 = 0.5$, and $\gamma_1 = 2$. The other parameters from master equation (12.3.1) are $\epsilon_2 = 0.5$, $\gamma_2 = 2$. (Time scale in arbitrary units.)

probability of counting a second photon at a time t after the initial detection of a photon. For a two level atom this function is identically zero for $t = 0$.

A generalization of this correlation function which we shall compute for the two atom system at hand is given by

$$\langle : I_i(t)I_j(0) :\rangle_s = \beta_i\beta_j \mathrm{Tr}\left\{\sigma_i^+\sigma_i^- V(t)\left\{\sigma_j^-\rho_s\sigma_j^+\right\}\right\}, \tag{12.3.2}$$

where $i, j = 1, 2$ (first or second atom), $V(t)$ is the two sided evolution operator, such that $V(t)\rho(t') = \rho(t+t')$, and ρ_s is the stationary solution of the master equation, as in Sect. 5.2. These functions can be interpreted as the conditional probability of counting a photon from atom i at time t after counting a photon from atom j. Their measurement would correspond to photon counting measurements on the output fields in the channels $(1-\epsilon_1)$ and ϵ_2 (i.e. $\beta_1 = (1-\epsilon_1)\gamma_1$ and $\beta_2 = \epsilon_2\gamma_2$); these fields are not contaminated with non-vacuum input fields, which simplifies the analysis.

In Fig. 12.6, where we plot the normalized intensity correlation functions,

$$g_{ij}^{(2)}(t) = \frac{\langle : I_i(t)I_j(0) :\rangle_s}{\langle I_i\rangle_s\langle I_j\rangle_s}. \tag{12.3.3}$$

The following features are notable:

i) A retardation of the evolution of $g_{22}^{(2)}(t)$ in comparison to that of $g_{11}^{(2)}(t)$. This is not a consequence of the time delay τ between the two atoms; rather it simply reflects the different rates of re-excitation experienced by the two atoms.

ii) A pronounced anticorrelation in $g_{21}^{(2)}(0)$ (i.e. $g_{21}^{(2)}(0) < 1$). This arises because of the antibunching. If a photon is detected from atom 1, then it is not available to excite atom 2. The anticorrelation is not perfect, since emission depends on the excitation of each atom—however, because of the antibunching, if a photon was counted from atom 1, then there was a reduced probability of there having been one in the immediate past to excite atom 2.

iii) This anticorrelation initially becomes more pronounced as t increases. This is because there will be no further photons to excite atom 2 in the time immediately following the emission, i.e., atom 1 is unable to provide a driving field for atom 2 until it has recovered some excitation. Eventually, of course, photons do appear, and $g_{21}^{(2)}(t)$ approaches a value representing zero correlation.

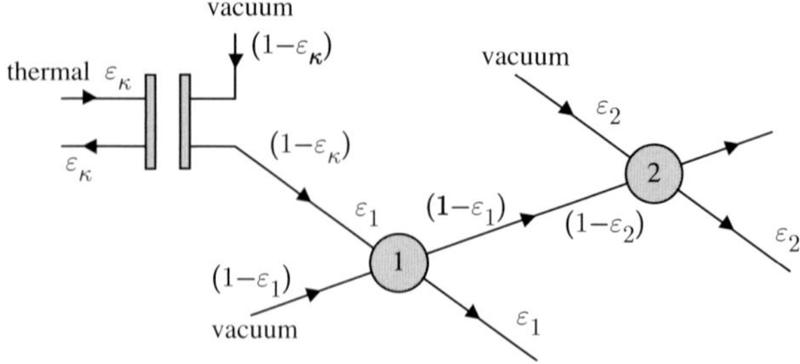

Fig. 12.7 Schematic diagram of the coupling between a thermally driven cavity and two atoms (1,2). The various input, output, and coupling channels are discussed in the text.

iv) In contrast, the correlation function $g^{(2)}_{12}(t)$ tends to mirror the behaviour of $g^{(2)}_{11}(t)$, simply reflecting the fact that excitation of atom 1 is provided by a coherent (Poissonian) light field.

Other measurable quantities, such as values of the atomic inversion and polarization can also be calculated, and are given in [12.4].

12.3.2 Incoherent Excitation of the Source Atom

a) Master Equation: We would like to test to what extent the antibunched nature of the light emitted from the first atom characterizes the nature of the photon counting from the second atom. To do this we shall show that we can produce a light beam from the first atom which has the same intensity and $g^{(2)}(t)$ as that emitted when the light incident is coherent, but is otherwise quite different.

Let us consider excitation of atom 1 by a thermal light field with finite bandwidth. The configuration we shall model is illustrated in Fig. 12.7 and the corresponding master equation is given by

$$\begin{aligned}\frac{d\rho}{dt} &= \kappa(\bar{N}+1)\left(2a\rho a^\dagger - a^\dagger a\rho - \rho a^\dagger a\right) + \kappa\bar{N}\left(2a^\dagger\rho a - aa^\dagger\rho - \rho aa^\dagger\right)\\ &\quad -\sqrt{2\kappa\eta_1\gamma_1}\,\left([\sigma_1^+, a\rho] + [\rho a^\dagger, \sigma_1^-]\right) + \tfrac{1}{2}\gamma_1\left(2\sigma_1^-\rho\sigma_1^+ - \sigma_1^+\sigma_1^-\rho - \rho\sigma_1^+\sigma_1^-\right)\\ &\quad -\sqrt{\eta_2\gamma_1\gamma_2}\,\left([\sigma_2^+, \sigma_1^-\rho] + [\rho\sigma_1^+, \sigma_2^-]\right) + \tfrac{1}{2}\gamma_2\left(2\sigma_2^-\rho\sigma_2^+ - \sigma_2^+\sigma_2^-\rho - \rho\sigma_2^+\sigma_2^-\right).\end{aligned} \tag{12.3.4}$$

The finite bandwidth thermal field is provided by the output from a cavity driven (through the input channel denoted by ϵ_κ) by a broadband, or "white noise", thermal field. The cavity output through channel $(1-\epsilon_\kappa)$ is coupled to atom 1 through channel ϵ_1. Again, this driving field is not incident upon atom 2. The coupling parameters are given by $\eta_1 = (1-\epsilon_\kappa)\epsilon_1$ and $\eta_2 = (1-\epsilon_1)(1-\epsilon_2)$.

Our choice of a finite bandwidth thermal field is motivated by the need to produce a $g^{(2)}(t)$ in which the initial time dependence is quadratic, as in the case of light from a coherently driven atom. A broadband thermal field would inevitably produce an initial linear t dependence. We shall see that by suitably adjusting the bandwidth of the driving field, we can closely match correlation functions for the two different

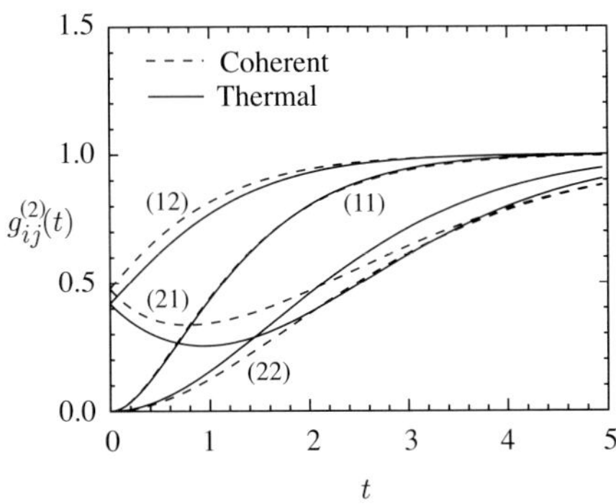

Fig. 12.8 Comparison of intensity correlation functions for coherent and thermal excitation of atom 1. The function $g^{(2)}_{11}(t)$ and the intensity $\eta_2\gamma_1\langle\sigma_1^+\sigma_1^-\rangle$ ($\eta_2 = (1-\epsilon_1)(1-\epsilon_2)$) have been matched for the two cases.

cases of coherent and incoherent excitation of the source atom, so that we may carry out the following comparison between the effects of different kinds of antibunched light.

b) Comparison with Coherent Excitation: The intensity of a light beam and fluctuations in the intensity are properties that one might typically measure, using photon counting, in order to achieve some degree of characterization of the light beam. In the case of antibunched light, both as a driving field (from atom 1) and as the emission induced (on atom 2) by this driving field, it is interesting to consider to what extent the photon counting properties of the induced emission are determined by those of the driving field. We can compare the cases of purely coherent, (12.3.1), and purely incoherent, (12.3.4), thermal excitation of atom 1 when the intensity of the light driving atom 2,

$$\eta_2\gamma_1\langle\sigma_1^+\sigma_1^-\rangle_s\,, \tag{12.3.5}$$

where $\eta_2 = (1-\epsilon_1)(1-\epsilon_2)$, and the intensity correlation function,

$$g^{(2)}_{11}(t) = \frac{\langle\sigma_1^+\sigma_1^+(t)\sigma_1^-(t)\sigma_1^-\rangle_s}{\langle\sigma_1^+\sigma_1^-\rangle_s^2}\,, \tag{12.3.6}$$

are as close as possible the same for both situations. The two different driving fields for atom 2 can then be regarded as having identical antibunching properties.

Such a comparison is presented in Fig. 12.8 where we plot the functions $g^{(2)}_{ij}(t)$ for coherent and thermal excitation. By a suitable choice of parameters, we are able to closely match $g^{(2)}_{11}(t)$ and $\eta_2\gamma_1\langle\sigma_1^+\sigma_1^-\rangle_s$ for the two cases (such that they would quite certainly be indistinguishable in an experiment). The curves describing $g^{(2)}_{ij}(t)$ for $(ij) = (12), (21), (22)$ are clearly different for the two cases, as are the respective intensities of light emitted from atom 2—values of the excited state population for atom 2 are $\langle\sigma_2^+\sigma_2^-\rangle_s = 0.0651$ (thermal), 0.0929 (coherent).

These results demonstrate that the photon counting properties of the light emitted by an atom driven with antibunched light depend on properties of the antibunched light other than those specified by $g^{(2)}_{11}(t)$ and $\eta_2\gamma_1\langle\sigma_1^+\sigma_1^-\rangle_s$. Indeed as shown in [12.4] the spectral properties of the antibunched driving field are quite different for the two cases.

12.4 Characterizing Non-Classical Light

The question which arises from the previous section is the following: How should one best characterize an arbitrary light beam in terms of a few parameters? In the previous section we showed how the photon counting properties of an experiment can depend on more than the photon counting properties of the incident light, and it is clear that a specification of the spectral properties alone would not be sufficient.

The clear message which comes from the cascaded systems technique is the following: *The best way of characterizing non-classical light is by means of the system used to produce it.* In general, it is not too difficult to give a quite accurate model of the production system, and this usually provides a few parameter specification of the light. The subsequent coupling of the output from such a system to the system under study is then relatively straightforward using the cascaded systems approach. This is in fact much simpler than to require the experimenter to measure a sufficient number of correlation functions and then to attempt to model light with the measured correlation functions.

12.5 Transmission of Quantum Information Through a Quantum Network

In this section we study as an application of cascaded quantum systems a model which implements transmission of quantum bits between two atoms stored in two high-Q cavities connected by an optical fibre [12.6, 12.7]. This problem is of significant relevance for quantum communication and quantum computing [12.8, 12.9].

12.5.1 Quantum Information

Over the last decade it has become apparent that information stored in a pure quantum states and processed in such a way as to preserve the purity of the state could well be the basis of a very much more powerful information processing technology than our current technology, which is based on physical concepts taken from classical physics. The elementary object of quantum information theory is the *qubit*, which is based on a two state system with basis states $|0\rangle, |1\rangle$. Unlike classical information, however, the quantum information to be stored in a qubit is not simply a 0 or a 1, but rather is in the form of any superposition of the form

$$|\phi\rangle = c_0|0\rangle + c_1|1\rangle \tag{12.5.1}$$

The goal of *quantum communication* is to send a qubit state $|\phi\rangle$, representing the *quantum information* we want to transmit, reliably from one place to another via a *quantum channel*.

Ideal quantum transmission of a qubit corresponds to a transformation

$$\left(c_0|0\rangle_1 + c_1|1\rangle_1\right) \otimes |0\rangle_2 \longrightarrow |0\rangle_1 \otimes \left(c_0|0\rangle_2 + c_1|1\rangle_2\right) \tag{12.5.2}$$

where we send an *unknown* superposition state stored in the first qubit via the quantum channel to the second qubit. Before the transmission, the second atom is prepared in a reference state, e.g., the state $|0\rangle_2$, while after the transmission the first atom will be left in $|0\rangle_1$. We cannot achieve quantum transmission by performing a measurement on the first qubit to read out its state, and then transmit the measurement result classically since this would destroy the superposition state. Instead we must find a scheme which preserves the quantum nature of (12.5.2).

In a more general context, (12.5.2) could be part of an operation in a *quantum network*. Such a quantum network would consist of spatially separated nodes connected by quantum communication channels. Each node is a quantum system that stores quantum information in quantum bits and processes this information locally using quantum gates, i.e., entanglement operations [12.8]. Exchange of information between the nodes of the network is accomplished via quantum channels. The possibility of combining local quantum processing with quantum transmission between the nodes of the network is relevant for a variety of applications ranging from entangled-state cryptography[12.9], teleportation[12.10] and purification [12.11], and is interesting from the perspective of distributed quantum computation.

12.5.2 Physical Implementation of a Quantum Network with Atoms and Photons

In a practical implementation of quantum communication the qubits would be stored in two level atoms, and the quantum channel would be represented by optical fibres, i.e., the atoms represent the *quantum memory* while the photons propagating in an optical fibre serve as a *data bus* to realise a quantum channel. This choice is motivated by the following advantages:

a) Nodes of a Quantum Network: Atoms and ions are particularly well suited for storing qubits reliably in long-lived internal states. In practice, these atomic levels could be, for example, the Zeeman ground states of an atom. (Alternative proposals have been made to implement quantum information processing operations using ion traps or cavity QED.)

b) Quantum Channels: Photons clearly represent the best qubit-carrier for fast and reliable communication over long distances, since fast and internal state preserving transportation of atoms or ions seems to be technically intractable.

12.5.3 Physical Idea behind Ideal Transmission

The basic idea is to utilize strong coupling between a high–Q optical cavity and the atoms forming a given node of the quantum network. The process than can be summarized as follows:

i) By applying laser beams, one first transfers the internal state of an atom at the first node to the optical state of the cavity mode.
ii) The generated photons leak out of the cavity, propagate as a wavepacket along the transmission line, and enter an optical cavity at the second node.
iii) Finally, the optical state of the second cavity is transferred to the internal state of an atom.

This arrangement has the form of a *cascaded quantum system* where a photon emerging from the first cavity drives the atom in the second cavity. Thus the theory of cascaded quantum systems provides the natural language to describe this process.

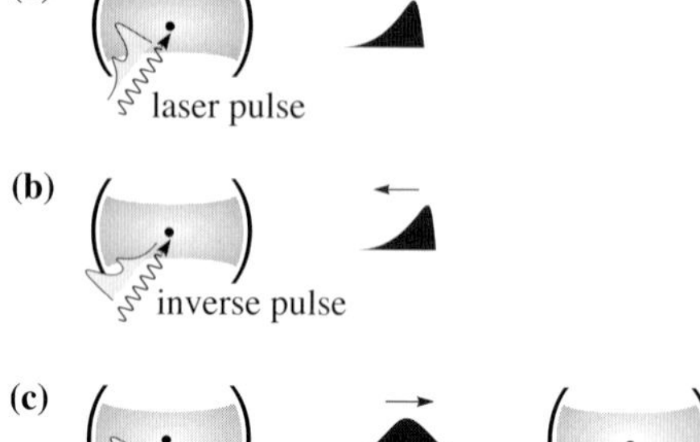

Fig. 12.9 **(a)** Cavity decay produces a photon wave packet. The photon inside the atom has been produced by laser excitation of an atom. **(b)** A "time reversed" cavity decay restores the atom in its original state. **(c)** Transfer of the quantum state from one atom to the other using a time-symmetric pulse.

a) The Problem of Photon Reflection from the Second Cavity: The central problem in the transmission described above is that the photon can be reflected from the second cavity, and thus the qubit will not be restored in the atom of the second cavity. In this case the transmission has failed. We must, therefore, develop a protocol which avoids the reflection from the second cavity.

Consider a photon which leaks out of an optical cavity and propagates away as a wavepacket as in Fig. 12.9a. Imagine that we were able to time reverse this wavepacket and send it back into the cavity as in Fig. 12.9b; then this would restore the original (unknown) superposition state of the atom, provided we would also reverse the timing of the laser pulses. If, on the other hand, we can drive the atom in a transmitting cavity in such a way that the outgoing pulse is already symmetric in time, the wavepacket entering a receiving cavity would mimic this time reversed process, thus generating the state of the first atom in the second one as in Fig. 12.9c.

In [12.6] a mathematical formulation of this idea was given which proposed a specific class of laser pulses which would achieve this goal. Finding a class of laser pulses which achieve these symmetric wave packets is a *quantum control* problem. This idea of time reversing the cavity decay is, of course, not the only possible solution to achieve ideal transmission.

b) A Cavity QED Model: The simplest possible configuration of quantum transmission between two nodes consists of two identical three level atoms 1 and 2, each with degenerate ground states. The atoms are strongly coupled to their respective cavity modes by the transition $g \leftrightarrow r$, as illustrated in Fig. 12.10, and couple to a separate strong laser field, treated classically, by the transition $e \leftrightarrow r$. The Hamiltonian describing the interaction of each atom with the corresponding cavity mode and the laser field, is:

$$\begin{aligned} H_i = {} & \hbar\omega_c a_i^\dagger a_i + \hbar\omega_0 |r\rangle_{i\,i}\langle r| + \hbar g\Big\{|r\rangle_{i\,i}\langle g|a_i + a_i^\dagger |g\rangle_{i\,i}\langle r|\Big\} \\ & + \tfrac{1}{2}\hbar\Omega_i(t)\Big\{\mathrm{e}^{-\mathrm{i}\omega_L t}|r\rangle_{i\,i}\langle e| + \mathrm{e}^{\mathrm{i}\omega_L t}|e\rangle_{i\,i}\langle r|\Big\} \quad (i = 1,2). \end{aligned} \tag{12.5.3}$$

Here, a_i is the destruction operator for cavity mode i with frequency ω_c. The atomic states $|g\rangle_i$, $|r\rangle_i$, and $|e\rangle_i$ refer to the three level system with excitation frequency ω_0 inside the first and second cavity ($i = 1, 2$) as in Fig. 12.10. The qubit is stored in a superposition of the two degenerate ground states. Thus in the present context we

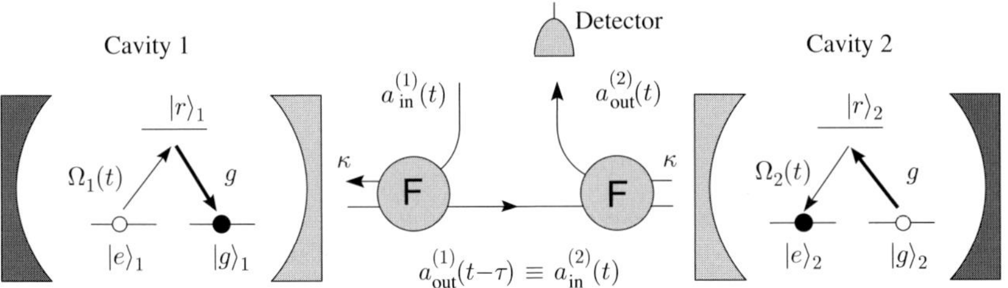

Fig. 12.10 Representation of unidirectional quantum transmission between two atoms in optical cavities connected by a quantized transmission line.

identify the two atomic ground states

$$|0\rangle_i \equiv |g\rangle_i, \quad |1\rangle_i \equiv |e\rangle_i \quad (i = 1,2) \tag{12.5.4}$$

with the qubits of (12.5.2).

The states $|e\rangle_i$ and $|g\rangle_i$ are coupled by a Raman transition, where a laser of frequency ω_L excites the atom from $|e\rangle_i$ to $|r\rangle_i$ with a time dependent Rabi frequency $\Omega_i(t)$, followed by a transition $|r\rangle_i \to |e\rangle_i$ which is accompanied by emission of a photon into the corresponding cavity mode, with coupling constant g. In order to suppress spontaneous emission from the excited state during the Raman process, the laser is strongly detuned from the atomic transition $|\Delta| \gg \Omega_{1,2}(t), g$ (with $\Delta = \omega_L - \omega_0$). In such a case, one can adiabatically eliminate the excited states $|r\rangle_i$. The new Hamiltonian for the dynamics of the two ground states becomes, in a rotating frame for the cavity modes at the laser frequency,

$$H_i = -\hbar\delta a_i^\dagger a_i - i\hbar g_i(t)\Big\{|e\rangle_i\,{}_i\langle g|a_i - a_i^\dagger|g\rangle_i\,{}_i\langle e|\Big\}. \quad (i = 1,2) \tag{12.5.5}$$

Here $\delta = \omega_L - \omega_c$ is the Raman detuning between the laser and the cavity mode frequencies. This Hamiltonian now has the form of a two level atom with "ground" state $|g\rangle_i$ and "excited" state $|e\rangle_i$, coupled to a quantized cavity mode with an effective coupling constant $g_i(t) = g\,\Omega_i(t)/(2\Delta)$.

We must now find time-dependent Rabi frequencies to accomplish *ideal quantum transmission* (12.5.2),

$$\Big\{c_g|g\rangle_1 + c_e|e\rangle_1\Big\}|g\rangle_2|0\rangle_{c_1}|0\rangle_{c_2}|\mathrm{vac}\rangle \to |g\rangle_1\Big\{c_g|g\rangle_2 + c_e|e\rangle_2\Big\}|0\rangle_{c_1}|0\rangle_{c_2}|\mathrm{vac}\rangle, \tag{12.5.6}$$

where $c_{g,e}$ are complex numbers. In (12.5.6), $|0\rangle_{c_1}$, $|0\rangle_{c_2}$ and $|\mathrm{vac}\rangle$ represent the vacuum state of the first and second cavity mode, and the free electromagnetic modes connecting the cavities. Transmission will occur by photon exchange via these modes.

12.5.4 Quantum Transmission in a Quantum Trajectory Picture

The present model is a particular example of a cascaded quantum system and can be described within the formalism developed in the present chapter. In particular this theory allows us to write down QSDEs for the system operators of the coupled cavities, or the corresponding master equation for the density operator. In the present context, it is convenient to reformulate the the present problem in the language of quantum trajectories for cascaded systems, as in Sect. 12.1.5. This will simplify the solution of our problem considerably.

a) Physical Picture: We first explain the physical picture underlying the quantum trajectory formulation. Let us consider a fictitious experiment where the output field of the second cavity is continuously monitored by a photodetector, illustrated in Fig. 12.10. The evolution of the quantum system under continuous observation, conditional to observing a particular trajectory of counts, can be described by a pure state wavefunction $|\psi_c(t)\rangle$ in the system Hilbert space, as formulated in Sect. 11.3.9. During the time intervals when no count is detected, this wavefunction evolves according to a Schrödinger equation with the non-Hermitian effective Hamiltonian

$$H_{\rm eff}(t) = H_1(t) + H_2(t) - \mathrm{i}\hbar\kappa\left(a_1^\dagger a_1 + a_2^\dagger a_2 + 2a_2^\dagger a_1\right) \tag{12.5.7}$$

with κ the decay rate of the cavities. The detection of a count at time t_r is associated with a quantum jump with jump operator

$$c = a_1 + a_2, \tag{12.5.8}$$

according to

$$|\psi_c(t_r + dt)\rangle \propto c|\psi_c(t_r)\rangle. \tag{12.5.9}$$

This form of the quantum jump operator follows from Sect. 12.1.5. The probability density for a jump (detector click) to occur during the time interval from $(t, t + dt)$ is given by $\langle\psi_c(t)|c^\dagger c|\psi_c(t)\rangle\, dt$.

b) Quantum Control in the Quantum Trajectory Language: We can now reformulate our quantum control problem: We wish to design the laser pulses in both cavities in such a way that ideal quantum transmission condition (12.5.6) is satisfied. A necessary condition for the time evolution is that a quantum jump (detector click, see Fig. 12.10) never occurs, i.e.,

$$c|\psi_c(t)\rangle \equiv (a_1 + a_2)|\psi_c(t)\rangle = 0 \quad \text{for all } t. \tag{12.5.10}$$

In the subspace defined by this condition, the effective Hamiltonian (12.5.7) will become a Hermitian operator—the system will remain in a "dark" state of the cascaded quantum system, which cannot emit any light into the outside world. Physically, this means that the wavepacket is not reflected from the second cavity.

We can now expand the state of the system consisting of the two atoms and the two cavity modes as

$$\begin{aligned}|\psi_c(t)\rangle = c_g|gg\rangle|00\rangle \\ +c_e\left[\alpha_1(t)|eg\rangle|00\rangle + \alpha_2(t)|ge\rangle|00\rangle + \beta_1(t)|gg\rangle|10\rangle + \beta_2(t)|gg\rangle|01\rangle\right]\end{aligned} \tag{12.5.11}$$

where we use the abbreviated notation $|gg\rangle|00\rangle \equiv |g\rangle_1|g\rangle_1|0\rangle_{c_1}|0\rangle_{c_2}$ *etc.*. Ideal quantum transmission (12.5.6) will occur for

$$\alpha_1(-\infty) = \alpha_2(+\infty) = 1 \tag{12.5.12}$$

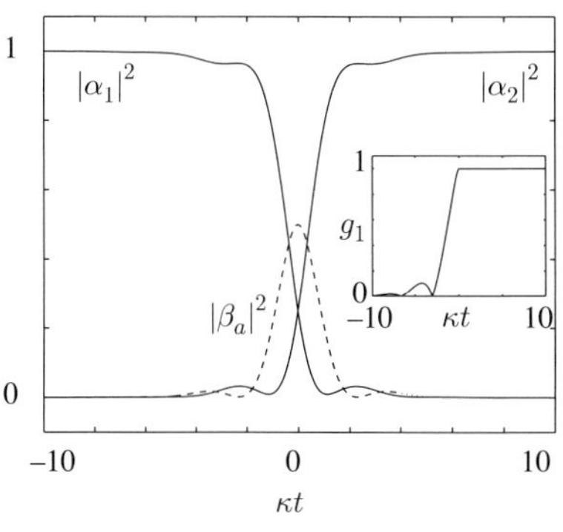

Fig. 12.11 Populations $\alpha_{1,2}(t)^2$ and $\beta_a(t)^2$ for a specific ideal transmission pulse $g_1(t) = g_2(-t)$ given in the inset (see [12.6] for details).

with all of the other coefficients being zero at the initial and final time. The first term on the right hand side of (12.5.11) does not change under the time evolution generated by H_{eff}. Defining symmetric and antisymmetric coefficients $\beta_{1,2} = (\beta_s \mp \beta_a)/\sqrt{2}$, we find the following *evolution equations*

$$\dot{\alpha}_1(t) = g_1(t)\beta_a(t)/\sqrt{2}\,, \tag{12.5.13}$$

$$\dot{\alpha}_2(t) = -g_2(t)\beta_a(t)/\sqrt{2}\,, \tag{12.5.14}$$

$$\dot{\beta}_a(t) = -g_1(t)\alpha_1(t)/\sqrt{2} + g_2(t)\alpha_2(t)/\sqrt{2}\,. \tag{12.5.15}$$

The *dark state condition* (12.5.10) implies $\beta_s(t) = 0$, and therefore

$$\dot{\beta}_s(t) = g_1(t)\alpha_1(t)/\sqrt{2} + g_2(t)\alpha_2(t)/\sqrt{2} + \kappa\beta_a(t) \equiv 0. \tag{12.5.16}$$

As well this, there is the normalization condition

$$|\alpha_1(t)|^2 + |\alpha_2(t)|^2 + |\beta_a(t)|^2 = 1. \tag{12.5.17}$$

The mathematical problem is now to find pulse shapes $\Omega_{1,2}(t) \propto g_{1,2}(t)$ such that the conditions (12.5.12,16) are fulfilled.

c) Solution: Solving these equations is a difficult problem, as imposing the conditions (12.5.12,16) on the solutions of the differential equations (12.5.13–15) gives functional relations for the pulse shape whose solution are not obvious. In [12.6] a class of solutions was constructed which satisfy the above conditions, based on the physical expectation that the time evolution in the second cavity should reverse the time evolution in the first one. Thus one looks for solutions satisfying the *symmetric pulse condition*:

$$g_2(t) = g_1(-t) \quad \text{for all } t. \tag{12.5.18}$$

As an illustration, Fig. 12.11 shows results of numerically integrating the full time-dependent Schrödinger equation with the effective Hamiltonian (12.5.7) for one of the laser pulses, shown in the inset to Fig. 12.11, constructed in this way. As the figure shows, the quantum transmission is indeed very close to ideal.

13. Supplement

In this supplement, we want to indicate briefly what areas of progress have been made in the subject since this edition was finalized in 1999, and give an indication of the available literature. There have been advances both in the understanding of the technical methodology involved in quantum stochastic processes, as well as in the extension of the methodology to new systems, most notably in the field of cold atom physics. The field nowadays is increasingly dominated by the physics of *atoms*, interacting either with themselves or with light, as opposed to the former dominance of the physics of *light*, i.e., quantum optics.

13.1 Laser Cooling

The practical implementation and theoretical understanding of laser cooling of neutral atoms is now a well-developed subject, as summarized in the recent book by *Metcalf* and *van der Straten* [1], as well as in the older book by *Minogin* and *Letokhov* [2]. The theoretical description of laser cooling is formulated in the language of master equations, and thus provides a prime example of applications of various techniques discussed in the present book. Laser cooling master equations (see Sect.11.5.4 for a simple example) describe the coupled internal (electronic) dynamics and external (centre of mass) dynamics of an atom, driven by laser light and undergoing spontaneous emission. The coupling arises from the recoil force of the photons involved in the absorption and emission of light. In many cases the different dynamical time scales of the internal and external dynamics allow an adiabatic elimination of the faster internal degrees of freedom, resulting in a master equation for the center of mass motion of the atoms alone. In addition, semiclassical approximations (paralleling those described in the context of laser theory in Sect.9.3.2) can often be applied to map this master equation to a Fokker-Planck equation describing the atomic motion. The techniques used are described in detail by *Dalibard* and *Cohen-Tannoudji* [3], *Minogin* and *Letokhov* [2], and *Cirac, Blatt, Zoller* and *Phillips* [4].

1 H. J. Metcalf and P. van der Straten, *Laser Cooling and Trapping*, *Graduate Texts in Contemporary Physics* (Springer Verlag, New York, 1999).

2 V. G. Minogin and V. Letokhov, *Laser Light Pressure on Atoms* (Gordon and Breach Scientific Publishers, Amsterdam, 1987).

3 J. Dalibard and C. Cohen-Tannoudji, J. Phys. B **18**, 1661 (1985).

4 J. I. Cirac, R. Blatt, P. Zoller, and W. D. Phillips, Phys. Rev. A **46**, 2668 (1992).

13.2 Bose-Einstein Condensation

The progress in laser cooling of atoms led in 1995 to the achievement of Bose-Einstein condensation in a vapour of Rubidium atoms [1], and a consequent rapid expansion in the appropriate theory, from both the point of view of traditional condensed matter theory, and also from the point of view of quantum optics. There is of course a strong analogy between photons and Bosonic atoms, with two major differences: Atoms are conserved, whereas photons are not; and in the condensed phase the interactions between atoms are never negligible, whereas photons only interact significantly with each other in non-linear media.

The basic theory of the Bose-Einstein condensate is available in the book of *Pethick* and *Smith* [2], and in the review article of *Dalfovo*, *Georgini* and *Stringari* [3], while a more up to date treatment is available in the book of *Pitaevskii* and *Stringari* [4]. The theory of irreversible processes, such as condensate growth and the damping of excitations is describable in a degree of approximation using similar techniques to those in this book, and we have done this in our series of papers on *Quantum Kinetic theory* [5, 6]. Other approaches have also been developed, more based on traditional techniques. *Zaremba*, *Griffin* and *Nikuni* [7] developed a method based on the techniques of traditional condensed matter physics; *Stoof* [8] used a Keldysh technique to develop functional Fokker-Planck equations; and *Walser* [9] developed a method based on the Bogoliubov's formulation of kinetic theory. The fact that the "heat bath" for a condensate is formed by the thermal atoms or quasiparticles, and cannot be precisely separated from the "system"—the condensate itself—is a central problem in all approaches, none of which is entirely satisfactory.

For a large class of problems it appears that the condensate can be treated as a classical field obeying the Gross-Pitaevskii equation, supplemented by added initial noise, in much the same way as discussed in Sect.4.5.2. This leads to the use of the *classical field* method based either explicitly or implicitly on the Wigner function for the Bosons, which has been developed by a number of workers [10–14].

1 M. H. Anderson *et al.*, Science **269**, 198 (1995).
2 C. J. Pethick and H. Smith, *Bose-Einstein condensation in dilute gases* (Cambridge, Cambridge, 2001).
3 F. Dalfovo, S. Giorgini, L. P. Pitaevskii, and S. Stringari, Rev. Mod. Phys. **71**, 463 (1999).
4 L. P. Pitaevskii and S. Stringari, *Bose-Einstein Condensation* (Oxford, Oxford, 2003).
5 C. W. Gardiner and P. Zoller, Phys. Rev. A **55**, 2902 (1997).
6 C. W. Gardiner and P. Zoller, Phys. Rev. A **58**, 536 (1998).
7 E. Zaremba, T. Nikuni, and A. Griffin, J. Low Temp. Phys. **116**, 277 (1999).
8 H. T. C. Stoof, J. Low Temp. Phys. **114**, 11 (1999).
9 R. Walser, J. Williams, J. Cooper, and M. Holland, Phys. Rev. A **59**, 3878 (1999).
10 M. J. Davis, S. A. Morgan, and K. Burnett, Phys. Rev. A **66**, 053618 (2002).
11 C. W. Gardiner, J. R. Anglin, and T. I. A. Fudge, J. Phys. B **35**, 1555 (2002).
12 C. W. Gardiner and M. J. Davis, J.Phys. B **36**, 4731 (2003).
13 A. Sinatra, C. Lobo, and Y. Castin, Phys. Rev. Lett **87**, 210404 (2001).
14 H. Schmidt *et al.*, J. Opt. B **5**, 96 (2003).

13.3 Phase Space Methods

Considerable progress has been made in the field of phase-space methods for quantum mechanical master equations, by two principal methods. One way, devised by *Deuar* and *Drummond* [1], is to modify the positive P-representation, taking advantage of the considerable flexibility afforded by the doubled phase space. An alternative formulation, due to *Carusotto*, *Castin* and *Dalibard* [2], is in terms of a kind of number state basis, which yields a somewhat similar formalism. The new methods have much better stability properties than can be achieved using the stochastic simulation methods of Sect.6.6.2, and the problem of "spikes" (as in Sect.6.6.6) is largely solved. However, most realistic problems are still too complex computationally for the new methods in their current form.

1 P. Deuar and P. D. Drummond, Phys. Rev. A **66**, 033812 (2002).
2 I. Carusotto, Y. Castin, and J. Dalibard, Phys. Rev. A **63**, 0223606 (2001).

13.4 Input-Output Theory for Fermions

The input-output methods of Chap.5 have now been adapted to beams of Fermions. There are treatments of some aspects of the problem in [1] and [2], while a treatment fully equivalent to the Boson input-output formalism of Chap.5 is given in [3]. The main difficulty not present in the case of Bosons is the correct treatment of the antisymmetrization required for *all* Fermions, including those possibly implicitly contained within the "system" which couples to the inputs and outputs.

1 H. B. Sun and G. J. Milburn, Phys. Rev. B **59**, 10748 (1999).
2 C. P. Search, S. Pötting, W. Zhang, and P. Meystre, Phys. Rev. A **66**, 043616 (2002).
3 C. W. Gardiner, cond-mat/0310542 (2003).

13.5 Quantum Feedback

This subject has been developed largely by *Wiseman* and *Milburn* [1, 2]. The main issues are very well covered in the paper by *Wiseman*, *Mancini* and *Wang* [3], which also gives a comprehensive set of references to other literature. A very recent paper with a good bibliography is that of *van Handel*, *Stockton* and *Mabuchi* [4].

As noted by *Wiseman*, *Mancini* and *Wang*, quantum feedback arises when the environment of an open quantum system is deliberately engineered so that information lost from the system into that environment comes back to affect the system again. Typically, the environment is large, and would be regarded at least in part as a classical system. In the case where the system dynamics are Markovian in the absence of feedback, the information lost to the environment can be treated as classical information—the measurement result. The feedback loop thus consists of a quantum system, a classical detector (which turns quantum information into classical information), and a classical actuator (which uses the classical information to affect the quantum system).

In general, quantum feedback is difficult to treat because any time delay or filtering in the feedback loop makes the system dynamics non-Markovian. A great simplification arises for Markovian feedback, where the measurement results are used immediately to alter the system state, and may then be forgotten. In this case the dynamics including feedback may be described by a master equation in the Lindblad form.

For non-Markovian feedback, the master equation approach of Wiseman and Milburn cannot be used. However, the formalism first used to derive the Wiseman-Milburn master equation, quantum trajectories (as in Chap.11), can be used. In the special case where the system has linear dynamics, the measurement is linear (e.g., homodyne detection), and the feedback dynamics is linear, the quantum trajectories including feedback can be solved analytically. In this case techniques based on quantum Langevin equations can be used. However, for nonlinear systems, a numerical solution of the non-Markovian quantum trajectories is the only recourse.

1 H. M. Wiseman and G. J. Milburn, Phys. Rev. Lett. **70**, 548 (1993).
2 H. M. Wiseman and G. Milburn, Phys. Rev. A **49**, 1350 (1994).
3 H. M. Wiseman, S. Mancini, and J. Wang, Phys. Rev. A **66**, 013807 (2002).
4 R. van Handel, J. K. Stockton, and H. Mabuchi, quant-ph/042136 (2004).

13.6 Non-Markovian QSDEs

Diósi, *Strunz* and *Gisin* [1, 2] have shown that a non-Markovian version of the stochastic Schrödinger equation of Sect.11.4.3 can be developed, and that this can be used to give a wavefunction simulation equivalent to the quantum Brownian motion master equation of Sect.3.6.1. The methodology is very reminiscent of that of the adjoint equation of Sect.3.5, which uses a density operator formalism, and which was much earlier used by *Parkins*, *Gardiner* and *Steyn-Ross* [3] for non-Markovian simulations of the damping of macroscopic quantum coherence.

1 L. Diosi and W. T. Strunz, Phys. Lett. **235A**, 569 (1997).
2 W. T. Strunz, L. Diósi, N. Gisin, and T. Yu, Phys. Rev. Lett. **83**, 4909 (1997).
3 A. S. Parkins, C. W. Gardiner, and M. L. Steyn-Ross, Z. Phys B **83**, 413 (1991).

13.7 Quantum Information

The recent establishment of the field of *quantum information*, *quantum computing* and *quantum communications* (see Sect.12.5.1) is one of the most significant and exciting developments in physics—it employs very subtle features of quantum mechanics, that have probably never before been experimentally tested, to offer the possibility of ultra-fast and highly secure information processing. However, because sensitive entangled states are an essential aspect of the implementation, there is a very strong susceptibility to decoherence—that is, to quantum and classical noise. A very significant part of the field is devoted to overcoming these problems.

For an introduction we refer to articles in the book edited by *Bouwmeester*, *Ekert* and *Zeilinger* [1], and to the comprehensive book on the subject by *Nielsen* and

Chuang [2]. In particular, we refer to the chapter covering *quantum operations*, which provides a link to the master equation description provided in the present book.

1 *The Physics of Quantum Information: Quantum Cryptography, Quantum Teleportation, Quantum Computation*, edited by D. Bouwmeester, A. Ekert, and A. Zeilinger (Springer Verlag, New York, 2000).

2 M. Nielsen and I. Chuang, *Quantum Computation and Quantum Information* (Cambridge University Press, Cambridge, 2000).

13.8 Implementing Quantum Information Using Atoms and Ions

During the past ten years there has been dramatic progress in gaining control of systems at the single quantum level, while suppressing the unwanted interactions with the environment which cause decoherence. These achievements, exemplified by the storage and laser cooling of single trapped ions and atoms, and by the manipulation of single photons in cavity quantum electrodynamics, have opened a new field: The engineering of interesting and useful quantum states. Currently, the frontier is moving towards building larger composite systems of a few atoms and photons, while still maintaining complete quantum control of the individual particles. The new physics to be studied in these systems is based on entangled states and ranges from fundamental aspects—testing quantum mechanics for larger and larger systems—to possible new applications such as quantum information processing and precision measurements.

The techniques described in this book are fundamental for the theoretical treatment of all of these systems. *Cirac* and *Zoller* have recently written a very accessible review [1], while more comprehensive reviews are by *Cirac*, *Duan* and *Zoller* [2], and by *Zoller*, *Cirac*, *Duan* and *García-Ripoll* [3].

a) Ion Traps

The technology for controlling and manipulating single (or few) ions has been very strongly developed for ultrahigh-precision spectroscopy and atomic clocks. In particular, ions can be trapped and cooled in such a way that:

i) They remain practically frozen in a specific region of space;

ii) Their internal states can be precisely manipulated using lasers and can be measured with practically 100% efficiency;

iii) They interact with each other very strongly due to the Coulomb repulsion, yet they can, at the same time, be decoupled from the environment very efficiently.

Ions stored and laser-cooled in an electromagnetic trap can be described in terms of a set of external and internal degrees of freedom. The external degrees of freedom are closely related to the center of mass motion of each ion; the internal degrees of freedom—responsible for the existence of a discrete energy level structure in each ion—relate to the motion of electrons within each ion and to the presence of electronic and nuclear spins. These levels can be used to store qubits provided they

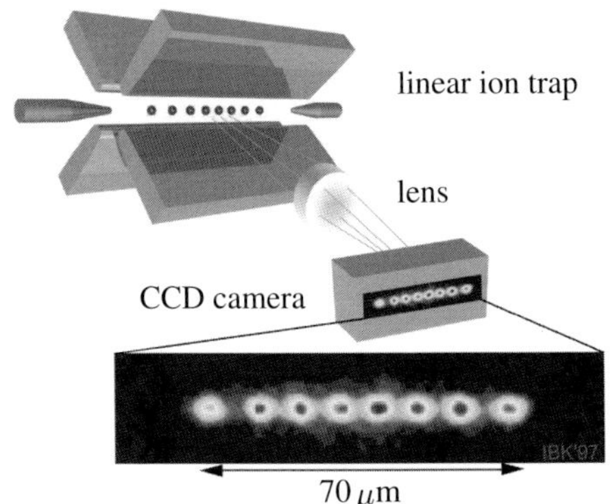

Fig. 13.1 A linear ion trap loaded with several ions. Their mutual repulsive Coulomb forces held them in a regular array, and allow communication of vibrational motion between ions. (Figure kindly supplied by Prof. Rainer Blatt.)

are very long-lived and suffer no decoherence, so that they are not disturbed during the computation—fortunately, this condition can be achieved.

A complete list of references on trapped ion can be found in the experimentally oriented review by *Leibfried*, *Blatt*, *Monroe* and *Wineland* [4], while a comprehensive review of the theoretical description is given in [2, 3].

b) Cavity Quantum Electrodynamics

Cavity quantum electrodynamics (CQED) realizes a situation where one or a few atoms interact strongly with a single quantized high-Q cavity mode, where the light field can be either in the optical or in the microwave domain. The coupling of atoms via the cavity mode can be used to engineer entanglement between the atoms. The underlying physics is described by the Jaynes-Cummings Hamiltonian similar to (12.5.3), which for a single two level atom has the form

$$H = \tfrac{1}{2}\hbar\omega_0\sigma_z + \hbar\omega_c a^\dagger a + \tfrac{1}{2}\hbar\left(\Omega(\mathbf{r})a^\dagger\sigma_- + \Omega(\mathbf{r})^*\sigma_+ a\right), \tag{13.8.1}$$

where the σ are the Pauli spin operators describing the two level atom, and $a^\dagger$ and a are the Boson creation and annihilation operators of the field mode. The term proportional to $\Omega(\mathbf{r})$ in (13.8.1) describes the coherent oscillatory exchange of energy between the atom and the field mode.

In optical CQED, dissipation in this system due to spontaneous emission of the atom in the excited state, and cavity decay, can be modelled by a master equation. From a formal point of view there is a close analogy between the ion trap models and the Jaynes-Cummings type models of CQED, where the role of the collective phonon modes of trapped ions is now taken by photons in the high-Q cavity mode. Thus, in many cases schemes for the engineering of entangled quantum states proposed for ion traps are readily translated to their CQED counterparts.

As discussed in Sect.12.5.2, CQED provides a natural interface between atoms as carriers of qubits and photons, which—once they leave the cavity—can be guided by optical fibers to transport qubits to distant locations. CQED thus serves as the paradigm for an optical interconnect between atoms as quantum memory and photons as carriers of qubits for quantum communication.

A good introduction to CQED is the book by *Berman* [5], while recent reviews on the subject include the articles by *Raimond*, *Brune* and *Haroche* [6], and *Mabuchi* and *Doherty* [7]. The theoretical description of the application to quantum computing is given in [2, 3].

There have also been significant developments on CQED in mesoscopic systems manufactured in solid state systems. The book by *Yamomoto* and *İmamoḡlu* [8] provides a valuable introduction to quantum optics in this context.

c) Cold Atoms in an Optical Lattice

Bose-Einstein condensates provide a source of a large number of ultracold atoms. As a result of their weak interactions, all atoms in a condensate occupy the single-particle ground state of the trapping potential, and the condensate is in a product state of the individual ground-state wavefunctions. Such a collection of atoms can be harnessed for quantum information processing by loading the atoms into an optical lattice. The result is an array of many identifiable qubits that can be entangled in a massively parallel operation. This scenario, devised by *Jaksch*, *Bruder*, *Briegel*, *Cirac*, *Gardiner* and *Zoller*, [9, 10] has recently been realized in the laboratory in a series of remarkable experiments [11, 12].

An optical lattice, generated by standing-wave laser fields, forms a periodic array of microtraps for cold atoms, and thus the energy levels of the atoms form Bloch bands, in the same way as electrons in a crystal. At the the low temperatures now achievable, atoms loaded in an optical lattice will occupy only the lowest energy Bloch band. The physics of these atoms can be understood in terms of a Hubbard model with Hamiltonian

$$H = -\sum_{\langle i,j\rangle} J_{ij} b_i^\dagger b_j + \frac{U}{2}\sum_i b_i^\dagger b_i^\dagger b_i b_i \tag{13.8.2}$$

where b_k and $b_k^\dagger$ are Bosonic annihilation and creation operators for atoms at each lattice site. The parameters J_{ij} are hopping matrix elements that connect two lattice sites via tunnelling, and U is the on-site interaction of atoms resulting from the collisional interactions (see Fig.13.2(a)). The distinguishing feature of this system is potential for time-dependent control of the hopping matrix elements (which correspond to the kinetic energy) and on-site interaction (the potential energy) through the intensity of the lattice laser.

Increasing the intensity deepens the lattice potential and suppresses the hopping; at the same time, it increases the atomic density at each lattice site and thus the on-site interaction. Increasing the intensity will, therefore, decrease the ratio of kinetic to potential energy, J_{ij}/U, and the system will eventually become strongly interacting. In the case of Bosonic atoms, the system will undergo a quantum phase transition from the condensate's superfluid state, with long-range phase coherence, to a Mott insulator state, which has no phase coherence.

In the Mott insulator regime, loading *exactly* one atom per lattice site can be achieved, an arrangement that provides a large number of identifiable atoms whose internal hyperfine or spin states can serve as qubits. Entanglement of these atomic qubits is obtained by combining the collisional interactions with a spin dependent

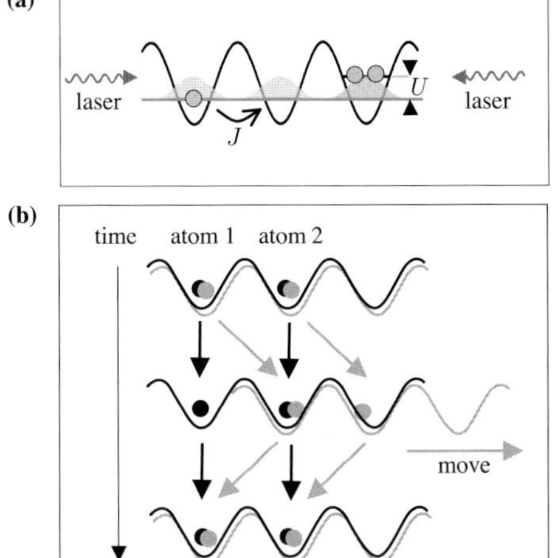

Fig. 13.2 (a) Illustration of the processes of hopping and on-site energy for the Bose-Hubbard Hamiltonian; (b) The different internal quantum states (grey and black) of the atoms feel different optical potentials (grey and black), which can be moved independently, producing quantum superpositions of collisions; (c) Construction of a quantum gate using these collisions.

optical lattice. Here, with appropriate choice of atomic states and laser configurations, the qubit states $|0\rangle$ and $|1\rangle$ see different lattice potentials. Atoms can thus be moved according to the state of the qubit. In particular, one can collide two atoms "by hand," as illustrated in Fig.13.2(b), so that the wavefunction component with the first atom in $|1\rangle$ and the second atom in $|0\rangle$ will pick up a collisional phase ϕ that entangles the atoms. For two adjacent atoms, this physical process realizes the two qubit gate required for quantum computation (see Fig.13.2(c)).

A comprehensive review of the theoretical description of the application to quantum computing is given in [2, 3].

d) Quantum Information Processing with Atomic Ensembles

Many quantum information protocols can be implemented simply by laser manipulation of atomic ensembles containing a large number of identical neutral atoms. The experimental candidate systems for the atomic ensembles can be either laser-cooled atoms confined in a magneto-optical trap, or room-temperature atoms contained in a glass cell with coated walls to avoid bad collisions. Quantum information is stored in the ground-state manifold of the atoms, such as in Zeeman sublevels with different atomic spins, or in some hyperfine atomic levels which are stable or metastable under optical transitions. Long coherence times of the relevant states have been observed in both of these kinds of experimental systems.

The motivation for using atomic ensembles instead of single particles for quantum information processing is threefold:

i) Laser manipulation of atomic ensembles without separate addressing of individual atoms is much easier than the laser manipulation of single particles;

ii) The quantum information encoded in atomic ensembles is robust against some practical noise. For instance, the loss of few atoms in a large atomic ensemble has negligible influence on the quantum information it carries;

iii) Perhaps most importantly, the use of the atomic ensembles provides a way of enhancing the signal to noise ratio by using the collective effects present in this system. We know that in physical implementations of quantum information protocols, the central problem is to enhance the signal to noise ratio—that is, to enhance the ratio of the magnitude of the coherent information processing to that of the decoherence process caused by the noisy coupling to the environment. For example, to achieve a high signal to noise ratio in CQED schemes, it is necessary both to achieve strong light-atom coupling by building a high finesse cavity around the atoms, and to enter the challenging strong coupling regime. However, for the case of atomic ensembles, the collective enhancement induced by the many-atom coherence enables the signal to noise ratio to be greatly increased by encoding quantum information into some collective excitations of the ensembles. As a result of this effect, quantum information processing is made possible with relatively simplified experimental systems, such as an atomic ensemble in a weak-coupling cavity or even in free space.

Lukin [13] has summarized recent progress with atomic ensembles with emphasis on quantum information aspects, as well as covering topics like *slow light*. The theoretical description is also given in [2, 3].

1 J. I. Cirac and P. Zoller, *New Frontiers in Quantum Information With Atoms and Ions*, Physics Today March 2004, 38, (2004).

2 J. I. Cirac, L. Duan, and P. Zoller, in *Experimental Quantum Computation and Information*, Vol. 148 of *International School of Physics Enrico Fermi*, edited by F. De Martini and C. Monroe (IOS Press, Amsterdam, 2002; also available as quant-ph/0405030).

3 P. Zoller, J. I. Cirac, L. M. Duan, and J. J. García-Ripoll, in *Quantum entanglement and information processing*, No. 79 in *les Houches summer school proceedings*, edited by D. Estève, J. Raimond, and J. Dalibard (Elsevier, Amsterdam, 2004; also available as quant-ph/0405025).

4 D. Leibfried, R. Blatt, C. Monroe, and D. Wineland, Rev. Mod. Phys. **75**, 281 (2003).

5 P. R. Berman, *Cavity Quantum Electrodynamics, Advances in Atomic, Molecular and Optical Physics, Supplement 2* (Academic Press, New York, 1994).

6 J.-M. Raimond, M. Brune, and S. Haroche, Rev. Mod. Phys. **73**, 565 (2001).

7 H. Mabuchi and A. C. Doherty, Science **298**, 1372 (2002).

8 Y. Yamamoto and A. İmamoḡlu, *Mesoscopic Quantum Optics* (Wiley, New York, 1999).

9 D. Jaksch, C. Bruder, J. I. Cirac, C. W. Gardiner and P. Zoller, Phys. Rev. Lett. **81**, 3108 (1998).

10 D. Jaksch. H.-J. Briegel, J.I. Cirac, C. W. Gardiner and P. Zoller, Phys. Rev. Lett. **82**, 1975 (1999).

11 M. Greiner, O. Mandel, T. Esslinger, T. W. Hänsch and I. Bloch, Nature **415**, 39 (2002).

12 O. Mandel, M. Greiner, A. Widera, T. Rom, T. W. Hänsch and I. Bloch, Nature **425**, 937 (2003).

13 M. D. Lukin, Rev. Mod. Phys. **75**, 457 (2003).

References

Chapter 1

1.1 M. Born: Z. Phys. **37**, 863 (1926)

1.2 V. Weisskopf, E. Wigner: Z. Phys. **63**, 54 (1930)

1.3 C. M. Caves, K. S. Thorne, R. W. P. Dreaver, V. D. Sandberg, M. Zimmerman: Rev. Mod. Phys **52**, 341 (1980)

1.4 H. Nyquist: Phys. Rev. **32**, 110 (1928)

1.5 J. B. Johnson: Phys. Rev. **32**, 97 (1928)

1.6 H. B. Callen, T. A. Welton: Phys. Rev. **83**, 34 (1951)

1.7 G. W. Ford, M. Kac, P. Mazur: J. Math. Phys **6**, 504 (1965)

1.8 R. H. Koch, D. van Harlingen, J. Clarke: Phys. Rev. B **26**, 74 (1982)

1.9 A. Schmid: J. Low Temp. Phys. **49**, 609 (1982)

1.10 P. A. M. Dirac: *The Principles of Quantum Mechanics* (4th Ed, Oxford 1960)

1.11 A. Einstein: Phys. Zeits. **18**, 121 (1917)

1.12 E. Fermi: *Lecture Notes on Quantum Mechanics* (Univ. Chicago, Chicago 1952)

1.13 W. Pauli: *Festschrift zum 60 Geburtstag A. Sommerfeld S.30.* (Leipzig, Hirzel 1928)

1.14 R. Benguria, M. Kac: Phys. Rev. Lett. **46**, 1 (1981)

1.15 H. Haken: Z. Phys. **181**, 96 (1964); M. Lax: Phys. Rev. **145**, 110 (1966); J. P. Gordon, L. R. Walker, W. Louisell: Phys. Rev. **130**, (1963)

1.16 J. von Neumann: *Mathematical Foundations of Quantum Mechanics* (Princeton University Press, Princeton 1955) Chap. IV

Chapter 2

2.1 J. von Neumann: *Mathematical Foundations of Quantum Mechanics* (Princeton University Press, Princeton 1955 Chap. IV)

2.2 C. M. Caves: Phys. Rev. D **33**, 1643 (1986)

2.3 J. B. Hartle: Am. J. Phys. **36**, 704 (1968)

2.4 R. C. Tolman: *The Principles of Statistical Mechanics* (Oxford University Press, Oxford 1938)

2.5 L. D. Landau, I. M. Lifshitz: *Statistical Physics* (Addison Wesley, Reading, Mass. 1974)

Chapter 3

3.1 R. Landauer: I. B. M. J. Res. Develop. **1**, 223 (1957)

3.2 G. W. Ford, M. Kac, P. Mazur: J. Math. Phys. **6**, 504 (1965)

3.3 N. G. van Kampen: K. Dansk. Vid. Sels. Mat-fys. Med. **26**, No. 15 (1951)

3.4 H. Lehmann, K. Symanzik, W. Zimmerman: Nuovo Cimento **1**, 1425 (1955)

3.5 C. M. Caves: Phys. Rev. D **26**, 1817 (1982)

3.6 B. Yurke and J. S. Denker: Phys. Rev. A. **29**, 1419 (1984)

3.7 M. J. Collett: *M. Sc. Thesis* (University of Waikato, Hamilton, New Zealand 1983)

3.8 I. S. Gradshteyn, I. M. Rhyzhik: *Tables of Integrals, Series, and Products* (Academic Press, New York 1980 p507, Et5883.)

3.9 G. N. Watson: *Theory of Bessel Functions* (2nd Ed. Cambridge University Press, New York 1966)

3.10 R. Kubo, M. Toda, N. Hashitsume: *Statistical Physics II* (Springer, Berlin 1985)

3.11 R. Kubo: J. Math. Phys. **4**, 174 (1963)

3.12 A. S. Parkins, C. W. Gardiner: Phys. Rev. A **37**, 3867 (1988)

3.13 H. Ritsch, P. Zoller: Phys. Rev. Lett. **61**, 1097 (1988)

3.14 N. G. van Kampen: *Stochastic Processes in Physics and Chemistry* (North Holland, Amsterdam 1981)

3.15 C. M. Savage, D. F. Walls: Phys. Rev. A **32**, 2316 (1985)

3.16 H. Haken: *Laser Theory* (Springer, Berlin 1984) See references on p51.

3.17 W. Louisell in : *Quantum Optics* (ed. S. Kay, A. Maitland, Academic New York, 1970)

Chapter 4

4.1 R. J. Glauber: Phys. Rev. **131**, 2766 (1963)

4.2 E. C. G. Sudarshan: Phys. Rev. Lett. **10**, 277 (1963)

4.3 E. P. Wigner: Phys. Rev. **40**, 749 (1932)

4.4 M. Abramowitz, I. Stegun: *Handbook of Mathematical Functions* (Dover, New York 1965)

4.5 J. R. Klauder, J. McKenna, D. G. Currie: J. Math. Phys. **6**, 734 (1965)

4.6 A. Schmid: J. Low Temp. Phys. **49**, 609 (1982)

4.7 R. H. Koch, D. van Harlingen, J. Clarke: Phys. Rev. Lett. **45**, 2132 (1980); Phys. Rev. B **26**, 74 (1982)

4.8 T. W. Marshall: Proc. Roy. Soc. Lond. A **276**, 475 (1963)

Chapter 5

5.1 N. G. van Kampen: J. Stat. Phys. **24**, 175 (1981)

5.2 I. R. Senitzky: Phys. Rev. **119**, 670 (1960); Phys. Rev. **124**, 642 (1961)

5.3 M. Lax: Phys. Rev. **129**, 2342 (1963)

5.4 H. Haken: *Laser Theory* (Springer, Berlin 1984)

5.5 W. Louisell in : *Quantum Optics* ed. S. Kay, A. Maitland (Academic, New York 1970)

5.6 W. Louisell: *Quantum Statistical Properties of Radiation* (Wiley, New York 1974)

5.7 H. Spohn: Rev. Mod. Phys. **52**, 569 (1980)

5.8 W. J. Munro, C. W. Gardiner: Phys. Rev. A **53**, 2633 (1996)

5.9 M. J. Collett, C. W. Gardiner: Phys. Rev. A **30**, 1386 (1984); C. W. Gardiner, M. J. Collett: Phys. Rev. A **31**, 3761 (1985)

5.10 M. J. Collett: *M. Sc. Thesis* (University of Waikato, Hamilton, New Zealand 1983)

5.11 A. Barchielli: Phys. Rev. A **34**, 1642 (1986)

Chapter 6

6.1 P. D. Drummond, C. W. Gardiner: J. Phys. A **13**, 2353 (1980)

6.2 R. J. Glauber: Phys. Rev. **131**, 2766 (1963)

6.3 P. Drummond, D. F. Walls: J. Phys. A **13**, 725 (1980)

6.4 M. L. Steyn: *M. Sc. Thesis* (University of Waikato, Hamilton, New Zealand 1978); M. Dörfle, A. Schenzle: Z. Phys. B **65**, 113 (1986); M. Wolinsky, H. J. Carmichael: Phys. Rev. Lett **60**, 1836 (1988)

6.5 A. M. Smith, C. W. Gardiner: Phys. Rev. A **39**, 3511 (1989)

6.6 A. Gilchrist, C. W. Gardiner, P. D. Drummond: Phys. Rev. A **55**, 3104 (1997)

6.7 R. Schack, A. Schenzle: Phys. Rev. A **44**, 682 (1991)

6.8 P. D. Drummond, I. K. Mortimer: J. Comp. Phys. **93**, 144 (1991)

6.9 M. Wolinsky, H. J. Carmichael: Phys. Rev. Lett. **60**, 1836 (1988)

6.10 L.-A. Wu, H. J. Kimble, J. Hall, H. Wu: Phys. Rev. Lett. **57**, 2520 (1986)

6.11 P. Kinsler, P. D. Drummond: Phys. Rev. Lett. **64**, 236 (1990); P. D. Drummond, P. Kinsler: Phys. Rev. A **40**, 4813 (1989)

6.12 R. Graham: Springer Tracts in Modern Physics **66**, 1 (1973)

6.13 P. D. Drummond: Phys. Rev. A **33**, 4462 (1986)

6.14 P. D. Drummond, K. J. McNeil, D. F. Walls: Opt. Acta **28**, 211 (1980)

Chapter 7

7.1 J. von Neumann: *U. S. Patent* 2 815 488

7.2 G. J. Milburn, D. F. Walls: Opt. Comm. **39**, 410 (1981); B. Yurke: Phys. Rev. A **29**, 408 (1984); M. J. Collett, C. W. Gardiner: Phys. Rev. A **30**, 1386 (1984)

7.3 D. F. Walls, M. J. Collett, G. J. Milburn: Phys. Rev. D **32**, 3208 (1985)

7.4 W. H. Zurek: Phys. Rev. D **24**, 1516 (1981); Phys. Rev. D **26**, 1862 (1982)

Chapter 8

8.1 H. Haken: Z. Phys. **181**, 210 (1964); M. Lax: Phys. Rev. **145**, 534 (1966); J. P. Gordon, L. R. Walker, W. Louisell: Phys. Rev. **130**, 806 (1963)

8.2 R. J. Glauber: Phys. Rev. **130**, 2529 (1963); Phys. Rev. **131**, 2766 (1963)

8.3 M. Planck: Verh. Deutsch. Phys. Ges. **2**, 202,237 (1900)

8.4 A. Einstein: Phys. Zeits. **18**, 121 (1917)

8.5 P. D. Drummond: Phys. Rev. A **35**, 4523 (1987)

8.6 M. D. Srinivas, E. B. Davies. Optica Acta **28**, 981 (1981)

8.7 L.-A. Wu, H. J. Kimble, J. L. Hall, H. Wu: Phys. Rev. Lett. **57**, 2520 (1986)

8.8 F. Ghielmetti: Phys. Lett. **12**, 210 (1964)

8.9 H. P. Yuen, J. H. Shapiro: IEEE Trans. Inf. Theory IT **26**, 78 (1980)

Chapter 9

9.1 L. A. Orozco, A. T. Rosenberger, H. J. Kimble: Phys. Rev. A **36**, 3248 (1987); L. A. Orozco, H. J. Kimble, A. T. Rosenberger, L. A. Lugiato, M. L. Asquini, M. Brambilla, L. M. Narducci: Phys. Rev. A **39**, 1235 (1989)

9.2 H. J. Carmichael, D. F. Walls: J. Phys. B **9**, 1199 (1976)

9.3 H. Haken: *Laser Theory* (Springer, Berlin 1984)

9.4 M. Lax: Phys. Rev. **145**, 110 (1966)

9.5 W. Louisell: *Quantum Statistical Properties of Radiation* (Wiley, New York 1973); J. P. Gordon: Phys. Rev. **161**, 367 (1967)

9.6 N. G. van Kampen: Can. J. Phys. **39**, 551 (1961)

9.7 R. Bonifacio, L. A. Lugiato: Opt. Commun. **19**, 172 (1976) ; R. Bonifacio, L. A. Lugiato: Lett. Nuovo Cimento **21**, 517 (1978)

9.8 P. D. Drummond, D. F. Walls: Phys. Rev. A **23**, 2563 (1981)

9.9 F. Haake, M. Lewenstein: Phys. Rev. A **22**, 1013 (1983)

9.10 H. Haken, H. Risken, W. Weidlich: Z. Physik **206**, 355 (1967); See also L. Lugiato: Nuovo Cimento B **50**, 89 (1979); P. D. Drummond, D. F. Walls: Phys. Rev. A **23**, 2563 (1981)

9.11 P. D. Drummond, D. F. Walls: Phys. Rev. A **23**, 2563 (1981) ; M. D. Reid, D. F. Walls: Phys. Rev. A **33**, 4465 (1986) ; M. D. Reid, D. F. Walls: Phys. Rev. A **34**, 4929 (1986)

Chapter 10

10.1 C. M. Caves, K. S. Thorne, R. W. P. Dreaver, V. D. Sandberg, M. Zimmerman: Rev. Mod. Phys **52**, 341 (1980)

10.2 M. J. Collett, C. W. Gardiner: Phys. Rev. A **30**, 1386 (1984); C. W. Gardiner, M. J. Collett: Phys. Rev. A **31**, 3761 (1985)

10.3 A. S. Parkins, C. W. Gardiner : Phys. Rev. A **37**, 3867 (1988)

10.4 H. Ritsch, P. Zoller: Phys. Rev. Lett. **61**, 1097 (1988)

10.5 L-A. Wu, H. J. Kimble, J. L. Hall, H. Wu: Phys. Rev. Lett. **57**, 2520 (1986)

10.6 A. S. Parkins, C. W. Gardiner: Phys. Rev. A **37**, 3867 (1988)

10.7 H. Ritsch, P. Zoller: Phys. Rev. A **38**, 4657 (1988)

Chapter 11

11.1 N. Bohr: Phil. Mag. **26**, 1, 476, 857 (1913)

11.2 W. Nagourney, J. Sandberg, H. Dehmelt: Phys. Rev. Lett. **56**, 2797 (1986); J. C. Bergquist, R. G. Hulet, W. M. Itano, D. J. Wineland: Phys. Rev. Lett. **57**, 1699 (1986); Th. Sauter, W. Neuhauser, R. Blatt, P. E. Toschek: Phys. Rev. Lett. **57** 1696 (1986)

11.3 B. R. Mollow: Phys. Rev. A **12**, 1919 (1975)

11.4 P. Zoller, M. Marte, D. F. Walls: Phys. Rev. A **35** (1987); R. Blatt, W. Ertmer, J. Hall, P. Zoller: Phys. Rev. A **34**, 3022 (1986)

11.5 A. Barchielli: Phys. Rev. A **34**, 1642 (1986); J. Phys. A **20**, 6341 (1987); Quant. Opt. **2**, 423 (1990); A. Barchielli, V. P. Belavkin: J. Phys. A **24**, 1495 (1991); A. Barchielli, A. M. Paganoni: Quant. Semiclass. Opt. **8**, 133 (1996)

11.6 C. W. Gardiner, A. S. Parkins, P. Zoller: Phys. Rev. A **46**, 4363 (1992); R. Dum, A. S. Parkins, P. Zoller, C. W. Gardiner: Phys. Rev. A **45**, 4879 (1992)

11.7 J. Dalibard, Y. Castin, K. Mølmer: Phys. Rev. Lett. **68**, 580 (1992); K. Mølmer, Y. Castin, J. Dalibard: J. Opt. Soc. Am. B, **10**, 524 (1993)

11.8 H. J. Carmichael: *An Open Systems Approach to Quantum Optics, Lecture Notes in Physics m18* (Springer, Berlin 1993)

11.9 N. Gisin: Phys. Rev. Lett. **52**, 1657 (1984); N. Gisin, Helv. Phys. Acta **62**, 363 (1989); N. Gisin, I. C. Percival: Phys. Lett. A **167**, 315 (1992); J. Phys. A **25**, 5677 (1992); J. Phys. A **26**, 2233 (1993); N. Gisin, P. L. Knight, I. C. Percival, R. C. Thompson, D. C. Wilson: J. Mod. Opt. **40**, 1663 (1993); N. Gisin: J. Mod. Opt. **40**, 2313 (1993); Y. Salama, N. Gisin: Phys. Lett. A **181**, 269 (1993)

11.10 M. B. Plenio, P. L. Knight: Rev. Mod. Phys. **70**, 101 (1998)

11.11 P. Goetsch, R. Graham: Phys. Rev. A **50**, 5242 (1994); P. Goetsch, R. Graham, F. Haake: Phys. Rev. A **51** 136 (1995)

11.12 P. Marte, R. Dum, R. Taïeb, P. D. Lett, P. Zoller: Phys. Rev. Lett. **71**, 1335 (1993); P. Marte, R. Dum, R. Taïeb, P. Zoller: Phys. Rev. A **47**, 1378 (1993)

11.13 M. Holland: Phys. Rev. Lett. **81**, 5117 (1998)

11.14 L. Diosi, N. Gisin, J. Halliwell, I. C. Percival: Phys. Rev. Lett. **74**, 203 (1995)

11.15 H. M. Wiseman: *Quantum Trajectories and Feedback* (Ph. D. Thesis, University of Queensland 1994); H. M. Wiseman, G. Milburn: Phys. Rev. A **47**, 642 (1993); Phys. Rev. A **47**, 1652 (1993)

11.16 R. Blatt, P. Zoller: Europ. J. Phys. **9**, 250 (1988); R. J. Cook: Prog. Opt. **28**, 361 (1990)

11.17 K. Mølmer, Y. Castin: Quant. Semiclass. Opt. **8**, 49 (1996)

11.18 R. Dum, P. Zoller: H. Ritsch, Phys. Rev. A **45**, 4879 (1992)

11.19 Y. Castin: K. Mølmer: Phys. Rev. Lett. **74**, 3772 (1995)

11.20 P. S. Jessen, C. Gerz, P. D. Lett, W. D. Phillips, S. L. Rolston, R. J. C. Spreeuw, C. I. Westbrook: Phys. Rev. Lett. **69**, 49 (1992)

11.21 P. Verkerk, B. Lounis, C. Salomon, C. Cohen-Tannoudji, J.-Y. Courtois, G. Grynberg: Phys. Rev. Lett. **68**, 3861 (1992); G. Grynberg B. Lounis, P. Verkerk, J.-Y. Courtois, C. Salomon: Phys. Rev. Lett. **70**, 2249 (1993)

11.22 A. Hemmerich, T. W. Hänsch: Phys. Rev. Lett. **70**, 410 (1993)

11.23 M. Holland, S. Marksteiner, P. Marte, P. Zoller: Phys. Rev. Lett. **76** 3683 (1996)

11.24 J. Dalibard, J.-M. Raimond, J. Zinn-Justin (*Ed.*): *Fundamental Systems in Quantum Optics; Proceedings of the Les Houches Summer School, Session LIII, 1990* (North-Holland, Amsterdam 1992)

11.25 A. P. Kazantsev, G. I. Surdutovich, V. P. Yakovlev: *Mechanical Action of Light on Atoms* (World Scientific, Singapore 1990)

11.26 Y. Castin, J. Dalibard: Europhys. Lett. **14**, 761 (1991)

11.27 J. Dalibard, C. Cohen-Tannoudji: J. Opt. Soc. Am. B **6**, 2023 (1989); J. Dalibard, C. Cohen-Tannoudji: J. Opt. Soc. Am. B **2**, 1707 (1985)

Chapter 12

12.1 M. I. Kolobov, I. V. Sokolov: Opt. Spektrosk. **62**, 112 (1987)

12.2 H. J. Carmichael: Phys. Rev. Lett. **70**, 2273 (1993)

12.3 C. W. Gardiner: Phys. Rev. Lett. **70**, 2269 (1993)

12.4 C. W. Gardiner, A. S. Parkins: Phys. Rev. A **50**, 1792 (1993)

12.5 P. L. Knight, D. T. Pegg: J. Phys. B **15**, 3211 (1982)

12.6 J. I. Cirac, P. Zoller, H. J. Kimble, H. Mabuchi: Phys. Rev. Lett. **78**, 3221 (1997)

12.7 S. J. van Enk, J. I. Cirac, P. Zoller: Science **279**, 205 (1998)

12.8 D. P. DiVincenzo: Science **270**, 255 (1995)

12.9 C. H. Bennett: Physics Today **48**, 24 (1995)

12.10 C. H. Bennett, G. Brassard, C. Crepeau, R. Josza, A. Peres, W. K. Wootters: Phys. Rev. Lett. **70**, 1895 (1993)

12.11 C H. Bennett, G. Brassard, S. Popescu, B. Schumacher, J. A. Smolin, W. K. Wootters: Phys. Rev. Lett. **76**, 722 (1996)

Bibliography

1960–1969

J. R. Klauder, E. C. G. Sudarshan, *Fundamentals of Quantum Optics* (Benjamin, New York 1968) The first book on quantum optics. Deals mainly with the description of quantum optical fields, rather than the evolution of damped systems, which was a subject then still in development.

1970–1979

S. M. Kay, A. Maitland ed., *Quantum Optics* (Proceedings of Scottish Universities Summer School) (Academic, New York 1970) A historical classic, in which most of main participants in the origin of quantum optics present the newly developed understanding of the laser and other quantum optical phenomena.

L. Mandel, E. Wolf, ed., *Selected Papers on Coherence and Fluctuations in Light* (2 Vols) (Dover, New York 1970) An invaluable collection which includes the main papers in which the theories of quantum damping and noise in the context of quantum optics were first developed. It can still be regarded as essential reading.

B. Davies, *Quantum Theory of Open Systems* (Academic, New York 1972) The book by Davies expresses the mathematician's view of quantum Markov processes. This is a rigorous book, and the value to physics lies in its demonstration of the self consistency of quantum Markov processes.

F. Haake, *Statistical Treatment of Open Systems by Generalized Master Equations* (Springer Tracts in Modern Physics, Vol. 66) (Springer, Berlin 1973) A summary of the state of the art (when written) of the master equation, with applications to quantum optics and solid state physics.

G. S. Agarwal, *Quantum Statistical Theories of Spontaneous Emission and their Relation to Other Approaches* (Springer Tracts in Modern Physics, Vol. 70) (Springer, Berlin 1974) Agarwal's book contains excellent early work on the use of phase space representations in quantum optical problems.

M. Sargent, M. O. Scully, W. E. Lamb, *Laser Physics* (Addison-Wesley, Reading, Mass. 1974) This book represents a viewpoint of laser and quantum damping theory as developed by Scully and Lamb in their theoretical treatment of the laser.

1980-1989

H. Spohn, *Kinetic Equations from Hamiltonian Dynamics: Markovian Limits, Rev. Mod. Phys.* **52**, *569 (1980)* This paper deals mainly with non-quantum systems, but the section on quantum matters gives an excellent summary of the rigorous results on the Markovian limit of quantum Hamiltonian systems.

R. Loudon, *Quantum Theory of Light* (Oxford University Press, Oxford 1983) An excellent book on the physics of quantum optics.

H. Haken, *Laser Theory* (Springer, Berlin 1970, 1984) The fundamental work on the laser. This contains the complete treatment of all kinds of laser, and goes into all the details thoroughly. It contains detailed treatments of phase space methods and master equation techniques as applied to the laser.

H. Haken, *Light* (2 vols.) (North Holland, Amsterdam 1981, 1985) A more tutorial treatment of quantum optics than Haken's monumental "Laser Theory", which covers material on lasers, optical bistability, interaction of laser light with atoms, as well as basic quantum mechanics.

R. Kubo, M. Toda, N. Hashitsume, *Statistical Physics II* (Nonequilibrium Statistical Mechanics) (Springer, Berlin 1985) Among the wide range of topics considered in this book is the treatment of both classical and quantum damping and noise theory. Here one can find how to treat such problems in the context of many body theory and solid state physics, rather than quantum optics.

W. Louisell, *Quantum Statistical Properties of Radiation* (Wiley, New York 1974, 1989) An early treatment of quantum optics, damping, noise and the master equation.

H. Risken, *The Fokker Planck Equation* (Springer, Berlin 1984, 1989) Although mainly a book on the classical Fokker-Planck equation, Risken's book contains valuable work on the semi-classical approach to the laser. The book is also valuable as a thorough handbook of methods of solution and approximation of the classical Fokker Planck equation, which is invaluable for those who wish to use phase space methods for treating master equations.

1990-1999

H. J. Carmichael, *An Open Systems Approach to Quantum Optics* (Lectures Presented at the Université Libre de Bruxelles, October 28 to November 4, 1991) (Springer, Berlin 1993) This has become the standard work on the stochastic Schrödinger equation and wavefunction simulation methods. It pays particular attention to the measurement theoretic interpretation of the stochastic processes generated.

W. Vogel, D.-G. Welsch, *Lectures on Quantum Optics* (Wiley, N.Y. 1994) The authors present quantum optics with a strong emphasis on the detailed analysis of the electromagnetic field.

D. F. Walls, G. J. Milburn, *Quantum Optics* (Springer, Berlin 1994) This book can be regarded as the expression of the "New Zealand School" of quantum optics, which has been so influential in quantum optics. Dan Walls had an influence on the whole field of quantum optics which extended into every aspect of the subject. The book, written with Gerard Milburn (an early student of Walls) emphasizes how much can be learned from relatively simple models, even in the context of a sophisticated theoretical point of view.

V. P. Belavkin, O. Hirota, R. L. Hudson ed., *Quantum Communications and Measurement* (Plenum, N.Y. 1995) This conference proceedings serves as a good introduction to the extent of the mathematical theory of quantum stochastic processes.

L. Mandel, E. Wolf, *Optical Coherence and Quantum Optics* (Cambridge U. P. , Cambridge 1995) A monumental work, which covers the complete field of quantum optics up to 1995, with a balance of experiment and theory.

A. Griffin, D. W. Snoke, S. Stringari (eds.), *Bose-Einstein Condensation* (Cambridge, Cambridge 1996) The proceedings of a conference held the year before the announcement of the creation of the first Bose-Einstein condennsate in 1995. Avery good summary of the state of the subject at that time.

U. Leonhardt, *Measuring the Quantum State of Light* (Cambridge Studies in Modern Optics) (Cambridge U.P., Cambridge 1997) A detailed description of this branch of quantum optics, including quantum tomography, a quantum description of optical instruments, and quantum measurement theory.

P. M. Radmore, S.M. Barnett, *Methods in Theoretical Quantum Optics* (Oxford Series in Optical and Imaging Sciences, 15) (Oxford U.P., Oxford 1997) This is a book on the mathematical methods used in quantum optics.

M. O. Scully, M. Zubairy, *Quantum Optics* (Cambridge U. P., Cambridge 1997) The summary of ideas which have had a unique influence on quantum optics.

C. Cohen-Tannoudji, J. Dupont-Roc, G. Grynberg, *Atom-Photon Interactions* (2 vols.) (Wiley, N.Y. 1994, 1998) A very thorough and extensive treatment of the physics of atoms and light by authors who are world leaders in the field.

H. J. Carmichael, *Statistical Methods in Quantum Optics* (2 vols) (Springer, Berlin 1999) Carmichael's treatment of quantum optics arises out of his background in the "New Zealand School", and is a very careful treatment with a strong emphasis on the physical meaning of the methodology used.

P. Meystre, M. Sargent III, *Elements of Quantum Optics* (3rd Ed.) (Springer, New York 1999) This covers the main practical theoretical aspects of quantum optics, in a reasonably simple exposition.

Y. Yamamoto, A. İmamoḡlu, *Mesoscopic Quantum Optics* (Wiley-Interscience, New York 1999) A quantum optics book with especial attention to the theory of semiconductor lasers and non-clasical light which can be produced this way.

2000-2003

D. Bouwmeester, A. Ekert, A. Zeilinger (eds.), *The Physics of Quantum Information: Quantum Cryptography, QuantumTeleportation, Quantum Computation* (Springer, Berlin, Heidelberg, New York 2000) A collection of mainly tutorial level articles by most of the workers in quantum information, presenting the field as it was in the year 2000.

M. Orszag, *Quantum Optics: Including Noise Reduction, Trapped Ions, Quantum Trajectories, and Decoherence* (Springer, Berlin, Heidelberg, New York 2000) A broad and up to date coverage of the basic laser-related phenomena, with a balance between experimental ant theoretical aspects.

P. Meystre, *Atom Optics* (Springer, Berlin, Heidelberg, New York 2001) The field of "Atom Optics" is broadly interpreted to cover laser cooling, propogation of atomic matter waves in linear and non-linear situations, and the elements of Bose-Einstein.

W. P. Schleich, *Quantum Optics in Phase Space* (John Wiley & Sons, New York 2001) This book presents quantum optics with a particular emphasis on phase space methods, the Wigner function in particular.

L. P. Pitaevskii, S. Stringari, *Bose-Einstein Condensation* (Oxford, Oxford 2003) Two of the leading figures in the field of Bose-Einstein condensation present the basics of the theory, as well as a review of the state of the field as of the year 2003, written from the point of view of condensed matter theory.

W. H. Zurek, *Decoherence, Einselection, and the Quantum origins of the Classical, Rev. Mod. Phys.* **75**, *715 (2003)* A thorough review and assessment of the foundations of quantum measurement theory using decoherence as thefundamental explanation for the existence of classical physics, the germ of which is reviewed in Sect.7.3 of this book.

Author Index

Subject Index

Printing: Mercedes-Druck, Berlin
Binding: Stein + Lehmann, Berlin